Grundlehren der mathematischen Wissenschaften 225

A Series of Comprehensive Studies in Mathematics

Kurt Schütte

Proof Theory

Translation from the German
by J. N. Crossley

Springer-Verlag
Berlin Heidelberg New York 1977

Kurt Schütte

Mathematisches Institut, Ludwig-Maximilians-Universität,
8000 München 2/Germany

Translator:

J. N. Crossley

Department of Mathematics, Monash University, Clayton,
Victoria 3168/Australia

Translation of the revised version of "Beweistheorie",
1st edition, 1960; Grundlehren der mathematischen
Wissenschaften, Band 103

AMS Subject Classification (1970): 02B10, 02B99, 02C15, 02Dxx,
02E99, 10N99

ISBN-13: 978-3-642-66475-5 e-ISBN-13: 978-3-642-66473-1
DOI: 10.1007/978-3-642-66473-1

Library of Congress Cataloging in Publication Data. Schütte, Kurt. Proof theory. (Grundlehren der mathematischen Wissenschaften; 225). Translation of Beweistheorie. Bibliography: p. Includes index. 1. Proof theory. I. Title. II. Series: Die Grundlehren der mathematischen Wissenschaften in Einzeldarstellungen; 225. QA9.54.S3813. 511'.3. 76-45768.

Softcover reprint of the hardcover 1st edition 1977

2141/3140-543210

Preface

This book was originally intended to be the second edition of the book "Beweistheorie" (Grundlehren der mathematischen Wissenschaften, Band 103, Springer 1960), but in fact has been completely rewritten. As well as classical predicate logic we also treat intuitionistic predicate logic. The sentential calculus properties of classical formal and semiformal systems are treated using positive and negative parts of formulas as in the book "Beweistheorie". In a similar way we use right and left parts of formulas for intuitionistic predicate logic.

We introduce the theory of functionals of finite types in order to present the Gödel interpretation of pure number theory. Instead of ramified type theory, type-free logic and the associated formalization of parts of analysis which we treated in the book "Beweistheorie", we have developed simple classical type theory and predicative analysis in a systematic way. Finally we have given consistency proofs for systems of Π_1^1-analysis following the work of G. Takeuti. In order to do this we have introduced a constructive system of notation for ordinals which goes far beyond the notation system in "Beweistheorie".

My hearty thanks are due to Professor J. N. Crossley for his translation of my German manuscript and for his careful checking of the proof sheets. I also wish to thank Dr. W. Pohlers and Dr. W. Buchholz for the help which they have given me technically and also in reading the proofs. I would especially like to thank Frau Ernst (Munich) and Ms. Vandenberg (Monash University) for their very careful typing of the text. I wish to thank Springer-Verlag for their kindness and help in the production of the book.

Munich, March 1977

Kurt Schütte

Preface

Table of Contents

Introduction 1

Part A. Pure Logic 5

Chapter I. Fundamentals 7

§ 1. Classical Sentential Calculus 7
1. Truth Functions 7
2. Sentential Forms 8
3. Complete Systems of Connectives 8
4. A Formal Language for the Sentential Calculus 9
5. Positive and Negative Parts of Formulas 10
6. Syntactic Characterization of Valid Formulas 12

§ 2. Formal Systems 13
1. Fundamentals 13
2. Deducible Formulas 14
3. Permissible Inferences 15
4. Sentential Properties of Formal Systems. 16
5. The Formal System CS of the Classical Sentential Calculus . . 17

Chapter II. Classical Predicate Calculus 19

§ 3. The Formal System CP 19
1. Primitive Symbols 19
2. Inductive Definition of the Formulas 20
3. P-Forms and N-Forms 20
4. Positive and Negative Parts of a Formula 20
5. Axioms 20
6. Basic Inferences 20

§ 4. Deducible Formulas and Permissible Inferences 21
1. Generalizations of the Axioms 21
2. Weak Inferences 22
3. Further Permissible Inferences 24
4. Defined Logical Connectives 25

§ 5. Semantics of Classical Predicate Calculus 26
1. Classical Models 27
2. The Consistency Theorem 28
3. The Completeness Theorem 28
4. The Satisfiability Theorem 32
5. Syntactic and Semantic Consequences 35

Chapter III. Intuitionistic Predicate Calculus 36

§ 6. Formalization of Intuitionistic Predicate Calculus 36
1. The Formal System IP1 36
2. The Formal System IP2 38
3. Left and Right Parts of Formulas 39
4. The Formal System IP3 40

§ 7. Deducible Formulas and Permissible Inferences in the System IP3 . 42
1. Generalizations of the Axioms 42
2. Weak Inferences 42
3. More Permissible Inferences 44
4. Special Features of Intuitionistic Logic 46
5. Properties of Negation 47
6. Syntactic Equivalence 47

§ 8. Relations between Classical and Intuitionistic Predicate Calculus . 48
1. Embedding IP3 in CP 48
2. Interpretation of CP in IP3 49

§ 9. The Interpolation Theorem 51
1. Interpolation Theorem for the System IP3 51
2. Interpolation Theorem for the System CP 54
3. Finitely Axiomatisable Theories 54
4. Beth's Definability Theorem 55

Chapter IV. Classical Simple Type Theory 56

§ 10. The Formal System CT 56
1. The Formal Language 56
2. Chains of Subterms 57
3. Axioms and Basic Inferences 60
4. Deducible Formulas and Permissible Inferences 60
5. The Cut Rule 62

§ 11. Deduction Chains and Partial Valuations 62
1. Definition of Deduction Chains 62
2. Partial Valuations 63
3. Principal Lemmata 64

§ 12. Semantics 65
1. Total Valuations over a System of Sets 65
2. Soundness Theorem 66

3. Extending a Partial Valuation 67
4. Completeness Theorem and Cut Rule 69

Part B. Systems of Arithmetic 71

Chapter V. Ordinal Numbers and Ordinal Terms 73

§ 13. Theory of Ordinals of the 1st and 2nd Number Classes 73
1. Order Types of Well-Ordered Sets 73
2. Axiomatic Characterization of the 1st and 2nd Number Classes . 74
3. Zero, Successor and Limit Numbers and Supremum 75
4. Ordering Functions 76
5. Addition of Ordinals 79
6. α-Critical Ordinals 81
7. Maximal α-Critical Ordinals 84

§ 14. A Notation System for the Ordinals $<\Gamma_0$ 86
1. Definition of Ordinal Terms 86
2. The Ordering of the Ordinal Terms 88
3. Addition of Ordinal Terms 89
4. Ordinal Terms $\phi\alpha\beta$. 90
5. Ordinal Terms ω^α, $\omega^\alpha \cdot n$ and $\omega^\alpha \cdot \beta$ 92
6. Ordinal Terms $n \cdot \beta$, 2^α and $2^\alpha \cdot \beta$ 94
7. Ordinal Terms ε_α, $\omega_n(\alpha)$ and ζ_n 95
8. A Mapping onto the Natural Numbers 96

Chapter VI. Functionals of Finite Type 98

§ 15. The System of Terms of Finite Type 98
1. Types 98
2. Terms 98
3. Interpretation 99
4. Reduction Procedure for Terms 100
5. Characterization of Numerals 101
6. Substitution Properties 101
7. The Normal Form of a Term 104

§ 16. Orders of Terms 105
1. Natural Sum and Natural Product of Ordinal Terms 105
2. Assigning Ordinal Terms to Terms 106
3. Estimates of the Order of Terms 108
4. Equality of Terms 112

§ 17. The Formal System FT of Functionals of Finite Type 113
1. The Formal Language 113
2. Deduction Procedures 114
3. The Consistency of the System FT 115
4. Fundamental Deduction Rules 117

5. Addition and Multiplication 120
6. The Indentity Functional I_τ and λ-Abstraction 121
7. The Predecessor Functional and the Arithmetic Difference . . 122
8. The Recursor 124
9. Simultaneous Recursion 126
10. The Characteristic Term of a Basic Formula 129

Chapter VII. Pure Number Theory 134

§ 18. The Formal System PN for Pure Number Theory 134
1. The Formal Language 134
2. The Deduction Procedure 135
3. Basic Properties of Deducibility 136
4. Properties of Negation. 138
5. Positive and Negative Parts of Formulas 142
6. The Consistency of the System PN 147

§ 19. Interpretation of PN in FT 148
1. Sequences of Terms of the System FT 148
2. The Formal System QFT 149
3. Interpreting Formulas 155
4. Interpretations of the Axioms of the System PN 157
5. Interpretations of the Basic Inferences in the System PN . . . 158

Part C. Subsystems of Analysis 165

Chapter VIII. Predicative Analysis 167

§ 20. Systems of Δ_1^1-Analysis 167
1. The Formal Language of Second Order Arithmetic 167
2. The Formal System DA 170
3. Deducible Formulas and Permissible Inference of the System DA 171
4. The Semi-Formal System DA* 174
5. Embedding DA in DA* 175
6. General Properties of Deduction in the System DA* 176
7. Subsystems of DA and DA* 176

§ 21. Deductions of Transfinite Induction. 177
1. Formalisation of Transfinite Induction 177
2. Deductions in EN 179
3. Deductions in EN*. 181
4. Deductions in EA and EA* 182
5. The Formula $\mathscr{R}$ $[P, Q, t]$ 184
6. Deductions in DA 188
7. Deductions in DA* 192

§ 22. The Semi-Formal System RA* for Ramified Analysis 197
1. The Formal Language 197

2. The Deduction Procedures 200
3. Weak Inferences 201
4. Elimination of Cuts 201
5. Further Properties of Deductions 204
6. Interpretations of EA* and DA* in RA* 207

§ 23. The Limits of the Deducibility of Transfinite Induction 209
1. Orders of Deductions of Induction in RA* 209
2. The Limiting Numbers of the Systems EN, EA and DA . . . 211
3. The Autonomous Ordinal Terms of the Systems EN*, EA* and DA* 213
4. The Autonomous Ordinal Terms of the System RA* 217
5. The Limits of Predicativity 220

Chapter IX. Higher Ordinals and Systems of Π_1^1-Analysis 221

§ 24. Normal Functions on a Segment $\mathbb{O}^*$ of the Ordinals 221
1. Axiomatic Characterization of the Segment $\mathbb{O}^*$ of the Ordinals . 221
2. Basic Properties of $\mathbb{O}^*$ 222
3. Definition of the Functions θ_α 224
4. Properties of the θ Functions 225
5. The Sets $\overline{In}(\alpha)$ and Functions $\bar{\theta}_\alpha$ 230

§ 25. A Notation System for Ordinals Based on the θ_α Functions . . . 234
1. The Set $\bar{\theta}(\Omega)$ of Ordinals 234
2. Sets of Coefficients 234
3. The Systems T* and OT* of Terms 236
4. Subsystems $\bar{\theta}(\tau)$ of $\bar{\theta}(\Omega)$ 238
5. The Ordinal Λ_0 239
6. Relations between Cr (α) and In (α) 240

§ 26. Level-Lowering Functions of the Ordinals 241
1. Basic Concepts 242
2. Properties of the Sets of Coefficients 242
3. The Ordinal Term $d_i\alpha$ 245
4. The Natural Sum 248
5. Deduction Functions 248

§ 27. The Formal System GPA for a Generalized Π_1^1-Analysis 252
1. The Formal Language 252
2. Axioms, Basic Inference and Substitution Inferences 253
3. Deductions 255
4. Orders of Normal Deductions 257
5. Transformations of Normal Deductions 258
6. Reducible Normal Deductions 259
7. Singular Normal Deductions 261
8. Reduction of a Suitable Cut 262
9. The Consistency of the System GPA 268
10. The Subsystem PA of GPA 268

§ 28. The Semi-Formal System PA* 269
1. Axioms and Basic Inferences of the System PA* 269
2. The Strength of a Formula 270
3. Basic Deductions in the System PA* 271
4. Embedding of PA in PA* 272
5. Elimination of Strong Cuts in PA* 273
6. Normal Deductions in the System PA* 275
7. Reducible Normal Deductions 277
8. Elimination of Cuts in PA* 278

§ 29. Proof of Well-Ordering 281
1. A Constructive Proof of Well-Ordering for Subsystems of $\bar{\theta}(\Omega)$. 281
2. The Formal System ID_n of n-Fold Iterated Inductive Definitions . 285
3. Formalization of the Proof of Well-Ordering of $\bar{\theta}(N)$ in ID_N . . 287
4. Embedding ID_n in a Subsystem of PA 289

Bibliography 293

Subject Index 297

Introduction

In mathematics, dealing with infinite number systems and other infinite sets brings a fundamental problem with it. For example consider the two statements:

I. Every even number greater than 2 can be represented as the sum of two primes.

II. There is an even number which is greater than 2 but cannot be represented as the sum of two primes.

According to classical logic the second statement is the contradictory of statement I. So far neither of these two statements can be proved. We cannot prove statement I by testing its validity for all even numbers since there are infinitely many such numbers. In order to prove statement I we must rather solve the following problem:

(I) To give a general procedure which assigns to each even number $g>2$ two primes $p(g)$ and $q(g)$ such that the equation

$$g=p(g)+q(g)$$

is satisfied for all $g>2$.

If one does not succeed in solving this problem then one must attempt to prove the opposite statement, II. Then one is dealing with an existential statement. If one takes the existential question seriously then in order to prove statement II one must solve the following problem:

(II) To give an even number $g_0>2$ which one can recognize as not being representable as the sum of two prime numbers.

We now ask whether either of the two statements I and II is fundamentally correct. Now the correctness of a mathematical statement can only be determined by a proof. There is no sense in accepting a mathematical truth which is not possibly provable. So we can only ask whether one of statements I and II is provable in principle. If one says this question is solvable in the affirmative then one must be certain that one of problems (I) or (II) really is solvable. But such certainty is not attainable. In fact one cannot even deduce from the assumption that problem (I) is unsolvable that a number g_0, which satisfies problem II, can actually be shown to exist. Neither can one deduce, from the assumption that problem (II) is unsolvable, that there is a general procedure for solving problem (I).

We therefore have no grounds for assuming that one of the two statements I and II is correct. The assumption would only be justified if a proof of I or II were given. But then the general situation regarding the provability of mathematical propositions would not be altered since there are always other cases where the question of provability of a statement or its contradictory remains open. Therefore there is no immediate sense in saying that the principle of *Tertium non datur*, which says that either a proposition or its logical contradictory is correct, be accepted as satisfactory for mathematical propositions.

This brief critique of *Tertium non datur* leads us to the basic problem of proof theory. We start from the premise that in mathematics *Tertium non datur* is not meaningful as a general basic principle. But if one gives up this principle as a method of proof then one loses the right to indirect proof procedures such as one cannot do without in classical mathematics. In order to guarantee the correctness of classical mathematics in a logically unobjectionable way there only remains the possibility of basing mathematical statements not on provability as such but on formal deducibility. This deducibility which may be regarded as a notional provability must be given axiomatically so that it is sufficient for the basic theorems of classical mathematics. In this way one does not need to interpret the concept of deducibility in a meaningful way, but only to establish its formal admissibility by means of a consistency proof. This is in fact Hilbert's programme: to establish the consistency of mathematics in its classical form.

Hilbert called the theory which solves this problem proof theory (*Beweistheorie*). The object of a proof-theoretic investigation is a mathematical theory which is precisely determined not only by its mathematical notions and axioms but also by the basic logical concepts and modes of inference which it allows. Now one must distinguish clearly between the laws of the theory which are to be investigated and those of the theory in which the investigation is to be carried out. The notions and theorems of the theory under investigation are represented in a formal system and treated purely formally without reference to their meaning, while the proof-theoretic investigation is concerned with the logical meaning of the notions and inference modes. Thus the formal theory is complemented by a meaningful metatheory (proof theory). This metatheory is also called "metalogic" or "metamathematics". This does not mean that one is extending the usual logic or mathematics, as is the case with metaphysics and physics. Here it only delineates the distinction between formalized mathematics and the logic in which the investigation is carried out.

If proof theory is to provide a foundation for a problematic area of mathematics then its own methods and notions and modes of inference must be restricted to those which lie outside this problematic area. The methods which Hilbert called "finitist" (*finit*) do lie outside. These consist of just elementary logical and mathematical relations of a combinatorial nature which are used as bases for all theoretical investigations and are generally acknowledged as admissible. In a finitist proof *Tertium non datur* may not be used for propositions concerning infinite totalities and existential propositions are only to be treated as valid if the object whose existence is asserted can in fact be constructed.

But Gödel's investigations [1] have shown that the strictest finitist methods are basically inadequate for carrying out the consistency proof required by Hilbert's programme. So proof theory needs not only the very strict finitist methods of a combinatorial nature but also higher level proof procedures. Thus we arrive at methods, first used by Gentzen [2], using induction which in fact goes beyond the usual complete (mathematical) induction but still has a constructive character.

In this book we give a systematic presentation of the most important results which have so far been achieved in the pursuit of Hilbert's programme. The main emphasis is on consistency proofs but other investigations, which are closely related in terms of methodology, are also carried out. We use inductive methods for the consistency proofs but do not admit *Tertium non datur* as a proof procedure. We need to forgo the constructive standpoint and use classical mathematics as a foundation despite its problems only for the semantic treatment of classical predicate calculus (§5) and classical simple type theory (§12). Likewise we first consider the necessary systems of ordinals (in §13 and §24) from the standpoint of classical mathematics but then (in §§14, 25 and 26) we define these and treat them constructively.

As in the first edition of this book we use the notions of "positive and negative parts" as a basis for formalization in the realm of classical logic. These are a sort of generalization of the notions of "succedent and antecedent formulas" of Gentzen's sequence calculus (Sequenzenkalkul). In intuitionistic logic we use instead the weaker notions of "left and right parts of a formula".

In Chapter I we collect together the general basic syntactic notions which are needed in the book and used for the formalization of classical propositional calculus.

Chapter II treats the constructive syntax (§§3, 4) and the non-constructive semantics (§5) of classical predicate calculus. Intuitionistic predicate calculus is developed in Chapter III but only syntactically (constructively) and related to classical predicate calculus. The Interpolation Theorem proved here (§9) also yields the corresponding theorem for classical predicate calculus.

In Chapter IV we first assemble the basic syntactic properties for classical simple type theory in §10 and then introduce partial valuations in §11 which are used to carry out the non-constructive proof, following M. Takahashi [1] and D. Prawitz [1], of Takeuti's Fundamental Conjecture (Eliminability of Cuts).

In Chapter V we introduce a notation system for a segment of the ordinals which is used later for the treatment of pure number theory (Chapter VI) and predicative analysis (Chapter VIII). This system of notations is defined in §13 in terms of an ordinary non-constructive approach in the classical theory of ordinal numbers and then defined as a constructive system of ordinal terms in §14. In this section too we constructively prove the necessary properties of functions on ordinal terms of this system (which are needed later) in an elementary way.

In Chapters VI and VII we develop the Godel interpretation of pure number theory in a system of functionals of finite type. Following W. Howard we prove (in §16) by transfinite induction up to ε_0 that the functionals introduced in §15 are calculable. In §17 we introduce a formal system FT for these functionals in the context of the positive implicational calculus. To this end Chapter VII treats, in

§18, an obvious formal system of pure number theory, and in §19 we carry out the interpretation of this system in the system FT.

In Chapter VIII we are concerned with the predicative limits of analysis and first we develop a formal and a semi-formal system of Δ_1^1-analysis whose languages are that of classical second order arithmetic and which have restricted comprehension axioms as opposed to classical mathematics. In §21 we give deductions of formalized transfinite induction in the spirit of S. Feferman in these systems and in subsystems of Δ_1^1-analysis. In §22 we introduce a semi-formal system for ramified analysis in which the system of Δ_1^1-analysis can be interpreted. This yields the limits for the deducibility of transfinite induction for the formal and semi-formal systems considered and demonstrates the predicativity of these systems.

In Chapter IX we prove the consistency of a formal system of Π_1^1-analysis following a proof of G. Takeuti [4]. For this we need a transfinite induction which goes far beyond those previously introduced. Here we do not use the ordinal diagrams of Takeuti but a system of notations for ordinal terms based on normal functions θ_α defined and studied by S. Feferman, P. Aczel and J. Bridge and constructively characterized by W. Buchholz. Using these functions θ_α which are introduced in §24 in the context of the classical theory of ordinals, we build a constructive system of notations in §25. The necessary properties of this system of notations which are needed for the consistency proof are developed in §§25 and 26. §27 gives the consistency proof for the formal system of Π_1^1-analysis and essentially this corresponds to the proof of Takeuti but in a somewhat different way. Similarly we give a consistency proof for a semi-formal system of Π_1^1-analysis in §28, but this is done using a weaker transfinite induction than in §27. Finally we give constructive proofs of well-ordering for the systems of ordinals which we have used and their subsystems in §29.

Part A

Pure Logic

Chapter I

Fundamentals

In mathematical logic sentences are represented by formulas. Those investigations which only depend on the formal structure of logical formulas and fixed formal rules of deduction are said to be *syntactic*. On the other hand those which depend on model-theoretic interpretations are said to be *semantic*. Syntactic investigations, which feature predominantly in proof theory, are usually carried out constructively whilst the semantic ones can only be carried out with constructive concepts and proof procedures in the simplest part of logic, that is, in the sentential (or propositional) calculus.

In §1 we start with a semantic characterization of classical sentential calculus and then develop an adequate syntax for it. In §2 we present the basic syntactic concepts which will be used in the subsequent chapters.

§1. Classical Sentential Calculus

1. Truth Functions

Truth values are denoted by t (*true*) and f (*false*). An n-place truth function is an assignment of one truth value (t or f) to each n-tuple of truth values. Classical sentential calculus can be presented semantically as the theory of these truth functions.

As there are exactly 2^n distinct n-tuples of truth values and each n-place truth function assigns just one of two possible values to each n-tuple there are exactly $2^{(2^n)}$ distinct n-place truth functions.

We use the following *connectives* as symbols for truth functions. $\top$ (*verum*) and $\bot$ (*falsum*) denote the 0-place truth functions with values $\top = t$ and $\bot = f$.

The connective $\neg$ (*negation*) denotes the one-place truth function with values $\neg(t) = f$ and $\neg(f) = t$.

The connectives $\vee$ (*disjunction*), $\wedge$ (*conjunction*) and $\to$ (*implication*) denote the 2-place truth functions with values as follows:

$$\vee(t, t) = \vee(t, f) = \vee(f, t) = t, \quad \vee(f, f) = f,$$
$$\wedge(t, t) = t, \quad \wedge(t, f) = \wedge(f, t) = \wedge(f, f) = f,$$
$$\to(t, t) = \to(f, t) = \to(f, f) = t, \quad \to(t, f) = f.$$

In the same way we could define arbitrarily many other connectives. But they will not be used in the sequel.

2. Sentential Forms

Using connectives and sentential variables, which we denote by $v, v_1, v_2, \ldots$, we construct *sentential forms* according to the following **inductive definition**:

SF1. Every sentential variable is a *sentential form*.

SF2. Every 0-place connective is a *sentential form*.

SF3. If ϕ is an n-place connective ($n \geqslant 1$) and $A_1, \ldots, A_n$ are *sentential forms* then $\phi(A_1, \ldots, A_n)$ is also a *sentential form*.

Such an inductive definition, like the ones we use constantly in the sequel, is to be interpreted as follows: a string of symbols has the inductively defined property if, and only if, it can be obtained by a finite number of applications of the defining rules.

Example. $\to(\wedge(v_1, v_2), \vee(\bot, \neg(v_1)))$ is a sentential form since the defining rules SF1–SF3 yield the following sentential forms:

1. v_1 and v_2 by SF1,
2. $\bot$ by SF2,
3. $\neg(v_1)$ by SF3 from 1,
4. $\wedge(v_1, v_2)$ by SF3 from 1,
5. $\vee(\bot, \neg(v_1))$ by SF3 from 2 and 3,
6. $\to(\wedge(v_1, v_2), \vee(\bot, \neg(v_1)))$ by SF3 from 4 and 5.

The semantics of the sentential calculus are determined by sentential valuations. Such a *valuation* is an assignment V of a truth value Vv to each sentential variable v.

For a given valuation V every sentential form A gets a truth value VA according to the following **inductive definition.**

V1. A sentential form which is a sentential variable v alone gets the truth value Vv given by the valuation V.

V2. $V\top = t$, $V\bot = f$.

V3. If ϕ is an n-place connective ($n \geqslant 1$) and $A_1, \ldots, A_n$ are sentential forms, then $V\phi(A_1, \ldots, A_n)$ is the truth value which the truth function represented by ϕ assigns to the n-tuple of truth values $(VA_1, \ldots, VA_n)$.

Example. For a valuation V where $Vv_1 = f$ and $Vv_2 = t$ we have $V\to(\wedge(v_1, v_2), \vee(\bot, \neg(v_1))) = \to(\wedge(f, t), \vee(f, \neg(f))) = \to(f, \vee(f, t)) = \to(f, t) = t$.

Two sentential forms A and B are said to be *semantically equivalent* if $VA = VB$ for every sentential valuation V.

3. Complete Systems of Connectives

A set M of connectives is said to be a *complete system of connectives* if to every sentential form A there is a semantically equivalent sentential form B all of whose connectives are in M.

Theorem 1.1. *The connectives* $\neg$, $\wedge$ *and* $\vee$ *form a complete system of connectives.*

Proof. Let A be an arbitrary sentential form. We prove by induction on the number of sentential variables in A that there is a sentential form B, semantically equivalent to A, in which the only connectives are $\neg$, $\wedge$ and $\vee$.

1. A contains no sentential variables. Then A has the same truth value VA for every valuation V. If $VA=t$, then A is semantically equivalent to the sentential form $\vee(v, \neg(v))$. If $VA=f$, then A is semantically equivalent to the sentential form $\wedge(v, \neg(v))$.

2. A contains the sentential variables $v_1, \ldots, v_n$ $(n \geqslant 1)$. Let A_1 (respectively, A_2) be the sentential form obtained from A by replacing v_n by $\top$ (respectively, $\bot$). By the induction hypothesis there are sentential forms B_1, B_2 semantically equivalent to A_1, A_2 in which the only connectives are $\neg$, $\wedge$ and $\vee$. But then A is semantically equivalent to the sentential form $\vee(\wedge(B_1, v_n), \wedge(B_2, \neg(v_n)))$.

Corollary 1.2. *The connectives* $\neg$ *and* $\vee$ *form a complete system of connectives.*

Proof. Immediate from Theorem 1.1 since every sentential form $\wedge(A, B)$ is semantically equivalent to the sentential form $\neg(\vee(\neg(A), \neg(B)))$.

Corollary 1.3. *The connectives* $\neg$ *and* $\rightarrow$ *form a complete system of connectives.*

Proof. Immediate from corollary 1.2 since every sentential form $\vee(A, B)$ is semantically equivalent to the sentential form $\rightarrow(\neg(A), B)$.

Corollary 1.4. *The connectives* $\bot$ *and* $\rightarrow$ *form a complete system of connectives.*

Proof. Immediate from Corollary 1.3 since every sentential form $\neg(A)$ is semantically equivalent to the sentential form $\rightarrow(A, \bot)$.

There are many other complete systems of connectives but we shall not go into them as we shall not use them in the sequel.

4. A Formal Language for the Sentential Calculus

In formalizing classical sentential calculus we need only consider connectives from a complete system of connectives since semantically equivalent sentential forms have the same logical interpretation. We now define a formal language with connectives $\bot$, $\neg$, $\wedge$, $\vee$ and $\rightarrow$. These connectives certainly form a complete system but of course they do not form a minimal complete system of connectives.

We assume we are given a countably infinite set of symbols which we call *sentential variables*. In fact we shall never need to consider these symbols themselves. We shall only use *syntactic symbols* $v, v_1, v_2, \ldots$ for the given symbols for sentential variables. The *formulas* of our formal language are given by the following **inductive definition.**

F1. Every sentential variable is a *formula*.

F2. $\perp$ is a *formula*.

F3. If A is a *formula* then $\neg A$ is a *formula*.

F4. If A and B are *formulas* then $(A \wedge B)$, $(A \vee B)$ and $(A \rightarrow B)$ are also *formulas*.

Formulas may be identified with particular sentential forms where $\neg A$ stands for $\neg(A)$ and the 2-place connectives are placed between rather than in front of their arguments following ordinary usage. The formulas $\perp$, $\neg A$, $(A \wedge B)$, $(A \vee B)$ and $(A \rightarrow B)$ are to be read "falsum", "not A", "A and B", "A or B" and "A implies B".

For any valuation V every formula gets a truth value as defined for sentential forms. Accordingly we have here the following particular defining rules for V2 and V3 (from page 8):

V2. $V\perp = f$,

V3.1. If $VA = t$ (respectively, $VA = f$), then $V\neg A = f$ (respectively, $V\neg A = t$).

V3.2. If $VA = t$ and $VB = t$, then $V(A \wedge B) = t$. Otherwise $V(A \wedge B) = f$.

V3.3. If $VA = t$ or $VB = t$, then $V(A \vee B) = t$. Otherwise $V(A \vee B) = f$.

V3.4. If $VA = f$ or $VB = t$, then $V(A \rightarrow B) = t$. Otherwise $V(A \rightarrow B) = f$.

A formula A is said to be *valid* if $VA = t$ for every sentential valuation V.

Given a formula A we can decide, in an elementary way, whether it is valid or whether there is a valuation under which it gets the value f. Indeed there are only finitely many possible combinations of truth values for the sentential variables occurring in A and in order to decide whether A is valid we only have to compute VA for these.

Our aim now is to characterize the valid formulas just by their structure and without considering valuations and then to employ such a syntactic characterization again for stronger logical systems where the semantics cannot be expressed in as elementary and constructive a way as they can for classical sentential calculus.

5. Positive and Negative Parts of Formulas

A formula A which is part of a formula F may have the property that the truth value VF is already determined by the truth value VA. We shall call A a *positive* (*negative*) *part* of F if $VA = t$ ($VA = f$) implies $VF = t$.

Examples. A is a positive part of the formulas $\neg\neg A$, $(A \vee B)$ and $(B \rightarrow A)$ and a negative part of the formulas $\neg A$, $\neg(B \wedge A)$, $(A \rightarrow B)$.

We establish a connection between the semantics and syntax of classical sentential calculus by using the notions of "*positive*" and "*negative parts*". We define these notions purely syntactically and without reference to valuations and then prove they have the intended semantic properties.

We use certain nominal forms in order to give the syntactic definition of positive and negative parts. By an *n-place simple nominal form* ($n \geqslant 1$) we mean a finite string of symbols in which besides the primitive symbols of our formal language (sentential variables, connectives and round brackets) the *nominal symbols* $*_1, \ldots, *_n$ occur just once. If $\mathscr{E}$ is an n-place simple nominal form ($n \geqslant 1$) and $r_1, \ldots, r_n$ are non-empty finite strings of symbols then $\mathscr{E}[r_1, \ldots, r_n]$ denotes the result of replacing the nominal symbols $*_1, \ldots, *_n$ in $\mathscr{E}$ by $r_1, \ldots, r_n$, respectively.

P-forms (*positive forms*) and *N-forms* (*negative forms*) are inductively defined as 1-place simple nominal forms as follows:

PNF1. $*_1$ is a *P-form*.

PNF2. If $\mathscr{P}$ is a *P-form* and B is a formula, then $\mathscr{P}[(*_1 \vee B)]$, $\mathscr{P}[(B \vee *_1)]$ and $\mathscr{P}[(B \rightarrow *_1)]$ are *P-forms* and $\mathscr{P}[\neg *_1]$ and $\mathscr{P}[(*_1 \rightarrow B)]$ are *N-forms*.

PNF3. If $\mathscr{N}$ is an *N-form* and B is a formula, then $\mathscr{N}[\neg *_1]$ and $\mathscr{N}[(*_1 \rightarrow \bot)]$ are *P-forms* and $\mathscr{N}[(*_1 \wedge B)]$ and $\mathscr{N}[(B \wedge *_1)]$ are *N-forms*.

We have the following consequences of this inductive definition:

PNF4. No P-form is an N-form.

PNF5. If $\mathscr{P}$ is a P-form and $\mathscr{E}$ is a P-form (N-form), then $\mathscr{P}[\mathscr{E}]$ is a P-form (N-form).

PNF6. If $\mathscr{E}$ is a P-form or an N-form and A is a formula, then $\mathscr{E}[A]$ is also a formula.

Definition. A formula A is said to be a *positive* (*negative*) *part* of a formula F, if there is a *P-form* (*N-form*) $\mathscr{E}$ such that $\mathscr{E}[A]$ is the formula F.

The rules PNF1–PNF3 yield the following **inductive characterization** of the *positive* and *negative parts* of a formula F.

PN1. F is a *positive part* of F.

PN2. If $\neg A$ is a *positive* (*negative*) *part* of F then A is a *negative* (*positive*) *part* of F.

PN3. If $(A \vee B)$ is a *positive part* of F, then A and B are both *positive parts* of F.

PN4. If $(A \wedge B)$ is a *negative part* of F, then A and B are both *negative parts* of F.

PN5. If $(A \rightarrow B)$ is a *positive part* of F, then A is a *negative part* and B is a *positive part* of F.

PN6. If $(A \rightarrow \bot)$ is a *negative part* of F, then A is a *positive part* of F.

Example. If F is the formula $\neg(\neg(v_1 \rightarrow v_2) \wedge (v_2 \vee \bot))$ then we get the positive and negative parts of F as follows:

1. By PN1 F is a positive part of F.
2. By 1 and PN2 $(\neg(v_1 \rightarrow v_2) \wedge (v_2 \vee \bot))$ is a negative part of F.
3. By 2 and PN4 $\neg(v_1 \rightarrow v_2)$ and $(v_2 \vee \bot)$ are negative parts of F.
4. By 3 and PN2 $(v_1 \rightarrow v_2)$ is a positive part of F.
5. By 4 and PN5 v_1 is a negative and v_2 a positive part of F.

These are all the positive and negative parts of F.

By an *NP-form* we mean a 2-place simple nominal form $\mathscr{Q}$ such that for every formula B $\mathscr{Q}[*_1, B]$ is an N-form and $\mathscr{Q}[B, *_1]$ is a P-form.

Examples. $(*_1 \rightarrow *_2)$, $(\neg(*_1 \wedge v) \vee (v \vee *_2))$ and $((*_2 \rightarrow \bot) \rightarrow (*_1 \rightarrow v))$ are NP-forms.

Convention. From now on $\mathscr{P}$, $\mathscr{P}_i$ will always denote P-forms, $\mathscr{N}$, $\mathscr{N}_i$ N-forms and $\mathscr{Q}$, $\mathscr{Q}_i$ NP-forms.

In the next few theorems we establish the semantic properties of positive and negative parts.

Theorem 1.5. *If $\mathscr{E}$ is a P-form (N-form) and V a sentential valuation with $VA=t$ ($VA=f$), then $V\mathscr{E}[A]=t$.*

Proof by induction on the length of $\mathscr{E}$.

1. $\mathscr{E}$ is the P-form $*_1$. Then $\mathscr{E}[A]$ is the formula A so by hypothesis $V\mathscr{E}[A]=t$.
2. $\mathscr{E}$ is a P-form $\mathscr{P}[(*_1 \vee B)]$, $\mathscr{P}[(B \vee *_1)]$ or $\mathscr{P}[(B \rightarrow *_1)]$. Now $VA=t$ implies $V(A \vee B)=V(B \vee A)=V(B \rightarrow A)=t$. By I.H. (induction hypothesis) we have $V\mathscr{E}[A]=t$.
3. $\mathscr{E}$ is an N-form $\mathscr{P}[\neg *_1]$ or $\mathscr{P}[(*_1 \rightarrow B)]$. Now $VA=f$ implies $V\neg A=V(A \rightarrow B)=t$. By I.H. we have $V\mathscr{E}[A]=t$.
4. $\mathscr{E}$ is a P-form $\mathscr{N}[\neg *_1]$ or $\mathscr{N}[(*_1 \rightarrow \bot)]$. Now $VA=t$ implies $V\neg A=V(A \rightarrow \bot)=f$. By I.H. we have $V\mathscr{E}[A]=t$.
5. $\mathscr{E}$ is an N-form $\mathscr{N}[(*_1 \wedge \mathrm{B})]$ or $\mathscr{N}[(B \wedge *_1)]$. Now $VA=f$ implies $V(A \wedge B)=V(B \wedge A)=f$. By I.H. we have $V\mathscr{E}[A]=t$.

By the definition of P-form and N-form all the cases have been treated.

Definition. A positive part is said to be *minimal* if it is not of the form $\neg A$, $(A \vee B)$ or $(A \rightarrow B)$. A negative part is said to be *minimal* if it is not of the form $\neg A$, $(A \wedge B)$ or $(A \rightarrow \bot)$.

Theorem 1.6. *If, under a sentential valuation V, every minimal positive part of a formula F takes the value f and every minimal negative part of F takes the value t, then every positive part of F takes the value f and every negative part of F takes the value t. (In particular, $VF=f$, since F is a positive part of F.)*

Proof by induction on the length of the positive (or negative) part C of the formula F. By hypothesis we need only treat the following cases which are those where C is not a minimal positive or negative part.

1. C is a positive (negative) part $\neg A$. Then A is a negative (positive) part of F. By I.H. $VA=t$ ($VA=f$) hence $VC=f$ ($VC=t$).
2. C is a positive part $(A \vee B)$ (negative part $(A \wedge B)$). Then A and B are positive (negative) parts of F. By I.H. $VA=B=f$ ($VA=VB=t$), hence $VC=f$ ($VC=t$).
3. C is a positive part $(A \rightarrow B)$. Then A is a negative and B a positive part of F. By I.H. $VA=t$ and $VB=f$, hence $VC=f$.
4. C is a negative part $(A \rightarrow \bot)$. Then A is a positive part of F. By I.H. $VA=f$, hence $VC=t$.

6. Syntactic Characterization of Valid Formulas

A formula is said to be *reducible* if it is of the form $\mathscr{P}[(A \wedge B)]$, $\mathscr{N}[(A \vee B)]$ or

$\mathcal{N}[(A \rightarrow B)]$ where, in the last case, B is not the formula $\perp$. Otherwise it is said to be *irreducible*.

Theorem 1.7. *An irreducible formula is valid if, and only if, it is of the form* $\mathcal{Q}[v, v]$ *or* $\mathcal{N}[\perp]$ *(that is, if it contains a sentential variable v both as a positive and as a negative part or has* $\perp$ *as a negative part).*

Proof. 1. Let V be an arbitrary sentential valuation. If $Vv = t$, then by Theorem 1.5 $V\mathcal{Q}[v, v] = t$, since v occurs as a positive part in $\mathcal{Q}[v, v]$. If $Vv = f$, then by Theorem 1.5 $V\mathcal{Q}[v, v] = t$, since v occurs as a negative part in $\mathcal{Q}[v, v]$. In either case, therefore, $V\mathcal{Q}[v, v] = t$. By Theorem 1.5 $V\mathcal{N}[\perp] = t$, since $V\perp = f$. Hence $\mathcal{Q}[v, v]$ and $\mathcal{N}[\perp]$ are valid formulas.

2. Let F be an irreducible formula which is not of the form $\mathcal{Q}[v, v]$ or $\mathcal{N}[\perp]$. We have to prove that F is not valid. Since F is irreducible every minimal positive part and every minimal negative part of F is either a sentential variable or the formula $\perp$. Since F is not of the form $\mathcal{N}[\perp]$ every minimal negative part of F is a sentential variable. Since F is not of the form $\mathcal{Q}[v, v]$ no sentential variable occurs both as a positive and as a negative part of F. Let V be the valuation such that $Vv = t$ if, and only if, the sentential variable v occurs as a negative part of F. Under this valuation V every minimal positive part of F takes the value f and every minimal negative part takes the value t. By Theorem 1.6 we have $VF = f$. Hence F is not valid.

Theorem 1.8. *The following hold for reducible formulas:*

(1) $\mathcal{P}[A \wedge B)]$ *is valid if, and only if,* $\mathcal{P}[A]$ *and* $\mathcal{P}[B]$ *are valid.*
(2) $\mathcal{N}[(A \vee B)]$ *is valid if, and only if,* $\mathcal{N}[A]$ *and* $\mathcal{N}[B]$ *are valid.*
(3) $\mathcal{N}[(A \rightarrow B)]$ *is valid if, and only if,* $\mathcal{N}[(A \rightarrow \perp)]$ *and* $\mathcal{N}[B]$ *are valid.*

Proof of (1). 1. Under a valuation V either $V(A \wedge B) = VA$ or $V(A \wedge B) = VB$, so $V\mathcal{P}[(A \wedge B)] = V\mathcal{P}[A]$ or $V\mathcal{P}[(A \wedge B)] = V\mathcal{P}[B]$. Consequently, if $\mathcal{P}[A]$ and $\mathcal{P}[B]$ are valid, then $\mathcal{P}[(A \wedge B)]$ is also valid.

2. Suppose $\mathcal{P}[A]$ is not valid. Then there is a valuation V such that $V\mathcal{P}[A] = f$. By Theorem 1.5 we have $VA = f$. But then $V(A \wedge B) = f$. Hence $V\mathcal{P}[(A \wedge B)] = V\mathcal{P}[A] = f$. Therefore $\mathcal{P}[(A \wedge B)]$ is not valid. Similarly, if $\mathcal{P}[B]$ is not valid then $\mathcal{P}[(A \wedge B)]$ is not valid.

The proof of (2) proceeds as for (1).

(3) follows from (2) since $\mathcal{N}[(A \rightarrow B)]$ is semantically equivalent to $\mathcal{N}[(\neg A \vee B)]$ and $\mathcal{N}[\neg A]$ is semantically equivalent to $\mathcal{N}[(A \rightarrow \perp)]$.

§2. Formal Systems

1. Fundamentals

The formal systems we are going to deal with are defined syntactically in four steps.

Σ1. Certain symbols (in general denumerably infinitely many) are introduced as *primitive symbols*.

Σ2. Certain finite strings of primitive symbols are selected as *formulas* by an inductive definition.

Σ3. Formulas of a particular sort (in general determined by formula schemata) are defined to be *axioms*.

Σ4. Certain configurations of the form

$$A_1, \dots, A_n \vdash B$$

where $A_1, \dots, A_n, B$ are formulas and in general $n=1$ or $n=2$, are defined to be *basic inferences*. A formula on the left of the symbol $\vdash$ is said to be a *premise* of the given basic inference. The formula on the right of the symbol $\vdash$ is said to be the *conclusion* of the given basic inference.

Re Σ1. We shall not in fact specify explicitly what all the primitive symbols are to be. We shall merely assume that the several primitive symbols are identifiable as such and that primitive symbols of different kinds are different from one another. (For example, a primitive symbol introduced as a connective shall not simultaneously be used as a propositional variable which is a primitive symbol.) In talking about primitive symbols we shall sometimes use the primitive symbols themselves (for example, connectives or round brackets) and sometimes use *syntactic symbols* (for example, $v, v_1, v_2, \dots$ for the propositional variables which are primitive symbols).

Re Σ2. Often, as well as formulas some further kinds of finite strings of primitive symbols are inductively defined. These are the *terms* and *expressions*. In order to define and characterise formulas we frequently use *nominal forms*.

By an *n-place nominal form* ($n \geqslant 1$) in a formal system Σ we mean a non-empty finite string of symbols which contains, in addition to the primitive symbols of the formal system Σ, at most the *nominal symbols* $*_1, \dots, *_n$. These nominal symbols are to be distinct from the primitive symbols of the formal system. In an n-place nominal form the nominal symbols $*_1, \dots, *_n$ may occur in one or more places but need not occur at all. We shall always denote nominal forms by script letters.

If $\mathscr{A}$ is an n-place nominal form ($n \geqslant 1$) and $r_1, \dots, r_n$ are non-empty finite strings of symbols, then $\mathscr{A}[r_1, \dots, r_n]$ denotes the string of symbols which results from the nominal form $\mathscr{A}$ when the nominal symbols $*_1, \dots, *_n$ are replaced at all occurrences in $\mathscr{A}$ by $r_1, \dots, r_n$, respectively.

We have already defined some nominal forms in §1 as P-forms, N-forms and NP-forms.

2. Deducible Formulas

Deducibility of formulas in a formal system is determined by the axioms and basic inferences according to the following **inductive definition**.

D1. Every axiom is *deducible*.

D2. If every premise of a basic inference is *deducible*, then the conclusion of the inference is *deducible*.

A proof (demonstration) of the deducibility of a formula F (according to D1 and D2) is said to be a *deduction* of F.

Sometimes we need to specify the deducibility of a formula more closely so to the formula we assign a natural number, which serves as the *order*, by the following **inductive definition**:

DO 1. Every axiom is *deducible with order* 0.

DO 2. If the premises $A_1, \ldots, A_n$ $(n \geq 1)$ of a basic inference are *deducible with orders* $m_1, \ldots, m_n$, respectively, then the conclusion of the inference *is deducible with order* $\max(m_1, \ldots, m_n) + 1$.

A formula may be deducible with many different orders. However, given that a formula is deducible we may always assume that it is deducible with some fixed order.

We write $\vdash F$ to indicate that the formula F is deducible.

We prove theorems about deducible formulas by *induction on the deduction*. This induction goes as follows.

First we prove that the theorem to be proved holds for every axiom. The second step is to prove the theorem holds for every conclusion of a basic inference under the induction hypothesis that the theorem holds for every premise of the given basic inference. When both these steps have been completed then it follows from the above inductive definition of deducibility that the theorem has been completely proved.

A formal system is said to be *consistent* if not every formula of the system is deducible. Consistency is the least condition one can place on a formal system if deducibility is to be meaningful since otherwise being deducible adds nothing to the notion of formula.

3. Permissible Inferences

A configuration

$$A_1, \ldots, A_n \vdash B$$

where $A_1, \ldots, A_n$ and B are formulas $(n \geq 1)$ is said to be a *permissible inference* if the following holds: if $A_1, \ldots, A_n$ are deducible then B is also deducible.

In particular every basic inference is a permissible inference. In the formal systems we treat in the sequel we consider many more inferences which, though not defined to be basic inferences, will be shown to be permissible. These will be used to establish completeness of the notion of deducibility.

A *weak inference* is a permissible inference $A \vdash B$ such that: if the formula A is deducible with order m, then the formula B is deducible with order $\leq m$.

An inference

$$A_1, \ldots, A_n \vdash B$$

may be permissible for several different reasons.

It may happen that B is deducible from the formulas $A_1, \ldots, A_n$ by using the axioms and further basic inferences. In this case we say that the given inference is *directly derivable*. Clearly a directly derivable inference remains permissible if we extend the notion of deducibility by adding axioms or basic inferences to the formal system.

But it may also happen that the inference can only be proved permissible by referring to the assumed deductions of the formulae $A_1, \ldots, A_n$. Such an inference, which is not directly derivable, may cease to be permissible if deducibility in the formal system is extended. In what follows we shall principally be concerned with permissible inferences of this second kind.

Example. Let the formulas of a formal system Σ be those of the sentential calculus defined in §1.4. Let the only axioms of the system Σ be those formulas of the form

$$(A \vee \neg A).$$

Let the only basic inferences of the system Σ be those inferences of the forms

$$A \vdash (A \vee B) \quad \text{and} \quad B \vdash (A \vee B).$$

Obviously

$$B \vdash ((A \vee B) \vee C)$$

is a directly derivable inference of the system Σ since one can deduce $(A \vee B)$ from B by a basic inference and then deduce $((A \vee B) \vee C)$ from $(A \vee B)$.

Every inference of the form

$$(A \vee A) \vdash A$$

is permissible in Σ since a formula $(A \vee A)$ can only be deduced from the basic inference $A \vdash (A \vee A)$. So the formula $(A \vee A)$ is only deducible if the formula A is deducible. Now if we extend the system Σ by adding all axioms of the form $(A \vee A)$, then it is no longer the case that every inference $(A \vee A) \vdash A$ is permissible since then, for example, the formula $(\bot \vee \bot)$ is deducible while the formula $\bot$ is not deducible in the extended system. Thus the inference $(A \vee A) \vdash A$ is permissible in Σ but not directly derivable.

4. Sentential Properties of Formal Systems

A formal system is said to be *sententially closed* (*s-closed* for short) if among its primitive symbols it contains the connectives from a complete set of connectives which are used for joining formulae as in the sentential calculus. All sentential connectives can be defined in such a formal system because of the completeness of the underlying system of connectives.

A formula in a formal system is said to be *s-valid* if it is obtained by substituting formulas of the formal system for the sentential variables in a valid formula A of the classical sentential calculus where every sentential variable which occurs in more than one place in A is always replaced by the same formula of the formal system.

Example. If Σ is a formal system containing $\rightarrow$ as a primitive symbol and $(A \rightarrow A)$ is a formula of the system Σ, then this formula is s-valid.

A formal system is said to be *s-complete* if it is s-closed and every s-valid formula of the system is deducible in the system.

Note that the notions of s-closure, s-validity and s-completeness defined above refer to the classical sentential calculus and not to the intuitionistic sentential calculus which forms a part of the intuitionistic predicate calculus we treat in chapter III.

A formal system in which every formula A has a negation $\neg A$ defined is said to be *s-consistent* if there is no formula A such that A and $\neg A$ are both deducible in the system. Every s-consistent system is obviously consistent. In many formal systems the converse is easily established, that consistency implies s-consistency. We shall only deal with formal systems which are s-consistent.

5. The Formal System CS of the Classical Sentential Calculus

Primitive symbols of the system CS: Denumerably infinitely many sentential variables, the connectives $\bot$, $\neg$, $\wedge$, $\vee$, $\rightarrow$ and round brackets.

The *formulas* of the system CS are defined as in §1.4 and P-forms, N-forms and NP-forms as in §1.5.

As in §1 we use the following *syntactic symbols*: v for sentential variables, A, B, F for formulas, $\mathscr{P}$ for P-forms, $\mathscr{N}$ for N-forms and $\mathscr{Q}$ for NP-forms.

The *axioms* of the system CS are all formulas of the forms:

(Ax. I) $\mathscr{Q}[v, v]$
(Ax. II) $\mathscr{N}[\bot]$

The *basic inferences* of the system CS are all inferences of forms:

$(I\wedge)$ $\mathscr{P}[A], \mathscr{P}[B] \vdash \mathscr{P}[(A \wedge B)]$
$(I\vee)$ $\mathscr{N}[A], \mathscr{N}[B] \vdash \mathscr{N}[(A \vee B)]$
$(I\rightarrow)$ $\mathscr{N}[(A \rightarrow \bot)], \mathscr{N}[B] \vdash \mathscr{N}[(A \rightarrow B)]$.

The next two theorems show that (syntactic) deducibility given by the above coincides with the semantic notion of validity.

Consistency Theorem. *Every deducible formula is valid.*

Proof by induction on the deduction. It follows from Theorem 1.5, that every axiom is valid. By Theorem 1.8, if both premises of a basic inference are valid then so is the conclusion. Hence every deducible formula is valid.

Completeness Theorem. *Every valid formula is deducible.*

Proof. Let F be a valid formula. Define the *reducibility degree* of F to be the number of connectives $\wedge$, $\vee$ and $\rightarrow$ occurring in the minimal positive and minimal negative parts of F. We prove by induction on the reducibility degree of F that the formula F is deducible.

1. F is an irreducible formula. Then by Theorem 1.7 F is an axiom and hence is deducible.

2. F is a reducible formula. Then F is of the form of the conclusion of a basic inference both of whose premises have smaller reducibility degree than F and hence by Theorem 1.8 are valid. By the induction hypothesis these premises are therefore deducible. But then F is deducible by means of the basic inference involved.

It follows from the consistency theorem and the completeness theorem that the system CS is s-consistent and s-complete.

Remark. For every formula of the system CS we can decide in an elementary way whether and how the formula is deducible in the system CS. For we can easily see whether a formula has the form of an axiom or of a conclusion of a basic inference and because every premise of a basic inference has a lower reducibility degree than the conclusion of the inference.

Chapter II

Classical Predicate Calculus

Predicate calculus forms an extension of the sentential calculus in two different regards.

1. As well as variables for arbitrary sentences we use variables ranging over an arbitrary (non-empty) object domain and arbitrary one- and many-place predicates over the object domain. So in addition to sentential variables we have as basic formulas, formulas of the form $p(a_1, \ldots, a_n)$ with the interpretation: "The objects $a_1, \ldots, a_n$ satisfy the n-place predicate p".

2. As well as the connectives $\wedge$ ("and"), $\vee$ ("or") ... of sentential calculus we use the *quantifiers* $\forall$ and $\exists$ which can be employed to form formulas of the following kinds:

$\forall x \mathscr{F}[x]$ interpreted "For all objects x, $\mathscr{F}[x]$ holds".

$\exists x \mathscr{F}[x]$ interpreted "There is an object x, such that $\mathscr{F}[x]$ holds".

We shall make the interpretation of these formulas more precise in §5 by an appropriate semantics but first we develop a pure syntax for classical predicate calculus. For this we shall use only the connectives $\perp$ and $\rightarrow$ and only the universal quantifier $\forall$ as primitive symbols since the remaining connectives and the existential quantifier $\exists$ can be defined from these in classical predicate calculus. We could in fact use instead the connectives of any other complete system and $\exists$ as primitive symbols. The choice of $\perp$, $\rightarrow$ and $\forall$ as primitive symbols turns out to be especially suitable for other parts of logic as well, in particular for intuitionistic type theory.

§3. The Formal System CP

1. *Primitive Symbols* of the system CP:

1.1. Countably infinitely many free and bound object variables, sentential variables and predicate symbols of each number of arguments $\geqslant 1$.

1.2. The logical symbols $\perp$, $\rightarrow$ and $\forall$.

1.3. Round brackets and the comma.

As syntactic variables we use

a, a_i for free object variables,

x, x_i for bound object variables,

v, v_i for sentential variables.

2. **Inductive Definition** of the *formulas* of the system CP.

2.1. Every sentential variable is a *formula*.

2.2. The symbol $\bot$ is a *formula*.

2.3. If p is an n-place predicate symbol ($n \geqslant 1$) and $a_1, \ldots, a_n$ are free object variables then $p(a_1, \ldots, a_n)$ is a *formula*.

2.4. If A and B are *formulas*, then $(A \rightarrow B)$ is also a *formula*.

2.5. If $\mathscr{F}[a]$ is a *formula* and x a bound object variable which does not occur in the 1-place nominal form $\mathscr{F}$, then $\forall x \mathscr{F}[x]$ is also a *formula*.

The formulas defined by 2.1–2.3 are said to be *atomic* (or *prime*) *formulas*.

As *syntactic symbols* we use $A, A_i, B, B_i, C, C_i, F, F_i, G, G_i$ for formulas, P, P_i for atomic formulas, $\mathscr{F}, \mathscr{F}_i$ for 1-place nominal forms such that $\mathscr{F}[a], \mathscr{F}_i[a]$ are formulas.

3. *P-forms* and *N-forms* are defined as in §1.5. In the system CS since the only connectives which occur as primitive symbols are $\bot$ and $\rightarrow$ we have just the following rules for the **inductive definition**:

3.1. $*_1$ is a *P-form*.

3.2 If $\mathscr{P}$ is a *P-form* and B a formula; then $\mathscr{P}[(B \rightarrow *_1)]$ is a *P-form* and $\mathscr{P}[(*_1 \rightarrow B)]$ is an *N-form*.

3.3. If $\mathscr{N}$ is an *N-form*, then $\mathscr{N}[(*_1 \rightarrow \bot)]$ is a *P-form*.

4. For *positive* and *negative parts* of a formula F, which are determined by P-forms and N-forms as in §1.5, we have the following **inductive characterization**:

4.1. F is a *positive part* of F.

4.2. If $(A \rightarrow B)$ is a *positive part* of F, then A is a *negative* and B a *positive part* of F.

4.3. If $(A \rightarrow \bot)$ is a *negative part* of F, then A is a *positive part* of F.

As in §1 we use *syntactic symbols* as follows: $\mathscr{P}, \mathscr{P}_i$ for P-forms, $\mathscr{N}, \mathscr{N}_i$ for N-forms and $\mathscr{Q}, \mathscr{Q}_i$ for NP-forms.

A positive part of a formula is said to be *minimal* if it is not of the form $(A \rightarrow B)$. A negative part is said to be *minimal* if it is not of the form $(A \rightarrow \bot)$.

$F \models^{S} G$ ("*G follows structurally from F*") denotes that every minimal positive part of F also occurs as a positive part of G and every minimal negative part of F also occurs as a negative part of G.

5. The *axioms* of the system CP are all formulas of the form

(Ax. I) $\mathscr{Q}[P, P]$
(Ax. II) $\mathscr{N}[\bot]$.

6. The *basic inferences* of the system CP are all inferences of the forms

(S1) $\mathscr{N}[(A \rightarrow \bot)], \mathscr{N}[B] \vdash \mathscr{N}[(A \rightarrow B)]$
(S2) $\mathscr{P}[\mathscr{F}[a]] \vdash \mathscr{P}[\forall x \mathscr{F}[x]]$ with a *condition on the variable a*
(S3) $(\forall x \mathscr{F}[x] \rightarrow \mathscr{N}[\mathscr{F}[a]]) \vdash \mathscr{N}[\forall x \mathscr{F}[x]]$.

The free object variable denoted by a in the premise of a basic inference (S2) is said to be the *eigenvariable* of the inference. The *condition on the variable* of an (S2) inference is: The eigenvariable must not occur in the conclusion.

The indicated positive or negative part in the conclusion of a basic inference is said to be the *principal part* of the given basic inference. In an (S1)-inference we may assume that its principal part is not of the form $(A \rightarrow \bot)$ since otherwise the conclusion would be the same as the first premise. Hence every principal part of a basic inference is a minimal positive or negative part of the conclusion.

No atomic formula is deducible in CP since an atomic formula cannot occur as an axiom nor as the conclusion of a basic inference. Hence the system CP is consistent. We shall see in §4 that it is also s-complete and s-consistent.

Remark. The notions of "negative part" and "positive part" may be regarded as generalizations of G. Gentzen's "antecedent formula" and "succedent formula". The generalizations turn out so that there is no need to have any structural inferences as basic inferences. In §4 we shall prove that the inferences corresponding to Gentzen's structural inferences are permissible. Then Theorem 4.6 corresponds to Gentzen's Hauptsatz on the elimination of the cut rule.

§4. Deducible Formulas and Permissible Inferences

1. Generalizations of the Axioms

Theorem 4.1. *Every formula of the form* $\mathscr{Q}[C, C]$ *is deducible.*

Proof by induction on the number of $\rightarrow$ and $\forall$ symbols occurring in C.

1. C is an atomic formula. Then $\mathscr{Q}[C, C]$ is an axiom.

2. C is a formula $(A \rightarrow B)$. Since $\mathscr{Q}$ is an NP-form so are $\mathscr{Q}[(*_2 \rightarrow \bot), (*_1 \rightarrow B)]$ and $\mathscr{Q}[*_1, (A \rightarrow *_2)]$. By I.H. (induction hypothesis) both of the formulas $\mathscr{Q}[(A \rightarrow \bot), (A \rightarrow B)]$ and $\mathscr{Q}[B, (A \rightarrow B)]$ are deducible. Hence $\mathscr{Q}[(A \rightarrow B), (A \rightarrow B)]$ is deducible by an S1-inference.

3. C is a formula $\forall x \mathscr{F}[x]$. We choose a free variable a which does not occur in $\mathscr{Q}[C, C]$. Since $\mathscr{Q}$ is an NP-form so is $(C \rightarrow \mathscr{Q})$. By I.H. the formula $(\forall x \mathscr{F}[x] \rightarrow \mathscr{Q}[\mathscr{F}[a], \mathscr{F}[a]])$ is therefore deducible. By an (S3)-inference we have $\mathscr{Q}[\forall x \mathscr{F}[x], \mathscr{F}[a]]$ and then by an (S2)-inference we have $\mathscr{Q}[\forall x \mathscr{F}[x], \forall x \mathscr{F}[x]]$.

Corollary to Theorem 4.1. *The system* CP *is s-complete.*

Proof. By Corollary 1.4 $\bot$ and $\rightarrow$ form a complete set of connectives. By §2.5 it follows that every s-valid formula which contains no connectives other than $\bot$ and $\rightarrow$ is deducible from formulas only of the forms $\mathscr{Q}[C, C]$ and $\mathscr{N}[\bot]$ using (S1)-inferences. By Theorem 4.1 it follows that CP is s-complete.

2. Weak Inferences

Theorem 4.2 (Substitution rule). *If a_1 does not occur in the nominal form $\mathscr{F}$, then*

$$\mathscr{F}[a_1] \vdash \mathscr{F}[a_2]$$

is a weak inference.

Proof by induction on the deduction. Suppose the formula $\mathscr{F}[a_1]$ is deduced with order m.

1. Suppose $\mathscr{F}[a_1]$ is an axiom. Then $\mathscr{F}[a_2]$ is also an axiom.

2. Suppose $\mathscr{F}[a_1]$ has been inferred by an (S1) or (S3) basic inference from F_i ($i=1, 2$ or $i=1$). F_i is a formula $\mathscr{F}_i[a_1]$ where a_1 does not occur in $\mathscr{F}_i$. By the induction hypothesis $\mathscr{F}_i[a_2]$ is deducible with order $<m$.

$\mathscr{F}_i[a_2]$ ($i=1, 2$ or $i=1$) $\vdash \mathscr{F}[a_2]$ is then an (S1) or (S3) basic inference. Hence $\mathscr{F}[a_2]$ is deducible with order $\leqslant m$.

3. Suppose $\mathscr{F}[a_1]$ has been inferred from F_1 by an (S2)-inference. We choose a free object variable a_0, different from a_1 and a_2 which does not occur in $\mathscr{F}$. Let F_0 be the formula obtained from F_1 by replacing the eigenvariable of the (S2)-inference, wherever it occurs in F_1, by a_0. By I.H. F_0 is deducible with order $<m$. F_0 is a formula $\mathscr{F}_0[a_1]$ where a_1 does not occur in $\mathscr{F}_0$. By I.H. $\mathscr{F}_0[a_2]$ is also deducible with order $<m$.
Now

$$\mathscr{F}_0[a_2] \vdash \mathscr{F}[a_2]$$

is an (S2)-inference with eigenvariable a_0. Hence $\mathscr{F}[a_2]$ is deducible with order $\leqslant m$.

Theorem 4.3 (Inversion rules). *The following are weak inferences*

(1) $\mathscr{N}[(A \to B)] \vdash \mathscr{N}[(A \to \bot)]$
(2) $\mathscr{N}[(A \to B)] \vdash \mathscr{N}[B]$
(3) $\mathscr{P}[\forall x \mathscr{F}[x]] \vdash \mathscr{P}[\mathscr{F}[a]]$.

1. *Proof* of (1) and (2) by induction on the deduction. Suppose $\mathscr{N}[(A \to B)]$ is deduced with order m. If B is the formula $\bot$ then the assertions are trivial since then the conclusion of (1) is the same as the premise and the conclusion of (2) is an axiom. Now let B be a formula different from $\bot$.

1.1. Suppose $\mathscr{N}[(A \to B)]$ is an axiom. Then $\mathscr{N}[(A \to \bot)]$ and $\mathscr{N}[B]$ are axioms.

1.2. Suppose $\mathscr{N}[(A \to B)]$ has been inferred by an (S1)-inference whose principal negative part is the part $((A \to B)$ shown. Then the premises $\mathscr{N}[(A \to \bot)]$ and $\mathscr{N}[B]$ are deducible with order $<m$.

1.3. Suppose $\mathscr{N}[(A \to B)]$ has been deduced in some other way by a basic inference from F_i ($i=1, 2$ or $i=1$). Then F_i is a formula $\mathscr{N}_i[(A \to B)]$ such that

$$\mathscr{N}_i[(A \to \bot)] \; (i=1, 2 \text{ or } i=1) \vdash \mathscr{N}(A \to \bot)$$

and

$\mathcal{N}_i[B]$ ($i=1,'2$ or $i=1$) $\vdash \mathcal{N}[B]$

are basic inferences. Hence the assertion holds by the induction hypothesis.

2. *Proof* of (3) by induction on the deduction. Suppose $\mathcal{P}[\forall x\mathcal{F}[x]]$ is deduced with order m.

2.1. Suppose $\mathcal{P}[\forall x\mathcal{F}[x]]$ is an axiom. Then $\mathcal{P}[F[a]]$ is also an axiom.

2.2. Suppose $\mathcal{P}[\forall x\mathcal{F}[x]]$ has been inferred by an (S2)-inference whose principal part is the given positive part $\forall x\mathcal{F}[x]$. The premise $\mathcal{P}[\mathcal{F}[a_0]]$ of this basic inference is deducible with order $m-1$ where a_0 is the eigenvariable of the (S2)-inference which does not occur in the nominal form $\mathcal{P}[\mathcal{F}]$. It follows from Theorem 4.2 that $\mathcal{P}[\mathcal{F}[a]]$ is deducible with order $<m$.

2.3. Suppose $\mathcal{P}[\forall x\mathcal{F}[x]]$ has been deduced in some other way by a basic inference. We choose a free object variable a_1 which is not the eigenvariable of the basic inference and which does not occur in the nominal form $\mathcal{P}[\mathcal{F}]$. As in case 1.3 it then follows from I.H. that the formula $\mathcal{P}[\mathcal{F}[a_1]]$ is deducible with order $\leqslant m$. By Theorem 4.2 $\mathcal{P}[\mathcal{F}[a]]$ is also deducible with order $\leqslant m$.

Theorem 4.4 (Structural rule). *If $F \models^{S} G$, then $F \vdash G$ is a weak inference.*

Proof by induction on the deduction. Suppose the formula F is deducible with order m.

1. Suppose F is an axiom. Then it follows from $F \models^{S} G$ that G is also an axiom.

2. Suppose F is a formula $\mathcal{N}_1[(A \to B)]$ which has been inferred from $\mathcal{N}_1[(A \to \bot)]$ and $\mathcal{N}_1[B]$ by an (S1)-inference. We replace every negative part $(A \to B)$ in $\mathcal{N}_1[(A \to \bot)]$ by $(A \to \bot)$ and every negative part $(A \to B)$ in $\mathcal{N}_1[B]$ by B. Then we obtain formulas F_1 and F_2 which by Theorem 4.3 are deducible with orders $<m$. It follows from $F \models^{S} G$ that G is a formula $\mathcal{N}_2[(A \to B)]$ and that $F_1 \models^{S} \mathcal{N}_2[(A \to \bot)]$ and $F_2 \models^{S} \mathcal{N}_2[B]$. By I.H. the formulas $\mathcal{N}_2[(A \to \bot)]$ and $\mathcal{N}_2[B]$ are therefore deducible with orders $<m$. It follows by an (S1)-inference that $\mathcal{N}_2[(A \to B)]$ is deducible with order $\leqslant m$.

3. Suppose F is a formula $\mathcal{P}_1[\forall x\mathcal{F}[x]]$ inferred from $\mathcal{P}_1[\mathcal{F}[a]]$ by an (S2)-inference. We choose a free object variable a_0 which occurs neither in F nor in G. In the formula $\mathcal{P}_1[\mathcal{F}[a]]$ we replace a by a_0 and every positive part $\forall x\mathcal{F}[x]$ by $\mathcal{F}[a_0]$. This gives a formula F_0 which, by Theorems 4.2 and 4.3, is deducible with order $<m$. It follows from $F \models^{S} G$ that G is a formula $\mathcal{P}_2[\forall x\mathcal{F}[x]]$ and $F_0 \models^{S} \mathcal{P}_2[\mathcal{F}[a_0]]$ holds. By I.H. the formula $\mathcal{P}_2[\mathcal{F}[a_0]]$ is deducible with order $<m$. It follows by an (S2)-inference that $\mathcal{P}_2[\forall x\mathcal{F}[x]]$ is deducible with order $\leqslant m$.

4. Suppose F is a formula $\mathcal{N}_1[\forall x\mathcal{F}[x]]$ inferred from $(\forall x\mathcal{F}[x] \to \mathcal{N}_1[\mathcal{F}[a]])$ by an (S3)-inference. It follows from $F \models^{S} G$ that G is a formula $\mathcal{N}_2[\forall x\mathcal{F}[x]]$ and $(\forall x\mathcal{F}[x] \to \mathcal{N}_1[\mathcal{F}[a]]) \models^{S} (\forall x\mathcal{F}[x] \to \mathcal{N}_2[\mathcal{F}[a]])$ holds. By I.H. the formula $(\forall x\mathcal{F}[x] \to \mathcal{N}_2[\mathcal{F}[a]])$ is therefore deducible with order $<m$. It follows by an (S3)-inference that $\mathcal{N}_2[\forall x\mathcal{F}[x]]$ is deducible with order $\leqslant m$.

3. Further Permissible Inferences

Theorem 4.5. $\mathcal{N}[\mathcal{F}[a]] \vdash \mathcal{N}[\forall x \mathcal{F}[x]]$ *is a permissible inference (provided x does not occur in* $\mathcal{F}$*).*

Proof. We have $\mathcal{N}[\mathcal{F}[a]] \models^{S} (\forall x \mathcal{F}[x] \to \mathcal{N}[\mathcal{F}[a]])$. From $\vdash \mathcal{N}[\mathcal{F}[a]]$ by Theorem 4.4 we therefore have $\vdash (\forall x \mathcal{F}[x] \to \mathcal{N}[\mathcal{F}[a]])$. By an (S3)-inference $\vdash \mathcal{N}[\forall x \mathcal{F}[x]]$ follows.

Theorem 4.6 (Cut rule).

$$\mathcal{P}[A], (A \to B) \vdash \mathcal{P}[B]$$

is a permissible inference.

Proof by induction on the number of $\to$ and $\forall$ symbols occurring in the formula A.

1. Suppose A is an atomic formula. For this case we have to do a subsidiary induction on the order of a deduction of the formula $\mathcal{P}[A]$.

1.1. Suppose $\mathcal{P}[A]$ is an axiom. Then either $\mathcal{P}[B]$ is also an axiom or $\mathcal{P}$ is a P-form $\mathcal{Q}[A, *_1]$. In the latter case $(A \to B) \models^{S} \mathcal{P}[B]$. Then by Theorem 4.4 from $\vdash (A \to B)$ we have $\vdash \mathcal{P}[B]$.

1.2. Suppose $\mathcal{P}[A]$ was obtained by a basic inference from F_i ($i=1, 2$ or $i=1$). Then by Theorem 4.2 we may assume that the basic inference has no free object variable which occurs in B as its eigenvariable. Then F_i is a formula of the form $\mathcal{P}_i[A]$ so

$$\mathcal{P}_i[B] \ (i=1, 2 \text{ or } i=1) \vdash \mathcal{P}[B]$$

is also a basic inference. Then $\vdash \mathcal{P}_i[B]$ follows from $\vdash \mathcal{P}_i[A]$ and $\vdash (A \to B)$ by the S.I.H. (subsidiary induction hypothesis). Using the given basic inference $\vdash \mathcal{P}[B]$ follows.

2. Suppose A is a formula $A_1 \to A_2$. By Theorem 4.3 from $\vdash ((A_1 \to A_2) \to B)$ we have $\vdash ((A_1 \to \bot) \to B)$ and $\vdash (A_2 \to B)$. Since $\mathcal{P}$ is a P-form so is $\mathcal{P}[(A_1 \to *_1)]$. Hence by the I.H. from $\vdash \mathcal{P}[(A_1 \to A_2)]$ and $\vdash (A_2 \to B)$ we have $\vdash \mathcal{P}[A_1 \to B)]$. By Theorem 4.4 there follows $\vdash (A_1 \to \mathcal{P}[B])$. Since $((*_1 \to \bot) \to B)$ is a P-form, from $\vdash ((A_1 \to \bot) \to B)$ and $\vdash (A_1 \to \mathcal{P}[B])$ by I.H. we have $\vdash ((\mathcal{P}[B] \to \bot) \to B)$. We have $((\mathcal{P}[B] \to \bot) \to B) \models^{S} \mathcal{P}[B]$. Therefore by Theorem 4.4 we have $\vdash \mathcal{P}[B]$.

3. Suppose A is a formula $\forall x \mathcal{F}[x]$. In this case we use a subsidiary induction on the order m of a deduction of the formula $(\forall x \mathcal{F}[x] \to B)$.

3.1. Suppose $(\forall x \mathcal{F}[x] \to B)$ is an axiom. Then $\mathcal{P}[B]$ is also an axiom.

3.2. Suppose $(\forall x \mathcal{F}[x] \to B)$ was inferred by an (S1) or (S2) basic inference from F_i ($i=1, 2$ or $i=1$). Then by Theorem 4.2 we may assume that the basic inference has no free object variable which occurs in $\mathcal{P}$ as its eigenvariable. Then F_i is a formula of the form $(\forall x \mathcal{F}[x] \to B_i)$ so

$$\mathcal{P}[B_i] \ (i=1, 2 \text{ or } i=1) \vdash \mathcal{P}[B]$$

is also an (S1) or (S2) basic inference. By the S.I.H. from $\vdash \mathscr{P}[\forall x \mathscr{F}[x]]$ and $\vdash (\forall x \mathscr{F}[x] \rightarrow B_i)$ we have $\vdash \mathscr{P}[B_i]$. Then by the given basic inference we have $\vdash \mathscr{P}[B]$.

3.3. Suppose $(\forall x \mathscr{F}[x] \rightarrow B)$ was inferred by an (S3)-inference from $(\forall x \mathscr{F}[x] \rightarrow (\mathscr{F}[a] \rightarrow B)$. By the S.I.H. we then have $\vdash \mathscr{P}[(\mathscr{F}[a] \rightarrow B)]$. $\vdash (\mathscr{F}[a] \rightarrow \mathscr{P}[B])$ then follows by Theorem 4.4. By Theorem 4.3 from $\vdash \mathscr{P}[\forall x \mathscr{F}[x]]$ we have $\vdash \mathscr{P}[\mathscr{F}[a]]$. Then using $\vdash (\mathscr{F}[a] \rightarrow \mathscr{P}[B])$ we have by the I.H. $\vdash \mathscr{P}[\mathscr{P}[B]]$. Now $\mathscr{P}[\mathscr{P}[B]] \models^{S} \mathscr{P}[B]$. Therefore by Theorem 4.4 we have $\vdash \mathscr{P}[B]$.

3.4. Suppose B is a formula $\mathscr{N}[\forall x_0 \mathscr{F}_0[x_0]]$ and $(\forall x \mathscr{F}[x] \rightarrow \mathscr{N}[\forall x_0 \mathscr{F}_0[x_0]])$ was inferred from $(\forall x_0 \mathscr{F}_0[x_0] \rightarrow (\forall x \mathscr{F}[x] \rightarrow \mathscr{N}[\mathscr{F}_0[a]]))$ by an (S3)-inference. By Theorem 4.4 $(\forall x \mathscr{F}[x] \rightarrow (\forall x_0 \mathscr{F}_0[x_0] \rightarrow \mathscr{N}[\mathscr{F}_0[a]]))$ is also deducible with order $<m$. Hence by the S.I.H. we have $\vdash \mathscr{P}[(\forall x_0 \mathscr{F}_0[x_0] \rightarrow \mathscr{N}[\mathscr{F}_0[a]])]$. By Theorem 4.4 we therefore have $\vdash (\forall x_0 \mathscr{F}_0[x_0] \rightarrow \mathscr{P}[\mathscr{N}[\mathscr{F}_0[a]]])$. Since $\mathscr{P}[\mathscr{N}]$ is also an N-form, using an (S3)-inference we have $\vdash \mathscr{P}[[\mathscr{N}[\forall x_0 \mathscr{F}_0[x_0]]]$.

4. Defined Logical Connectives

The connectives $\neg$, $\wedge$, $\vee$ and the existential quantifier $\exists$ are defined as follows in the system CP:

$$\begin{aligned} \neg A &:= (A \rightarrow \bot) \\ (A \wedge B) &:= ((A \rightarrow (B \rightarrow \bot)) \rightarrow \bot) \\ (A \vee B) &:= ((A \rightarrow \bot) \rightarrow B) \\ \exists x \mathscr{F}[x] &:= (\forall x(\mathscr{F}[x] \rightarrow \bot) \rightarrow \bot). \end{aligned}$$

These connectives we have just defined satisfy the defining rules given in §1.5 for positive and negative parts as we now show.

If $\neg A$ (that is $(A \rightarrow \bot)$) is a *positive* (or *negative*) *part* of a formula F, then A is a *negative* (or *positive*) *part* of F.

If $((A \rightarrow (B \rightarrow \bot)) \rightarrow \bot)$ is a *negative part* of F, then $(A \rightarrow (B \rightarrow \bot))$ is a *positive part*, A is a *negative part*, $(B \rightarrow \bot)$ a *positive part* and B a *negative part* of F. So we have: If $(A \wedge B)$ is a *negative part* of F, then A and B are also *negative parts* of F.

If $((A \rightarrow \bot) \rightarrow B)$ is a *positive part* of F, then $(A \rightarrow \bot)$ is a *negative part*, A a *positive part* and B a *positive part* of F.

So we have: If $(A \vee B)$ is a *positive part* of F, then A and B are also *positive parts* of F.

As we have defined negation, we therefore have:

Theorem 4.7. *The formal system* CP *is s-consistent.*

Proof. By Theorem 4.6 and the definition of $\neg$

$$A, \neg A \vdash \bot$$

is a permissible inference. Since the formula $\bot$ is not deducible in CP it therefore follows that there is no formula A such that A and $\neg A$ are both deducible in CP.

The proof-theoretic properties of $\wedge$, $\vee$ and $\exists$ are given by the following theorem.

Theorem 4.8. *The following inferences are permissible:*

(1) $\mathscr{P}[A], \mathscr{P}[B] \vdash \mathscr{P}[(A \wedge B)]$
(2) $\mathscr{N}[A_i] \vdash \mathscr{N}[(A_1 \wedge A_2)] \quad (i=1,2)$
(3) $\mathscr{P}[A_i] \vdash \mathscr{P}[(A_1 \vee A_2)] \quad (i=1,2)$
(4) $\mathscr{N}[A], \mathscr{N}[B] \vdash \mathscr{N}[(A \vee B)]$
(5) $\mathscr{P}[\mathscr{F}[a]] \vdash \mathscr{P}[\exists x \mathscr{F}[x]]$
(6) $\mathscr{N}[\mathscr{F}[a]] \vdash \mathscr{N}[\exists x \mathscr{F}[x]]$,

if a does not occur in the conclusion (condition on variables).

Proof. (1) For every formula C we have $\mathscr{P}[C] \models^{\mathrm{S}} \mathscr{P}[((C \to \bot) \to \bot)]$. Therefore by Theorem 4.4 from $\vdash \mathscr{P}[A]$ and $\vdash \mathscr{P}[B]$ we have $\vdash \mathscr{P}[((A \to \bot) \to \bot)]$ and $\vdash \mathscr{P}[((B \to \bot) \to \bot)]$. Since $\mathscr{P}[(*_1 \to \bot)]$ is an N-form, we obtain by an (S1)-inference $\vdash \mathscr{P}[((A \to (B \to \bot)) \to \bot)]$, that is $\vdash \mathscr{P}[(A \wedge B)]$.

(2) Follows from Theorem 4.4 since $\mathscr{N}[A_i] \models^{\mathrm{S}} \mathscr{N}[(A_1 \wedge A_2)]$ $(i=1,2)$ by the definition of $\wedge$.

(3) Follows from Theorem 4.4 since $\mathscr{P}[A_i] \models^{\mathrm{S}} \mathscr{P}[(A_1 \vee A_2)]$ $(i=1,2)$ by the definition of $\vee$.

(4) We have $\mathscr{N}[A] \models^{\mathrm{S}} \mathscr{N}[((A \to \bot) \to \bot)]$. By Theorem 4.4 from $\vdash \mathscr{N}[A]$ we therefore have $\vdash \mathscr{N}[((A \to \bot) \to \bot)]$. Using $\vdash \mathscr{N}[B]$ by an (S1)-inference we have $\vdash \mathscr{N}[((A \to \bot) \to B)]$, that is $\mathscr{N}[(A \vee B)]$.

(5) By Theorem 4.4 from $\vdash \mathscr{P}[\mathscr{F}[a]]$ we have $\vdash \mathscr{P}[((\mathscr{F}[a] \to \bot) \to \bot)]$. Since $\mathscr{P}[(*_1 \to \bot)]$ is an N-form Theorem 4.5 yields $\vdash \mathscr{P}[(\forall x(\mathscr{F}[x] \to \bot) \to \bot)]$, that is $\mathscr{P}[\exists x \mathscr{F}[x]]$.

(6) From $\vdash \mathscr{N}[\mathscr{F}[a]]$ by Theorem 4.4 we have $\vdash \mathscr{N}[((\mathscr{F}[a] \to \bot) \to \bot)]$. $\mathscr{N}[(*_1 \to \bot)]$ is a P-form. If a does not occur in $\mathscr{N}[\mathscr{F}]$, then by an (S2)-inference we have $\vdash \mathscr{N}[(\forall x(\mathscr{F}[x] \to \bot) \to \bot)]$, that is $\vdash \mathscr{N}[\exists x \mathscr{F}[x]]$.

Remark. All the proofs we have given in this section which show that certain inferences are permissible are constructive. That is to say: In each case, if deductions of the formulas $A_1, \ldots, A_n$ are given, then one can obtain a deduction of the formula B from the proof of the permissibility of the inference

$$A_1, \ldots, A_n \vdash B.$$

§5. Semantics of Classical Predicate Calculus

Although all the notions we have used so far and all the proofs we have given were strictly constructive (*finitist* in the sense of D. Hilbert), when it comes to classical predicate calculus we have to deal with non-constructive notions.

1. Classical Models

A *model* $\mathcal{M} = \langle \mathcal{D}, V \rangle$ consists of a non-empty set $\mathcal{D}$ (the *domain of individuals* of the model) and a function V (the *valuation* of the model) defined on the free object variables, predicate letters and propositional variables as follows:

V1. For each free object variable a, Va is an element of the set $\mathcal{D}$.

V2. For each n-place predicate letter p ($n \geqslant 1$), Vp is a set of n-tuples $(\alpha_1, \ldots, \alpha_n)$ of elements $\alpha_1, \ldots, \alpha_n$ from the set $\mathcal{D}$.

V3. For each sentential variable v, Vv is one of the truth-values t, f.

By a *$\mathcal{D}$-formula* we mean a string of symbols which is the result of replacing some free object variables in a formula of the system CP by names of elements from the set $\mathcal{D}$.

We use the following as *syntactic symbols*:

α, α_i for names of elements of the set $\mathcal{D}$,

e, e_i for free object variables and names of elements of the set $\mathcal{D}$,

A', B', F', $\forall x \mathcal{F}'[x]$ for $\mathcal{D}$-formulas.

For a model $\mathcal{M} = \langle \mathcal{D}, V \rangle$ the valuation V has properties V1–V3. We now extend V by definition so that every $\mathcal{D}$-formula F' also gets a truth-value VF'.

V4. If $\alpha \in \mathcal{D}$, then $V\alpha := \alpha$.

V5. $V\bot := f$.

V6. If p is an n-place predicate letter ($n \geqslant 1$) then $Vp(e_1, \ldots, e_n) := t$ if $(Ve_1, \ldots, Ve_n) \in Vp$. Otherwise, $Vp(e_1, \ldots, e_n) = f$.

V7. If $VA' = f$ or $VB' = t$, then $V(A' \to B') := t$. Otherwise $V(A' \to B') := f$.

V8. If $V\mathcal{F}'[\alpha] = t$ for all $\alpha \in \mathcal{D}$ then $V\forall x \mathcal{F}'[x] := t$. Otherwise $V\forall x \mathcal{F}'[x] := f$.

We note in particular that every formula F gets a truth-value VF, since formulas are just particular $\mathcal{D}$-formulas.

VF' is defined by induction on the number of $\to$ and $\forall$ symbols which occur in the $\mathcal{D}$-formula F'. But this is non-constructive since we cannot always calculate the value VF' as the following example shows.

Example. Let $\mathcal{D}$ be the set of positive integers. Let p be a 4-place predicate letter and set Vp to be the set of quadruples $(\alpha_1, \ldots, \alpha_4)$ of positive integers such that $\alpha_1^{\alpha_4+2} + \alpha_2^{\alpha_4+2} \neq \alpha_3^{\alpha_4+2}$. Let F be the formula $\forall x_1 \forall x_2 \forall x_3 \forall x_4 p(x_1, x_2, x_3, x_4)$. Then in the model $\langle \mathcal{D}, V \rangle$ we have $VF = t$ if, and only if, Fermat's last theorem is true, that is if for every $n \geqslant 3$ there are no positive integers $\alpha_1, \alpha_2, \alpha_3$ such that $\alpha_1^n + \alpha_2^n = \alpha_3^n$. But Fermat's conjecture has not yet been fully established.

In the definition of VF' for a $\mathcal{D}$-formula F' the connectives $\bot$ and $\to$ are treated as they were in §1 in the definition of VA for a formula A of the sentential calculus. Consequently, corresponding to Theorem 1.5 we have:

PN-Lemma. If $\mathcal{E}$ is a P-form (or an N-form) and $VA = t$ (or $VA = f$) for a valuation V in a model, then $V\mathcal{E}[A] = t$.

A formula F is said to be *true in a model* $\mathcal{M} = \langle \mathcal{D}, V \rangle$ if $VF = t$. We also say in this case that $\mathcal{M}$ is a *model of the formula F*.

A formula is said to be *valid* if it is true in every model. It is said to be *satisfiable* if there is some model in which it is true.

2. The Consistency Theorem

Theorem 5.1 (Consistency Theorem). *Every deducible formula is valid.*

Proof by induction on the deduction. Let F be a deducible formula and $\mathcal{M}=\langle\mathcal{D}, V\rangle$ an arbitrary model. We have to prove that $VF=t$.

1. Suppose F is an axiom $\mathcal{Q}[P, P]$ or $\mathcal{N}[\bot]$. By the PN-lemma we have $VF=t$.

2. Suppose F is a formula $\mathcal{N}[(A\to B)]$ inferred from $\mathcal{N}[(A\to\bot)]$ and $\mathcal{N}[B]$ by an (S1)-inference. By I.H. we have $V\mathcal{N}[(A\to\bot)]=t$ and $V\mathcal{N}[B]=t$. If $VB=t$, then $V(A\to B)=VB$ and $V\mathcal{N}[(A\to B)]=V\mathcal{N}[B]=t$. If $VB=f$, then $V(A\to B)=V(A\to\bot)$ and $V\mathcal{N}[(A\to B)]=V\mathcal{N}[(A\to\bot)]=t$. So in either case $V\mathcal{N}[(A\to B)]=t$.

3. Suppose F is a formula $\mathcal{P}[\forall x\mathcal{F}[x]]$ inferred from $\mathcal{P}[\mathcal{F}[a]]$ by an (S2)-inference. If $V\forall x\mathcal{F}[x]=t$, then $V\mathcal{P}[\forall x\mathcal{F}[x]]=t$ by the PN-lemma. Now suppose there exists $\alpha\in\mathcal{D}$ with $V\mathcal{F}[\alpha]=f$. We define a new model $\mathcal{M}'=\langle\mathcal{D}, V'\rangle$ where $V'a=\alpha$ and otherwise V' agrees with V on free object variables, predicate letters and sentential variables. Then, since the eigenvariable a does not occur in $\mathcal{P}[\mathcal{F}]$ we obtain

$$V'\mathcal{F}[a]=V\mathcal{F}[\alpha]=f$$
$$V'\forall x\mathcal{F}[x]=V\forall x\mathcal{F}[x]=f$$

and

$$V\mathcal{P}[\forall x\mathcal{F}[x]]=V'\mathcal{P}[\forall x\mathcal{F}[x]]=V'\mathcal{P}[\mathcal{F}[a]].$$

Since the premise $\mathcal{P}[\mathcal{F}[a]]$ of the (S2)-inference is valid by I.H., we have $V'\mathcal{P}[\mathcal{F}[a]]=t$. Hence $V\mathcal{P}[\forall x\mathcal{F}[x]]=t$.

4. Suppose F is a formula $\mathcal{N}[\forall x\mathcal{F}[x]]$ inferred from $(\forall x\mathcal{F}[x]\to\mathcal{N}(\mathcal{F}[a]])$ by an (S3)-inference. If $V\forall x\mathcal{F}[x]=f$, then $V\mathcal{N}[\forall x\mathcal{F}[x]]=t$ by the PN-lemma. Now suppose $V\forall x\mathcal{F}[x]=t$. Then $V\mathcal{F}[a]=t$ too and it follows that $V\mathcal{N}[\forall x\mathcal{F}[x]]=V\mathcal{N}[\mathcal{F}[a]]=V(\forall x\mathcal{F}[x]\to\mathcal{N}[\mathcal{F}[a]])$. By I.H. $V(\forall x\mathcal{F}[x]\to\mathcal{N}[\mathcal{F}[a]])=t$. Hence we also have $V\mathcal{N}[\forall x\mathcal{F}[x]]=t$.

3. The Completeness Theorem

In order to prove that, conversely, every valid formula is deducible we need several auxiliary notions.

First we let

$$\alpha_0, \alpha_1, \alpha_2, \ldots$$

be an enumeration of all the free object variables.

A part of a formula F is said to be *reducible* if it is a positive part of the form $\forall x\mathcal{F}[x]$ or a negative part of the form $\forall x\mathcal{F}[x]$ or $(A\to B)$ provided B is not the formula $\perp$. A formula is said to be *reducible* if it has at least one reducible part, otherwise it is said to be *irreducible*. According to this definition a formula F is irreducible if, and only if, every minimal positive part and every minimal negative part of F is an atomic formula. The *reducibility degree* of a formula F is the number of $\rightarrow$ and $\forall$ symbols occurring in the reducible parts of F.

By the *distinguished part* of a reducible formula F we mean that reducible part of F which occurs furthest to the right.

A *D-chain* (*deduction chain*) for a formula F is a sequence of formulas

$$F_0, F_1, F_2, \ldots,$$

constructed as follows:

D1. The initial formula F_0 of the D-chain is the formula F.

D2. If the formula F_n of the D-chain is an axiom or an irreducible formula, then it is the last formula of the D-chain. We then say that the D-chain has *length n*.

D3. If the formula F_n of the D-chain is a reducible formula which is not an axiom then F_n has an immediate successor F_{n+1} in the D-chain determined as follows by $F_0, \ldots, F_n$.

D3.1. If F_n is a formula $\mathcal{N}[(A \rightarrow B)]$ with distinguished negative part $(A \rightarrow B)$, then F_{n+1} is either the formula $\mathcal{N}[(A \rightarrow \perp)]$ or the formula $\mathcal{N}[B]$.

D3.2. If F_n is a formula $\mathcal{P}[\forall x \mathcal{F}[x]]$ with distinguished positive part $\forall x \mathcal{F}[x]$, then F_{n+1} is the formula $\mathcal{P}[\mathcal{F}[\alpha_i]]$ where i is the least number such that α_i does not occur in F_n.

D3.3. If F_n is a formula $\mathcal{N}[\forall x \mathcal{F}[x]]$ with distinguished negative part $\forall x \mathcal{F}[x]$, then F_{n+1} is the formula $(\forall x \mathcal{F}[x] \rightarrow \mathcal{N}[\mathcal{F}[\alpha_i]])$, where i is the least number such that $\mathcal{F}[\alpha_i]$ does not occur as a negative part in any of the formulas among $F_0, \ldots, F_n$ if the nominal symbol $*_1$ occurs in $\mathcal{F}$.

D-chains are therefore formed inversely to the basic inferences.

We shall shortly prove:

Principal Syntactic Lemma. *If every D-chain of a formula F contains an axiom then the formula F is deducible.*

Principal Semantic Lemma. *If there is a D-chain of a formula F which contains no axiom, then there is a model $\langle \mathbb{N}, V \rangle$ with domain the natural numbers $\mathbb{N}$ such that $VF = f$.*

From these two lemmata we obtain:

Theorem 5.2 (Completeness Theorem). *Every valid formula is deducible.*

Proof. Let F be a valid formula. By the principal semantic lemma every D-chain of F contains an axiom. From the principal syntactic lemma it follows that F is deducible.

Theorem 5.3 (Löwenheim-Skolem). *Every satisfiable formula has a model whose domain is the natural numbers.*

Proof. Let F be a satisfiable formula. Then the formula $(F \to \bot)$ is not valid, so by Theorem 5.1, not deducible. From the principal syntactic lemma it follows that there is a D-chain for $(F \to \bot)$ which contains no axiom. By the principal semantic lemma it follows that there is a model $\mathcal{M} = \langle \mathbb{N}, V \rangle$ such that $V(F \to \bot) = f$. Then $VF = t$ so F is true in the model $\mathcal{M}$.

In order to prove the principal syntactic lemma we need the following version of **König's Lemma**:

If a formula F has infinitely many D-chains then there is an infinite D-chain for F.

Proof of the lemma. An infinite D-chain

$$F_0, F_1, F_2, \ldots$$

is obtained from F as follows. Let F_0 be the formula F. By hypothesis there are infinitely many D-chains beginning with F_0. Now assume $F_0, \ldots, F_n$ have been determined in such a way that infinitely many D-chains begin with $F_0, \ldots, F_n$. Then F_n is a reducible formula which is not an axiom.

1. Suppose F_n is a formula $\mathcal{N}[(A \to B)]$ with distinguished negative part $(A \to B)$. Then only one of the two formulas $\mathcal{N}[(A \to \bot)]$ and $\mathcal{N}[B]$ can occur in D-chain after F_n. Let F_{n+1} be the formula $\mathcal{N}[(A \to \bot)]$ if infinitely many D-chains begin with $F_0, \ldots, F_n, \mathcal{N}[(A \to \bot)]$. Otherwise let F_{n+1} be the formula $\mathcal{N}[B]$.

2. Suppose F_n has a distinguished part of the form $\forall x \mathcal{F}[x]$. Then the formula F_{n+1} following F_n is uniquely determined by $F_0, \ldots, F_n$.

It follows from our assumption in each case that there are infinitely many D-chains which begin with $F_0, \ldots, F_{n+1}$. Since this process of going from F_n to F_{n+1} never stops, it determines an infinite D-chain of F. (The proof is non-constructive since the choice of F_{n+1} depends on infinitely many successors.)

Proof of the principal syntactic lemma. Let F be a formula such that every D-chain of F contains an axiom. Then such an axiom is the last formula of such a D-chain. Hence every D-chain of F is finite. From König's lemma it follows that there are only finitely many D-chains. So these have a maximal length m. We prove by induction on m–n that every formula F_n which occurs as the n-th formula in a D-chain of F is deducible. If F_n is the last formula of a D-chain of F then F_n is an axiom. Otherwise F_n is of the form of a conclusion of a basic inference whose premises come after F_n in D-chains of F. By I.H. these premises are deducible. Consequently F_n is also deducible. In particular the formula F_0, that is, F, is deducible, as was to be shown.

Proof of the principal semantic lemma. Let

$$\text{(DC)} \qquad F_0, F_1, \ldots, F_n, \ldots$$

be a D-chain which contains no axiom where F_0 is the formula F. We prove some lemmata about this D-chain.

Lemma 1. *If an atomic formula occurs as a positive (or as a negative) part in a formula F_n of the D-chain, then it also occurs in every formula F_m ($m \geqslant n$) of the D-chain as a positive (or as a negative) part.*

Proof. This follows immediately from the definition of the D-chain.

Lemma 2. *If a formula F_n of the D-chain has a reducible positive (or reducible negative) part C, then the D-chain contains a reducible formula F_m ($m \geqslant n$) with distinguished positive (negative) part C.*

Proof by induction on the number k of $\rightarrow$ and $\forall$ symbols occurring in the reducible parts of F_n to the right of C. If C is the distinguished part of F_n then the assertion is satisfied with $m=n$. Otherwise F_n is followed by a formula F_{n+1} with positive (or negative) part C which contains less than $k \rightarrow$ and $\forall$ symbols in reducible parts to the right of C. But then the assertion follows from the induction hypothesis.

Lemma 3. *If a formula F_n of the D-chain has a negative part $\forall x\mathscr{F}[x]$, then the D-chain contains infinitely many reducible formulas with distinguished negative part $\forall x\mathscr{F}[x]$.*

Proof. It follows from the definition of D-chain that the formula $\forall x\mathscr{F}[x]$ occurs as a negative part in every formula F_m ($m \geqslant n$) in the D-chain. Therefore every formula in the D-chain is reducible. Since the D-chain contains no axiom it follows that the D-chain is infinite. The assertion now follows from Lemma 2.

We say that a formula is a *positive (or negative) part of* (DC) if it occurs as a positive (or negative) part in some formula of the D-chain (DC). We now define a *model* $\mathscr{M}=\langle \mathbb{N}, V\rangle$ as follows:

1. $V\alpha_i := i$
2. If p is an n-place predicate letter ($n \geqslant 1$) then Vp is the set of n-tuples $(i_1, \ldots, i_n)$ of natural numbers such that $p(\alpha_{i_1}, \ldots, \alpha_{i_n})$ is a negative part of (DC).
3. If v is a negative part of (DC) then set $Vv := t$. Otherwise set $Vv := f$.

The following lemmata hold for this model.

Lemma 4. *If P is an atomic formula then $VP=t$ if, and only if, P is a negative part of* (DC).

Proof. Since the D-chain (DC) contains no axiom, $\perp$ is not a negative part of (DC). For the remaining atomic formulas the assertion follows immediately from the definition of the model.

Lemma 5. *If C is a positive (or a negative) part of* (DC) *then $VC=f$ (or $VC=t$).*

Proof by induction on the length of C.

1. Suppose C is an atomic formula. Since the D-chain (DC) contains no axiom $\mathscr{Q}[C, C]$ it follows from Lemma 1 that C is not both a positive and a negative part of (DC). The assertion then follows from Lemma 4.

2. Suppose C is a formula $(A \to B)$. If $(A \to B)$ is a positive part of (DC), then A is a negative and B a positive part of (DC). Then by I.H. $VA = t$ and $VB = f$. It follows that $V(A \to B) = f$. If $(A \to \perp)$ is a negative part of (DC), then A is a positive part of (DC). Therefore by I.H. $VA = f$. It follows that $V(A \to \perp) = t$. If $(A \to B)$ is a negative part of (DC) but B is not the formula $\perp$, then it follows from Lemma 2 and the definition of D-chain that $(A \to \perp)$ or B is a negative part of (DC). Then by I.H., $VA = f$ or $VB = t$. Hence $V(A \to B) = t$.

3. Suppose C is a formula $\forall x \mathscr{F}[x]$. If $\forall x \mathscr{F}[x]$ is a positive part of (DC), then it follows from Lemma 2 and the definition of the D-chain that some formula $\mathscr{F}[\alpha_i]$ is a positive part of (DC). Then by I.H. $V\mathscr{F}[\alpha_i] = f$. Hence $V\forall x \mathscr{F}[x] = f$. If $\forall x \mathscr{F}[x]$ is a negative part of (DC), then it follows from Lemma 3 and the definition of the D-chain that every formula $\mathscr{F}[\alpha_i]$ is a negative part of (DC). Then by I.H., $V\mathscr{F}[\alpha_i] = t$ for all $i \in \mathbb{N}$. Hence $V\forall x \mathscr{F}[x] = t$.

In particular from Lemma 5 we have $VF = f$, since F is a positive part of (DC). This completes the proof of the principal semantic lemma.

Remark. The proof of the completeness theorem is non-constructive since it draws on the non-constructive notion of validity and König's lemma which has a non-constructive proof. On the other hand the D-chains were constructively defined and it follows from the proof that for every deducible formula one has a *standard deduction* which is given, constructively, by the D-chains.

4. The Satisfiability Theorem

For brevity we write

$$A_1, \ldots, A_n \to B \quad (n \geqslant 0)$$

for

$$(A_1 \to (\cdots \to (A_n \to B) \ldots)).$$

In the case $n = 0$ this means the formula B.

A set S of formulas is said to be *inconsistent* if there is a finite subset $\{A_1, \ldots, A_n\}$ of S such that the formula

$$A_1, \ldots, A_n \to \perp$$

is deducible. Otherwise the set S is said to be *consistent*.

A set S of formulas is said to be *satisfiable* if there is a model $\mathscr{M}$ in which every formula in the set S is true. Then we also say that S is *satisfiable in* $\mathscr{M}$ or that $\mathscr{M}$ is a *model of the set* S.

Consistency is a syntactic notion, satisfiability a semantic one.

An inconsistent set of formulas is not satisfiable since if a formula $A_1, \ldots, A_n \to \perp$ is deducible then by the consistency theorem it follows that for every model $\mathcal{M}=\langle\mathcal{D}, V\rangle$: $V(A_1,\ldots,A_n\rightarrow\bot)=t$ and hence $VA_i=f$ for at least one formula A_i from the set S.

Now, in order to prove that every consistent set of formulas is satisfiable, we proceed as for the proof of the completeness theorem. We again start with an enumeration

$$\alpha_0, \alpha_1, \alpha_2, \ldots$$

of all free object variables.

Let S be a non-empty set of formulas. We assume that no formula in S contains a free object variable α_i with odd index i. (In order to ensure this, one can replace each free object variable α_i which occurs in a formula of a set S_1 by α_{2i}. One then obtains a set of formulas S_2 of the desired sort which is satisfiable if, and only if, S_1 is satisfiable.)

The set S is at most denumerably infinite since we are only considering countably infinitely many primitive symbols. Now we assume we are given an infinite sequence of formulas

$$A_0, A_1, A_2, \ldots$$

which contains all and only formulas from the set S. (This sequence must contain some formula infinitely many times if the set S of formulas is finite.)

From this sequence of formulas we define a *D-chain for S* as a sequence of formulas

$$F_0, F_1, F_2, \ldots,$$

formed as follows:

DS1. The initial formula F_0 of the D-chain is the formula $(A_0\rightarrow\bot)$.

DS2. If a formula F_n of the D-chain is an axiom, then it is the last formula of the D-chain. We then say that the D-chain has *length n*.

DS3. If a formula F_n of the D-chain is not an axiom then the D-chain has as the immediate successor F_{n+1} of F_n a formula $(A_{n+1}\rightarrow G_n)$, where G_n is determined as follows by $F_0,\ldots,F_n$.

DS3.1. If F_n is an irreducible formula, then G_n is the formula F_n.

DS3.2. If F_n is a reducible formula $\mathcal{N}[(A\rightarrow B)]$ with distinguished negative part $(A\rightarrow B)$, then G_n is either the formula $\mathcal{N}[(A\rightarrow\bot)]$ or the formula $\mathcal{N}[B]$.

DS3.3. If F_n is a reducible formula $\mathcal{P}[\forall x\mathcal{F}[x]]$ with distinguished positive part $\forall x\mathcal{F}[x]$, then G_n is the formula $\mathcal{P}[\mathcal{F}[\alpha_{2i+1}]]$ where i is the least number such that α_{2i+1} does not occur in F_n.

DS3.4. If F_n is a reducible formula $\mathcal{N}[\forall x\mathcal{F}[x]]$ with distinguished negative part $\forall x\mathcal{F}[x]$, then G_n is the formula $(\forall x\mathcal{F}[x]\rightarrow\mathcal{N}[\mathcal{F}[\alpha_i]])$ where i is the least number such that $\mathcal{F}[\alpha_i]$ does not occur as a negative part in any of the formulas $F_0,\ldots,F_n$ if the nominal symbol $*_1$ occurs in $\mathcal{F}$.

Principal syntactic lemma for sets of formulas. *If every D-chain for S is finite, then the set S is inconsistent.*

Proof. It follows from König's lemma that there are only finitely many D-chains for S. Let m be the maximal length of such a D-chain. We prove by induction on $m-n$:

If a formula F_n occurs in the n-th place of a D-chain for S, then the formula

(1) $A_m, \ldots, A_{n+1} \rightarrow F_n$

is deducible. (If $m=n$, then (1) is the formula F_m.)

If F_n is the last formula of a D-chain for S, then F_n is an axiom. Then the formula (1) is also an axiom. Otherwise following F_n in the D-chains for S there are the formulas $(A_{n+1} \rightarrow G_n^{(i)})$ ($i=1, 2$ or $i=1$).

By I.H. the formula

(2) $A_m, \ldots, A_{n+2} \rightarrow (A_{n+1} \rightarrow G_n^{(i)})$

is deducible. It follows by a basic inference (possibly using structural inferences and Theorem 4.4), that the formula (1) is also deducible. (The condition on variables for (S2)-inferences is not violated since only free object variables with even index occur in formulas of the set S.)

It now follows by taking $n=0$ that the formula

$$A_m, \ldots, A_0 \rightarrow \perp$$

is deducible. Hence the set S is inconsistent.

Principal semantic lemma for sets of formulas. *If there is an infinite D-chain for S, then the set S of formulas is satisfiable in a model $\langle \mathbb{N}, V \rangle$ with domain the natural numbers $\mathbb{N}$.*

The proof proceeds just as for the D-chain of a formula. One defines the model $\langle \mathbb{N}, V \rangle$ as before and has the same Lemmata 1–5. Since every formula A_i of the set S occurs as a negative part in the D-chain, it follows that $VA_i = t$ for all $A_i \in S$.

Theorem 5.4 (Satisfiability Theorem). *Every consistent set of formulas is satisfiable in a model whose domain is the natural numbers $\mathbb{N}$.*

Proof. If S is a consistent set of formulas, then by the principal syntactic lemma there is an infinite D-chain for S. The assertion now follows by the principal semantic lemma.

Theorem 5.5 (Compactness Theorem). *If every finite subset of a set S of formulas is satisfiable, then S is satisfiable.*

Proof. If a finite set $\{A_1, \ldots, A_n\}$ is satisfiable, then the formula

$$A_1, \ldots, A_n \rightarrow \perp$$

is not valid and hence by the consistency theorem not deducible. It therefore follows from the hypothesis of our theorem, that the set S is consistent. Then by Theorem 5.4 it is satisfiable.

Remark. Theorem 5.5 establishes the compactness of a set of models in a suitable topology.

5. Syntactic and Semantic Consequences

We define the notions of *syntactic consequence* $S \vdash F$ and *semantic consequence* $S \models F$ for a (possibly empty) set S of formulas as follows:

$S \vdash F$ ("*F follows syntactically from S*") means: There is a finite (possibly empty) subset $\{A_1, \ldots, A_n\}$ of S, such that the formula

$$A_1, \ldots, A_n \to F \quad (n \geqslant 0)$$

is deducible. (If S is empty, then this says that F is deducible.)

$S \models F$ ("*F follows semantically from S*") means: Every model of the set S is also a model of the formula F. (If S is empty, then this says that F is valid.)

We now prove that these two notions of consequence coincide.

Theorem 5.6 (Strong Consistency Theorem). *$S \vdash F$ implies $S \models F$.*

Proof. By hypothesis there is a finite subset $\{A_1, \ldots, A_n\}$ of S such that the formula

$$A_1, \ldots, A_n \to F$$

is deducible, therefore by the consistency theorem it is valid. So for every model $\mathcal{M} = \langle \mathcal{D}, V \rangle$ of S we have $V(A_1, \ldots, A_n \to F) = t$ and $VA_i = t\ (i = 1, \ldots, n)$. Therefore $VF = t$. Hence we have $S \models F$.

Theorem 5.7 (Strong Completeness Theorem). *$S \models F$ implies $S \vdash F$.*

Proof. From $S \models F$ it follows that the set $S \cup \{(F \to \bot)\}$ of formulas is not satisfiable. By the satisfiability theorem it follows this set is inconsistent. Hence there is a finite subset $\{A_1, \ldots, A_n\}$ of S such that the formula

$$A_1, \ldots, A_n \to ((F \to \bot) \to \bot)$$

is deducible. From this formula and the deducible formula

$$((F \to \bot) \to \bot) \to F$$

by Theorem 4.6 it follows that the formula

$$A_1, \ldots, A_n \to F$$

is also deducible. Hence $S \vdash F$.

Chapter III

Intuitionistic Predicate Calculus

Intuitionistic logic is based on constructive inferences which are not covered as in classical logic by truth values. This leads immediately to theories in which the truth of a proposition is either not determined semantically or is not in general decidable.

A calculus for intuitionistic predicate calculus was first developed by A. Heyting in order to use it to delimit the predicate logic which is adequate for the intuitionism of L. E. J. Brouwer. Corresponding semantic characterizations of this logic as were later given by E. W. Beth and S. Kripke, are essentially more complicated than the semantics of the classical predicate calculus.

In what follows we do not start from such semantics, instead we only define intuitionistic predicate calculus syntactically by means of axioms and basic inferences which are motivated in a natural way by logical inferences. As before the symbols $\perp$, $\wedge$, $\vee$ and $\rightarrow$ denote *connectives*, but they are no longer conceived as symbols for truth functions.

§6. Formalization of Intuitionistic Predicate Calculus

1. The Formal System IP1

In intuitionistic predicate calculus the connectives $\wedge$, $\vee$ and the existential quantifier $\exists$ cannot be defined in terms of $\perp$, $\rightarrow$ and $\forall$ as in classical predicate calculus.

We therefore use as *primitive symbols*: The primitive symbols of the system CP and the symbols $\wedge$, $\vee$ and $\exists$.

Consequently in addition to the defining rules for formulas of the system CP we have the following defining rules for *formulas*:

2.4′. If A and B are formulas, then $(A \wedge B)$ and $(A \vee B)$ are also formulas.

2.5′. If $\mathscr{F}[a]$ is a formula and x is a bound object variable which does not occur in the 1-place nominal form $\mathscr{F}$, then $\exists x \mathscr{F}[x]$ is also a formula.

Atomic formulas are defined as in the system CP.

We use the same syntactic symbols for free and bound object variables, propositional variables, formulas and atomic formulas as in the system CP.

For brevity we write

$$A_1 \to \cdots \to A_n \to B \quad (n \geqslant 1)$$

for

$$(A_1 \to (\cdots \to (A_n \to B)\ldots)).$$

We denote by Γ, Γ_i, Δ, Δ_i finite (possibly empty) sequences of formulas. If $\Gamma = A_1, \ldots, A_n$ $(n \geqslant 1)$, then $\Gamma \to B$ denotes the formula

$$A_1 \to \cdots \to A_n \to B.$$

If Γ is empty, then $\Gamma \to B$ denotes the formula B.

$\Gamma \subset \Delta$ means that every formula in the sequence Γ is also a formula in the sequence Δ. A formula which occurs several times in Γ however, need only occur once in Δ. (In particular $\Gamma \subset \Delta$ holds for all Δ when Γ is empty.)

For implication we first take as basic only the following *axioms* and *rules of inference*:

(Ax. 1) $A \to A$.
(Str) $\Gamma \to B \vdash \Delta \to B$ if $\Gamma \subset \Delta$ holds.
(Çut) $\Gamma \to A,\ A \to B \vdash \Gamma \to B$.

This says that a formula $\Gamma \to B$ should only be deducible if one can get from the formulas in the sequence Γ to the formula B in a trivial way. We call a (Str) inference a *structural inference* and a (Cut) inference a *cut* (with *cut formula* A).

Restricting to formulas built up only from propositional variables by using the connective $\to$ and using (Ax. 1), (Str) and (Cut) gives the *positive implicational calculus* of P. Bernays.

Now for the connective $\perp$ we add the *axiomschema*

(Ax. 2) $\perp \to A$ (*ex falso quodlibet*)

and we fix the syntactic use of the symbols $\wedge$, $\vee$, $\forall$ and $\exists$ by two *basic inferences* each depending on whether these symbols are introduced in a left or right component of a formula.

($\wedge$l) $A \to B \to C \vdash (A \wedge B) \to C$
($\wedge$r) $\Gamma \to A,\ \Gamma \to B \vdash \Gamma \to (A \wedge B)$
($\vee$l) $A \to C,\ B \to C \vdash (A \vee B) \to C$
($\vee$r) $\Gamma \to A_i \vdash \Gamma \to (A_1 \vee A_2)$ $(i = 1, 2)$
($\forall$l) $\mathscr{F}[a] \to C \vdash \forall x \mathscr{F}[x] \to C$
($\forall$r) $\Gamma \to \mathscr{F}[a] \vdash \Gamma \to \forall x \mathscr{F}[x]$ } if a does not occur
($\exists$l) $\mathscr{F}[a] \to C \vdash \exists x \mathscr{F}[x] \to C$ } in the conclusion
($\exists$r) $\Gamma \to \mathscr{F}[a] \vdash \Gamma \to \exists x \mathscr{F}[x]$.

We call the system with axioms (Ax. 1), (Ax. 2) and basic inferences (Str), (Cut), ($\wedge$ l), ..., ($\exists$r) the formal system IP1 of intuitionistic predicate calculus.

2. The Formal System IP2

Among the basic inferences of the system IP1 the cuts

$$\Gamma \to A,\ A \to B \vdash \Gamma \to B$$

are the only inferences where a formula A may occur in the premises which need bear no formal relation to the conclusion. Thus the deducibility of a formula $\Gamma \to B$ may, because of a cut, depend in a roundabout way on an arbitrary formula A. However, G. Gentzen arrived at just as straightforward a pattern of inference for intuitionistic as for classical predicate calculus in which in every premise of an inference only subformulas of the conclusion occur (where one regards $\mathscr{F}[a]$ as a subformula of $\forall x \mathscr{F}[x]$ and $\exists x \mathscr{F}[x]$). One obtains a straightforward pattern of inference if one takes a suitable basic rule of inference for implication so that the cut is no longer needed as a basic inference but nevertheless still can be proved permissible.

One rule of inference for introducing implication, which corresponds to the straightforward pattern of inference, is ($\to$ l) $\Gamma \to A,\ B \to C \vdash \Gamma \to (A \to B) \to C$. Such an ($\to$ l)-inference is permissible in the system IP1 as one sees as follows:

1) $(A \to B) \to A \to B$ is an (Ax. 1),
2) $A \to (A \to B) \to B$ follows from 1) by a structural inference,
3) $\Gamma \to A$ is the first premise of the ($\to$ l)-inference,
4) $\Gamma \to (A \to B) \to B$ follows from 3) and 2) by a cut,
5) $B \to C$ is the second premise of the ($\to$ l)-inference,
6) $\Gamma \to (A \to B) \to C$ follows from 4) and 5) by a cut.

If one adds ($\to$ l) as a basic rule of inference then (Ax. 1) can be restricted to

(Ax. 1A) $P \to P$

where P is an atomic formula, since then every formula $A \to A$ is deducible (without the use of cuts). Thus, for example, from $A \to A$ and $B \to B$ by an ($\to$ l)-inference we get $A \to (A \to B) \to B$ and by a structural inference $(A \to B \to (A \to B)$. (Ax. 2) may also be restricted to

(Ax. 2A) $\perp \to P$.

We denote by IP2 the formal system with axioms (Ax. 1A), (Ax. 2A) and basic inferences (Str), ($\to$l), ($\wedge$ l), ..., ($\exists$r). This system has the desired property that only subformulas of the conclusion occur in each premise of a basic inference. This has certain advantages for syntactic investigations. For example it is immediately apparent that the system IP2 is consistent since no atomic formula has the form of an axiom or a conclusion of a basic inference of the system IP2 it follows that no atomic formula is deducible in IP2. In order to prove this is also true for IP1 one has to show that an atomic formula cannot be shown to be deducible using a cut. This is not immediately obvious, since the conclusion of a cut can be an atomic formula.

In IP2 it turns out that the formulas $A \to A$ and $\bot \to A$ are deducible and the cut is a permissible inference whence it follows that precisely the same formulas are deducible as in IP1. We shall not carry out the proof of the deducibility of $A \to A$ and $\bot \to A$ and of the permissibility of cut in IP2 since we shall pass to another formal system IP3 and give the corresponding proof for it in §7.

3. Left and Right Parts of Formulas

The notions of "positive" and "negative part" of a formula which we used in classical predicate calculus refer to valuations of formulas and truth values and are therefore not available for intuitionistic predicate calculus. Instead of using these for the basic inferences we turn to certain left and right parts of formulas which are defined as follows. Every formula can be uniquely written in the form

$$\Gamma \to B$$

where B is not of the form $(B_1 \to B_2)$. The *left parts* of such a formula are all formulas of the sequence Γ and the *right parts* are the formula B and, if $\Gamma = A_1, \ldots, A_n$ $(n \geqslant 1)$, so too are $A_n \to B$ and all formulas $A_i \to \cdots \to A_n \to B$ $(i = 1, \ldots, n-1)$. Logical symbols are introduced into a left part of the conclusion by $(\to \mathrm{l})$, $(\wedge \mathrm{l})$, $(\vee \mathrm{l})$, $(\forall \mathrm{l})$ and $(\exists \mathrm{l})$ basic inferences, and into a right part of the conclusion by $(\wedge r)$, $(\vee \mathrm{r})$, $(\forall \mathrm{r})$ and $(\exists \mathrm{r})$ basic inferences.

In order to be able to do without the $(\wedge \mathrm{l})$ inference as a basic inference we make the notion of "left part" somewhat wider in that we also regard A and B as left parts of a formula F if $(A \wedge B)$ is a left part of F.

In order to characterize left and right parts of a formula we use *L-forms* (*left part forms*) and *R-forms* (*right part forms*) which are inductively defined as 1-place nominal forms as follows:

LRF 1. $*_1$ is an *R-form*.

LRF 2. If $\mathscr{R}$ is an *R-form* and B is a formula, then $\mathscr{R}[(*_1 \to B)]$ is an *L-form* and $\mathscr{R}[(B \to *_1)]$ is an *R-form*.

LRF 3. If $\mathscr{L}$ is an *L-form* and B a formula then $\mathscr{L}[(*_1 \wedge B)]$ and $\mathscr{L}[(B \wedge *_1)]$ are *L-forms*.

We have the following consequences of this inductive definition:

LRF 4. No R-form is an L-form.

LRF 5. If $\mathscr{R}$ is an R-form and $\mathscr{U}$ is an L-form (or an R-form) then $\mathscr{R}[\mathscr{U}]$ is an L-form (or an R-form).

LRF 6. If $\mathscr{U}$ is an L-form or an R-form and A is a formula then $\mathscr{U}[A]$ is also a formula.

We now define: A formula A is said to be a *left part* (or a *right part*) of a formula F if there is an L-form (or an R-form) $\mathscr{U}$ such that $\mathscr{U}[A]$ is the formula F. From this definition by means of the defining rules LRF 1–LRF 3 we obtain the following **inductive characterization** of *left* and *right parts* of a formula F.

LR 1. F is a *right part* of F.

LR 2. If $(A \to B)$ is a *right part* of F, then A is a *left part* and B a *right part* of F.

LR 3. If $(A \wedge B)$ is a *left part* of F, then A and B are also *left parts* of F.

As corollaries we have:

LR 4. If F is not an implication, that is not a formula of the form $(A \to B)$, then F is the unique right part of F.

LR 5. If F is an implication $(A \to B)$, then the right parts of F are the formula F and the right parts of B.

LR 6. Every formula F has precisely one right part which is not an implication. We call this the *end part* of the formula F.

LR 7. The end part of a formula F is also the end part of every right part of F.

By a *D-form* (double form) we mean a 2-place nominal form $\mathscr{D}$ such that $\mathscr{D}[*_1, B]$ is an L-form and $\mathscr{D}[A, *_1]$ is an R-form.

For *syntactic symbols* we use

$\mathscr{L}, \mathscr{L}_i$ for L-forms,
$\mathscr{R}, \mathscr{R}_i$ for R-forms,
$\mathscr{D}, \mathscr{D}_i$ for D-forms.

A left part of a formula is said to be *minimal* if it is not of the form $(A \wedge B)$.

We generalize the structural inference (Str) thus: $F \stackrel{\text{S}}{\vdash} G$ ("*G follows structurally from F*") means that F and G have the same end part and every minimal left part of F also occurs as a left part of G. (Obviously $\Gamma \to B \stackrel{\text{S}}{\vdash} \Delta \to B$ holds if $\Gamma \to B \vdash \Delta \to B$ is an (Str)-inference.)

4. The Formal System IP3

We obtain a formal system for intuitionistic predicate calculus in which neither cuts nor structural inferences are needed if the axiom schemata and basic rules of inference of the system IP 2 are modified in such a way that they are invariant under structural inferences in the following sense: Every formula which follows from an axiom by a structural inference is likewise to be an axiom and every formula which follows from the conclusion of a basic inference by a structural inference is to be the conclusion of a corresponding basic inference whose premises are to be derivable from the premises of the first basic inference.

We get such an invariance under (Str)-inferences if, for example we replace the axiom schema (Ax. 1A) by

$$\Gamma_1 \to P \to \Gamma_2 \to P$$

and the basic rule of inference ($\to$ l) by

$$\Gamma_1 \to (A \to B) \to \Gamma_2 \to A,\ \Gamma_1 \to B \to \Gamma_2 \to C \vdash \Gamma_1 \to (A \to B) \to \Gamma_2 \to C.$$

In order to get this invariance with respect to $F \overset{S}{\models} G$ too we replace (Ax. 1A) by $\mathscr{D}[P, P]$ and $(\rightarrow 1)$ by

$$\mathscr{D}[(A \rightarrow B), A], \mathscr{D}[B, C] \vdash \mathscr{D}[(A \rightarrow B), C]$$

We correspondingly modify the remaining axioms and basic inferences of the system IP2.

The formal system IP3 of the type we want has the same formal language as the systems IP1 and IP2.

The *axioms* of the system IP3 are all formulas of the following forms

(Ax. I) $\mathscr{D}[P, P]$
(Ax. II) $\mathscr{L}[\perp]$.

The *basic inferences* of the system IP3 are all inferences of the following sort:

$(\rightarrow \mathrm{L})$ $\mathscr{D}[(A \rightarrow B), A], \mathscr{D}[B, C] \vdash \mathscr{D}[(A \rightarrow B), C]$
$(\wedge \mathrm{R})$ $\mathscr{R}[A], \mathscr{R}[B] \vdash \mathscr{R}[(A \wedge B)]$
$(\vee \mathrm{L})$ $\mathscr{L}[A], \mathscr{L}[B] \vdash \mathscr{L}[(A \vee B)]$
$(\vee \mathrm{R})$ $\mathscr{R}[A_i] \vdash \mathscr{R}[(A_1 \vee A_2)]$ $(i=1, 2)$
$(\forall \mathrm{L})$ $\mathscr{L}[(\forall x \mathscr{F}[x] \wedge \mathscr{F}[a])] \vdash \mathscr{L}[\forall x \mathscr{F}[x]]$
$(\forall \mathrm{R})$ $\mathscr{R}[\mathscr{F}[a]] \vdash \mathscr{R}[\forall x \mathscr{F}[x]]$ } with a *condition on*
$(\exists \mathrm{L})$ $\mathscr{L}[\mathscr{F}[a]] \vdash \mathscr{L}[\exists x \mathscr{F}[x]]$ } *variables*
$(\exists \mathrm{R})$ $\mathscr{R}[\mathscr{F}[a]] \vdash \mathscr{R}[\exists x \mathscr{F}[x]]$.

The free object variable denoted by a in the premises of the $(\forall \mathrm{R})$ and $(\exists \mathrm{L})$ inferences is said to be the *eigenvariable* of the given inference. The *condition on variables* for these inferences is: The eigenvariable must not occur in the conclusion.

The left part denoted by $(A \rightarrow B)$ in the conclusion of an $(\rightarrow \mathrm{L})$-inference is called the *principal part*, the right part of the conclusion denoted by C is called the *secondary part* of the given basic inference. For any other basic inference the *principal part* is the left or right part of its conclusion shown. In every case the principal part of a basic inference is either a minimal left part or the end part of the conclusion. For $\rightarrow$ one only has a basic L-inference with principal left part, for $\wedge$ only a basic R-inference with principal right part and for $\vee$, $\forall$, $\exists$ one basic L-inference and one basic R-inference.

By the *degree* of a formula we mean the number of $\rightarrow$, $\wedge$, $\vee$, $\forall$ and $\exists$ symbols occurring in the formula.

It follows easily from the results of the next section that precisely the same formulas are deducible in the system IP3 as in the system IP1. However, the system IP3 has the advantage that induction on deductions is particularly simple and transparent since here the derivation is direct and also no structural inference occurs as a basic inference.

§7. Deducible Formulas and Permissible Inferences in the System IP3

1. Generalizations of the Axioms

Theorem 7.1. *Every formula $\mathscr{D}[C, C]$ is deducible.*

Proof by induction on the degree of the formula C.

1. Suppose C is an atomic formula. Then $\mathscr{D}[C, C]$ is an axiom.

2. Suppose C is a formula $(A \to B)$. Since $\mathscr{D}[(A \to B), (*_1 \to *_2)]$ and $\mathscr{D}[*_1, (A \to *_2)]$ are D-forms, by the I.H. (induction hypothesis) $\vdash \mathscr{D}[(A \to B), (A \to A)]$ and $\vdash \mathscr{D}[B, (A \to B)]$. Using an $(\to \mathrm{L})$-inference with the D-form $\mathscr{D}[*_1, (A \to *_2)]$ we get $\vdash \mathscr{D}[(A \to B), (A \to B)]$.

3. Suppose C is a formula $(A \wedge B)$. Since $\mathscr{D}[(*_1 \wedge B), *_2]$ and $\mathscr{D}[(A \wedge *_1), *_2]$ are D-forms, by I.H. we have $\vdash \mathscr{D}[(A \wedge B), A]$ and $\vdash \mathscr{D}[(A \wedge B), B]$. Since $\mathscr{D}[(A \wedge B), *_1]$ is an R-form, by an $(\wedge \mathrm{R})$-inference we obtain $\vdash \mathscr{D}[(A \wedge B), (A \wedge B)]$.

4. Suppose C is a formula $(A \vee B)$. By I.H. we have $\vdash \mathscr{D}[A, A]$ and $\vdash \mathscr{D}[B, B]$. Using $(\vee \mathrm{R})$-inferences we get $\vdash \mathscr{D}[A, (A \vee B)]$ and $\vdash \mathscr{D}[B, (A \vee B)]$. Then by an $(\vee \mathrm{L})$-inference we have $\vdash \mathscr{D}[(A \vee B), (A \vee B)]$.

5. Suppose C is a formula $\forall x \mathscr{F}[x]$ or $\exists x \mathscr{F}[x]$. We choose a free object variable a which does not occur in $\mathscr{D}[C, C]$. By I.H. we have $\vdash \mathscr{D}[(\forall x \mathscr{F}[x] \wedge \mathscr{F}[a]), \mathscr{F}[a]]$ and $\vdash \mathscr{D}[\mathscr{F}[a], \mathscr{F}[a]]$. Using an $(\forall \mathrm{L})$-inference and an $(\exists \mathrm{R})$-inference we have $\vdash \mathscr{D}[\forall x \mathscr{F}[x], \mathscr{F}[a]]$ and $\vdash \mathscr{D}[\mathscr{F}[a], \exists x \mathscr{F}[x]]$. Using an $(\forall \mathrm{R})$-inference and an $(\exists \mathrm{L})$-inference we obtain $\vdash \mathscr{D}[\forall x \mathscr{F}[x], \forall x \mathscr{F}[x]]$ and $\vdash \mathscr{D}[\exists x \mathscr{F}[x], \exists x \mathscr{F}[x]]$.

2. Weak Inferences

Theorem 7.2 (Substitution rule). *If a_1 does not occur in the nominal form $\mathscr{F}$ then*

$$\mathscr{F}[a_1] \vdash \mathscr{F}[a_2]$$

is a weak inference.

The proof proceeds by induction on the deduction as for Theorem 4.2.

Theorem 7.3. *The following are weak inferences for the elimination of logical symbols*

$(\to \mathrm{E})$ $\mathscr{L}[(A_1 \to A_2)] \vdash \mathscr{L}[A_2]$

$(\wedge \mathrm{E})$ $\mathscr{R}[(A_1 \wedge A_2)] \vdash \mathscr{R}[A_i]$ $(i=1, 2)$

$(\vee \mathrm{E})$ $\mathscr{L}[(A_1 \vee A_2)] \vdash \mathscr{L}[A_i]$ $(i=1, 2)$

$(\forall \mathrm{E})$ $\mathscr{R}[\forall x \mathscr{F}[x]] \vdash \mathscr{R}[\mathscr{F}[a]]$

$(\exists \mathrm{E})$ $\mathscr{L}[\exists x \mathscr{F}[x]] \vdash \mathscr{L}[\mathscr{F}[a]]$.

1. *Proof* for $(\to \mathrm{E})$, $(\wedge \mathrm{E})$ and $(\vee \mathrm{E})$. Suppose $\mathscr{U}[(A_1 \times A_2)]$ is a formula $\mathscr{L}[(A_1 \to A_2)]$, $\mathscr{R}[(A_1 \wedge A_2)]$ or $\mathscr{L}[(A_1 \vee A_2)]$ which is deducible with order n.

We prove by induction on the deduction that $\mathcal{U}[A_i]$ is then deducible with order $\leqslant n$. For this let $i=2$ if $\mathcal{U}[(A_1 \times A_2)]$ is the formula $\mathcal{L}[(A_1 \to A_2)]$, otherwise $i=1$ or $i=2$.

1.1. Suppose $\mathcal{U}[(A_1 \times A_2)]$ is an axiom. Then $\mathcal{U}[A_i]$ is also an axiom.

1.2. Suppose $\mathcal{U}[(A_1 \times A_2)]$ was obtained by a basic inference whose principal part is the left or right part $(A_1 \times A_2)$ shown. Then $\mathcal{U}[A_i]$ is a premise of this basic inference and therefore is deducible with order $<n$.

1.3. Suppose $\mathcal{U}[(A_1 \times A_2)]$ was obtained by an $(\to \mathrm{L})$-inference whose secondary part contains the part $(A_1 \times A_2)$ shown. The premises of this $(\to \mathrm{L})$-inference are of the form $\mathcal{D}[(B_1 \to B_2), B_1]$ and $\mathcal{D}[B_2, \mathcal{U}_0[(A_1 \times A_2)]]$ where $\mathcal{U}[A_i]$ is the formula $\mathcal{D}[(B_1 \to B_2), \mathcal{U}_0[A_i]]$. Since $\mathcal{D}[B_2, \mathcal{U}_0[(A_1 \times A_2)]]$ is deducible with order $<n$, by I.H. $\mathcal{D}[B_2, \mathcal{U}_0[A_i]]$ is also deducible with order $<n$. Since $\mathcal{D}[(B_1 \to B_2), B_1]$ is also deducible with order $<n$ it follows by an $(\to \mathrm{L})$-inference that $\mathcal{D}[(B_1 \to B_2), \mathcal{U}_0[A_i]]$ is deducible with order $\leqslant n$.

1.4. Suppose $\mathcal{U}[(A_1 \times A_2)]$ was obtained in some other way by a basic inference. Then every premise of this basic inference is of such a form $\mathcal{U}_j[(A_1 \times A_2)]$ that

$$\mathcal{U}_j[A_i] \ (j=1, 2 \text{ or } j=1) \vdash \mathcal{U}[A_i]$$

is also a basic inference. By I.H. $\mathcal{U}_j[A_i]$ is deducible with order $<n$. Using the given basic inference it follows that $\mathcal{U}[A_i]$ is deducible with order $\leqslant n$.

2. *Proof* for $(\forall \mathrm{E})$ and $(\exists \mathrm{E})$. Suppose $\mathcal{U}[Qx\mathcal{F}[x]]$ is a formula $\mathcal{R}[\forall x\mathcal{F}(x]]$ or $\mathcal{L}[\exists x\mathcal{F}[x]]$, which is deducible with order n. We prove by induction on the deduction that $\mathcal{U}[\mathcal{F}[a]]$ is then deducible with order $\leqslant n$.

2.1. Suppose $\mathcal{U}[Qx\mathcal{F}[x]]$ is an axiom. Then $\mathcal{U}[\mathcal{F}[a]]$ is also an axiom.

2.2. Suppose $\mathcal{U}[Qx\mathcal{F}[x]]$ was obtained by a basic inference whose principal part is the left or right part $Qx\mathcal{F}[x]$ shown. The premise $\mathcal{U}[\mathcal{F}[a_1]]$ of this basic inference is deducible with order $<n$. Since the eigenvariable a_1 of the basic inference does not occur in the nominal form $\mathcal{U}[\mathcal{F}]$ it follows by Theorem 7.2 that $\mathcal{U}[\mathcal{F}[a]]$ is also deducible with order $<n$.

2.3. Suppose $\mathcal{U}[Qx\mathcal{F}[x]]$ was obtained in some other way by a basic inference. We choose a free object variable a_1 which does not occur in the nominal form $\mathcal{U}[\mathcal{F}]$ and is not the eigenvariable of the basic inference. By I.H. it follows as in 1.3 and 1.4 that $\mathcal{U}[\mathcal{F}[a_1]]$ is deducible with order $\leqslant n$. Then by Theorem 7.2 $\mathcal{U}[\mathcal{F}[a]]$ is also deducible with order $\leqslant n$.

Theorem 7.4 (Structural rule). *If $F \overset{S}{\vdash} G$ then $F \vdash G$ is a weak inference.*

Proof by induction on the deduction. Suppose F is deducible with order n and $F \overset{S}{\vdash} G$ holds.

1. Suppose F is an axiom. Then from $F \overset{S}{\vdash} G$ it follows that G is also an axiom.

2. Suppose F is a formula $\mathcal{D}_1[(A \to B), C]$ which was obtained by an $(\to \mathrm{L})$-inference from $\mathcal{D}_1[(A \to B), A]$ and $\mathcal{D}_1[B, C]$. Then G is a formula $\mathcal{D}_2[(A \to B), E]$ where E is the end part of F. Let $\mathcal{D}_0[B, E]$ be the formula obtained from $\mathcal{D}_1[B, C]$

by replacing every left part $(A \to B)$ of this formula by B. By Theorem 7.3 $\mathscr{D}_0[B, E]$ is deducible with order $<n$. From $F \models^{S} G$ we have $\mathscr{D}_1[(A \to B), A] \models^{S} \mathscr{D}_2[(A \to B), A]$ and $\mathscr{D}_0[B, E] \models^{S} \mathscr{D}_2[B, E]$. Then by I.H. $\mathscr{D}_2[(A \to B, A]$ and $\mathscr{D}_2[B, E]$ are deducible with orders $<n$. Using an $(\to \mathrm{L})$-inference it follows that $\mathscr{D}_2[(A \to B), E]$ is deducible with order $\leqslant n$.

3. Suppose F is a formula $\mathscr{R}_1[(A_1 \wedge A_2)]$ which was obtained by an $(\wedge \mathrm{R})$-inference from $\mathscr{R}_1[A_1]$ and $\mathscr{R}_1[A_2]$. Then G is a formula $\mathscr{R}_2[(A_1 \wedge A_2)]$ where $\mathscr{R}_1[A_i] \models^{S} \mathscr{R}_2[A_i]$ $(i=1, 2)$. By I.H. $\mathscr{R}_2[A_i]$ is deducible with order $<n$. Using an $(\wedge \mathrm{R})$-inference it follows that $\mathscr{R}_2[(A_1 \wedge A_2)]$ is deducible with order $\leqslant n$.

4. Suppose F is a formula $\mathscr{L}_1[(A_1 \vee A_2)]$ which was obtained by an $(\vee \mathrm{L})$-inference from $\mathscr{L}_1[A_1]$ and $\mathscr{L}_1[A_2]$. Then G is a formula $\mathscr{L}_2[(A_1 \vee A_2)]$. For $i=1, 2$ let F_i be the formula obtained by replacing every left part $(A_1 \vee A_2)$ in $\mathscr{L}_1[A_i]$ by A_i. By Theorem 7.3 F_i is deducible with order $<n$. From $F \models^{S} G$ we have $F_i \models^{S} \mathscr{L}_2[A_i]$. Therefore by I.H. $\mathscr{L}_2[A_i]$ is deducible with order $<n$. Using an $(\vee \mathrm{L})$-inference it follows that $\mathscr{L}_2[(A_1 \vee A_2)]$ is deducible with order $\leqslant n$.

5. Suppose F was obtained by an $(\vee \mathrm{R})$-inference, $(\forall \mathrm{L})$-inference or $(\exists \mathrm{R})$-inference. Then the assertion follows as in case 3.

6. Suppose F was obtained by a $(\forall \mathrm{R})$-inference or $(\exists \mathrm{L})$-inference with eigenvariable a. We choose a free object variable a_1 which occurs in neither F nor G. Let F_1 be the formula obtained by replacing the eigenvariable a in the premise of the basic inference by a_1. By Theorem 7.2 F_1 is deducible with order $<n$. $F_1 \vdash F$ is a basic inference with eigenvariable a_1. Now the assertion follows as in case 3 or 4.

3. More Permissible Inferences

Theorem 7.5. $\mathscr{L}[\mathscr{F}[a]] \vdash \mathscr{L}[\forall x \mathscr{F}[x]]$ (provided x does not occur in $\mathscr{F}$) *is a permissible inference.*

Proof. We have $\mathscr{L}[\mathscr{F}[a]] \models^{S} \mathscr{L}[(\forall x \mathscr{F}[x] \wedge \mathscr{F}[a])]$. Therefore by Theorem 7.4 from $\vdash \mathscr{L}[\mathscr{F}[a]]$ we have $\vdash \mathscr{L}[(\forall x \mathscr{F}[x] \wedge \mathscr{F}[a])]$. Using a $(\forall \mathrm{L})$-inference we get $\vdash \mathscr{L}[\forall x \mathscr{F}[x]]$.

Theorem 7.6 (Cut rule). $\mathscr{R}[A], (A \to B) \vdash \mathscr{R}[B]$ *is a permissible inference.*

Proof by induction on the degree of the formula A.

1. Suppose A is a formula $(A_1 \wedge A_2)$. By Theorem 7.3 from $\vdash \mathscr{R}[(A_1 \wedge A_2)]$ we have $\vdash \mathscr{R}[A_1]$ and $\vdash \mathscr{R}[A_2]$. By Theorem 7.4 from $\vdash((A_1 \wedge A_2) \to B)$ we have $\vdash(A_2 \to (A_2 \to B))$. By the I.H. using $\vdash \mathscr{R}[A_1]$ we have $\vdash \mathscr{R}[(A_2 \to B)]$. By Theorem 7.4 we have $\vdash A_2 \to \mathscr{R}[B]$. By I.H. using $\vdash \mathscr{R}[A_2]$ we get $\vdash \mathscr{R}[\mathscr{R}[B]]$. By Theorem 7.4 there follows $\vdash \mathscr{R}[B]$.

2. Suppose A is an atomic formula or a formula $(A_1 \vee A_2)$ or $\exists x \mathscr{F}[x]$. In this case we perform a subsidiary induction on the order of a deduction of $\mathscr{R}[A]$.

2.1. Suppose $\mathscr{R}[A]$ is an axiom.

2.1.1. If $\mathscr{R}[A]$ is an (Ax. I) then A is an atomic formula and $\mathscr{R}[B]$ is a formula $\mathscr{D}[A, B]$. By Theorem 7.4 from $\vdash(A \to B)$ we have $\vdash \mathscr{D}[A, B]$.

2.1.2. If $\mathscr{R}[A]$ is an (Ax. II) then $\mathscr{R}[B]$ is also an (Ax. II).

2.2. Suppose $\mathscr{R}[A]$ was inferred by a basic R-inference. Then A is either $(A_1 \vee A_2)$ or $\exists x\mathscr{F}[x]$.

2.2.1. If A is the formula $(A_1 \vee A_2)$ then the basic R-inference has a premise $\mathscr{R}[A_i]$ ($i=1$ or $i=2$). By Theorem 7.3 from $\vdash((A_1 \vee A_2) \rightarrow B)$ we have $\vdash(A_i \rightarrow B)$. By I.H. using $\mathscr{R}[A_i]$ we have $\vdash\mathscr{R}[B]$.

2.2.2. If A is the formula $\exists x\mathscr{F}[x]$ then the basic R-inference has a premise $\mathscr{R}[\mathscr{F}[a]]$. By Theorem 7.3 from $\vdash(\exists x\mathscr{F}[x] \rightarrow B)$ we have $\vdash(\mathscr{F}[a] \rightarrow B)$. By I.H. using $\vdash\mathscr{R}[F[a]]$ we have $\vdash\mathscr{R}[B]$.

2.3. Suppose $\mathscr{R}[A]$ was inferred by an $(\rightarrow$ L)-inference. This inference has as premises $\mathscr{D}[(C_1 \rightarrow C_2), C_1]$ and $\mathscr{D}[C_2, \mathscr{R}_0[A]]$ where $\mathscr{R}$ is the R-form $\mathscr{D}[(C_1 \rightarrow C_2), \mathscr{R}_0]$. From $\vdash\mathscr{D}[C_2, \mathscr{R}_0[A]]$ and $\vdash(A \rightarrow B)$ by the S.I.H. (subsidiary induction hypothesis) we obtain $\vdash\mathscr{D}[C_2, \mathscr{R}_0[B]]$. By an $(\rightarrow$ L)-inference using $\vdash\mathscr{D}[(C_1 \rightarrow C_2), C_1]$ we have $\vdash\mathscr{D}[(C_1 \rightarrow C_2), \mathscr{R}_0[B]]$, i.e. $\vdash\mathscr{R}[B]$.

2.4. Suppose $\mathscr{R}[A]$ was inferred by an $(\vee$ L)-inference, $(\forall$L)-inference or $(\exists$L)-inference. In the case of an $(\exists$L)-inference we may assume by Theorem 7.2 that the eigenvariable of this inference does not occur in B. Then each premise of the basic L-inference is of the form $\mathscr{R}_i[A]$ so

$$\mathscr{R}_i[B] \; (i=1, 2 \text{ or } i=1) \vdash\mathscr{R}[B]$$

is also a basic L-inference. From $\vdash\mathscr{R}_i[A]$ and $\vdash(A \rightarrow B)$ by the S.I.H. we therefore have $\vdash\mathscr{R}_i[B]$. Using the given basic L-inference $\vdash\mathscr{R}[B]$ follows.

3. Suppose A is a formula $(A_1 \rightarrow A_2)$ or $\forall x\mathscr{F}[x]$. For this case we use a subsidiary induction on the order of a deduction of $(A \rightarrow B)$.

3.1. Suppose $(A \rightarrow B)$ is an axiom. Since A is a minimal left part of $(A \rightarrow B)$ which is not an atomic formula then $\mathscr{R}[B]$ is also an axiom.

3.2. Suppose $(A \rightarrow B)$ was inferred by a basic inference with principal term A.

3.2.1. Suppose A is the formula $(A_1 \rightarrow A_2)$. Then the basic inference has premises $((A_1 \rightarrow A_2) \rightarrow \mathscr{R}_0[A_1])$ and $(A_2 \rightarrow \mathscr{R}_0[C])$ where B is the formula $\mathscr{R}_0[C]$. By the S.I.H. from $\vdash\mathscr{R}[(A_1 \rightarrow A_2)]$ and $\vdash((A_1 \rightarrow A_2) \rightarrow \mathscr{R}_0[A_1])$ we have $\vdash\mathscr{R}[\mathscr{R}_0[A_1]]$. By I.H. from $\vdash\mathscr{R}[(A_1 \rightarrow A_2)]$ and $\vdash(A_2 \rightarrow \mathscr{R}_0[C])$ we have $\vdash\mathscr{R}[(A_1 \rightarrow \mathscr{R}_0[C])]$. By Theorem 7.4 we have $\vdash(A_1 \rightarrow \mathscr{R}[\mathscr{R}_0[C]])$. Using $\vdash\mathscr{R}[\mathscr{R}_0[A_1]]$ there follows by I.H. $\vdash\mathscr{R}[\mathscr{R}_0[\mathscr{R}[\mathscr{R}_0[C]]]]$. By Theorem 7.4 we have $\vdash\mathscr{R}[\mathscr{R}_0[C]]$, that is, $\vdash\mathscr{R}[B]$.

3.2.2. Suppose A is the formula $\forall x\mathscr{F}[x]$. Then the basic inference has a premise $((\forall x\mathscr{F}[x] \wedge \mathscr{F}[a]) \rightarrow B)$. By Theorem 7.4 the formula $(\forall x\mathscr{F}[x] \rightarrow (\mathscr{F}[a] \rightarrow B))$ is also deducible with an order smaller than that of the formula $(\forall x\mathscr{F}[x] \rightarrow B)$. Using $\vdash\mathscr{R}[\forall x\mathscr{F}[x]]$ therefore, $\vdash\mathscr{R}[(\mathscr{F}[a] \rightarrow B)]$ follows by the S.I.H. By Theorem 7.4 there follows $\vdash(\mathscr{F}[a] \rightarrow \mathscr{R}[B])$. From $\vdash\mathscr{R}[\forall x\mathscr{F}[x]]$ by Theorem 7.3 we have $\vdash\mathscr{R}[\mathscr{F}[a]]$. Using $\vdash(\mathscr{F}[a] \rightarrow \mathscr{R}[B])$, by I.H. we have $\vdash\mathscr{R}[\mathscr{R}[B]]$. By Theorem 7.4 we have $\vdash\mathscr{R}[B]$.

3.3 Suppose $(A \rightarrow B)$ was inferred by a basic inference whose principal part is contained in B. Then by Theorem 7.2 we may assume that an eigenvariable of the basic inference is not a free object variable occurring in $\mathscr{R}$. Then each premise of the basic inference is of the form $(A \rightarrow B_i)$ so

$$\mathscr{R}[B_i] \; (i=1, 2 \text{ or } i=1) \vdash\mathscr{R}[B]$$

is a basic inference. From $\vdash \mathscr{R}[A]$ and $\vdash (A \to B_i)$ by the S.I.H. we have $\vdash \mathscr{R}[B_i]$. Using the given basic inference we have $\vdash \mathscr{R}[B]$.

Corollary to Theorem 7.6. *The formal system* IP3 *is s-consistent, that is, there is no formula A such that A and $(A \to \bot)$ are both deducible in* IP3.

Proof. By Theorem 7.6

$$A, (A \to \bot) \vdash \bot$$

is a permissible inference in the system IP3. Were A and $(A \to \bot)$ deducible then $\bot$ would also be deducible. But this is not the case since $\bot$ is not an axiom nor does it occur as the conclusion of a basic inference.

Remark. As will be shown in §8, every formula deducible in IP3 is also deducible in CP. Hence the s-consistency of the system IP3 also follows from the s-consistency of the system CP.

4. Special Features of Intuitionistic Logic

Theorem 7.7. 1. *A formula $(A \vee B)$ is deducible only if A is deducible or B is deducible.*

2. *A formula $\exists x \mathscr{F}[x]$ is deducible only if there is a free object variable a such that $\mathscr{F}[a]$ is deducible.*

Proof. 1. $(A \vee B)$ is not an axiom. The only basic inferences with $(A \vee B)$ as conclusion are the $(\vee R)$-inferences with premise A or B. Therefore $\vdash (A \vee B)$ holds only if $\vdash A$ or $\vdash B$ holds.

2. $\exists x \mathscr{F}[x]$ is not an axiom. The only basic inferences with $\exists x \mathscr{F}[x]$ as conclusion are the $(\exists R)$-inferences with a premise $\mathscr{F}[a]$. Therefore $\vdash \exists x \mathscr{F}[x]$ holds only if there is an a with $\vdash \mathscr{F}[a]$.

Corollaries to Theorem 7.7. *The formulas* $(v \vee (v \to \bot))$ and $(\forall x\, p(x) \vee \exists x (p(x) \to \bot))$, *where p is a one-place predicate letter, which are valid in classical predicate calculus are not deducible in* IP3.

Proof. 1. Were $(v \vee (v \to \bot))$ deducible then by Theorem 7.7 v or $(v \to \bot)$ would be deducible. But this is not the case since neither v nor $(v \to \bot)$ occurs as an axiom or as the conclusion of a basic inference.

2. Were $(\forall x\, p(x) \vee (\exists x (p(x) \to \bot))$ deducible then by Theorem 7.7 $\forall x\, p(x)$ or $\exists x (p(x) \to \bot)$ would be deducible. By Theorems 7.3 and 7.7 $p(a)$ or $(p(a) \to \bot)$ would then be deducible for some free object variable a. But this is not the case since neither $p(a)$ nor $(p(a) \to \bot)$ occurs as an axiom or as the conclusion of a basic inference.

5. Properties of Negation

As in the system CP we define $\neg A := (A \to \bot)$.

Theorem 7.8. *$\mathscr{D}[A, B] \vdash \mathscr{D}[\neg B, \neg A]$ is a permissible inference.*

Proof. By Theorem 7.4 from $\mathscr{D}[A, B]$ we have $\vdash \mathscr{D}[(B \to \bot), (A \to B)]$. By an $(\to \mathrm{L})$-inference using the axiom $\mathscr{D}[\bot, (A \to \bot)]$ we have $\vdash \mathscr{D}[(B \to \bot), (A \to \bot)]$, hence $\vdash \mathscr{D}[\neg B, \neg A]$.

Theorem 7.9. *The following formulas are deducible*

(1) $(A \to \neg\neg A)$
(2) $(\neg\neg\bot \to \bot)$
(3) $(\neg\neg\neg A \to \neg A)$
(4) $((\neg A \to A) \to \neg\neg A)$

Proof. 1. By Theorem 7.1 we have $\vdash((A \to \bot) \to (A \to \bot))$. By Theorem 7.4 there follows $\vdash(A \to ((A \to \bot) \to \bot)$, that is, $\vdash(A \to \neg\neg A)$.

2. From the axioms $(((\bot \to \bot) \to \bot) \to (\bot \to \bot))$ and $(\bot \to \bot)$ by an $(\to \mathrm{L})$-inference we have $\vdash(((\bot \to \bot) \to \bot) \to \bot)$, that is, $\vdash(\neg\neg\bot \to \bot)$.

3. By (1) we have $\vdash(A \to \neg\neg A)$. By Theorem 7.8 there follows $\vdash(\neg\neg\neg A \to \neg A)$.

4. By Theorem 7.1 we have $\vdash((\neg A \to A) \to (\neg A \to \neg A))$. From (1) we have $\vdash(A \to (\neg A \to \bot))$. By an $(\to \mathrm{L})$-inference there follows $\vdash((\neg A \to A) \to (\neg A \to \bot)$, that is, $\vdash((\neg A \to A) \to \neg\neg A)$.

6. Syntactic Equivalence

Let $\vdash^{\mathrm{I}} F$ denote that the formula F is deducible in the system IP3.
Let $A \overset{\mathrm{I}}{\sim} B$ denote that $(A \to B)$ and $(B \to A)$ are deducible in IP3.
Let $A \overset{\mathrm{I}}{\approx} B$ denote that $A \vdash^{\mathrm{S}} B$ and $B \vdash^{\mathrm{S}} A$ hold in IP3.

Theorem 7.10. *$\overset{\mathrm{I}}{\sim}$ is an equivalence relation which is compatible with $\vdash^{\mathrm{I}}$, $\overset{\mathrm{I}}{\approx}$ and $\to$, that is, the following hold:*

(1) *$A \overset{\mathrm{I}}{\sim} A$.*
(2) *$A \overset{\mathrm{I}}{\sim} B$ implies $B \overset{\mathrm{I}}{\sim} A$.*
(3) *$A \overset{\mathrm{I}}{\sim} B$ and $B \overset{\mathrm{I}}{\sim} C$ imply $A \overset{\mathrm{I}}{\sim} C$.*
(4) *$A \overset{\mathrm{I}}{\sim} B$ and $\vdash^{\mathrm{I}} A$ imply $\vdash^{\mathrm{I}} B$.*
(5) *$A \overset{\mathrm{I}}{\approx} B$ implies $A \overset{\mathrm{I}}{\sim} B$.*
(6) *$A \overset{\mathrm{I}}{\sim} B$ implies $(A \to C) \overset{\mathrm{I}}{\sim} (B \to C)$ and $(C \to A) \overset{\mathrm{I}}{\sim} (C \to B)$.*

Proof. (1) holds by Theorem 7.1. (2) holds by the definition of $\overset{\mathrm{I}}{\sim}$. (3) and (4) follow from Theorem 7.6.

From $\vdash(A \to A)$ and $A \models^{S} B$ by Theorem 7.4 we have $\vdash(A \to B)$. This gives (5).

From $\vdash(A \to B)$ we have $\vdash(A \to (C \to B))$. Using $\vdash((C \to A) \to (C \to C))$ we obtain $\vdash((C \to A) \to (C \to B))$ by an $(\to \mathrm{L})$-inference. From $\vdash(B \to A)$ there follows $\vdash((A \to C) \to (B \to A))$. Using $\vdash(C \to (B \to C))$ we obtain $\vdash((A \to C) \to (B \to C))$. This gives (6).

§8. Relations between Classical and Intuitionistic Predicate Calculus

1. Embedding IP3 in CP

We now regard every formula of the system IP3 as also being a formula of the system CP where $\wedge$, $\vee$ and $\exists$ are defined as in §4.4. We need the following lemmata in order to show that every formula deducible in IP3 is also deducible in CP.

Lemma 1. *Every R-form is a P-form, every L-form is an N-form.*

Proof by induction on the length of the R-form or L-form.

1. The R-form $*_1$ is a P-form.
2. If $\mathscr{R}$ is a P-form then: $\mathscr{R}[(*_1 \to B)]$ is an N-form, $\mathscr{R}[(B \to *_1)]$ is a P-form.
3. If $\mathscr{L}$ is an N-form then: $\mathscr{L}[(*_1 \to \bot)]$ is a P-form, $\mathscr{L}[((*_1 \to (B \to \bot)) \to \bot)]$ is an N-form, $\mathscr{L}[((B \to *_1) \to \bot)]$ is a P-form and $\mathscr{L}[((B \to (*_1 \to \bot)) \to \bot)]$ is an N-form. By the definition of $\wedge$, $\mathscr{L}[(*_1 \wedge B)]$ and $\mathscr{L}[(B \wedge *_1)]$ are therefore N-forms.

Lemma 2. *Every D-form is an NP-form.*

Proof. This follows from Lemma 1.

Lemma 3. *The following are permissible inferences in the system* CP:

(1) $\mathscr{Q}[(A \to B), A], \mathscr{Q}[B, C] \vdash \mathscr{Q}[(A \to B), C]$,
(2) $\mathscr{N}[\forall x \mathscr{F}[x] \wedge \mathscr{F}[a])] \vdash \mathscr{N}[\forall x \mathscr{F}[x]]$.

Proof. 1. By Theorem 4.3, from $\vdash \mathscr{Q}[(A \to B), A]$ by an inversion inference we have $\vdash \mathscr{Q}[(A \to \bot), A]$. By Theorem 4.4, using a structural inference we obtain $\vdash \mathscr{Q}[(A \to \bot), C]$. From this and $\vdash \mathscr{Q}[B, C]$ an (S1)-inference gives $\vdash \mathscr{Q}[(A \to B), C]$.

2. Since $(\forall x \mathscr{F}[x] \wedge \mathscr{F}[a])$ is defined in PC to be the formula $((\forall x \mathscr{F}[x] \to (\mathscr{F}[a] \to \bot)) \to \bot)$, from $\vdash \mathscr{N}[(\forall x \mathscr{F}[x] \wedge \mathscr{F}[a])]$ by Theorem 4.4 a structural inference we have $\vdash((\bot \to \bot) \to (\forall x \mathscr{F}[x] \to \mathscr{N}[\mathscr{F}[a]]))$. Since $(\bot \to \bot)$ is deducible by Theorem 4.6 we have $\vdash(\forall x \mathscr{F}[x] \to \mathscr{N}[\mathscr{F}[a]])$. By an (S3)-inference we have $\vdash \mathscr{N}[\forall x \mathscr{F}[x]]$.

Theorem 8.1. *Every formula deducible in* IP3 *is deducible in* CP.

Proof. By Lemmata 1 and 2 every axiom of the system IP3 is an axiom of the system CP. By Lemmata 2 and 3 every ($\rightarrow$ L)-inference is permissible in CP. By Lemma 1 every ($\forall$R)-inference is an (S2)-inference. By Lemmata 1 and 3 every ($\forall$L)-inference is permissible in CP. By Lemma 1 and Theorem 4.8 all the remaining basic inferences of the system IP3 are permissible in CP.

2. Interpretation of CP in IP3

Inductive definition of a formula $\bar{F}$ for each formula F of the system CP.

1. $\overline{\perp} := \perp$
2. $\bar{P} := \neg\neg P$ for every other atomic formula P
3. $\overline{(A \rightarrow B)} := (\bar{A} \rightarrow \bar{B})$
4. If $\overline{\mathscr{F}[a]} = \overline{\mathscr{F}}[a]$, where a does not occur in the nominal forms $\mathscr{F}$ and $\overline{\mathscr{F}}$ then $\overline{\forall x \mathscr{F}[x]} := \forall x \overline{\mathscr{F}}[x]$.

Lemma 4. $(\neg\neg\bar{A} \rightarrow \bar{A})$ *is deducible in* IP3.

Proof by induction on the degree of the formula A.

1. Suppose A is an atomic formula. Then we have $\vdash^{I}(\neg\neg\bar{A} \rightarrow \bar{A})$ by Theorem 7.9 (2) and (3).

2. Suppose A is a formula $(A_1 \rightarrow A_2)$. By Theorem 7.1 we have $\vdash^{I}((\bar{A}_1 \rightarrow \bar{A}_2) \rightarrow (\bar{A}_1 \rightarrow \bar{A}_2))$. By two applications of Theorem 7.8 we have $\vdash^{I}(\neg\neg(\bar{A}_1 \rightarrow \bar{A}_2) \rightarrow (\bar{A}_1 \rightarrow \neg\neg\bar{A}_2))$. By I.H. we have $\vdash^{I}(\neg\neg\bar{A}_2 \rightarrow \bar{A}_2)$. By Theorem 7.6 there follows $\vdash^{I}(\neg\neg(\bar{A}_1 \rightarrow \bar{A}_2) \rightarrow (\bar{A}_1 \rightarrow \bar{A}_2))$ and therefore $\vdash^{I}(\neg\neg\bar{A} \rightarrow \bar{A})$.

3. Suppose A is a formula $\forall x \mathscr{F}[x]$. Suppose $\overline{\mathscr{F}[a]} = \overline{\mathscr{F}}[a]$, where a occurs in neither $\mathscr{F}$ nor $\overline{\mathscr{F}}$. By Theorem 7.1 we have $\vdash^{I}(\overline{\mathscr{F}}[a] \rightarrow \overline{\mathscr{F}}[a])$. Therefore by Theorem 7.5 we have $\vdash^{I}(\forall x \overline{\mathscr{F}}[x] \rightarrow \overline{\mathscr{F}}[a])$. Two applications of Theorem 7.8 yield $\vdash^{I}(\neg\neg\forall x \overline{\mathscr{F}}[x] \rightarrow \neg\neg\overline{\mathscr{F}}[a])$. By I.H. we have $\vdash^{I}(\neg\neg\overline{\mathscr{F}}[a] \rightarrow \overline{\mathscr{F}}[a])$. By Theorem 7.6 therefore $\vdash^{I}(\neg\neg\forall x \overline{\mathscr{F}}[x] \rightarrow \overline{\mathscr{F}}[a])$. From an ($\forall$R)-inference we have $\vdash^{I}(\neg\neg\forall x \overline{\mathscr{F}}[x] \rightarrow \forall x \overline{\mathscr{F}}[x])$ and therefore $\vdash^{I}(\neg\neg\bar{A} \rightarrow \bar{A})$.

Lemma 5. $(\neg\bar{A} \rightarrow \bar{B}) \overset{I}{\sim} (\neg\bar{B} \rightarrow \bar{A})$.

Proof. By Theorem 7.9 (1) and Lemma 4 we have $\bar{B} \overset{I}{\sim} \neg\neg\bar{B}$. Therefore $(\neg\bar{A} \rightarrow \bar{B} \overset{I}{\sim} (\neg\bar{A} \rightarrow \neg\neg\bar{B})$. Since $(\neg\bar{A} \rightarrow \neg\neg\bar{B}) \overset{I}{\approx} (\neg\bar{B} \rightarrow \neg\neg\bar{A})$ holds we have $(\neg\bar{A} \rightarrow \bar{B}) \overset{I}{\sim} (\neg\bar{B} \rightarrow \neg\neg\bar{A})$. Using $\neg\neg\bar{A} \overset{I}{\sim} \bar{A}$ we obtain $(\neg\bar{A} \rightarrow \bar{B}) \overset{I}{\sim} (\neg\bar{B} \rightarrow \bar{A})$.

Lemma 6. $(\neg\bar{A} \rightarrow \bar{A}) \overset{I}{\sim} \bar{A}$.

Proof. By Theorem 7.9 (4) we have $\vdash^{I}((\neg\bar{A} \rightarrow \bar{A}) \rightarrow \neg\neg\bar{A})$, and by Lemma 4 $\vdash^{I}(\neg\neg\bar{A} \rightarrow \bar{A})$. Therefore by Theorem 7.6 we have $\vdash^{I}((\neg\bar{A} \rightarrow \bar{A}) \rightarrow \bar{A})$. By Theorem 7.1 we also have $\vdash^{I}(\bar{A} \rightarrow (\neg\bar{A} \rightarrow \bar{A}))$.

Lemma 7. *For every P-form $\mathscr{P}$ there is an R-form $\mathscr{R}$ such that $\overline{\mathscr{P}[A]} \overset{\perp}{\sim} \mathscr{R}[\bar{A}]$ holds for every formula A.*

Proof by induction on the length of $\mathscr{P}$.

1. Suppose $\mathscr{P}$ is the P-form $*_1$. Then the assertion holds for $\mathscr{R} := *_1$.

2. Suppose $\mathscr{P}$ is a P-form $\mathscr{P}_0[(B \to *_1)]$. Then by I.H. there is an R-form $\mathscr{R}_0$ such that $\overline{\mathscr{P}_0[C]} \overset{\perp}{\sim} \mathscr{R}_0[\bar{C}]$ for every formula C. Now $\mathscr{P}[A]$ is $\mathscr{P}_0[(B \to A)]$ hence $\overline{\mathscr{P}[A]} \overset{\perp}{\sim} \mathscr{R}_0[\overline{(B \to A)}]$, that is $\overline{\mathscr{P}[A]} \overset{\perp}{\sim} \mathscr{R}_0[(\bar{B} \to \bar{A})]$. The assertion is therefore satisfied by $\mathscr{R} := \mathscr{R}_0[(\bar{B} \to *_1)]$.

3. Suppose $\mathscr{P}$ is a P-form $\mathscr{N}[(*_1 \to \perp)]$. Then $\mathscr{N}$ is an N-form $\mathscr{P}_0[(*_1 \to B)]$ and therefore $\mathscr{P}$ is the P-form $\mathscr{P}_0[((*_1 \to \perp) \to B)]$. By I.H. there is an R-form $\mathscr{R}_0$ such that $\overline{\mathscr{P}_0[C]} \overset{\perp}{\sim} \mathscr{R}_0[\bar{C}]$ for every formula C. Now $\mathscr{P}[A]$ is $\mathscr{P}_0[((A \to \perp) \to B)]$ hence $\overline{\mathscr{P}[A]} \overset{\perp}{\sim} \mathscr{R}_0[\overline{((A \to \perp) \to B)}]$, that is $\overline{\mathscr{P}[A]} \overset{\perp}{\sim} \mathscr{R}_0[((\bar{A} \to \perp) \to \bar{B})]$. Since $\mathscr{R}_0[((\bar{A} \to \perp) \to \bar{B})] \overset{\perp}{\approx} (\neg A \to \mathscr{R}_0[\bar{B}])$ it follows that $\overline{\mathscr{P}[A]} \overset{\perp}{\sim} (\neg A \to \mathscr{R}_0[\bar{B}])$. By Lemma 5 we have $\overline{\mathscr{P}[A]} \overset{\perp}{\sim} (\neg \mathscr{R}_0[\bar{B}] \to \bar{A})$. The assertion is therefore satisfied by $\mathscr{R} := (\neg \mathscr{R}_0[\bar{B}] \to *_1)$.

Lemma 8. *For every N-form $\mathscr{N}$ there is an L-form $\mathscr{L}$ such that $\overline{\mathscr{N}[A]} \overset{\perp}{\sim} \mathscr{L}[\bar{A}]$ for every formula A.*

Proof. $\mathscr{N}$ is an N-form $\mathscr{P}[(*_1 \to B)]$. By Lemma 7 there is an R-form $\mathscr{R}$ such that $\overline{\mathscr{N}[A]} \overset{\perp}{\sim} \mathscr{R}[(\bar{A} \to \bar{B})]$. The assertion is therefore satisfied by $\mathscr{L} := \mathscr{R}[(*_1 \to \bar{B})]$.

Theorem 8.2 (Interpretation Theorem). *If F is deducible in* CP *then $\bar{F}$ is deducible in* IP3.

Proof by induction on the deduction.

1. Suppose F is an axiom $\mathscr{Q}[P, P]$. By Lemmata 7 and 8 there is an R-form $\mathscr{R}$ and an L-form $\mathscr{L}$ such that $\bar{F} \overset{\perp}{\sim} \mathscr{R}[\bar{P}]$ and $\bar{F} \overset{\perp}{\sim} \mathscr{L}[\bar{P}]$. Hence $(\bar{P} \to \mathscr{R}[\bar{P}]) \overset{\perp}{\sim} (\bar{P} \to \mathscr{L}[\bar{P}])$. By Theorem 7.1 we have $\vdash^{\perp} (\bar{P} \to \mathscr{R}[\bar{P}])$. Hence $\vdash^{\perp} (\bar{P} \to \mathscr{L}[\bar{P}])$. From Theorem 7.4 we have $\vdash^{\perp} \mathscr{L}[\bar{P}]$. Since $\bar{F} \overset{\perp}{\sim} \mathscr{L}[\bar{P}]$, we have $\vdash^{\perp} \bar{F}$.

2. Suppose F is an axiom $\mathscr{N}[\perp]$. By Lemma 8 there is an L-form $\mathscr{L}$ such that $\bar{F} \overset{\perp}{\sim} \mathscr{L}[\perp]$. Since $\mathscr{L}[\perp]$ is an axiom of the system IP3, we have $\vdash^{\perp} \bar{F}$.

3. Suppose F is a formula $\mathscr{N}[(A \to B)]$ obtained by an (S1)-inference from $\mathscr{N}[A \to \perp]$ and $\mathscr{N}[B]$. By Lemma 8 there is an L-form $\mathscr{L}$ such that $\bar{F} \overset{\perp}{\sim} \mathscr{L}[(\bar{A} \to \bar{B})]$, $\overline{\mathscr{N}[(A \to \perp)]} \overset{\perp}{\sim} \mathscr{L}[(\bar{A} \to \perp)]$ and $\overline{\mathscr{N}[B]} \overset{\perp}{\sim} \mathscr{L}[\bar{B}]$. By I.H. therefore $\vdash^{\perp} \mathscr{L}[(\bar{A} \to \perp)]$ and $\vdash^{\perp} \mathscr{L}[\bar{B}]$. By Theorem 7.4 we have $\vdash^{\perp} ((\bar{A} \to \bar{B}) \to (\neg \bar{A} \to \mathscr{L}[(\bar{A} \to \bar{B})]))$ and $\vdash^{\perp} (\bar{B} \to (\neg \bar{F} \to \mathscr{L}[(\bar{A} \to \bar{B})]))$. Using $\bar{F} \overset{\perp}{\sim} \mathscr{L}[(\bar{A} \to \bar{B})]$ we have $\vdash^{\perp} ((\bar{A} \to \bar{B}) \to (\neg \bar{A} \to \bar{F}))$ and $\vdash^{\perp} (\bar{B} \to (\neg \bar{F} \to \bar{F}))$. By Lemma 5 we have $\vdash^{\perp} ((\bar{A} \to \bar{B}) \to (\neg \bar{F} \to \bar{A}))$. Using an $(\to \mathrm{L})$-inference we obtain $\vdash^{\perp} ((\bar{A} \to \bar{B}) \to (\neg \bar{F} \to \bar{F}))$. By Lemma 6 we have $\vdash^{\perp} ((\bar{A} \to \bar{B}) \to \bar{F})$. Using $\bar{F} \overset{\perp}{\sim} \mathscr{L}[(\bar{A} \to \bar{B})]$ we obtain $\vdash^{\perp} ((\bar{A} \to \bar{B}) \to \mathscr{L}[(\bar{A} \to \bar{B})]$. By Theorem 7.4 we have $\vdash^{\perp} \mathscr{L}[(\bar{A} \to \bar{B})]$, hence $\vdash^{\perp} \bar{F}$.

4. Suppose F is a formula $\mathscr{P}[\forall x \mathscr{F}[x]]$ obtained by an (S2)-inference from $\mathscr{P}[\mathscr{F}[a]]$. By Lemma 7 there is an R-form $\mathscr{R}$ such that $\bar{F} \overset{\perp}{\sim} \mathscr{R}[\overline{\forall x \mathscr{F}[x]}]$ and

$\overline{\mathscr{P}[\mathscr{F}[a]]} \stackrel{\mathrm{I}}{\sim} \mathscr{R}[\overline{\mathscr{F}}[a]]$. By I.H. we have $\vdash^{\mathrm{I}} \mathscr{R}[\overline{\mathscr{F}}[a]]$. Now a does not occur in $\mathscr{R}[\overline{\mathscr{F}}]$. By an ($\forall$R)-inference we have $\vdash^{\mathrm{I}} \mathscr{R}[\forall x \overline{\mathscr{F}}[x]]$, hence $\vdash^{\mathrm{I}} \overline{F}$.

5. Suppose F is a formula $\mathscr{N}[\forall x \mathscr{F}[x]]$ obtained by an (S3)-inference from $(\forall x \mathscr{F}[x] \to \mathscr{N}[\mathscr{F}[a]])$. By Lemma 8 there is an L-form $\mathscr{L}$ such that $\overline{F} \stackrel{\mathrm{I}}{\sim} \mathscr{L}[\forall x \overline{\mathscr{F}}[x]]$ and $\overline{(\forall x \mathscr{F}[x] \to \mathscr{N}[\mathscr{F}[a]])} \stackrel{\mathrm{I}}{\sim} (\forall x \overline{\mathscr{F}}[x] \to \mathscr{L}[\overline{\mathscr{F}}[a]])$. By I.H. we have $\vdash^{\mathrm{I}} (\forall x \overline{\mathscr{F}}[x] \to \mathscr{L}[\overline{\mathscr{F}}[a]])$. By Theorem 7.4 we have $\vdash^{\mathrm{I}} \mathscr{L}[(\forall x \overline{\mathscr{F}}[x] \wedge \overline{\mathscr{F}}[a])]$. By an ($\forall$L)-inference we have $\vdash^{\mathrm{I}} \mathscr{L}[\forall x \overline{\mathscr{F}}[x]]$, hence $\vdash^{\mathrm{I}} \overline{F}$.

Let $\vdash^{\mathrm{C}} F$ denote that the formula F is deducible in the system CP.
Let $A \stackrel{\mathrm{C}}{\sim} B$ denote that $(A \to B)$ and $(B \to A)$ are deducible in CP.

Theorem 8.3. $\overline{A} \stackrel{\mathrm{C}}{\sim} A$.

Proof by induction on the degree of the formula A.

$(((*_1 \to \bot) \to \bot) \to *_2)$ and $(*_1 \to ((*_2 \to \bot) \to \bot))$ are NP-forms. Consequently $(\neg\neg P \to P)$ and $(P \to \neg\neg P)$ are axioms of the system CP. Hence $\overline{P} \stackrel{\mathrm{C}}{\sim} P$ for every atomic formula P. For the remaining formulas the assertion follows from the induction hypothesis since two formulas are classically equivalent if they only differ from each other by classically equivalent components.

§9. The Interpolation Theorem

As a sort of converse of the cut rule we have an interpolation theorem which was first formulated and proved for classical predicate calculus by W. Craig. Here we prove the corresponding theorem for the formal system IP3 and then show that this yields the interpolation theorem for the system CP.

1. Interpolation Theorem for the System IP3

If X is a formula or a nominal form then we write $\langle X \rangle$ for the set of free object variables, sentential variables and predicate letters occurring in X.

Interpolation Theorem. *For every formula $\mathscr{R}[B]$ deducible in* IP3 *there is a formula A with the following properties:*

1. *$\mathscr{R}[A]$ and $(A \to B)$ are deducible in* IP3.
2. *A contains only free object variables, sentential variables and predicate letters which occur in both $\mathscr{R}$ and B, that is to say, $\langle A \rangle \subseteq \langle \mathscr{R} \rangle \cap \langle B \rangle$.*

We call such a formula an *interpolant between $\mathscr{R}$ and B.*

Proof of the interpolation theorem. Suppose the formula $\mathscr{R}[B]$ was deduced with order n. We prove, by induction on the order n of the deduction, that there is an interpolant between $\mathscr{R}$ and B.

1. Suppose $\mathscr{R}[B]$ is an axiom.

1.1. Suppose B is an axiom. Then $\mathscr{R}[(\perp \to \perp)]$ and $((\perp \to \perp) \to B)$ are axioms. Then $(\perp \to \perp)$ is an interpolant between $\mathscr{R}$ and B.

1.2. Suppose $\mathscr{R}[B]$ is an (Ax. I) and B is not an axiom. Then $\mathscr{R}$ is an R-form $\mathscr{D}[P, *_1]$ where P is the end part of B. In this case $\mathscr{R}[P]$ and $(P \to B)$ are axioms. P occurs in $\mathscr{R}$ as well as in B. Hence P is an interpolant between $\mathscr{R}$ and B.

1.3. Suppose $\mathscr{R}[B]$ is an (Ax. II) and B is not an axiom. Then $\perp$ is an interpolant between $\mathscr{R}$ and B.

2. Suppose $\mathscr{R}[B]$ was obtained by a basic inference whose principal part is contained in B. Each premise of this basic inference is of the form $\mathscr{R}[B_i]$ $(i=1, 2$ or $i=1)$. By I.H. there exist A_i such that $\vdash \mathscr{R}[A_i]$, $\vdash (A_i \to B_i)$ and $\langle A_i \rangle \subseteq \langle \mathscr{R} \rangle \cap \langle B_i \rangle$.

2.1. Suppose the basic inference was an $(\to \mathrm{L})$-inference, $(\wedge \mathrm{R})$-inference or $(\vee \mathrm{L})$-inference. These have two premises $\mathscr{R}[B_i]$ $(i=1, 2)$ with $\langle B_i \rangle \subseteq \langle B \rangle$. Therefore $\langle (A_1 \wedge A_2) \rangle \subseteq \langle \mathscr{R} \rangle \cap \langle B \rangle$. By an $(\wedge \mathrm{R})$-inference and by using Theorem 7.4, from $\vdash \mathscr{R}[A_i]$ and $\vdash (A_i \to B_i)$ we obtain $\vdash \mathscr{R}[(A_1 \wedge A_2)]$ and $\vdash ((A_1 \wedge A_2) \to B_i)$ $(i=1, 2)$. In the present case

$$((A_1 \wedge A_2) \to B_i) \ (i=1, 2) \vdash ((A_1 \wedge A_2) \to B)$$

is also a basic inference. Hence we have $\vdash ((A_1 \wedge A_2) \to B)$. Therefore $(A_1 \wedge A_2)$ is an interpolant between $\mathscr{R}$ and B.

2.2. Suppose the basic inference was an $(\vee \mathrm{R})$-inference, $(\forall \mathrm{R})$-inference or $(\exists \mathrm{L})$-inference. This has one premise $\mathscr{R}[B_1]$. If the basic inference has an eigenvariable then this does not occur in $\mathscr{R}$. In every case we have $\langle \mathscr{R} \rangle \cap \langle B_1 \rangle \subseteq \langle \mathscr{R} \rangle \cap \langle B \rangle$. Hence $\langle A_1 \rangle \subseteq \langle \mathscr{R} \rangle \cap \langle B \rangle$. In the present situation we also have that

$$(A_1 \to B_1) \vdash (A_1 \to B)$$

is a basic inference. So from $\vdash (A_1 \to B_1)$ we have $\vdash (A_1 \to B)$. Since $\vdash \mathscr{R}[A_1]$ also holds, A_1 is an interpolant between $\mathscr{R}$ and B.

2.3. Suppose the basic inference was an $(\forall \mathrm{L})$-inference or an $(\exists \mathrm{R})$-inference. These have one premise $\mathscr{R}[B_1]$. Then

$$(A_1 \to B_1) \vdash (A_1 \to B)$$

is also a basic inference. So from $\vdash (A_1 \to B_1)$ we have $\vdash (A_1 \to B)$. There is a free object variable a such that $\langle B_1 \rangle \subseteq \langle B \rangle \cup \{a\}$. If a occurs in B then $\langle A_1 \rangle \subseteq \langle \mathscr{R} \rangle \cap \langle B \rangle$. Since we also have $\vdash \mathscr{R}[A_1]$, A_1 is therefore an interpolant between $\mathscr{R}$ and B. Now assume that a does not occur in B. Then A_1 is a formula $\mathscr{F}_1[a]$ such that a does not occur in the 1-place nominal form $\mathscr{F}_1$. By an $(\exists \mathrm{R})$-inference and an $(\exists \mathrm{L})$-inference, from $\vdash \mathscr{R}[\mathscr{F}_1[a]]$ and $\vdash (\mathscr{F}_1[a] \to B)$ we have $\vdash \mathscr{R}[\exists x \mathscr{F}_1[x]]$ and $\vdash (\exists x \mathscr{F}_1[x] \to B)$. Then $\langle \exists x \mathscr{F}_1[x] \rangle \subseteq \langle \mathscr{R} \rangle \cap \langle B \rangle$. Hence $\exists x \mathscr{F}_1[x]$ is an interpolant between $\mathscr{R}$ and B.

3. Suppose $\mathscr{R}[B]$ was obtained by a basic inference whose principal part is not contained in B. Then this basic inference is an L-inference.

3.1. Suppose the basic inference was an $(\to \mathrm{L})$-inference whose secondary part contains the formula B. This inference has two premises $\mathscr{D}[(C_1 \rightarrow C_2), C_1]$ and $\mathscr{D}[C_2, \mathscr{R}_0[B]]$, where $\mathscr{R}$ is the R-form $\mathscr{D}[(C_1 \rightarrow C_2), \mathscr{R}_0]$. By I.H. for the second premise there is a formula A such that $\vdash \mathscr{D}[C_2, \mathscr{R}_0[A]]$, $\vdash (A \rightarrow B)$ and $\langle A\rangle \subseteq \langle \mathscr{D}[C_2, \mathscr{R}_0]\rangle \cap \langle B\rangle$. Hence $\langle A\rangle \subseteq \langle \mathscr{R}\rangle \cap \langle B\rangle$.. By an $(\rightarrow$ L)-inference from $\vdash \mathscr{D}[(C_1 \rightarrow C_2), C_1]$ and $\vdash \mathscr{D}[C_2, \mathscr{R}_0[A]]$ we have $\vdash \mathscr{D}[(C_1 \rightarrow C_2), \mathscr{R}_0[A]]$, that is, $\vdash \mathscr{R}[A]$. Hence A is an interpolant between $\mathscr{R}$ and B.

3.2. Suppose the basic inference was an $(\rightarrow$ L)-inference whose secondary term does not contain the formula B. This inference has two premises $\mathscr{D}[(C_1 \rightarrow C_2), \mathscr{R}_0[C_1]]$ and $\mathscr{D}[C_2, \mathscr{R}_0[B_0]]$, where B is the formula $\mathscr{R}_0[B_0]$ and $\mathscr{R}$ is the R-form $\mathscr{D}[(C_1 \rightarrow C_2), *_1]$. Since the first premise $\mathscr{D}[(C_1 \rightarrow C_2), \mathscr{R}_0[C_1]]$ is deducible with order $<n$, by Theorem 7.4 $\mathscr{R}_0[\mathscr{D}[(C_1 \rightarrow C_2), C_1]]$ is also deducible with order $<n$. By I.H. there is therefore a formula A_1 such that $\vdash \mathscr{R}_0[A_1]$, $\vdash (A_1 \rightarrow \mathscr{D}[(C_1 \rightarrow C_2), C_1])$ and $\langle A_1\rangle \subseteq \langle \mathscr{R}_0\rangle \cap \langle \mathscr{D}[(C_1 \rightarrow C_2), C_1]\rangle$. Since the second premise is also deducible with order $<n$, by I.H. there is also a formula A_2 such that $\vdash \mathscr{D}[C_2, A_2]$, $\vdash (A_2 \rightarrow \mathscr{R}_0[B_0])$ and $\langle A_2\rangle \subseteq \langle \mathscr{D}[C_2, *_1]\rangle \cap \langle \mathscr{R}_0[B_0]\rangle$. Since $\langle \mathscr{D}[(C_1 \rightarrow C_2), C_1]\rangle \subseteq \langle \mathscr{R}\rangle$, $\langle \mathscr{D}[C_2, *_1]\rangle \subseteq \langle \mathscr{R}\rangle$ and $\mathscr{R}_0[B_0] = B$, it follows that $\langle (A_1 \rightarrow A_2)\rangle \subseteq \langle \mathscr{R}\rangle \cap \langle B\rangle$. By Theorem 7.4 from $\vdash (A_1 \rightarrow \mathscr{D}[(C_1 \rightarrow C_2), C_1])$ and $\vdash \mathscr{D}[C_2, A_2]$ we have $\vdash \mathscr{D}[(C_1 \rightarrow C_2), (A_1 \rightarrow C_1)]$ and $\vdash \mathscr{D}[C_2, (A_1 \rightarrow A_2)]$. Using an $(\rightarrow$ L)-inference we have $\vdash \mathscr{D}[(C_1 \rightarrow C_2), (A_1 \rightarrow A_2)]$, that is, $\vdash \mathscr{R}[(A_1 \rightarrow A_2)]$. By Theorem 7.4 from $\vdash \mathscr{R}_0[A_1]$ we have $\vdash ((A_1 \rightarrow A_2) \rightarrow \mathscr{R}_0[A_1])$. This together with $\vdash (A_2 \rightarrow \mathscr{R}[B_0])$ by an $(\rightarrow$ L)-inference gives $\vdash ((A_1 \rightarrow A_2) \rightarrow \mathscr{R}_0[B_0])$, that is, $\vdash ((A_1 \rightarrow A_2) \rightarrow B)$. Hence $(A_1 \rightarrow A_2)$ is an interpolant between $\mathscr{R}$ and B.

3.3. Suppose the basic inference was an $(\vee$L)-inference. This inference has two premises $\mathscr{D}[C_i, B]$ $(i = 1, 2)$ where $\mathscr{R}$ is the R-form $\mathscr{D}[(C_1 \vee C_2), *_1]$. By I.H. there are formulas A_i such that $\vdash \mathscr{D}[C_i, A_i]$, $\vdash (A_i \rightarrow B)$ and $\langle A_i\rangle \subseteq \langle \mathscr{D}[C_i, *_1]\rangle \cap \langle B\rangle$. Hence $\langle (A_1 \vee A_2)\rangle \subseteq \langle \mathscr{R}\rangle \cap \langle B\rangle$. By an $(\vee$R)-inference from $\vdash \mathscr{D}[C_i, A_i]$ we have $\vdash \mathscr{D}[C_i, (A_1 \vee A_2)]$ $(i = 1, 2)$ and by an $(\vee$L)-inference $\vdash \mathscr{D}[(C_1 \vee C_2), (A_1 \vee A_2)]$, that is, $\vdash \mathscr{R}[(A_1 \vee A_2)]$. By an $(\vee$L)-inference from $\vdash (A_i \rightarrow B)$ $(i = 1, 2)$ we have $\vdash ((A_1 \vee A_2) \rightarrow B)$. Hence $(A_1 \vee A_2)$ is an interpolant between $\mathscr{R}$ and B.

3.4. Suppose the basic inference was an $(\forall$L)-inference. This inference has a premise $\mathscr{D}[(\forall x \mathscr{F}[x] \wedge \mathscr{F}[a]), B]$ where $\mathscr{R}$ is the R-form $\mathscr{D}[\forall x \mathscr{F}[x], *_1]$. By I.H. there is a formula A_1 such that $\vdash \mathscr{D}[(\forall x \mathscr{F}[x] \wedge \mathscr{F}[a], A_1])$, $\vdash (A_1 \rightarrow B)$ and $\langle A_1\rangle \subseteq \langle \mathscr{D}[(\forall x \mathscr{F}[x] \wedge \mathscr{F}[a]), *_1]\rangle \cap \langle B\rangle$. Consequently $\langle A_1\rangle \subseteq (\langle \mathscr{R}\rangle \cup \{a\}) \cap \langle B\rangle$. By an $(\forall$L)-inference from $\vdash \mathscr{D}[(\forall x \mathscr{F}[x] \wedge \mathscr{F}[a]), A_1]$ we have $\vdash \mathscr{D}[\forall x \mathscr{F}[x], A_1]$, that is, $\vdash \mathscr{R}[A_1]$. If a occurs in $\mathscr{R}$, then $\langle A_1\rangle \subseteq \langle \mathscr{R}\rangle \cap \langle B\rangle$ and hence A_1 is an interpolant between $\mathscr{R}$ and B. Now assume that a does not occur in $\mathscr{R}$. A_1 is a formula $\mathscr{F}_1[a]$, where a does not occur in $\mathscr{F}_1$. By an $(\forall$R)-inference and by Theorem 7.5 from $\vdash \mathscr{R}[\mathscr{F}_1[a]]$ and $\vdash (\mathscr{F}_1[a] \rightarrow B)$ we have $\vdash \mathscr{R}[\forall x \mathscr{F}_1[x]]$ and $\vdash (\forall x \mathscr{F}_1[x] \rightarrow B)$. Then $\langle \forall x \mathscr{F}_1[x]\rangle \subseteq \langle \mathscr{R}\rangle \cap \langle B\rangle$. Hence $\forall x \mathscr{F}_1[x]$ is an interpolant between $\mathscr{R}$ and B.

3.5. Suppose the basic inference was an $(\exists$L)-inference. This inference has one premise $\mathscr{D}[\mathscr{F}[a], B]$ where $\mathscr{R}$ is the R-form $\mathscr{D}[\exists x \mathscr{F}[x], *_1]$. By I.H. there is a formula A such that $\vdash \mathscr{D}[\mathscr{F}[a], A]$, $\vdash (A \rightarrow B)$ and $\langle A\rangle \subseteq \langle \mathscr{D}[\mathscr{F}[a], *_1]\rangle \cap \langle B\rangle$. The eigenvariable a of the $(\exists$L)-inference does not occur in B. Therefore it does not occur in A either. Consequently $\langle A\rangle \subseteq \langle \mathscr{R}\rangle \cap \langle B\rangle$, and by an $(\exists$L)-inference from

$\vdash\mathscr{D}[\mathscr{F}[a], A]$ we have $\vdash\mathscr{D}[\exists x\mathscr{F}[x], A]$, that is, $\vdash\mathscr{R}[A]$. Hence A is an interpolant between $\mathscr{R}$ and B.

Remark. The above proof is constructive. It shows how to construct an interpolant between $\mathscr{R}$ and B from a given deduction of $\mathscr{R}[B]$.

2. Interpolation Theorem for the System CP

We are here content with the following simple version:

Interpolation Theorem. *For every formula* $(A \to C)$ *deducible in* CP *there is a formula* B *with the following properties:*

1. $(A \to B)$ *and* $(B \to C)$ *are deducible in* CP.
2. $\langle B\rangle \subseteq \langle A\rangle \cap \langle C\rangle$.

We call such a formula B an *interpolant between* A *and* C.

This theorem can be proved directly in CP. But we can derive it as follows from the interpolation theorem for the system IP3.

By Theorem 8.2 from $\vdash^{\mathrm{C}}(A \to C)$ we have $\vdash^{\mathrm{I}}(\bar{A} \to \bar{C})$. By the interpolation theorem for the system IP3 it follows that there is a formula B such that $\vdash^{\mathrm{I}}(\bar{A} \to B)$, $\vdash^{\mathrm{I}}(B \to \bar{C})$ and $\langle B\rangle \subseteq \langle \bar{A}\rangle \cap \langle \bar{C}\rangle$. By Theorem 8.1 we have $\vdash^{\mathrm{C}}(\bar{A} \to B)$ and $\vdash^{\mathrm{C}}(B \to \bar{C})$. By Theorem 8.3 we have $\vdash^{\mathrm{C}}(A \to \bar{A})$ and $\vdash^{\mathrm{C}}(\bar{C} \to C)$. By Theorem 4.6 therefore $\vdash^{\mathrm{C}}(A \to B)$ and $\vdash^{\mathrm{C}}(B \to C)$. From $\langle B\rangle \subseteq \langle \bar{A}\rangle \cap \langle \bar{C}\rangle$ we have $\langle B\rangle \subseteq \langle A\rangle \cap \langle C\rangle$. Hence B is an interpolant between A and C.

We now give two applications of the interpolation theorem.

3. Finitely Axiomatisable Theories

We consider mathematical theories which can be formulated in the language of the predicate calculus. Such a theory is said to be *finitely axiomatisable* if it can be based on finitely many formulas of predicate calculus as axioms. We can then take the conjunction of these formulas to obtain a single formula A. The theory $\mathscr{T}(A)$ for which A is the axiom is determined in predicate calculus as follows.

The free object variables, sentential variables and predicate letters occurring in A denote the *relevant concepts* of the theory $\mathscr{T}(A)$, namely the underlying objects, sentences and predicates which are axiomatically characterized by the formula A.

A formula B expresses a classically (or intuitionistically) *provable theorem* of the theory $\mathscr{T}(A)$ if the formula $(A \to B)$ is deducible in the system CP (or IP3).

The theory $\mathscr{T}(A)$ is said to be classically (or intuitionistically) *consistent* if the formula $(A \to \bot)$ is not deducible in CP (or IP3).

A theory $\mathscr{T}(B)$ is said to be *more general* than a theory $\mathscr{T}(A)$ if the formula $(A \to B)$ is deducible. In this case every provable theorem of the theory $\mathscr{T}(B)$ is also provable in the theory $\mathscr{T}(A)$.

The interpolation theorem says: *If C is provable in a theory $\mathscr{T}(A)$ then C is also provable in a more general theory $\mathscr{T}(B)$ whose relevant concepts are only those relevant concepts of $\mathscr{T}(A)$ which occur in C.* That is: Those relevant concepts of $\mathscr{T}(A)$ which do not occur in C are not needed for the proof of C in $\mathscr{T}(A)$.

This holds not only for theories in classical predicate calculus but also for those in intuitionistic predicate calculus.

4. Beth's Definability Theorem

The simplest version of the definability theorem runs:

If the formula

(1) $((\mathscr{A}[p] \wedge \mathscr{A}[q]) \rightarrow \forall x(p(x) \rightarrow q(x)))$

is deducible then there is a one-place nominal form $\mathscr{F}$, in which neither p nor q occurs, such that the formulas

(2) $(\mathscr{A}[p] \rightarrow \forall x(p(x) \rightarrow \mathscr{F}[x]))$
(3) $(\mathscr{A}[p] \rightarrow \forall x(\mathscr{F}[x] \rightarrow p(x)))$

are deducible. Here p and q are to be two distinct one-place predicate letters which do not occur in the nominal form $\mathscr{A}$.

That is to say: If the uniqueness condition (1) is deducible for $\mathscr{A}[p]$ then a predicate is defined which can be expressed in predicate calculus by a nominal form $\mathscr{F}$. By (2) and (3) this predicate is uniquely determined by the nominal form $\mathscr{A}$.

This definability theorem holds not only for classical predicate calculus but also for intuitionistic predicate calculus. Following G. Kreisel it can be proved from the interpolation theorem as follows.

Since the formula (1) is deducible so too is the formula

$$((\mathscr{A}[p] \wedge p(a)) \rightarrow (\mathscr{A}[q] \rightarrow q(a))).$$

Here a is a free object variable which does not occur in the nominal form $\mathscr{A}$. By the interpolation theorem there is therefore a formula A, in which neither p nor q occurs, such that the formulas

$$((\mathscr{A}[p] \wedge p(a)) \rightarrow A),$$
$$(A \rightarrow (\mathscr{A}[q] \rightarrow q(a)))$$

are deducible. We may replace q in the last formula by p. A is a formula $\mathscr{F}[a]$ where a does not occur in $\mathscr{F}$. Therefore we have

(4) $\vdash((\mathscr{A}[p] \wedge p(a)) \rightarrow \mathscr{F}[a])$,
(5) $\vdash(\mathscr{F}[a] \rightarrow (\mathscr{A}[p] \rightarrow p(a)))$.

From (4) it follows that (2) is deducible and from (5) it follows that (3) is deducible.

Versions of the definability theorem for many-place predicates can also be formulated and proved in analogous fashion.

Chapter IV

Classical Simple Type Theory

Like predicate calculus, simple type theory deals with a non-empty domain of basic objects. But it deals not only with predicates over this domain, but also with arbitrary predicates of predicates. Further, quantification is permitted not only over the basic domain but also over predicates of all types.

We denote the type of the basic objects by 0, the type of statements (formulas) by 1 and the type of predicates with arguments of types $\tau_1, \ldots, \tau_n$ by $(\tau_1, \ldots, \tau_n)$. The primitive logical symbols we use are just $\rightarrow$ and $\forall$ from which we can define $\perp$, $\neg$, $\wedge$, $\dot{\vee}$ and $\exists$. We use the abstraction symbol λ to form the predicate $\lambda x_1^{\tau_1} \ldots x_n^{\tau_n} \mathscr{A}[\chi_1^{\tau_1}, \ldots, x_n^{\tau_n}]$ of type $(\tau_1, \ldots, \tau_n)$. This predicate holds for those and only those elements $e_1^{\tau_1}, \ldots, e_n^{\tau_n}$ of types $\tau_1, \ldots, \tau_n$ for which the statement (formula) $\mathscr{A}[e_1^{\tau_1}, \ldots, e_n^{\tau_n}]$ holds (is deducible). In addition we use symbols for particular basic objects and for functions on the basic domain.

We first develop classical simple type theory purely syntactically using only constructive methods and then define a corresponding semantics which can only be adequately developed by using non-constructive methods.

§10. The Formal System CT

1. The Formal Language

1. **Inductive definition** of the *types*.

1.1. 0 and 1 are *types*.

1.2. If $\tau_1, \ldots, \tau_n$ $(n \geqslant 1)$ are *types*, then $(\tau_1, \ldots, \tau_n)$ is also a *type*.

We use

σ, σ_i, τ_i as *syntactic symbols* for types and
τ for sequences $\tau_1, \ldots, \tau_n$ $(n \geqslant 1)$ of types.

2. *Primitive symbols* of the system CT.

2.1. Denumerably infinitely many free and bound variables of each type.

2.2. A non-empty set of object symbols.

2.3. Certain function letters with specified numbers $\geqslant 1$ of arguments.

2.4. The symbols $\rightarrow$, $\forall$ and λ.

2.5. Round brackets and comma.

For *syntactic symbols* we use

a^σ, $a_i^{\tau_i}$ for free variables of types σ, τ_i,
x^σ, $x_i^{\tau_i}$ for bound variables of types σ, τ_i,
a^τ for sequences $a_1^{\tau_1}, \ldots, a_n^{\tau_n}$ ($n \geqslant 1$) of free variables,
x^τ for sequences $x_1^{\tau_1}, \ldots, x_n^{\tau_n}$ ($n \geqslant 1$) of pairwise distinct bound variables.

3. **Inductive definition** of *terms* in the system CT and their types.

3.1. Every free variable of type σ is a *term* of type σ.

3.2 Every object symbol is a *term* of type 0.

3.3. If ϕ is an n-place function letter ($n \geqslant 1$) and $t_1^0, \ldots, t_n^0$ are *terms* of type 0, then $\phi(t_1^0, \ldots, t_n^0)$ is a *term* of type 0.

3.4. If $t_1^{\tau_1}, \ldots, t_n^{\tau_n}(n \geqslant 1)$ and $t_{01}^{(\tau_1, \ldots, \tau_n)}$ are *terms* of types $\tau_1, \ldots, \tau_n$ and $(\tau_1, \ldots, \tau_n)$, then $t_0^{(\tau_1, \ldots, \tau_n)}(t_1^{\tau_1}, \ldots, t_n^{\tau_n})$ is a *term* of type 1. (We sometimes denote this by $t_0^{(\tau)}(t^\tau)$ for brevity where $\tau = \tau_1, \ldots, \tau_n$ and $t^\tau = (t_1^{\tau_1}, \ldots, t_n^{\tau_n}.)$

3.5. If A and B are *terms* of type 1 then $(A \to B)$ is also a *term* of type 1.

3.6. If $\mathscr{F}[a^\sigma]$ is a *term* of type 1 and x^σ is a bound variable which does not occur in the nominal form $\mathscr{F}$ then $\forall x^\sigma \mathscr{F}[x^\sigma]$ is also a *term* of type 1.

3.7. If $\mathscr{A}[a_1^{\tau_1}, \ldots, a_n^{\tau_n}]$ ($n \geqslant 1$) is a *term* of type 1 and $x_1^{\tau_1}, \ldots, x_n^{\tau_n}$ are pairwise distinct bound variables which do not occur in the n-place nominal form $\mathscr{A}$ then $\lambda x_1^{\tau_1} \ldots x_n^{\tau_n} \mathscr{A}[x_1^{\tau_1}, \ldots, x_n^{\tau_n}]$ is a *term* of type $(\tau_1, \ldots, \tau_n)$. (We sometimes denote this by $\lambda x^\tau \mathscr{A}[x^\tau]$ for brevity where $x^\tau = x_1^{\tau_1}, \ldots, x_n^{\tau_n}$.)

The terms of type 1 are called *formulas*. Free variables of type 1 and formulas of the form $a_0^{(\tau_1, \ldots, \tau_n)}(t_1^{\tau_1}, \ldots, t_n^{\tau_n})$ where $a_0^{(\tau_1, \ldots, \tau_n)}$ is a free variable, are called *atomic formulas*.

For *syntactic symbols* we use

t^σ, $t_i^{\tau_i}$ for terms of types σ, τ_i,
t^τ for sequences $t_1^{\tau_1}, \ldots, t_n^{\tau_n}$ ($n \geqslant 1$) of terms,
A, A_i, B, B_i, C, C_i, F, F_i, G, G_i for formulas,
P, P_i for atomic formulas,
$\mathscr{F}$, $\mathscr{F}_i$ for 1-place nominal forms such that $\forall x^\sigma \mathscr{F}[x^\sigma]$, $\forall x^\sigma \mathscr{F}_i[x^\sigma]$ are formulas,
$\mathscr{A}$, $\mathscr{A}_i$ for n-place nominal forms ($n \geqslant 1$) such that $\lambda x^\tau \mathscr{A}[x^\tau]$, $\lambda x^\tau \mathscr{A}_i[x^\tau]$ are terms of type (τ) where $\tau = \tau_1, \ldots, \tau_n$.

We identify terms which only differ in the choice of the bound variables occurring in them. When we write $\mathscr{F}[t^\sigma]$ or $\mathscr{A}[t^\tau]$ we assume that the bound variables which occur in t^σ and t^τ are chosen so that they do not also occur in the nominal form $\mathscr{F}$ or $\mathscr{A}$ (respectively).

Corollaries: 1. *If* $\forall x^\sigma \mathscr{F}[x^\sigma]$ *is a formula and* t^σ *a term of type* σ, *then* $\mathscr{F}[t^\sigma]$ *is a formula.*

2. *If* $\lambda x^\tau \mathscr{A}[x^\tau]$ *is a term of type* $(\tau) = (\tau_1, \ldots, \tau_n)$ *and* $t^\tau = t_1^{\tau_1}, \ldots, t_n^{\tau_n}$ *is a sequence of terms of types* $\tau_1, \ldots, \tau_n$ *then* $\mathscr{A}[t^\tau]$ *is a formula.*

2. Chains of Subterms

We need to give formulas a rank for our proofs by induction. We define this rank with the aid of chains of subterms.

We call terms of type 0 and free variables *prime terms*.

We define the *subterms* of a term as follows:

ST1. A prime term has no subterms.

ST2. An atomic formula $a_0^{(\tau_1,\dots,\tau_n)}(t_1^{\tau_1},\dots,t_n^{\tau_n})$ has $t_1^{\tau_1},\dots,t_n^{\tau_n}$ as subterms.

ST3. A formula $\lambda x^{\tau}\mathscr{A}[x^{\tau}](t^{\tau})$ has $\mathscr{A}[t^{\tau}]$ as subterm.

ST4. A formula $(A \rightarrow B)$ has A and B as subterms.

ST5. A formula $\forall x^{\sigma}\mathscr{F}[x^{\sigma}]$ has as subterms every formula $\mathscr{F}[a^{\sigma}]$ (for every free variable a^{σ} of type σ).

ST6. A term $\lambda x^{\tau}\mathscr{A}[x^{\tau}]$ has as subterms every formula $\mathscr{A}[a^{\tau}]$ (for every sequence $a^{\tau}=a_1^{\tau_1},\dots,a_n^{\tau_n}$ of free variables of types $\tau_1,\dots,\tau_n$).

By a *subterm chain* of a term t^{σ} we mean a sequence

$$t_0^{\tau_0},\ t_1^{\tau_1},\ t_2^{\tau_2},\dots$$

of terms formed as follows:

SC1. The initial term $t_0^{\tau_0}$ of the subterm chain is the term t^{σ}.

SC2. If a term $t_i^{\tau_i}$ in the subterm chain is not a prime term, then the subterm chain contains as immediate successor of $t_i^{\tau_i}$ a subterm $t_{i+1}^{\tau_{i+1}}$ of $t_i^{\tau_i}$.

SC3. If a term $t_i^{\tau_i}$ in a subterm chain is a prime term then it is the last term in the subterm chain. We then say that the subterm chain has *length i*.

A term t^{σ} is said to be *regular* if all subterm chains of t^{σ} have a maximal finite length m. We define this number m to be the *rank* Rt^{σ} of a regular term t^{σ}.

In the following we suppose $\mathfrak{s}$ is a 1-place nominal form such that $\mathfrak{s}[a^{\sigma}]$ is a term.

Lemma 1. *If $\mathfrak{s}[a_1^{\sigma}]$ is a regular term, then $\mathfrak{s}[a_2^{\sigma}]$ is a regular term with the same rank as $\mathfrak{s}[a_1^{\sigma}]$.*

Proof. The lemma follows by induction on the rank of $\mathfrak{s}[a_1^{\sigma}]$.

Lemma 2. *If every subterm of a term t^{σ} is regular, then t^{σ} is regular.*

Proof. If a term has infinitely many regular subterms (by ST5 or ST6), then by Lemma 1 these have the same rank, m say. This yields the assertion since then t^{σ} has rank $m+1$.

We define the *height* of a type σ to be the number of round brackets which occur in σ.

Lemma 3. *The following conditions are sufficient for a term $\mathfrak{s}[t^{\sigma}]$ to be regular:*

(1) *$\mathfrak{s}[a^{\sigma}]$ and t^{σ} are regular terms.*

(2) *For every type τ_i of height less than σ we have: If $\mathscr{W}[a_i^{\tau_i}]$ and $t_i^{\tau_i}$ are regular terms, then $\mathscr{W}[t_i^{\tau_i}]$ is a regular term.*

(3) *For every regular term $\mathfrak{s}_i[a^{\sigma}]$ of smaller rank than $\mathfrak{s}[a^{\sigma}]$, $\mathfrak{s}_i[t^{\sigma}]$ is a regular term.*

Proof. 1. Suppose $\mathscr{o}$ is the nominal form $*_1$. Then $\mathscr{o}[t^\sigma]$ is the term t^σ which is regular by condition (1).

2. Suppose $\mathscr{o}$ is a nominal form $*_1(\mathscr{o}_1, \dots, \mathscr{o}_n)$ $(n \geq 1)$, where $\mathscr{o}_1, \dots, \mathscr{o}_n$ are 1-place nominal forms. Then $\sigma = (\tau_1, \dots, \tau_n)$ where $\mathscr{o}_i[t^\sigma]$ is of type τ_i. For $i = 1, \dots, n$, $\mathscr{o}_i[a^\sigma]$ is a regular term of smaller rank than $\mathscr{o}[a^\sigma]$. It follows by condition (3) that $\mathscr{o}_i[t^\sigma]$ is regular.

2.1. Suppose t^σ is a prime term. Then $\mathscr{o}[t^\sigma]$ has the subterms $\mathscr{o}_1[t^\sigma], \dots, \mathscr{o}_n[t^\sigma]$. Since these are regular, $\mathscr{o}[t^\sigma]$ is regular.

2.2. Suppose t^σ is a term $\lambda x^\tau \mathscr{A}[x^\tau]$ where $\tau = \tau_1, \dots, \tau_n$. Then t^σ has a subterm $\mathscr{A}[a^\tau]$ where $a^\tau = a_1^{\tau_1}, \dots, a_n^{\tau_n}$. This subterm is regular since t^σ is regular by condition (1). For $i = 1, \dots, n$, τ_i has a smaller height than σ. Since $\mathscr{o}_i[t^\sigma]$ is regular it follows by condition (2) that $\mathscr{A}[\mathscr{o}_1[t^\sigma], \dots, \mathscr{o}_n[t^\sigma]]$ is regular. This is the only subterm of $\mathscr{o}[t^\sigma]$. Hence $\mathscr{o}[t^\sigma]$ is regular.

3. In every other case every subterm of $\mathscr{o}[t^\sigma]$ is of the form $\mathscr{o}_i[t^\sigma]$, where $\mathscr{o}_i[a^\sigma]$ is a subterm of $\mathscr{o}[a^\sigma]$. Since $\mathscr{o}_i[a^\sigma]$ is then a regular term of smaller rank than $\mathscr{o}[a^\sigma]$, it follows by condition (3) that every subterm $\mathscr{o}_i[t^\sigma]$ of $\mathscr{o}[t^\sigma]$ is regular. Hence $\mathscr{o}[t^\sigma]$ is also regular.

Lemma 4. *If $\mathscr{o}[a^\sigma]$ and t^σ are regular terms then $\mathscr{o}[t^\sigma]$ is also a regular term.*

Proof. It follows from Lemma 3 by induction on the rank of the regular term $\mathscr{o}[a^\sigma]$ that $\mathscr{o}[t^\sigma]$ is regular if conditions (1) and (2) of Lemma 3 hold. The assertion now follows by induction on the height of the type σ.

We define the *length* of a term t^σ to be the number of primitive symbols which occur in t^σ.

Theorem 10.1. *Every term t^σ is regular.*

Proof by induction on the length of the term t^σ.

1. Suppose t^σ is a formula $\lambda x^\tau \mathscr{A}[x^\tau](t^\tau)$ where $t^\tau = t_1^{\tau_1}, \dots, t_n^{\tau_n}$. Then the terms $\mathscr{A}[a^\tau]$ and $t_1^{\tau_1}, \dots, t_n^{\tau_n}$ have smaller lengths than t^σ. By the induction hypothesis these are therefore regular. By Lemma 4 it follows that the term $\mathscr{A}[t^\tau]$ is also regular. This is the only subterm of t^σ. Hence t^σ is regular.

2. In every other case every subterm of t^σ has a smaller length than t^σ and therefore by the induction hypothesis is regular. Hence t^σ is regular by Lemma 2.

Corollary to Theorem 10.1: *Every formula has a computable finite rank.*

Theorem 10.2. *The following hold for ranks of formulas:*

(1) $RA < R(A \rightarrow B)$ *and* $RB < R(A \rightarrow B)$.

(2) $R\mathscr{F}[a^\sigma] < R\forall x^\sigma \mathscr{F}[x^\sigma]$.

(3) $R\mathscr{A}[t^\tau] < R\lambda x^\tau \mathscr{A}[x^\tau](t^\tau)$.

Proof. Immediate from the definition of the rank of a formula and the definition of subterm of a term.

3. Axioms and Basic Interferences

We set

$$\perp :=\forall x^1 x^1$$

and define *P-forms*, *N-forms*, *NP-forms*, *positive parts*, *negative parts*, *minimal positive and negative parts* and

$$F \overset{S}{\models} G$$

as in §3.

As before we use the following *syntactic symbols*:

$\mathscr{P}, \mathscr{P}_i$ for P-forms,
$\mathscr{N}, \mathscr{N}_i$ for N-forms,
$\mathscr{E}, \mathscr{E}_i$ for P-forms and N-forms,
$\mathscr{Q}, \mathscr{Q}_i$ for NP-forms.

The *axioms* of the system CT are all formulas of the form $\mathscr{Q}[P, P]$.

The *basic inferences* of the system CT are:

(S1) $\mathscr{N}[(A \to \forall x^1 x^1)], \quad \mathscr{N}[B] \vdash \mathscr{N}[(A \to B)]$,
if B is not a formula $\forall x^1 x^1$.

(S2) $\mathscr{P}[\mathscr{F}[a^\sigma]] \vdash \mathscr{P}[\forall x^\sigma \mathscr{F}[x^\sigma]]$ (with a *condition on the variable*).

(S3) $(\mathscr{F}[t^\sigma] \to \mathscr{N}[\forall x^\sigma \mathscr{F}[x^\sigma]]) \vdash \mathscr{N}[\forall x^\sigma \mathscr{F}[x^\sigma]]$.

(S4) $\mathscr{E}[\mathscr{A}[t^\tau]] \vdash \mathscr{E}[\lambda x^\tau \mathscr{A}[x^\tau](t^\tau)]$.

The free variable denoted by a^σ in the premise of an (S2)-inference is called the *eigenvariable* of the inference. It must not occur in the conclusion. The minimal positive or negative part in the conclusion of a basic inference is said to be the *principal part* of the given basic inference.

4. Deducible Formulas and Permissible Inferences

Theorem 10.3. *Every formula of the form $\mathscr{Q}[C, C]$ is deducible.*

Proof by induction on the rank of the formula C in the same way as for Theorem 4.1. For the case that C is a formula $\lambda x^\tau \mathscr{A}[x^\tau](t^\tau)$ we start from the assumption that, by I.H., $\mathscr{Q}[\mathscr{A}[t^\tau], \mathscr{A}[t^\tau]]$ is deducible. Using (S4)-inferences we have $\vdash \mathscr{Q}[C, C]$.

Theorem 10.4. *Every formula $\mathscr{N}[\forall x^1 x^1]$ is deducible.*

Proof. $\mathscr{N}$ is an N-form $\mathscr{P}[(*_1 \to B)]$. By Theorem 10.3 the formula $(B \to \mathscr{P}[(\forall x^1 x^1 \to B)])$ is deducible. Using an (S3)-inference we have $\vdash \mathscr{P}[(\forall x^1 x^1 \to B)]$, that is, $\vdash \mathscr{N}[\forall x^1 x^1]$.

Corollary to Theorems 10.3 and 10.4. *The formal system* CT *is sententially complete.*

Proof. The sentential completeness follows from Theorem 10.4 in the same way as for the system CP (see §4.1) since the connective $\perp$ is defined by $\forall x^1 x^1$ and hence P-forms and N-forms are defined as in CP.

Theorem 10.5. (Particular substitution rule). *If a_1^σ does not occur in the nominal form $\mathscr{F}$, then*

$$\mathscr{F}[a_1^\sigma] \vdash \mathscr{F}[a_2^\sigma]$$

is a weak inference.

Proof. Like that for Theorem 4.2.

Theorem 10.6 (Inversion rules). *The following are weak inferences.*
(1) $\mathscr{N}[(A \to B)] \vdash \mathscr{N}[(A \to \forall x^1 x^1)]$.
(2) $\mathscr{N}[(A \to B)] \vdash \mathscr{N}[B]$.
(3) $\mathscr{P}[\forall x^\sigma \mathscr{F}[x^\sigma]] \vdash \mathscr{P}[\mathscr{F}[a^\sigma]]$.
(4) $\mathscr{F}[\lambda x^\tau \mathscr{A}[x^\tau](t^\tau)] \vdash \mathscr{F}[\mathscr{A}[t^\tau]]$.

Proof. Like that for Theorem 4.3.

Theorem 10.7 (Structural rule of inference). *If $F \models^{S} G$ holds, then $F \vdash G$ is a weak inference.*

Proof. Like that for Theorem 4.4.

Theorem 10.8. $\mathscr{N}[\mathscr{F}[t^\sigma]] \vdash \mathscr{N}[\forall x^\sigma \mathscr{F}[x^\sigma]]$ *is a permissible inference.*

Proof. Like that for Theorem 4.5.

Theorem 10.9 (General substitution rule). *If a^σ does not occur in the nominal form $\mathscr{F}$, then*

$$\mathscr{F}[a^\sigma] \vdash \mathscr{F}[t^\sigma]$$

is a permissible inference.

Proof by induction on the deduction.

1. Suppose $\mathscr{F}[a^\sigma]$ is an axiom. Then $\mathscr{F}[t^\sigma]$ is a formula of the form $\mathscr{Q}[C, C]$, which is deducible by Theorem 10.3.

2. Suppose $F[a^\sigma]$ was obtained by a basic inference. Then by Theorem 10.5 we can assume that the basic inference does not have any free variable occurring in t^σ as its eigenvariable. Then the assertion follows from the induction hypothesis.

The logical connectives, $\wedge$, $\vee$ and $\exists$ can be defined as in the system CP and for these one obtains the permissible inferences as given in Theorem 4.8.

5. The Cut Rule

While hitherto all the proofs that certain inferences are permissible have been constructive, the *cut rule*

$$\mathscr{P}[A], (A \to B) \vdash \mathscr{P}[B]$$

cannot be shown to be permissible for the system CT by using constructive methods.

The permissibility of the cut rule in the system CT corresponds to the fundamental conjecture from which G. Takeuti [1] has constructively proved the consistency of a formal system of classical analysis.

But by K. Gödel [1] the consistency of a formal system of classical analysis cannot be proved by constructive methods. Hence the permissibility of the cut rule in CT is not constructively provable.

The permissibility of the cut rule was first proved by W. Tait [1] for second order predicate calculus (a subsystem of classical simple type theory) and then by M. Takahashi [1] and D. Prawitz [1] for the full classical simple type theory. These proofs are related to a semantic equivalent of the syntactic fundamental conjecture of Takeuti which was developed by K. Schütte [4] using partial valuations.

In §12 we introduce suitable semantics and prove the completeness theorem using the methods of Takahashi and Prawitz. From this it follows that the cut rule is permissible. Before doing this we establish in §11 the relation between deducibility and the existence of partial valuations which we need for the completeness proof.

§11. Deduction Chains and Partial Valuations

1. Definition of Deduction Chains

We define deduction chains in the same way as for the system CP (see §5.3). We first assume that for each type σ an enumeration

$$\alpha_0^\sigma, \alpha_1^\sigma, \alpha_2^\sigma, \ldots$$

of all free variables of type σ and an enumeration

$$s_0^\sigma, s_1^\sigma, s_2^\sigma, \ldots$$

of all terms of type σ are given.

The *reducible parts* of a formula F are defined to be the positive parts of F which are of the form $\forall x^\sigma \mathscr{F}[x^\sigma]$ or $\lambda x^\tau \mathscr{A}[x^\tau](t^\tau)$ and the negative parts of F which are of the form $\lambda x^\tau \mathscr{A}[x^\tau](t^\tau)$ or $(A \to B)$ where B is not a formula $\forall x^1 x^1$. A formula is said to be *reducible* if it has at least one reducible part. By the *distinguished part* of a reducible formula F we mean the reducible part of F which occurs furthest to the right in F.

By the *critical parts* of a formula F we mean the negative parts of F which are of the form $\forall x^\sigma \mathscr{F}[x^\sigma]$. A formula is said to be *critical* if it is not reducible and has at least one critical part.

A formula is said to be *primitive* if it is neither reducible nor critical. This is obviously the case if, and only if, every minimal positive part and every minimal negative part of the formula is an atomic formula.

By a *D-chain* (*deduction chain*) *of a formula F* we mean a sequence of formulas

$$F_0, F_1, F_2, \ldots,$$

formed as follows:

D1. The initial formula F_0 of the D-chain is the formula F.

D. If a formula F_m of the D-chain is an axiom or an atomic formula, then it is the last formula of the D-chain. We then say that the D-chain has *length m*.

D3. If a formula F_m of the D-chain is a reducible or critical formula but not an axiom, then F_m has an immediate successor F_{m+1} in the D-chain, where F_{m+1} is determined by F_m as follows:

D3.1. If F_m is a reducible formula $\mathscr{N}[(A \to B)]$ with distinguished negative part $(A \to B)$, then F_{m+1} is either the formula $\mathscr{N}[(A \to \forall x^1 x^1)]$ or the formula $\mathscr{N}[B]$.

D3.2. If F_m is a reducible formula $\mathscr{P}[\forall x^\sigma \mathscr{F}[x^\sigma]]$ with distinguished positive part $\forall x^\sigma \mathscr{F}[x^\sigma]$, then F_{m+1} is the formula $\mathscr{P}[\mathscr{F}[\alpha_i^\sigma]]$ where i is the least number such that α_i^σ does not occur in F_m.

D3.3. If F_m is a reducible formula $\mathscr{E}[\lambda x^\tau \mathscr{A}[x^\tau](t^\tau)]$ with distinguished positive or negative part $\lambda x^\tau \mathscr{A}[x^\tau](t^\tau)$ then F_{m+1} is the formula $\mathscr{E}[\mathscr{A}[t^\tau]]$.

D3.4. If F_m is a critical formula with critical parts

$$\forall x_1^{\sigma_1} \mathscr{F}_1[x_1^{\sigma_1}], \ldots, \forall x_k^{\sigma_k} \mathscr{F}_k[x_k^{\sigma_k}],$$

then F_{m+1} is the formula

$$(A_1 \to (\ldots \to (A_{(m+1)\cdot k} \to F_m)\ldots)),$$

where $A_{i+j.k} := \mathscr{F}_i[s_j^{\sigma_i}]$ $(i=1,\ldots,k;\, j=0,\ldots,m)$.

2. Partial Valuations

By a *partial valuation* we mean a function V which assigns a truth value $VF_i = t$ or $VF_i = f$ to certain formulas F_i (but which may not be defined for all formulas) such that the following conditions are satisfied:

V1. If $V(A \to B) = t$, then $VA = f$ or $VB = t$. (In this case VB or VA may be undefined.)

V2. If $V(A \to B) = f$, then $VA = t$ and $VB = f$.

V3. If $V\forall x^\sigma \mathscr{F}[x^\sigma]=t$, then $V\mathscr{F}[t^\sigma]=t$ for every term t^σ of type σ.

V4. If $V\forall x^\sigma \mathscr{F}[x^\sigma]=f$, then there is a free variable a^σ of type σ such that $V\mathscr{F}[a^\sigma]=f$. (In this case $V\mathscr{F}[t^\sigma]$ need not be defined for every term t^σ of type σ.)

V5. If $V\lambda x^\tau \mathscr{A}[x^\tau](t^\tau)$ is defined then $V\mathscr{A}[t^\tau]=V\lambda x^\tau \mathscr{A}[x^\tau](t^\tau)$.

Remark. These partial valuations correspond to the semivaluations in Schütte [4], Takahashi [1] and Prawitz [1].

3. Principal Lemmata

As in §5.3 we have:

Principal Syntactic Lemma. *If every D-chain of a formula F contains an axiom then the formula F is deducible.*

Proof. In this case every D-chain of F is finite. Hence by König's lemma F has only finitely many D-chains. These constitute a deduction of the formula F.

Principal Semantic Lemma. *If there is a D-chain of formula F which contains no axiom then there is a partial valuation V such that* $VF=f$.

Proof. Suppose

(DC) $\quad F_0, F_1, F_2, \ldots$

is a D-chain which contains no axiom where F_0 is the formula F. We now prove, in analogy with §5.3:

Lemma 1. *If an atomic formula occurs as a positive (or negative) part in a formula* F_n *of the D-chain, then it also occurs as a positive (or negative) part in every formula* F_m $(m \geqslant n)$ *of the D-chain.*

Proof. This follows immediately from the definition of D-chain.

Lemma 2. *If a formula* F_n *of the D-chain has a reducible positive (or reducible negative) part C, then the D-chain contains a reducible formula* F_m $(m \geqslant n)$ *with distinguished positive (or negative) part C.*

Proof. By induction on the sum of the ranks of the reducible parts of F_n which occur in F_n to the right of C.

Lemma 3. *If a formula* F_n *of the D-chain has a negative part* $\forall x^\sigma \mathscr{F}[x^\sigma]$ *then the D-chain contains infinitely many critical formulas with negative part* $\forall x^\sigma \mathscr{F}[x^\sigma]$.

Proof. Under the hypothesis for Lemma 3 it suffices to prove that the D-chain contains a critical formula F_m $(m>n)$ with negative part $\forall x^\sigma \mathscr{F}[x^\sigma]$. We call the sum of the ranks of the reducible parts of a formula the *reducibility rank* of the formula. By hypothesis the formula F_n is either reducible or critical. The D-chain therefore contains a formula F_{n+1} which also contains the formula $\forall x^\sigma \mathscr{F}[x^\sigma]$ as a negative part. We prove our assertion by induction on the reducibility rank of the formula F_{n+1}. If F_{n+1} is a critical formula then the assertion is satisfied with $m=n+1$. Otherwise the D-chain contains a formula F_{n+2} which has a smaller reducibility rank than F_{n+1} and also contains the formula $\forall x^\sigma \mathscr{F}[x^\sigma]$ as a negative part. In this case the assertion follows from the induction hypothesis.

Now we set $VC=t$ (or $VC=f$) if C occurs in the D-chain (DC) as a negative (or positive) part.

Lemma 4. *The function V satisfies conditions* V1–V5 *for a partial valuation.*

Proof. 1. Suppose $V(A \rightarrow B)=t$, then $(A \rightarrow B)$ is a negative part in (DC). If B is a formula $\forall x^1 x^1$, then A is a positive part in (DC) and therefore $VA=f$. Otherwise it follows from Lemma 2 and the definition of D-chain that A is a positive or B is a negative part in (DC), and therefore $VA=f$ or $VB=t$.

2. Suppose $V(A \rightarrow B)=f$, then $(A \rightarrow B)$ is a positive part in (DC). But then A is a negative and B a positive part in (DC), so $VA=t$ and $VB=f$.

3. Suppose $V\forall x^\sigma \mathscr{F}[x^\sigma]=t$, then $\forall x^\sigma \mathscr{F}[x^\sigma]$ is a negative part in (DC). From Lemma 3 and the definition of D-chain it follows that, for every term t^σ of type σ, $\mathscr{F}[t^\sigma]$ is a negative part in (DC). Hence $V\mathscr{F}[t^\sigma]=t$ for all t^σ of type σ.

4. Suppose $V\forall x^\sigma \mathscr{F}[x^\sigma]=f$, then $\forall x^\sigma \mathscr{F}[x^\sigma]$ is a positive part in (DC). From Lemma 2 and the definition of D-chain it follows that there is a variable a^σ of type σ such that $\mathscr{F}[a^\sigma]$ is a positive part in (DC), and therefore $V\mathscr{F}[a^\sigma]=f$.

5. Suppose $V\lambda x^\tau \mathscr{A}[x^\tau](t^\tau)$ is defined. Then $\lambda x^\tau \mathscr{A}[x^\tau](t^\tau)$ is a positive (or negative) part in (DC). From Lemma 2 and the definition of D-chain it follows that $\mathscr{A}[t^\tau]$ is a positive (or negative) part in (DC). Hence $V\mathscr{A}[t^\tau]=V\lambda x^\tau \mathscr{A}[x^\tau](t^\tau)$.

Since the D-chain (DC) contains no axiom it follows from Lemma 1 that no atomic formula occurs both as a positive and as a negative part in (DC). Therefore there is no atomic formula P for which $VP=t$ and $VP=f$. From Lemma 4 it follows by induction on the rank of the formula C that there is no formula C for which $VC=t$ and $VC=f$. Hence V is a partial valuation. Hence $VF=f$, since F occurs as a positive part in (DC).

§12. Semantics

1. Total Valuations over a System of Sets

By a *system M of sets* we mean a collection of non-empty sets M^σ for each type σ, such that M^{σ_1} and M^{σ_2} are disjoint if $\sigma_1 \neq \sigma_2$.

For a given system M of sets we introduce the following terminology.

The *basic M-terms* of type σ are the names of the elements of M^σ.

The *M-terms* of type σ are the expressions which result from terms of type σ

when all the free variables occurring in them are replaced by basic M-terms of the same type.

The *M-formulas* are the *M-terms* of type 1.

By an *M-variant* of a formula F we mean an M-formula which is obtained from F by replacing the free variables occurring in F by M-terms of the same type where a free variable which occurs more than once in F is replaced at all occurrences by the same M-term.

As *syntactic symbols* we use

c^σ, $c_i^{\tau_i}$ for basic M-terms of type σ, τ_i,

c^τ for sequences $c_1^{\tau_1}, \ldots, c_n^{\tau_n}$ $(n \geq 1)$ of basic M-terms,

u^σ, $u_i^{\tau_i}$ for M-terms of type σ, τ_i,

u^τ for sequences $u_1^{\tau_1}, \ldots, u_n^{\tau_n}$ $(n \geq 1)$ of M-terms,

A', B', C', F', G', $\forall x^\sigma \mathcal{F}'[x^\sigma]$ for M-formulas and

$\lambda x^\tau \mathcal{A}'[x^\tau]$ for M-terms of type $(\tau) = (\tau_1, \ldots, t_n)$.

By a *total valuation* over a system M of sets we mean a function W which assigns a truth value $WF' = t$ or $WF' = f$ to each M-formula F' such that the following hold:

W1. $W(A' \to B') = t$ if, and only if, $WA' = f$ or $WB' = t$.

W2. $W\forall x^\sigma \mathcal{F}'[x^\sigma] = t$ if, and only if, $W\mathcal{F}'[u^\sigma] = t$ for every M-term u^σ of type σ.

W3. $W\lambda x^\tau \mathcal{A}'[x^\tau](u^\tau) = W\mathcal{A}'[u^\tau]$.

A formula F is said to be *true* in a total valuation W over M if $WF' = t$ for every M-variant F' of F.

A formula is said to be *valid* if it is true in every total valuation over a system of sets.

Remark. An adequate semantics for the formal system CT is provided by total valuations but not by set-theoretic models since there is no extensionality axiom in CT. The semantics given here are therefore weaker than the semantics given by *general models* which L. Henkin [1] showed is adequate for classical simple type theory with extensionality.

2. Soundness Theorem

We denote by $\mathcal{P}'$, $\mathcal{N}'$ and $\mathcal{Q}'$ the nominal forms obtained from P-forms, N-forms and NP-forms when the free variables in them are replaced by basic M-terms of the same types.

For every total valuation W over M we obviously have:

PN-Lemma. *If $WA' = t$ (or $WA' = f$), then $W\mathcal{P}'[A'] = t$ (or $W\mathcal{N}'[A'] = t$).*

Theorem 12.1 (Soundness Theorem). *Every deducible formula is valid.*

Proof. Let F be a deducible formula, W a total valuation over a system M of sets and F' an M-variant of F. We prove that $WF' = t$ by induction on the order of a deduction of F.

1. Suppose F is an axiom $\mathcal{Q}[P, P]$. Then F' is an M-formula $\mathcal{Q}'[A', A']$. By the PN-lemma we have $WF' = t$.

2. Suppose F is a formula $\mathcal{N}[(A \to B)]$ obtained by an (S1)-inference from $\mathcal{N}[(A \to \forall x^1 x^1)]$ and $\mathcal{N}[B]$. Then F' is an M-formula $\mathcal{N}'[(A' \to B')]$. If $WA' = t$, then $W(A' \to B') = WB'$. Since, by I.H., $W\mathcal{N}'[B'] = t$, we have $WF' = t$. If $WA' = f$, then $W(A' \to B') = t = W(A' \to \forall x^1 x^1)$. Since, by I.H., $W\mathcal{N}'[(A' \to \forall x^1 x^1)] = t$, we have $WF' = t$.

3. Suppose F is a formula $\mathcal{P}[\forall x^\sigma \mathcal{F}[x^\sigma]]$ obtained by an (S2)-inference from $\mathcal{P}[\mathcal{F}[a^\sigma]]$. Then F' is an M-formula $\mathcal{P}'[\forall x^\sigma F'[x^\sigma]]$. If $W\forall x^\sigma \mathcal{F}'[x^\sigma] = t$, then $WF' = t$ by the PN-lemma. If $W\forall x^\sigma \mathcal{F}'[x^\sigma] = f$, then there is an M-term u^σ such that $W\mathcal{F}'[u^\sigma] = f$. $\mathcal{P}'[\mathcal{F}'[u^\sigma]]$ is an M-variant of $\mathcal{P}[\mathcal{F}[a^\sigma]]$. Hence, by I.H., $W\mathcal{P}'[\mathcal{F}'[u^\sigma]] = t$. Therefore $WF' = t$.

4. Suppose F is a formula $\mathcal{N}[\forall x^\sigma \mathcal{F}[x^\sigma]]$ obtained by an (S3)-inference from $(\mathcal{F}[t^\sigma] \to \mathcal{N}[\forall x^\sigma \mathcal{F}[x^\sigma]])$. Then F' is an M-formula $\mathcal{N}'[\forall x^\sigma \mathcal{F}'[x^\sigma]]$. By I.H., we have $W(\mathcal{F}'[u^\sigma] \to \mathcal{N}'[\forall x^\sigma \mathcal{F}'[x^\sigma]]) = t$ for an M-term u^σ. If $W\mathcal{F}'[u^\sigma] = t$, then $WF' = t$. If $W\mathcal{F}'[u^\sigma] = f$, then $W\forall x^\sigma \mathcal{F}'[x^\sigma] = f$ too. Then $WF' = t$ by the PN-lemma.

5. Suppose F is a formula $\mathcal{E}[\lambda x^\tau \mathcal{A}[x^\tau](t^\tau)]$ obtained by an (S4)-inference from $\mathcal{E}[\mathcal{A}[t^\tau]]$. Then F' is an M-formula $\mathcal{E}'[\lambda x \mathcal{A}'[x^\tau](u^\tau)]$. By I.H., $W\mathcal{E}'[\mathcal{A}'[u^\tau]] = t$. Since $W\lambda x^\tau \mathcal{A}'[x^\tau](u^\tau) = W\mathcal{A}'[u^\tau]$ it follows that $WF' = t$.

3. Extending a Partial Valuation

Let V be a partial valuation. We shall define a total valuation W over a suitable system M of sets such that no formula F for which $VF = f$ is true in the total valuation W. We follow D. Prawitz [1].

Inductive definition of a set Pt^σ (of *possible valuations*) for each term t^σ (by induction on the height (see p. 58) of the type σ).

P1. $Pt^0 := \{t^0\}$.

P2. If Vt^1 is defined, let $Pt^1 := \{Vt^1\}$. Otherwise let $Pt^1 := \{t, f\}$.

P3. For $\tau = \tau_1, \ldots, \tau_n$ $(n \geqslant 1)$ $p_0 \in Pt_0^{(\tau)}$ if, and only if, the following conditions are satisfied:

P3.1. p_0 is a set of n-tuples

$$(\langle t_1^{\tau_1}, p_1 \rangle, \ldots, \langle t_n^{\tau_n}, p_n \rangle)$$

such that $p_i \in Pt_i^{\tau_i}$ $(i = 1, \ldots, n)$.

P3.2. If $Vt_0^{(\tau)}(t_1^{\tau_1}, \ldots, t_n^{\tau_n})$ is defined and $p_i \in Pt_i^{\tau_i}$ $(i = 1, \ldots, n)$, then

$$(\langle t_1^{\tau_1}, p_1 \rangle, \ldots, \langle t_n^{\tau_n}, p_n \rangle) \in p_0$$

if, and only if, $Vt_0^{(\tau)}(t_1^{\tau_1}, \ldots, t_n^{\tau_n}) = t$.

For every type σ let M^σ be the set of ordered pairs $\langle t^\sigma, p \rangle$ with $p \in Pt^\sigma$. Obviously M^σ is non-empty. So the sets M^σ form a system M of sets. We now consider this system M and use the syntactic symbols introduced in §12.1.

If u^σ is an M-term of type σ let $u^{\sigma *}$ be the term of type σ obtained from u^σ by replacing each basic M-term $\langle t_i^{\tau_i}, p_i\rangle$ by the term $t_i^{\tau_i}$.

Inductive definition of the *degree* du^σ of an M-term u^σ.

d1. $du^\sigma := 0$ if $\sigma = 0$ or u^σ is a basic M-term.

d2. $du_0^{(\tau)}(u_1^{\tau_1}, \ldots, u_n^{\tau_n}) := \max(du_0^{(\tau)}, du_1^{\tau_1}, \ldots, du_n^{\tau_n}) + 1$.

d3. $d(A' \to B') := \max(dA', dB') + 1$.

d4. $d\forall x^\sigma \mathscr{F}'[x^\sigma] := d\mathscr{F}'[c^\sigma] + 1$, where c^σ is a basic M-term.

d5. $d\lambda x^\tau \mathscr{A}'[x^\tau] := d\mathscr{A}'[c^\tau] + 1$, where c^τ is a sequence of basic M-terms.

Inductive definition of Wu^σ for each M-term u^σ (by induction on the degree du^σ).

W1. $Wu^0 := u^{0*}$.

W2. $W\langle t^\sigma, p\rangle := p$ for $p \in Pt^\sigma$. (If $\sigma = 0$, we have $p = t^0$, hence $W\langle t^0, p\rangle = t^0 = \langle t^0, p\rangle^*$ by W2, in accordance with W1.)

W3. $Wu_0^{(\tau)}(u_1^{\tau_1}, \ldots, u_n^{\tau_n}) := t$, if

$$(\langle u_1^{\tau_1 *}, Wu_1^{\tau_1}\rangle, \ldots, \langle u_n^{\tau_n *}, Wu_n^{\tau_n}\rangle) \in Wu_0^{(\tau)}.$$

Otherwise let $Wu_0^{(\tau)}(u_1^{\tau_1}, \ldots, u_n^{\tau_n}) := f$.

W4. $W(A' \to B') := t$, if $WA' = f$ or $WB' = t$. Otherwise let $W(A' \to B') := f$.

W5. $W\forall x^\sigma \mathscr{F}'[x^\sigma] := t$, if $W\mathscr{F}'[c^\sigma] = t$ holds for every basic M-term c^σ of type σ. Otherwise let $W\forall x^\sigma \mathscr{F}'[x^\sigma] := f$.

W6. For $\tau = \tau_1, \ldots, \tau_n$ $(n \geqslant 1)$ let $W\lambda x^\tau \mathscr{A}'[x^\tau]$ be the set of n-tuples $(c_1^{\tau_1}, \ldots, c_n^{\tau_n})$ satisfying the condition $W\mathscr{A}'[c_1^{\tau_1}, \ldots, c_n^{\tau_n}] = t$.

Lemma 1. $Wu^\sigma \in Pu^{\sigma *}$.

Proof by induction on the degree du^σ.

1. Suppose $\sigma = 0$. Then we have $Wu^0 = u^{0*}$ and $Pu^{0*} = \{u^{0*}\}$, so $Wu^0 \in Pu^{0*}$.

2. Suppose u^σ is a basic M-term $\langle t^\sigma, p\rangle$. Then $Wu^\sigma = p \in Pt^\sigma$ and $u^{\sigma *} = t^\sigma$, so $Wu^\sigma \in Pu^{\sigma *}$.

Suppose $\sigma = 1$ and u^1 is not a basic M-term. By definition $Wu^1 \in \{t, f\}$. We now only need to prove that: If Vu^{1*} is defined, then $Wu^1 = Vu^{1*}$.

3.1. Suppose u^1 is an M-formula $u_0^{(\tau)}(u_1^{\tau_1}, \ldots, u_n^{\tau_n})$. Then by I.H. $Wu_0^{(\tau)} \in Pu_0^{(\tau)*}$ and $Wu_i^{\tau_i} \in Pu_i^{\tau_i *}$ $(i = 1, \ldots, n)$. If Vu^{1*} is defined then, by the definition of $Pu_0^{(\tau)*}$,

$$(\langle u_1^{\tau_1 *}, Wu_1^{\tau_1}\rangle, \ldots, \langle u_n^{\tau_n *}, Wu_n^{\tau_n}\rangle) \in Wu_0^{(\tau)}$$

if, and only if, $Vu^{1*} = t$. Hence in this case $Wu^1 = Vu^{1*}$.

3.2. Suppose u^1 is an M-formula $(A' \to B')$. Then by I.H. $WA' \in PA'^*$ and $WB' \in PB'^*$. If $Vu^{1*} = t$, then $VA'^* = f$ or $VB'^* = t$. Then by I.H. $WA' = f$ or $WB' = t$ and therefore $Wu^1 = t$. If $Vu^{1*} = f$, then $VA'^* = t$ and $VB'^* = f$. Then by I.H. $WA' = t$ and $WB' = f$ and therefore $Wu^1 = f$.

3.3. Suppose u^1 is an M-formula $\forall x^{\sigma_0} \mathscr{F}'[x^{\sigma_0}]$. Then u^{1*} is a formula $\forall x^{\sigma_0} \mathscr{F}[x^{\sigma_0}]$. By I.H. $W\mathscr{F}'[c^{\sigma_0}] \in P\mathscr{F}[c^{\sigma_0 *}]$ for every basic M-term of type σ_0. If $Vu^{1*} = t$ then $V\mathscr{F}[t^\sigma] = t$ holds for every term t^{σ_0} of type σ_0. Then by I.H.

$W\mathscr{F}'[c^{\sigma_0}]=t$ for every basic M-term c^{σ_0} of type σ_0 and therefore $Wu^1=t$. If $Vu^{1*}=f$, then there is a free variable a^{σ_0} of type σ_0 such that $V\mathscr{F}[a^{\sigma_0}]=f$. Then by I.H. $W\mathscr{F}'[c^{\sigma_0}]=f$ for some basic M-term c^{σ_0} of type σ_0 and therefore $Wu^1=f$.

4. Suppose u^σ is an M-term $\lambda x^\tau \mathscr{A}'[x^\tau]$ where $\sigma=(\tau)=(\tau_1,\dots,\tau_n)$ $(n\geq 1)$. Then $u^{\sigma*}$ is a term $\lambda x^\tau \mathscr{A}[x^\tau]$. If $Vu^{\sigma*}(c_1^{\tau_1*},\dots,c_n^{\tau_n*})$ is defined then $V\mathscr{A}[c_1^{\tau_1*},\dots,c_n^{\tau_n*}]=Vu^{\sigma*}(c_1^{\tau_1*},\dots,c_n^{\tau_n*})$. Then by I.H. $W\mathscr{A}'[c_1^{\tau_1},\dots,c_n^{\tau_n}]=V\mathscr{A}[c_1^{\tau_1*},\dots,c_n^{\tau_n*}]$. In this case by the definition of Wu^σ

$$(c_1^{\tau_1},\dots,c_n^{\tau_n})\in Wu^\sigma$$

if, and only if, $Vu^{\sigma*}(c_1^{\tau_1*},\dots,c_n^{\tau_n*})=t$. Hence we have $Wu^\sigma\in Pu^{\sigma*}$ by the definition of $Pu^{\sigma*}$.

Lemma 2. *If $\mathfrak{s}$ is a 1-place nominal form such that $\mathfrak{s}[u^\sigma]$ is an M-term, then $W\mathfrak{s}[u^\sigma]=W\mathfrak{s}[\langle u^{\sigma*}, Wu^\sigma\rangle]$.*

Proof. By Lemma 1 $Wu^\sigma\in Pu^{\sigma*}$ and therefore $\langle u^{\sigma*}, Wu^\sigma\rangle$ is a basic M-term of type σ. If $\mathfrak{s}=*_1$ then the assertion holds by the definition of $W\langle u^{\sigma*}, Wu^\sigma\rangle$. In the other cases the assertion follows by induction on the length of the nominal form $\mathfrak{s}$.

Lemma 3. *$W\forall x^\sigma\mathscr{F}'[x^\sigma]=t$ if, and only if, $W\mathscr{F}'[u^\sigma]=t$ for every M-term u^σ of type σ.*

Proof. If $W\mathscr{F}'[u^\sigma]=t$ for every M-term u^σ of type σ then $W\forall x^\sigma\mathscr{F}'[x^\sigma]=t$ by definition. If $W\forall x^\sigma\mathscr{F}'[x^\sigma]=t$, then $W\mathscr{F}'[(u^{\sigma*}, Wu^\sigma\rangle]=t$ and by Lemma 2 $W\mathscr{F}'[u^\sigma]=t$ for every M-term u^σ of type σ.

Lemma 4. $W\lambda x^\tau\mathscr{A}'[x^\tau](u^\tau)=W\mathscr{A}'[u^\tau]$.

Proof. Let $c_i^{\tau_i}:=\langle u_i^{\tau_i*}, Wu_i^{\tau_i}\rangle$ $(i=1,\dots,n)$ and $c^\tau:=c_1^{\tau_1},\dots,c_n^{\tau_n}$. By definition $W\lambda x^\tau\mathscr{A}'[x^\tau](u^\tau)=t$ if, and only if, $(c_1^{\tau_1},\dots,c_n^{\tau_n})\in W\lambda x^\tau\mathscr{A}'[x^\tau]$. This is the case if, and only if, $W\mathscr{A}'[c^\tau]=t$ and then by Lemma 2 $W\mathscr{A}'[u^\tau]=t$.

Theorem 12.2. *The restriction of W to M-formulas is a total valuation over M under which no formula F is true for which $VF=f$.*

Proof. By the definition of W and Lemmata 3 and 4 W is a total valuation over M. Suppose $VF=f$. Replace each free variable a occurring in F by a basic M-term $\langle^\sigma, p\rangle$ such that $p\in Pa^\sigma$ to obtain an M-formula F'. Then F' is an M-variant of F with $F'^*=F$. By Lemma 1, $WF'\in PF$. Since $VF=f$, we have $WF'=f$. Hence F is not true under the total valuation W.

4. Completeness Theorem and Cut Rule

Theorem 12.3 (Completeness Theorem). *Every valid formula is deducible.*

Indirect proof. Suppose F is not deducible. Then by the principal syntactic lemma (§11.3) there is a D-chain of F which contains no axiom and therefore by the principal semantic lemma (§11.3) there is a partial valuation V with $VF=f$. It follows from Theorem 12.2 that F is not valid.

Theorem 12.4 (Cut rule). *$\mathscr{P}[A]$, $(A \to B) \vdash \mathscr{P}[B]$ is a permissible inference.*

Proof. Suppose the formulas $\mathscr{P}[A]$ and $(A \to B)$ are deducible, then they are valid by Theorem 12.1. Let W be a total valuation over a system M of sets. If $\mathscr{P}'[A']$ and $(A' \to B')$ are M-variants of the formulas $\mathscr{P}[A]$ and $(A \to B)$ then $W\mathscr{P}'[A']=t$ and $W(A' \to B')=t$. If $WB'=t$, then $W\mathscr{P}'[B']=t$ by the PN-lemma (p. 66). If $WB'=f$, then since $W(A' \to B')=t$, we have $WA'=f$. Then from $W\mathscr{P}'[A']=t$ we have $W\mathscr{P}'[B']=t$. In any case, therefore, $W\mathscr{P}'[B']=t$. Hence $\mathscr{P}[B]$ is valid and, by Theorem 12.3, deducible.

Part B

Systems of Arithmetic

Chapter V

Ordinal Numbers and Ordinal Terms

In this chapter we develop a constructive system of ordinals which we shall use in §16 and Chapter VIII for the proof-theoretic treatment of pure number theory and predicative analysis. In §13 we start from a non-constructive presentation of the classical theory of ordinals in which we take as basis a corresponding axiomatic characterization of the set $\mathbb{O}$, of all finite and denumerably infinite ordinals where, however, we do not go very far into set theory but use sets in a naive way. In this non-constructive framework we develop a hierarchy of normal functions and critical ordinals which determine a certain initital segment of the ordinals (in fact up to Γ_0). This initial segment of the ordinals will be constructively characterized in §14 (by ordinal terms) and all the properties of this system of ordinal terms which we need later will be developed constructively. The full system of these ordinal terms (up to Γ_0) is first used in Chapter VIII (for predicative analysis) while a subsystem (up to ε_0) occurs in §16 (for pure number theory).

§13. Theory of Ordinals of the 1st and 2nd Number Classes

1. Order Types of Well-Ordered Sets

Definition. A set M is said to be *ordered* by a relation R, if, for all $a, b, c \in M$ we have:

1. Not $a\ R\ a$ (*Irreflexiveness*)
2. $a\ R\ b$ and $b\ R\ c$ imply $a\ R\ c$ (*Transitivity*)
3. $a\ R\ b$ or $a=b$ or $b\ R\ a$ (*Connection*)

Corollary of 1–3: *If $a, b \in M$ then precisely one of $a\ R\ b$, $a=b$, $b\ R\ a$ holds.*

M is said to be *well-ordered* (by R) if we also have:

4. Every non-empty subset $M_0 \subseteq M$ contains a first element with respect to R. That is: there exists $a_0 \in M_0$ such that $a_0 = x$ or $a_0\ R\ x$ for all $x \in M_0$.

We denote a set M, well-ordered by R, by (M, R). If $a \in M$ we let $(M, R) \mid a$ be the set of those $x \in M$ such that $x\ R\ a$ with the relation R restricted to this set. $(M, R) \mid a$ is again a well-ordered set.

Two well-ordered sets (M_1, R_1), (M_2, R_2) are said to be *isotonic* if there is an order preserving map of M_1 onto M_2, that is, a map $f: M_1 \to M_2$ such that $f(M_1) = M_2$ and

$$a\, R_1\, b \Rightarrow f(a)\, R_2\, f(b)$$

for $a, b \in M_1$. Such a map is bijective. Isotonism is an *equivalence relation.* The equivalence classes of well-ordered sets under isotonism are called *order types.*

We set $(M_1, R_1) < (M_2, R_2)$ if there is $a_2 \in M_2$ such that (M_1, R_1) and $(M_2, R_2) \mid a_2$ are isotonic. This relation is compatible with isotonism. That is: if $(M_1, R_1), (M_2, R_2)$ are isotonic and $(M_3, R_3), (M_4, R_4)$ are isotonic then $(M_1, R_1) < (M_3, R_3)$ implies $(M_2, R_2) < (M_4, R_4)$. We can therefore make the following definition for order types α_1, α_2: $\alpha_1 < \alpha_2$ if, and only if, $(M_1, R_1) < (M_2, R_2)$ for well-ordered sets (M_1, R_1), (M_2, R_2) of order types α_1, α_2, respectively. The totality of order types of well-ordered sets is then well-ordered by this relation $<$.

Ordinal numbers (or *ordinals*) serve as representatives of order types of well-ordered sets. In the context of axiomatic set theory (say Zermelo-Fraenkel or Bernays-Gödel) they are defined as specific sets which are well-ordered by the $\in$-relation in such a way that every order type of a well-ordered set has precisely one ordinal number belonging to it. The totality of these ordinal numbers is well-ordered by the $\in$-relation which corresponds to the $<$-relation on order types here.

The ordinal numbers which belong to order types of finite well-ordered sets form the 1st Cantor number class. These ordinals correspond to the natural numbers. The ordinals which belong to order types of denumerably infinite well-ordered sets form the 2nd Cantor number class.

2. Axiomatic Characterization of the 1st and 2nd Number Classes

In this section we restrict our attention to the ordinals of the 1st and 2nd number classes and we represent these by a set $\mathbb{O}$ which is characterized as a set of ordinals by the three axioms below. These axioms uniquely determine, up to isomorphism, the set of ordinals of the 1st and 2nd number classes as it arises in the context of an axiomatic set theory. Here we are sticking to an axiomatic determination of the concept of set, in general in what follows we shall be dealing with a naive concept of set.

Ax. I. *$\mathbb{O}$ is a set well-ordered by a relation $<$.*

Ax. II. *Every bounded subset of $\mathbb{O}$ is denumerable.* That is: if, given $M \subset \mathbb{O}$ there exists $\alpha \in \mathbb{O}$ such that $\xi < \alpha$ for all $\xi \in M$, then M is a finite or denumerably infinite set.

Ax. III. *Every denumerable subset of $\mathbb{O}$ is bounded.* That is: for each finite or denumerably infinite set $M \subset \mathbb{O}$ there exists $\alpha \in \mathbb{O}$ such that $\xi < \alpha$ for all $\xi \in M$.

Corollary: $\mathbb{O}$ is an infinite, but not denumerable set.

We call the elements of $\mathbb{O}$ *ordinal numbers* (*ordinals*, for short). We denote them by small Greek letters. We write $\alpha \leqslant \beta$ for $\alpha < \beta$ or $\alpha = \beta$, $\alpha > \beta$ for $\beta < \alpha$ and $\alpha \geqslant \beta$ for $\beta \leqslant \alpha$. All the definitions introduced in this section refer to the axiomatically given set $\mathbb{O}$.

If $\mathscr{A}(\alpha)$ is a statement about an arbitrary ordinal α then the property $\mathscr{A}$ of ordinals is said to be *progressive* if, for all ordinals α, $\mathscr{A}(\alpha)$ can be deduced from the hypothesis that $\mathscr{A}(\xi)$ holds for all $\xi < \alpha$.

Induction Theorem. *If $\mathscr{A}$ is progressive then $\mathscr{A}(\alpha)$ holds for all ordinals α.*

Proof. Suppose $\mathscr{A}(\alpha)$ did not hold for every ordinal α. Then the set of ordinals ξ such that $\mathscr{A}(\xi)$ is false is non-empty. By Ax. I this set must contain a least element. This contradicts the hypothesis that $\mathscr{A}$ is progressive.

The induction theorem shows that in order to prove a theorem $\mathscr{A}(\alpha)$ about ordinals it is sufficient to prove $\mathscr{A}$ is progressive. That is: it is sufficient to prove $\mathscr{A}(\alpha)$ for arbitrary $\alpha \in \mathbb{O}$ under the

Induction hypothesis: $\mathscr{A}(\xi)$ holds for all $\xi < \alpha$.

This proof principle for $\mathscr{A}(\alpha)$ is called *transfinite induction* (on α). We shall frequently apply this transfinite induction below to prove theorems about ordinals of the set $\mathbb{O}$.

3. Zero, Successor and Limit Numbers and Supremum

By Ax. I there is a unique smallest ordinal. We denote this by 0.

For each ordinal α there is a unique smallest ordinal $> \alpha$.

Proof. $\{\alpha\}$ is a denumerable subset of $\mathbb{O}$. By Ax. III it follows that the set of $\xi > \alpha$ is non-empty. By Ax. I this set contains a unique smallest element. We call this the *successor* α' of α.

The following hold for this definition of α':

$$\alpha < \alpha'$$
$$\alpha < \beta \Rightarrow \alpha' \leqslant \beta$$
$$\alpha < \beta' \Rightarrow \alpha \leqslant \beta.$$

An ordinal β is said to be a *successor number* if there is an $\alpha \in \mathbb{O}$ with $\alpha' = \beta$. An ordinal is said to be a *limit number* if it is neither 0 nor a successor number.

For every limit number β the following holds:

$$\alpha < \beta \Rightarrow \alpha' < \beta.$$

Proof. $\alpha < \beta$ implies $\alpha' \leqslant \beta$. If β is a limit number then $\alpha' \neq \beta$, hence $\alpha' < \beta$.

For every denumerable set $M \subset \mathbb{O}$ there is a unique smallest ordinal α such that $\xi \leqslant \alpha$ for all $\xi \in M$. (This follows from Ax. III and Ax. I.) We call this ordinal α the *supremum* sup M of the set M. The definition yields

$$\xi \in M \Rightarrow \xi \leqslant \sup M$$
$$\xi \leqslant \beta \textit{ for all } \xi \in M \Rightarrow \sup M \leqslant \beta$$
$$\beta < \sup M \Rightarrow \textit{there exists } \xi \in M \textit{ with } \beta < \xi.$$

Theorem 13.1. *If M is a non-empty denumerable subset of $\mathbb{O}$ which has no maximal element then* sup M *is a limit number.*

Proof. Since M has no maximal element, sup $M \notin M$. For all $\xi \in M$ it follows that $\xi < \sup M$. Since M is non-empty it follows that sup $M \neq 0$. If $\alpha' \leqslant \sup M$ then $\alpha < \sup M$. Then there exists $\xi \in M$ such that $\alpha < \xi$. It follows that $\alpha' \leqslant \xi < \sup M$. Therefore sup M is neither 0 nor a successor number, and hence is a limit number.

Definition. The set $\mathbb{N}$ of *finite ordinals* is the smallest subset of $\mathbb{O}$ which contains 0 and is such that if α is in the subset then so is its successor α'.

This set $\mathbb{N}$ which consists of the ordinals $0, 0', 0'', \ldots$, corresponds to the set of natural numbers. Accordingly we set $1 := 0'$, $2 := 0''$, ... Obviously $\mathbb{N}$ is a denumerably infinite set which contains no maximal element and no limit number.

Lemma 1. $\alpha < \beta,\ \beta \in \mathbb{N} \Rightarrow \alpha \in \mathbb{N}$.

Proof by transfinite induction on β. If $\alpha < \beta$ and $\beta \in \mathbb{N}$ then it follows that there exists $\beta_0 \in \mathbb{N}$ with $\beta_0' = \beta$. Then $\alpha \leqslant \beta_0$. If $\alpha = \beta_0$, then $\alpha \in \mathbb{N}$. If $\alpha < \beta_0$ then $\alpha \in \mathbb{N}$ by the induction hypothesis since $\beta_0 < \beta$.

Definition. $\omega := \sup \mathbb{N}$.

Theorem 13.2. *ω is the smallest number and $\alpha < \omega$ holds if, and only if, $\alpha \in \mathbb{N}$.*

Proof. By Theorem 13.2 ω is a limit number since $\mathbb{N}$ contains no maximal element. For all $\alpha \in \mathbb{N}$ it follows that $\alpha < \omega$. Conversely, if $\alpha < \omega$, then there exists $\beta \in \mathbb{N}$ with $\alpha < \beta$. By Lemma 1, $\alpha \in \mathbb{N}$. Hence $\alpha < \omega$ if, and only if, $\alpha \in \mathbb{N}$. It follows from this that ω is the smallest limit number, since $\mathbb{N}$ contains no limit number.

4. Ordering Functions

For $\alpha \in \mathbb{O}$ let $\mathbb{O}(\alpha)$ be the set of all $\xi < \alpha$. $\mathbb{O}(0)$ is the empty set, and by Theorem 13.2 $\mathbb{O}(\omega) = \mathbb{N}$. Every set $\mathbb{O}(\alpha)$ is denumerable by Ax. II.

Definition. A set $A \subseteq \mathbb{O}$ is said to be an $\mathbb{O}$-*segment* if

$$\xi < \eta, \quad \eta \in A \Rightarrow \xi \in A.$$

The whole set $\mathbb{O}$ is an $\mathbb{O}$-segment. An $\mathbb{O}$-segment which is $\neq \mathbb{O}$ is said to be a *proper* $\mathbb{O}$-segment. Obviously every $\mathbb{O}(\alpha)$ is a proper $\mathbb{O}$-segment. The converse also holds:

Every proper $\mathbb{O}$*-segment is an* $\mathbb{O}(\alpha)$.

Proof. If $A \neq \mathbb{O}$ then there is a least ordinal $\alpha \notin A$. If A is an $\mathbb{O}$-segment then $A = \mathbb{O}(\alpha)$.

Definition. A function $f: A \to B$ is said to be an *ordering function* of a set $B \subseteq \mathbb{O}$ if the following holds:

1. The domain A of the function f is an $\mathbb{O}$-segment.
2. The function f is *strictly monotone*, that is, for all $\xi < \eta \in A$ we have $f(\xi) < f(\eta)$.
3. The image of the function f is the set B.

Remark. If B is empty then the *empty function* $\varnothing \to \varnothing$ is an ordering function of B. (The empty set $\varnothing$ is an $\mathbb{O}$-segment.)

Theorem 13.3. *If* $f: A \to B$ *is an ordering function of* B *then* $\alpha \leqslant f(\alpha)$ *for all* $\alpha \in A$.

Proof by transfinite induction on α. For all $\xi < \alpha$, $f(\xi) < f(\alpha)$ and by the induction hypothesis $\xi \leqslant f(\xi)$, therefore $\xi < f(\alpha)$. Hence $\alpha \leqslant f(\alpha)$.

Lemma 2. *Every set* $B \subseteq \mathbb{O}$ *has at most one ordering function.*

Proof. Let $f_i: A_i \to B$ $(i = 1,2)$ be ordering functions of B. We prove by transfinite induction on α: If $\alpha \in A_1$, then $\alpha \in A_2$ and $f_1(\alpha) = f_2(\alpha)$. Now $\alpha \in A_1$ implies $\xi \in A_1$ for all $\xi < \alpha$ since A_1 is an $\mathbb{O}$-segment. By the induction hypothesis it follows that $\xi \in A_2$ and $f_1(\xi) = f_2(\xi)$ for all $\xi < \alpha$. By property 3 it follows that $\alpha \in A_2$. By properties 1 and 2 it follows that $f_1(\alpha)$ is the least element of B not equal to any $f_1(\xi)$ for $\xi < \alpha$ and similarly for $f_2(\alpha)$. Hence $f_1(\alpha) = f_2(\alpha)$. Therefore $A_1 \subseteq A_2$. Similarly $A_2 \subseteq A_1$. It follows that $A_1 = A_2$ and $f_1(\alpha) = f_2(\alpha)$ for all $\alpha \in A_1$ and therefore $f_1 = f_2$.

Definition. By a *proper segment* of a set $B \subset \mathbb{O}$ we mean a set of the form $B(\beta) := B \cap \mathbb{O}(\beta)$ for some $\beta \in B$.

Lemma 3. *If every proper segment of a set* $B \subset \mathbb{O}$ *has an ordering function then* B *also has an ordering function.*

Proof. Let $f_\beta: A_\beta \to B(\beta)$ be an ordering function of $B(\beta)$ for each proper segment $B(\beta)$ of B. By Ax. II the set $B(\beta)$ is denumerable. Since f_β is bijective, A_β is also denumerable and therefore a proper $\mathbb{O}$-segment. Therefore to each ordinal $\beta \in B$ there is an ordinal $g(\beta)$ such that $A_\beta = \mathbb{O}(g(\beta))$. g is a map from B into $\mathbb{O}$.

(1) *If* $\beta_1, \beta_2 \in B$ *and* $\beta_1 < \beta_2$ *then* $g(\beta_1) < g(\beta_2)$.

Proof. f_{β_2} is a map of $\mathbb{O}(g(\beta_2))$ onto $B(\beta_2)$. Since $\beta_1 \in B(\beta_2)$, there exists $\alpha < g(\beta_2)$ such that $f_{\beta_2}(\alpha) = \beta_1$. The restriction of f_{β_2} to $\mathbb{O}(\alpha)$ is an ordering function of $B(\beta_1)$. Since $f_{\beta_1}: \mathbb{O}(g(\beta_1)) \to B(\beta_1)$ is also an ordering function, it follows by Lemma 2 that $\mathbb{O}(\alpha) = \mathbb{O}(g(\beta_1))$ and therefore $g(\beta_1) = \alpha < g(\beta_2)$.

(2) *The image of fhe function g is an* $\mathbb{O}$*-segment.*

Proof. Let $\beta \in B$ and $\alpha < g(\beta)$. We have to prove that $\alpha \in g(B)$. Since f_β is a map of $\mathbb{O}(g(\beta))$ onto $B(\beta)$ and $\alpha < g(\beta)$ there exists $\beta_0 \in B(\beta)$ with $f_\beta(\alpha) = \beta_0$. The restriction of f_β to $\mathbb{O}(\alpha)$ is an ordering function of $B(\beta_0)$. Since $f_{\beta_0}: \mathbb{O}(g(\beta_0)) \to B(\beta_0)$ is also an ordering function, so by Lemma 2 $\alpha = g(\beta_0)$ and therefore $\alpha \in g(B)$.

From (1) and (2) g is an order preserving map of B onto an $\mathbb{O}$-segment A. The inverse of g is an order preserving map of A onto B and therefore an ordering function of B.

Theorem 13.4. *Every set* $B \subseteq \mathbb{O}$ *has exactly one ordering function.*

Proof. If $B(\beta)$ is a proper segment of B then every proper segment of $B(\beta)$ is a proper segment $B(\beta_0)$ of B for some $\beta_0 < \beta$. It follows from Lemma 3 by transfinite induction on β that every proper segment $B(\beta)$ of B has an ordering function. By Lemma 3, B then has an ordering function. By Lemma 2 this is uniquely determined by B.

Definitions. A set $B \subseteq \mathbb{O}$ is said to be *closed* if, for every non-empty denumerable set M,

$$M \subseteq B \Rightarrow \sup M \in B.$$

An ordering function $f: A \to B$ is said to be *continuous* if A is closed and, for every non-empty denumerable set $U \subseteq A$,

$$\sup f(U) = f(\sup U).$$

Theorem 13.5. *The ordering function of a set* $B \subseteq \mathbb{O}$ *is continuous if, and only if,* B *is closed.*

Proof. Let $f: A \to B$ be the ordering function of B.

1. Suppose f is continuous. For every non-empty denumerable set $M \subseteq B$ there is a non-empty denumerable set $U \subseteq A$ such that $f(U) = M$. Since f is continuous, $\sup M = f(\sup U)$ and therefore $\sup M \in B$. Hence B is closed.

2. Suppose B is closed. Let U be a non-empty denumerable subset of A. Then $f(U)$ is a non-empty denumerable subset of B. Since B is closed, $\sup f(U) \in B$. Hence there exists $\alpha \in A$ such that $f(\alpha) = \sup f(U)$. It follows that for all $\xi \in U$, $f(\xi) \leqslant f(\alpha)$ and therefore $\xi \leqslant \alpha$. Hence $\sup U \leqslant \alpha$. Since A is an $\mathbb{O}$-segment it follows that $\sup U \in A$. Hence A is closed. For all $\xi \in U$, $\xi \leqslant \sup U$, therefore $f(\xi) \leqslant f(\sup U)$. It follows that $\sup f(U) \leqslant f(\sup U)$. Now $\sup U \leqslant \alpha$ implies $f(\sup U) \leqslant f(\alpha) = \sup f(U)$. Hence $\sup f(U) = f(\sup U)$ and therefore f is continuous.

Definition. An ordering function $f: A \to B$ is said to be a *normal function* if $A = \mathbb{O}$ and f is continuous.

Theorem 13.6. *The ordering function of a set* $B \subseteq \mathbb{O}$ *is a normal function if, and only if,* B *is closed and unbounded.*

Proof. By Ax. II and Ax. III a subset of $\mathbb{O}$ is bounded if, and only if, it is denumerable. It follows that if $f: A \to B$ is an ordering function, then the set B is unbounded if, and only if, A is unbounded, but then $A = \mathbb{O}$. This together with Theorem 13.5 yields the assertion.

5. Addition of Ordinals

For $\alpha \in \mathbb{O}$ let B_α be the set of all ordinals $\geqslant \alpha$ and f_α the ordering function of B_α. Obviously the set B_α is closed and unbounded. So by Theorem 13.6 f_α is a normal function. In this way $f_\alpha(\beta)$ is defined for all $\alpha, \beta \in \mathbb{O}$. We write $\alpha + \beta$ for $f_\alpha(\beta)$. This definition of addition immediately yields

(+1) $\alpha \leqslant \alpha + \beta$
(+2) $\beta < \gamma \Rightarrow \alpha + \beta < \alpha + \gamma$ (*right strict monotonicity*)
(+3) If $\alpha \leqslant \beta$ *there is a unique* ξ *such that* $\alpha + \xi = \beta$.

Since f_α is a continuous function we have

(+4) $\alpha + \sup U = \sup(\alpha + U)$

for every non-empty denumerable set $U \subset \mathbb{O}$ where we write $\alpha + U$ for $\{\alpha + \beta : \beta \in U\}$.

If β is a limit number then $\mathbb{O}(\beta)$ is a non-empty denumerable set and $\sup \mathbb{O}(\beta) = \beta$. (+4) therefore yields in particular

(+5) $\alpha + \beta = \sup(\alpha + \mathbb{O}(\beta))$ *for every limit number* β.

We also easily obtain

(+6) $\alpha + 0 = \alpha$
(+7) $\alpha + \beta' = (\alpha + \beta)'$
(+8) $\beta \leqslant \alpha + \beta$ (by Theorem 13.3).

Since $B_0 = \mathbb{O}$, f_0 is the identity and hence

(+9) $0 + \beta = \beta$.

We also have *associativity*

(+10) $(\alpha + \beta) + \gamma = \alpha + (\beta + \gamma)$.

Proof. The ordering function for the set $B_{\alpha+\beta}$ of all ordinals $\geqslant \alpha + \beta$ is such that $f_{\alpha+\beta}(\gamma) = (\alpha + \beta) + \gamma$. Let $C_{\alpha,\beta} := \{\alpha + (\beta + \gamma) : \gamma \in \mathbb{O}\}$ and $g_{\alpha,\beta}(\gamma) := \alpha + (\beta + \gamma)$. It follows from (+2) that the function $g_{\alpha,\beta}$ is strictly monotone. Its image is the set $C_{\alpha,\beta}$. Hence $g_{\alpha,\beta}$ is the ordering function of $C_{\alpha,\beta}$. It follows from (+1)–(+3) that

$\xi \in C_{\alpha,\beta}$ if, and only if, $\alpha+\beta \leqslant \xi$. Hence $C_{\alpha,\beta}=B_{\alpha+\beta}$. Therefore $f_{\alpha+\beta}$ and $g_{\alpha,\beta}$ are ordering functions for the same set $B_{\alpha+\beta}$. By Theorem 13.4 it follows that $f_{\alpha+\beta}=g_{\alpha,\beta}$ and hence $(\alpha+\beta)+\gamma=\alpha+(\beta+\gamma)$.

We also have *left weak monotonicity*

$$(+11)\quad \alpha \leqslant \beta \Rightarrow \alpha+\gamma \leqslant \beta+\gamma.$$

Proof. It follows from $\alpha \leqslant \beta$ by $(+3)$ that there exists ξ with $\alpha+\xi=\beta$. By $(+8)$ $\gamma \leqslant \xi+\gamma$. From $(+2)$ and $(+10)$ it follows that $\alpha+\gamma \leqslant \alpha+(\xi+\gamma)=(\alpha+\xi)+\gamma=\beta+\gamma$.

From $(+6)$ and $(+7)$ one sees that addition of finite ordinals corresponds to addition of natural numbers. If α, $\beta \in \mathbb{N}$ then $\alpha+\beta \in \mathbb{N}$.

Remark. Addition of ordinals is not commutative, since for example we have: $0'+\omega=\sup(0'+\mathbb{N})=\sup \mathbb{N}=\omega$ and $\omega<\omega'=\omega+0'$ and therefore $0'+\omega<\omega+0'$.

Also left strict monotonicity fails since for example $0<0'$ but $0+\omega=\omega=0'+\omega$.

Definition. An ordinal α is said to be an *additive principal number* if $\alpha \neq 0$ and $\xi+\alpha=\alpha$ for all $\xi<\alpha$.

Obviously $1:=0'$ is the smallest additive principal number.

ω is the smallest additive principal number >1.

Proof. If $\xi<\omega$ then $\xi \in \mathbb{N}$ and $\xi+\omega=\sup(\xi+\mathbb{N})=\sup \mathbb{N}=\omega$. Hence ω is an additive principal number. If $1<\alpha<\omega$ then $\alpha \in \mathbb{N}$ and $1+\alpha=\alpha+1>\alpha$ so α is not an additive principal number. Hence ω is the smallest additive principal number >1.

If α is an additive principal number then the set $\mathbb{O}(\alpha)$ is closed under addition. That is: $\xi<\alpha$ and $\eta<\alpha$ imply $\xi+\eta<\alpha$.

Proof. If $\eta<\alpha$ then $\xi+\eta<\xi+\alpha$. If α is an additive principal number and $\xi<\alpha$ then $\xi+\alpha=\alpha$, and therefore $\xi+\eta<\alpha$.

Lemma 4. *The set of additive principal numbers is closed and unbounded.*

Proof of unboundedness. Let α be an arbitrary ordinal. We have to prove that there is an additive principal number $>\alpha$. Let $\beta_0:=\alpha'$, $\beta_{n+1}:=\beta_n+\beta_n$, $M:=\{\beta_n : n \in \mathbb{N}\}$ and $\beta=\sup M$. Then $\beta_n<\beta_{n+1}$ and $\alpha<\beta_0<\beta$ and therefore $\beta \neq 0$. If $\xi<\beta$ then there exists $n \in \mathbb{N}$ such that $\xi<\beta_n$. It follows that for all $m \geqslant n$, $\xi+\beta_m \leqslant \beta_m+\beta_m=\beta_{m+1}<\beta$. Hence $\sup(\xi+M) \leqslant \beta$. From $(+8)$ and $(+4)$ it follows that $\beta \leqslant \xi+\beta=\sup(\xi+M) \leqslant \beta$. Hence $\xi+\beta=\beta$ for all $\xi<\beta$. β is therefore an additive principal number $>\alpha$.

Proof of closure. Let M be a non-empty denumerable set of additive principal numbers. Then $\sup M \neq 0$. We have to prove that $\sup M$ is an additive principal number. If $\xi<\sup M$ then there exists $\alpha \in M$ such that $\xi<\alpha$. For all $\beta \in M$ which are $\geqslant \alpha$ it follows that $\xi<\beta$ and, since β is an additive principal number, $\xi+\beta=\beta$. Hence $\xi+\sup M=\sup M$ for all $\xi<\sup M$. Therefore $\sup M$ is an additive principal number.

Definition. Let $\alpha \mapsto \omega^\alpha$ be the ordering function of the additive principal numbers.

It follows from Lemma 4 and Theorem 13.6 that this function is a normal function so ω^α is defined for all $\alpha \in \mathbb{O}$ and we have:

(ω1) $0 < \omega^\alpha$

(ω2) $\xi < \omega^\alpha \Rightarrow \xi + \omega^\alpha = \omega^\alpha$

(ω3) $\alpha < \beta \Rightarrow \omega^\alpha < \omega^\beta$

(ω4) *For every additive principal number β there exists α such that $\omega^\alpha = \beta$.*

By continuity we have

(ω5) $\omega^{\sup U} = \sup \omega^U$

for every non-empty denumerable set $U \subset \mathbb{O}$. In particular we have

(ω6) $\omega^\beta = \sup \omega^{O(\beta)}$ *for every limit number β.*

From (ω2) and (ω3) we have

(ω7) $\alpha < \beta \Rightarrow \omega^\alpha + \omega^\beta = \omega^\beta$.

Since 1 is the smallest additive principal number and ω is the next smallest additive principal number we have

(ω8) $\omega^0 = 1$

(ω9) $\omega^1 = \omega$.

Theorem 13.7 (Cantor Normal Form). *For every ordinal $\alpha \neq 0$ there are unique $\alpha_1 \geqslant \cdots \geqslant \alpha_n$ $(n \geqslant 1)$ such that $\alpha = \omega^{\alpha_1} + \cdots + \omega^{\alpha_n}$.*

Proof of existence by transfinite induction on α.

1. Suppose α is an additive principal number. By (ω4) there is then an α_1 such that $\alpha = \omega^{\alpha_1}$.

2. Suppose α is not an additive principal number. Then there exists $\xi < \alpha$ such that $\xi + \alpha \neq \alpha$. It follows that $\xi \neq 0$ and $\alpha < \xi + \alpha$. Then there exists $\eta \neq 0$ such that $\eta < \alpha$ and $\alpha = \xi + \eta$. By the induction hypothesis there are $\alpha_1 \geqslant \cdots \geqslant \alpha_m$ $(m \geqslant 1)$ such that $\xi = \omega^{\alpha_1} + \cdots + \omega^{\alpha_m}$ and there exist $\beta_1 \geqslant \cdots \geqslant \beta_n (n \geqslant 1)$ such that $\eta = \omega^{\beta_1} + \cdots + \omega^{\beta_n}$. If we had $\alpha_i < \beta_1$ for all $i = 1, \ldots, m$ then it would follow from (ω7) that $\xi + \eta = \eta$ contradicting $\eta < \alpha$. Hence there is a largest number $k \leqslant m$ such that $\alpha_k \geqslant \beta_1$. It follows that $\alpha_1 \geqslant \cdots \geqslant \alpha_k \geqslant \beta_1 \geqslant \cdots \geqslant \beta_n$ and by (ω7)

$$\alpha = \omega^{\alpha_1} + \cdots + \omega^{\alpha_k} + \omega^{\beta_1} + \cdots + \omega^{\beta_n}.$$

Uniqueness proof. Suppose $\alpha = \omega^{\alpha_1} + \cdots + \omega^{\alpha_m} = \omega^{\beta_1} + \cdots + \omega^{\beta_n}$ where $\alpha_1 \geqslant \cdots \geqslant \alpha_m$ $(m \geqslant 1)$ and $\beta_1 \geqslant \cdots \geqslant \beta_n$ $(n \geqslant 1)$. We prove by induction on m that $m = n$ and $\alpha_i = \beta_i$ for all $i = 1, \ldots, m$. It follows from the properties of additive principal numbers that $\alpha < \omega^{\alpha_1'}$, $\alpha < \omega^{\beta_1'}$, $\alpha_1 < \beta_1'$ and $\beta_1 < \alpha_1'$. Hence $\alpha_1 \leqslant \beta_1$ and $\beta_1 \leqslant \alpha_1$ and therefore $\alpha_1 = \beta_1$. Hence either $m = n = 1$ or $m > 1$, $n > 1$ and $\omega^{\alpha_2} + \cdots + \omega^{\alpha_m} = \omega^{\beta_2} + \cdots + \omega^{\beta_n}$. The assertion now follows from the induction hypothesis.

6. α-Critical Ordinals

Definition. We define a set $\mathrm{Cr}(\alpha) \subset \mathbb{O}$ and a function ϕ_α for each ordinal α inductively as follows:

1. Cr (0) *is the set of additive principal numbers.*
2. $\phi_\alpha : A_\alpha \to \mathrm{Cr}(\alpha)$ *is the ordering function of* Cr (α).
3. *If* $\alpha \neq 0$, Cr (α) *is the set of common fixed points of all functions* ϕ_ξ *with* $\xi < \alpha$.

That is: $\eta \in \mathrm{Cr}(\alpha)$ if, and only if, $\eta \in A_\xi$ and $\phi_\xi(\eta) = \eta$ for all $\xi < \alpha$.

We call the elements of Cr (α) the *α-critical ordinals*. We write $\phi\alpha\beta$ instead of $\phi_\alpha(\beta)$ for $\beta \in A_\alpha$.

Corollaries. 1. $\alpha < \beta \Rightarrow \mathrm{Cr}(\beta) \subseteq \mathrm{Cr}(\alpha)$.
2. *Every ordinal* $\phi\alpha\beta$ *is an additive principal number.*
3. $\phi 0 \beta = \omega^\beta$.

Lemma 5. *The set* Cr (α) *is closed and unbounded.*

Proof by transfinite induction on α. If $\alpha = 0$ the assertion holds by Lemma 4. Now suppose $\alpha \neq 0$. If $\xi < \alpha$ it follows from the induction hypothesis by Theorem 13.6 that ϕ_ξ is a normal function and therefore $A_\xi = \mathbb{O}$.

Proof of unboundedness. Let β be an arbitrary ordinal. We have to show that there is an α-critical ordinal $> \beta$. Let $\gamma_0 := \beta'$, $\gamma_{n+1} := \sup\{\phi\xi\gamma_n : \xi < \alpha\}$, $U := \{\gamma_n : n \in N\}$ and $\gamma := \sup U$. Then $\beta < \gamma_0 \leqslant \gamma$. If $\xi < \alpha$ then $\phi\xi\gamma_n \leqslant \gamma_{n+1} \leqslant \gamma$ so $\sup \phi\xi U \leqslant \gamma$. Therefore $\phi\xi\gamma = \gamma$ for all $\xi < \alpha$. Hence γ is an α-critical ordinal $> \beta$.

Proof of closure. Let M be a non-empty denumerable subset of Cr (α). Then for all $\xi < \alpha$ and $\eta \in M$ we have $\phi\xi\eta = \eta$. Since ϕ_ξ is continuous it follows that $\phi\xi(\sup M) = \sup M$ for all $\xi < \alpha$ and therefore $\sup M \in \mathrm{Cr}(\alpha)$. Hence Cr ($\alpha$) is closed.

Theorem 13.8. *For every ordinal* α, ϕ_α *is a normal function and* $A_\alpha = \mathbb{O}$.

Proof. This follows from Lemma 5 and Theorem 13.6.

Theorem 13.9. $\phi\alpha_1\beta_1 = \phi\alpha_2\beta_2$ *holds if, and only if, one of the following holds:*
1. $\alpha_1 < \alpha_2$ *and* $\beta_1 = \phi\alpha_2\beta_2$,
2. $\alpha_1 = \alpha_2$ *and* $\beta_1 = \beta_2$,
3. $\alpha_2 < \alpha_1$ *and* $\phi\alpha_1\beta_1 = \beta_2$.

Proof. $\phi\alpha_2\beta_2 \in \mathrm{Cr}(\alpha_2)$. If $\alpha_1 < \alpha_2$ then $\phi\alpha_1(\phi\alpha_2\beta_2) = \phi\alpha_2\beta_2$ therefore $\phi\alpha_1\beta_1 = \phi\alpha_2\alpha_2$ if, and only if, $\beta_1 = \phi\alpha_2\beta_2$. The case of $\alpha_2 < \alpha_1$ is similar. For $\alpha_1 = \alpha_2$ the assertion is trivial.

Corollaries. 1. $\eta \in \mathrm{Cr}(\alpha') \Leftrightarrow \phi\alpha\eta = \eta$,
2. *If* β *is a limit ordinal then* Cr (β) *is the intersection* $\bigcap_{\alpha<\beta} \mathrm{Cr}(\alpha)$ *of all* Cr (α) *with* $\alpha < \beta$.

Theorem 13.10. $\phi\alpha_1\beta_1 < \phi\alpha_2\beta_2$ *holds if, and only if, one of the following holds:*
1. $\alpha_1 < \alpha_2$ *and* $\beta_1 < \phi\alpha_2\beta_2$,
2. $\alpha_1 = \alpha_2$ *and* $\beta_1 < \beta_2$,
3. $\alpha_2 < \alpha_1$ and $\phi\alpha_1\beta_1 < \beta_2$.

Proof. $\phi\alpha_2\beta_2 \in \mathrm{Cr}(\alpha_2)$. If $\alpha_1 < \alpha_2$, then $\phi\alpha_1(\phi\alpha_2\beta_2) = \phi\alpha_2\beta_2$, therefore $\phi\alpha_1\beta_1 < \phi\alpha_2\beta_2$ if, and only if, $\beta_1 < \phi\alpha_2\beta_2$. The case of $\alpha_2 < \alpha_1$ is similar. For $\alpha_1 = \alpha_2$ the assertion follows from the fact that ϕ_{α_1} is an ordering function.

Theorem 13.11. $\alpha \leqslant \phi\alpha 0$.

Proof by transfinite induction on α. If $\xi < \alpha$ then $\phi\xi 0 < \phi\xi(\phi\alpha 0) = \phi\alpha 0$. By the induction hypothesis it follows that $\xi < \phi\alpha 0$ for all $\xi < \alpha$. Hence $\alpha \leqslant \phi\alpha 0$.

Corollary. $\beta \in \mathrm{Cr}(\alpha) \to \alpha \leqslant \beta$.

Proof. If $\beta \in \mathrm{Cr}(\alpha)$ then there exists η such that $\beta = \phi\alpha\eta$. It follows that $\alpha \leqslant \phi\alpha 0 \leqslant \phi\alpha\eta = \beta$.

Theorem 13.12. *For every additive principal number γ there exist unique α and $\beta < \gamma$ such that $\gamma = \phi\alpha\beta$.*

Existence proof. By Theorem 13.11, $\gamma \leqslant \phi\gamma 0$. Since $0 < \gamma$, $\gamma < \phi\gamma\gamma$. Hence there is a least ordinal $\alpha \leqslant \gamma$ such that $\gamma \neq \phi\alpha\gamma$. If $\alpha \neq 0$ then $\phi\xi\gamma = \gamma$ for all $\xi < \alpha$ and therefore $\gamma \in \mathrm{Cr}(\alpha)$. By the hypothesis $\gamma \in \mathrm{Cr}(0)$. In any case therefore $\gamma \in \mathrm{Cr}(\alpha)$. Therefore there exists β such that $\gamma = \phi\alpha\beta$. Since $\gamma \neq \phi\alpha\gamma$ it follows that $\gamma < \phi\alpha\gamma$ and $\beta < \gamma$.

Uniqueness proof. If $\beta_1 < \gamma$, $\beta_2 < \gamma$ and $\gamma = \phi\alpha_1\beta_1 = \phi\alpha_2\beta_2$ then by Theorem 13.9 $\alpha_1 = \alpha_2$ and $\beta_1 = \beta_2$.

Definition. An ordinal α is said to be *strongly critical* if $\alpha \in \mathrm{Cr}(\alpha)$.

Theorem 13.13. *α is strongly critical if, and only if, $\phi\alpha 0 = \alpha$.*

Proof. If $\alpha \in \mathrm{Cr}(\alpha)$ then there exists β such that $\alpha = \phi\alpha\beta$. Now from $\alpha \leqslant \phi\alpha 0$ (by Theorem 13.11) it follows that $\beta = 0$ and hence $\phi\alpha 0 = \alpha$. Conversely: if $\phi\alpha 0 = \alpha$ then $\alpha \in \mathrm{Cr}(\alpha)$.

Theorem 13.14. *The set of strongly critical ordinal numbers is closed and unbounded.*

Proof of unboundedness. Let α be an arbitrary ordinal. We have to prove that there is a strongly critical ordinal $> \alpha$. Let $\beta_0 := \alpha'$, $\beta_{n+1} := \phi\beta_n 0$, $U := \{\beta_n : n \in \mathbb{N}\}$ and $\beta := \sup U$. Then $\alpha < \beta_0 \leqslant \beta$ and by Theorem 13.11 $\beta_n \leqslant \beta_{n+1}$. If $\xi < \beta$ then there exists $n \in \mathbb{N}$ such that $\xi < \beta_n$. It follows that for all $m \geqslant n$ we have $\xi < \beta_m$ and

$$\phi\xi\beta_{m+1} = \phi\xi(\phi\beta_m 0) = \phi\beta_m 0 \leqslant \beta.$$

From this it follows that $\sup \phi\xi U \leqslant \beta$. Since ϕ_ξ is a normal function it follows that $\phi\xi\beta = \beta$ for all $\xi < \beta$, and therefore $\beta \in \mathrm{Cr}(\beta)$. Hence β is a strongly critical ordinal $> \alpha$.

Proof of closure. Let M be a non-empty denumerable set of strongly critical ordinals. We have to prove that sup M is strongly critical. If $\xi < \sup M$ there exists $\alpha \in M$ such that $\xi < \alpha$. It follows that $\phi\xi\beta = \beta$ for every $\beta \in M$ which is $\geqslant \alpha$ since every β is strongly critical. Hence $\phi\xi(\sup M) = \sup M$ for all $\xi < \sup M$ and therefore $\sup M \in \mathrm{Cr}(\sup M)$. Thus sup M is strongly critical.

Following S. Feferman we denote the *least strongly critical ordinal* by Γ_0.

7. Maximal α-Critical Ordinals

Definition. An ordinal γ is said to be *maximal α-critical* if $\gamma \in \mathrm{Cr}(\alpha)$ and $\gamma \notin \mathrm{Cr}(\xi)$ for all $\xi > \alpha$.

Obviously an ordinal γ is maximal α-critical if, and only if, there exists $\beta < \gamma$ such that $\gamma = \phi\alpha\beta$. It therefore follows from Theorem 13.12 that for every additive principal number γ there is an ordinal α such that γ is maximal α-critical.

We define $\psi\alpha\beta$ in such a way that ψ_α, where $\psi_\alpha(\beta) := \psi\alpha\beta$, is the ordering function of the maximal α-critical ordinals.

Definition. If there exists β_0 and $n \in N$ such that $\beta = \beta_0 + n$ and $\phi\alpha\beta_0 = \beta_0$ let $\psi\alpha\beta := \phi\alpha\beta'$. Otherwise let $\psi\alpha\beta := \phi\alpha\beta$.

Lemma 6. *If $\psi\alpha\beta = \phi\alpha\beta^*$ then $\beta^* < \phi\alpha\beta^*$.*

Proof. By definition $\beta^* = \beta$ or $\beta^* = \beta'$. If $\phi\alpha\beta = \beta$ then $\beta^* = \beta'$. If $\beta^* = \beta'$ then $\beta \leqslant \phi\alpha\beta < \phi\alpha\beta^*$ and $0' \leqslant \phi\alpha\beta < \phi\alpha\beta^*$. Since $\phi\alpha\beta^*$ is an additive principal number it follows that $\beta^* = \beta + 0' < \phi\alpha\beta^*$. On the other hand, if $\beta^* = \beta$ then $\beta < \phi\alpha\beta$ and therefore $\beta^* < \phi\alpha\beta^*$.

Lemma 7. *$\psi\alpha\beta$ is a maximal α-critical ordinal.*

Proof. $\psi\alpha\beta = \phi\alpha\beta^*$ where $\beta^* = \beta$ or $\beta^* = \beta'$. By Lemma 6 $\beta^* < \phi\alpha\beta^*$. Consequently $\psi\alpha\beta$ is a maximal α-critical ordinal.

Lemma 8. *$\beta_1 < \beta_2$ implies $\psi\alpha\beta_1 < \psi\alpha\beta_2$.*

Proof. $\psi\alpha\beta_i = \phi\alpha\beta_i^*$ where $\beta_i^* = \beta_i$ or $\beta_i^* = \beta_i'$ $(i = 1, 2)$. Now $\beta_1 < \beta_2$ implies $\beta_i^* \leqslant \beta_2$. If $\beta_1^* = \beta_2$ then $\beta_1^* = \beta_1'$ and $\beta_1 = \beta_0 + n$ where $\phi\alpha\beta_0 = \beta_0$ and $n \in N$. But then $\beta_2 = \beta_0 + n'$ and $\beta_2^* = \beta_2'$. In every case therefore $\beta_1^* < \beta_2^*$. Hence $\phi\alpha\beta_1^* < \phi\alpha\beta_2^*$ and therefore $\psi\alpha\beta_1 < \psi\alpha\beta_2$.

Theorem 13.15. *For each additive principal number γ there are unique α and β such that $\gamma = \psi\alpha\beta$.*

Existence proof. By Theorem 13.12 there exist α and $\beta_1 < \gamma$ such that $\gamma = \phi\alpha\beta_1$. If $\psi\alpha\beta_1 = \phi\alpha\beta_1$ then the assertion is satisfied with $\beta := \beta_1$. Otherwise there exist

β_0 and $n \in N$ such that $\beta_1 = \beta_0 + n$ and $\phi\alpha\beta_0 = \beta_0$. Since $\beta_1 < \phi\alpha\beta_1$ it follows that $n \neq 0$. Therefore there is an $m \in \mathbb{N}$ such that $m' = n$. Putting $\beta := \beta_0 + m$ we obtain $\psi\alpha\beta = \phi\alpha\beta' = \gamma$.

Uniqueness proof. Suppose $\psi\alpha_1\beta_1 = \psi\alpha_2\beta_2$. By definition there exist β_i^* such that $\psi\alpha_i\beta_i = \phi\alpha_i\beta_i^*$ $(i=1,2)$. By Lemma 6 $\beta_i^* < \phi\alpha_i\beta_i^*$ $(i=1,2)$. Then by Theorem 13.9 $\phi\alpha_1\beta_1^* = \phi\alpha_2\beta_2^*$ implies $\alpha_1 = \alpha_2$. By Lemma 8 it follows that $\beta_1 = \beta_2$.

Theorem 13.16. *The function ψ_α given by $\psi_\alpha(\beta) := \psi\alpha\beta$ is the ordering function of the maximal α-critical numbers.*

Proof. This follows from Lemmata 7, 8 and Theorem 13.15.

Remark. By Lemmata 7 and 8 the set of maximal α-critical ordinals is unbounded. But by Theorem 13.5 it is not closed since its ordering function ψ_α is not continuous as the following shows.

If $\xi < \phi\alpha' 0$ then by Theorem 13.10 $\phi\alpha\xi < \phi\alpha' 0$ and hence $\phi\alpha\xi \notin \mathrm{Cr}(\alpha')$ and $\xi < \phi\alpha\xi$. It follows that there is no $\xi < \phi\alpha' 0$ with $\phi\alpha\xi = \xi$ and hence $\psi\alpha\xi = \phi\alpha\xi$ by the definition of ψ_α. Setting $U := \mathbb{O}(\phi\alpha' 0)$ it follows that $\sup \psi\alpha U = \sup \phi\alpha U$. Since ϕ_α is continuous and $\sup U = \phi\alpha' 0$ it follows that $\sup \psi\alpha U = \phi\alpha' 0$. But $\psi\alpha(\phi\alpha' 0) = \phi\alpha(\phi\alpha' 0)' > \phi\alpha' 0$ and therefore $\psi\alpha(\sup U) > \sup \psi\alpha U$. Hence ψ_α is not continuous.

Theorem 13.17. 1. *If $\gamma = \psi\alpha\beta$ then $\alpha < \gamma$ if, and only if, γ is not a strongly critical ordinal.*

2. *For all β, $\beta < \psi\alpha\beta$.*

Proof. $\psi\alpha\beta = \phi\alpha\beta^*$ where $\beta^* = \beta$ or $\beta^* = \beta'$.

1.1. Suppose γ is a strongly critical ordinal. By Theorem 13.13 we then have $\gamma = \phi\gamma 0$. In this case $\psi\gamma 0 = \phi\gamma 0$. But then $\gamma = \psi\gamma 0 = \psi\alpha\beta$ implies $\gamma = \alpha$ by Theorem 13.15.

1.2. Suppose γ is not a strongly critical ordinal. Then $\gamma < \phi\gamma 0$ by Theorem 13.11 and 13.13. By Theorem 13.10 $\gamma = \phi\alpha\beta^* < \phi\gamma 0$ implies $\alpha < \gamma$.

2. By Lemma 6 $\beta^* < \phi\alpha\beta^* = \psi\alpha\beta$. It follows that $\beta \leqslant \beta^* < \psi\alpha\beta$.

Theorem 13.18. *$\psi\alpha_1\beta_1 < \psi\alpha_2\beta_2$ holds if, and only if, one of the following holds:*

1. $\alpha_1 < \alpha_2$ *and* $\beta_1 < \psi\alpha_2\beta_2$,
2. $\alpha_1 = \alpha_2$ *and* $\beta_1 < \beta_2$,
3. $\alpha_2 < \alpha_1$ *and* $\psi\alpha_1\beta_1 \leqslant \beta_2$.

Proof. 1. Suppose $\alpha_1 < \alpha_2$ and $\beta_1 < \psi\alpha_2\beta_2$. Then $\psi\alpha_2\beta_2$ is an additive principal number > 1. Therefore $\beta_1 < \psi\alpha_2\beta_2$ implies $\beta_1' < \psi\alpha_2\beta_2$ and then $\psi\alpha_1\beta_1 \leqslant \phi\alpha_1\beta_1' < \phi\alpha_1(\psi\alpha_2\beta_2)$. Since $\psi\alpha_2\beta_2 \in \mathrm{Cr}(\alpha_2)$ and $\alpha_1 < \alpha_2$ we have $\phi\alpha_1(\psi\alpha_2\beta_2) = \psi\alpha_2\beta_2$. Therefore $\psi\alpha_1\beta_1 < \psi\alpha_2\beta_2$.

2. Suppose $\alpha_1 = \alpha_2$ and $\beta_1 < \beta_2$. Then by Lemma 8 $\psi\alpha_1\beta_1 < \psi\alpha_2\beta_2$.

3. Suppose $\psi\alpha_1\beta_1 \leqslant \beta_2$. By Theorem 13.17 $\beta_2 < \psi\alpha_2\beta_2$. Therefore $\psi\alpha_1\beta_1 < \psi\alpha_2\beta_2$.

4. Suppose none of cases 1–3 holds. Then either $\alpha_1=\alpha_2$ and $\beta_1=\beta_2$ so $\psi\alpha_1\beta_1=\psi\alpha_2\beta_2$ or one of cases 1–3 for $\psi\alpha_2\beta_2<\psi\alpha_1\beta_1$ holds.

§14. A Notation System for the Ordinals $<\Gamma_0$

1. Definition of Ordinal Terms

By Theorems 13.7, 13.15 and 13.17 every ordinal $\neq 0$ which is not strongly critical can be expressed in terms of $+$ and ψ. Accordingly we introduce a set T of *terms* denoting the ordinals $<\Gamma_0$ and at the same time we formally fix their *lengths* in the following way.

Inductive Definition of the *set* T *of terms.*

1. The symbol 0 is a term in the set T with length 0.
2. If $\alpha_1,\ldots,\alpha_n,\beta_1,\ldots,\beta_n\ (n\geqslant 1)$ are terms in the set T whose lengths are $L\alpha_1,\ldots,L\alpha_n,L\beta_1,\ldots,L\beta_n$ then $(\alpha_1,\beta_1)\ldots(\alpha_n,\beta_n)$ is a term in the set T of length

$$2\cdot\max(L\alpha_1,\ldots,L\alpha_n,L\beta_1,\ldots,L\beta_n)+n.$$

We denote the *length* of a term $\alpha\in$ T by $L\alpha$. This length is a natural number.

Interpretation. 1. The term 0 denotes the ordinal 0.

2. If $\alpha_1,\ldots,\alpha_n,\beta_1,\ldots,\beta_n\ (n\geqslant 1)$ are terms in the set T which denote ordinals $\bar\alpha_1,\ldots,\bar\alpha_2,\bar\beta_1,\ldots,\bar\beta_n$ then the term $(\alpha_1,\beta_1)\ldots(\alpha_n,\beta_n)$ denotes the ordinal $\psi\bar\alpha_1\bar\beta_1+\cdots+\psi\bar\alpha_n\bar\beta_n$.

These terms in the set T do not give a system of unique notations for ordinals. For example the terms

$$(0,(0,0))\quad\text{and}\quad(0,0)(0,(0,0))$$

both denote the same ordinal $\omega=\psi 0(\psi 00)=\psi 00+\psi 0(\psi 00)$ according to the given interpretation.

In order to obtain a system of unique notations for the ordinals $<\Gamma_0$ we single out a subset of OT of T. At the same time we inductively define the $<$-relation.

Inductive Definition of a set OT$\subset$T of *ordinal terms* and a relation $<$ on OT.

1. The symbol 0 is an ordinal term in the set OT.
2. If $\alpha_1,\ldots,\alpha_n,\beta_1,\ldots,\beta_n\ (n\geqslant 1)$ are ordinal terms in the set OT and $(\alpha_{i+1},\beta_{i+1})\leqslant(\alpha_i,\beta_i)$ for all $i=1,\ldots,n-1$ then $(\alpha_1,\beta_1)\ldots(\alpha_n,\beta_n)$ is an ordinal term in the set OT.

We call the ordinal terms $(\alpha,\beta)\in$ OT *principal terms.*

3. If $\alpha\in$ OT then $\alpha<0$ does *not* hold.
4. If $\beta\in$ OT then $0<\beta$ if, and only if, $\beta\neq 0$.

5. If (α_1, β_1), $(\alpha_2, \beta_2) \in \mathrm{OT}$ then $(\alpha_1, \beta_1) < (\alpha_2, \beta_2)$ exactly when one of the following holds:

5.1. $\alpha_1 < \alpha_2$ *and* $\beta_1 < (\alpha_2, \beta_2)$,

5.2. $\alpha_1 = \alpha_2$ *and* $\alpha_1 < \beta_2$,

5.3. $\alpha_2 < \alpha_1$ *and* $(\alpha_1, \beta_1) \leqslant \beta_2$.

6. If $\gamma = \gamma_1 \ldots \gamma_m \in 0T$ and $\delta = \delta_1 \ldots \delta_n \in 0T$ where $\gamma_1, \ldots, \gamma_m$, $\delta_1, \ldots, \delta_n$ $(m \geqslant 1, n \geqslant 1, m+n > 2)$ are principal terms then $\gamma < \delta$ exactly when one of the following holds:

6.1. $m < n$ and $\gamma_i = \delta_i$ for all $i = 1, \ldots, m$.

6.2. There exists $j \leqslant \min(m, n)$ such that $\gamma_j < \delta_j$ and $\gamma_i = \delta_i$ for all $i = 1, \ldots, j-1$.

$\alpha = \beta$ means that α and β are identical terms. $\alpha \leqslant \beta$ means that $\alpha < \beta$ or $\alpha = \beta$. We also write $\alpha > \beta$ for $\beta < \alpha$ and $\alpha \geqslant \beta$ for $\beta \leqslant \alpha$.

In the sequel only the elements of the inductively given set OT are called *ordinal terms*.

It is clear from §13 that the inductive definition of the $<$-relation corresponds to that in the interpretation we gave of the ordinal terms as notations for ordinals $<\Gamma_0$. The principal terms denote the additive principal numbers $<\Gamma_0$.

In this section we started out from the point of view of the non-constructive theory of ordinals developed in §13. Now the ordinal terms are to be understood without reference to their interpretations and simply as formal strings of symbols for which we have a constructively defined $<$-relation. All the theorems about ordinal terms in this section will be proved constructively using only the above inductive definition. In this way we lay a constructive foundation for the proof theoretic applications in the next chapter.

We first confirm that we are dealing with decidable properties of ordinal terms. It is obviously decidable for each string of symbols whether it is a term of the set T. By the following theorem it is also decidable whether a string of symbols is an ordinal term of the system OT.

Theorem 14.1. (1) *It is decidable whether* $\alpha < \beta$, $\alpha = \beta$ *or* $\beta < \alpha$ *for all* $\alpha, \beta \in \mathrm{OT}$.

(2) *If* $\gamma \in \mathrm{T}$ *it is decidable whether* $\gamma \in \mathrm{OT}$.

Proof by induction on the maximum of $L\alpha$, $L\beta$ and $L\gamma$ and within that by induction on $L\alpha + L\beta$.

(1) For $\xi, \eta \in \mathrm{OT}$ let $E(\xi, \eta)$ denote: It is decidable whether $\xi < \eta$, $\xi = \eta$ or $\eta < \xi$. We prove $E(\alpha, \beta)$ for $\alpha, \beta \in \mathrm{OT}$.

1. Suppose α or β is 0. Then $E(\alpha, \beta)$ is trivial.

2. $\alpha = (\alpha_1, \alpha_2)$, $\beta = (\beta_1, \beta_2)$. Then by the induction hypotheses we have $E(\alpha_1, \beta_1)$, $E(\alpha_2, \beta_2)$, $E(\alpha_2, \beta)$ and $E(\alpha, \beta_2)$. Now $E(\alpha, \beta)$ follows by the inductive definition of the $<$-relation.

3. $\alpha = \alpha_1 \ldots \alpha_m$, $\beta = \beta_1 \ldots \beta_n$ where $\alpha_1, \ldots, \alpha_m$, $\beta_1, \ldots, \beta_n$ $(m \geqslant 1, n \geqslant 1, m+n > 2)$ are principal terms. Then by the induction hypotheses $E(\alpha_i, \beta_i)$ for all $i \leqslant \min(m, n)$. If $\alpha_i = \beta_i$ for all $i \leqslant \min(m, n)$ then $\alpha < \beta$, $\alpha = \beta$ or $\beta < \alpha$ according as $m < n$, $m = n$ or $n < m$. If there is a least index $j \leqslant \min(m, n)$ such that $\alpha_j \neq \beta_j$ then $\alpha < \beta$ or $\beta < \alpha$ according as $\alpha_j < \beta_j$ or $\beta_j < \alpha_j$.

(2) If $\gamma=0$ then $\gamma\in OT$. Otherwise $\gamma=(\alpha_1,\beta_1)\dots(\alpha_n,\beta_n)$ $(n\geqslant 1)$. By the main induction hypothesis it is decidable whether $\alpha_i\in$ OT and $\beta_i\in$ OT for all $i=1,\dots,n$. If this is the case, then we also have $(\alpha_i,\beta_i)\in$ OT. If $n>1$ then by the induction hypotheses it is decidable whether $(\alpha_{i+1},\beta_{i+1})\leqslant(\alpha_i,\beta_i)$ for $i=1,\dots,n-1$. Hence $\gamma\in$ OT is decidable.

In the following parts of this section small Greek letters will always denote ordinal terms in the system OT.

2. The Ordering of the Ordinal Terms

Theorem 14.2. *The ordinal terms are linearly ordered by the* $<$*-relation, that is, for all* $\gamma,\gamma_1,\gamma_2,\gamma_3\in$ OT *we have:*

(1) *Not* $\gamma<\gamma$

(2) $\gamma_1<\gamma_2$ *or* $\gamma_1=\gamma_2$ or $\gamma_2<\gamma_1$

(3) $\gamma_1<\gamma_2$ *and* $\gamma_2<\gamma_3$ *imply* $\gamma_1<\gamma_3$.

Proof. (1) follows at once by induction on $L\gamma$ and (2) by induction on $L\gamma_1+L\gamma_2$. Using (2) we prove (3) by induction on $L\gamma_1+L\gamma_2+L\gamma_3$. We only give the details of the proof of (3) for the case $\gamma_i=(\alpha_i,\beta_i)$ $(i=1,2,3)$ since in the other cases the assertion follows easily from the induction hypothesis. The hypotheses $\gamma_1<\gamma_2$ and $\gamma_2<\gamma_3$ produce one of the following seven cases:

1. $\alpha_1<\alpha_2$, $\beta_1<\gamma_2$ and $\alpha_2\leqslant\alpha_3$.

By I.H. $\alpha_1<\alpha_2$ and $\alpha_2\leqslant\alpha_3$ yield $\alpha_1<\alpha_3$.
By I.H. $\beta_1<\gamma_2$ and $\gamma_2<\gamma_3$ yield $\beta_1<\gamma_3$.
$\alpha_1<\alpha_3$ and $\beta_1<\gamma_3$ yield $\gamma_1<\gamma_3$.

2. $\alpha_1<\alpha_2$, $\beta_1<\gamma_2$ and $\alpha_3<\alpha_2$, $\gamma_2\leqslant\beta_3$.

By (2) either $\alpha_1<\alpha_3$ or $\alpha_1=\alpha_3$ or $\alpha_3<\alpha_1$.

2.1. Suppose $\alpha_1<\alpha_3$.

By I.H. $\beta_1<\gamma_2$ and $\gamma_2<\gamma_3$ imply $\beta_1<\gamma_3$.
$\alpha_1<\alpha_3$ and $\beta_1<\gamma_3$ imply $\gamma_1<\gamma_3$.

2.2. Suppose $\alpha_1=\alpha_3$.

By I.H. $\beta_1<\gamma_2$ and $\gamma_2\leqslant\beta_3$ imply $\beta_1<\beta_3$.
$\alpha_1=\alpha_3$ and $\beta_1<\beta_3$ imply $\gamma_1<\gamma_3$.

2.3. Suppose $\alpha_3<\alpha_1$.

By I.H. $\gamma_1<\gamma_2$ and $\gamma_2\leqslant\beta_3$ imply $\gamma_1<\beta_3$.
$\alpha_3<\alpha_1$ and $\gamma_1<\beta_3$ imply $\gamma_1<\gamma_3$.

3. $\alpha_1=\alpha_2$, $\beta_1<\beta_2$ and $\alpha_2<\alpha_3$, $\beta_2<\gamma_3$.

Then $\alpha_1<\alpha_3$. By I.H. $\beta_1<\beta_2$ and $\beta_2<\gamma_3$ imply $\beta_1<\gamma_3$.
$\alpha_1<\alpha_3$ and $\beta_1<\gamma_3$ imply $\gamma_1<\gamma_3$.

4. $\alpha_1=\alpha_2$, $\beta_1<\beta_2$ and $\alpha_2=\alpha_3$, $\beta_2<\beta_3$.

Then $\alpha_1=\alpha_3$. By I.H. $\beta_1<\beta_2$ and $\beta_2<\beta_3$ imply $\beta_1<\beta_3$.
$\alpha_1=\alpha_3$ and $\beta_1<\beta_3$ imply $\gamma_1<\gamma_3$.

5. $\alpha_2\leqslant\alpha_1$ and $\alpha_3<\alpha_2$, $\gamma_2\leqslant\beta_3$.

By I.H. $\alpha_3<\alpha_2$ and $\alpha_2\leqslant\alpha_1$ imply $\alpha_3<\alpha_1$.

By I.H. $\gamma_1<\gamma_2$ and $\gamma_2\leqslant\beta_3$ imply $\gamma_1<\beta_3$.
$\alpha_3<\alpha_1$ and $\gamma_1<\beta_3$ imply $\gamma_1<\gamma_3$.

6. $\alpha_2<\alpha_1$, $\gamma_1\leqslant\beta_2$ and $\alpha_2=\alpha_3$, $\beta_2<\beta_3$.

Then $\alpha_3<\alpha_1$. By I.H. $\gamma_1\leqslant\beta_2$ and $\beta_2<\beta_3$ imply $\gamma_1<\beta_3$. $\alpha_3<\alpha_1$ and $\gamma_1<\beta_3$ imply $\gamma_1<\gamma_3$.

7. $\alpha_2<\alpha_1$, $\gamma_1\leqslant\beta_2$ and $\alpha_2<\alpha_3$, $\beta_2<\gamma_3$.

By I.H. $\gamma_1\leqslant\beta_2$ and $\beta_2<\gamma_3$ imply $\gamma_1<\gamma_3$.

Theorem 14.3. (1) $\beta<(\alpha, \beta)$
(2) $\alpha<(\alpha, \beta)$.

Proof of (1) by induction on $L\beta$.

1. $\beta=0$. Then the assertion is trivial.
2. $\beta=(\beta_1, \beta_2)$. By I.H. we then have $\beta_2<\beta$ and $\beta_2<(\alpha, \beta_2)$. Now $\beta_2<\beta$ implies $(\alpha, \beta_2)<(\alpha, \beta)$. Hence $\beta_2<(\alpha, \beta)$.

2.1. Suppose $\beta_1<\alpha$. Since $\beta_2<(\alpha, \beta)$ we then have $\beta<(\alpha, \beta)$.
2.2. Suppose $\beta_1=\alpha$. Since $\beta_2<\beta$ we then have $\beta<(\alpha, \beta)$.
2.3. Suppose $\alpha<\beta_1$. Since $\beta\leqslant\beta$ we have $\beta<(\alpha, \beta)$.

3. $\beta=\beta_1\dots\beta_n$ $(n\geqslant 2)$ where $\beta_1, \dots, \beta_n$ are principal terms. By I.H. $\beta_1<(\alpha, \beta_1)$. Since $\beta_1<\beta$ it follows that $\beta_1<(\alpha, \beta)$. Hence $\beta<(\alpha, \beta)$.

Proof of (2) by induction on $L\alpha$.

1. $\alpha=0$. Then the assertion is trivial.
2. $\alpha=(\alpha_1, \alpha_2)$. By I.H. we then have $\alpha_1<\alpha$ and $\alpha_2<(\alpha_2, \beta)$. By (1) $\alpha_2<\alpha$ and $\beta<(\alpha, \beta)$. Hence $(\alpha_2, \beta)<(\alpha, \beta)$. Therefore $\alpha_2<(\alpha, \beta)$. Now $\alpha_1<\alpha$ implies $\alpha<(\alpha, \beta)$.
3. $\alpha=\alpha_1\dots\alpha_n$ $(n\geqslant 2)$ where $\alpha_1, \dots, \alpha_n$ are principal terms. By I.H. $\alpha_1<(\alpha_1, \beta)$. By (1) $\beta<(\alpha, \beta)$. Since $\alpha_1<\alpha$ it follows that $(\alpha_1, \beta)<(\alpha, \beta)$. Hence $\alpha_1<(\alpha, \beta)$. It follows that $\alpha<(\alpha, \beta)$.

Remark. We require ordinal terms in the succeeding chapters in order to carry out induction over OT with respect to the $<$-relation. That is to say: we use the principle of proof by transfinite induction over the segment of the ordinals $<\Gamma_0$ represented by OT. However, the well-ordering of OT, on which this transfinite induction is based, cannot be proved in as elementary a way as Theorem 14.2. Later we shall establish transfinite induction for OT, using stronger methods step by step.

3. Addition of Ordinal Terms

Definition of $\alpha+\beta\in 0T$ and $\alpha'\in$ OT for $\alpha, \beta\in$ OT.

1. $\alpha+0:=\alpha$ and $0+\beta:=\beta$.
2. If $\alpha=\alpha_1\dots\alpha_m$ and $\beta=\beta_1\dots\beta_n$ with principal terms $\alpha_1, \dots, \alpha_m$, $\beta_1, \dots, \beta_n$ $(m\geqslant 1, n\geqslant 1)$ set

2.1. $\alpha+\beta:=\beta$, if $\alpha_1<\beta_1$,
2.2. $\alpha+\beta:=\alpha_1\dots\alpha_k\beta_1\dots\beta_n$, if k is the largest index $\leqslant m$ such that $\beta_1\leqslant\alpha_k$.

3. $\alpha':=\alpha+(0, 0)$.

The following properties are easily established for this addition of ordinal terms.

$\alpha \leqslant \alpha + \beta$ *and* $\beta \leqslant \alpha + \beta$,
$\beta < \gamma \Rightarrow \alpha + \beta < \alpha + \gamma$,
$\alpha \leqslant \beta \Rightarrow \alpha + \gamma \leqslant \beta + \gamma$,
$\alpha + (\beta + \gamma) = (\alpha + \beta) + \gamma$.
If $\alpha = \alpha_1 \ldots \alpha_n$ $(n \geqslant 2)$ *where* $\alpha_1, \ldots, \alpha_n$ *are principal terms* $\alpha = \alpha_1 + \cdots + \alpha_n$.

α' is the *successor* of α since we obviously have

$$\alpha < \alpha'$$
$$\alpha < \beta \Rightarrow \alpha' \leqslant \beta$$
$$\alpha < \beta' \Rightarrow \alpha \leqslant \beta.$$

The principal terms have the properties of additive principal numbers since the definition of addition yields

$$\xi < (\alpha, \beta) \Rightarrow \xi + (\alpha, \beta) = (\alpha, \beta)$$
$$\xi < (\alpha, \beta), \eta < (\alpha, \beta) \Rightarrow \xi + \eta < (\alpha, \beta).$$

Definition of $-\alpha + \beta$ for $\alpha \leqslant \beta$.

1. $-\alpha + \alpha := 0$
2. $-0 + \beta := \beta$
3. If $\alpha = \alpha_1 \ldots \alpha_m < \beta = \beta_1 \ldots \beta_n$ with principal terms $\alpha_1, \ldots, \alpha_m, \beta_1, \ldots, \beta_n$ $(m \geqslant 1, n \geqslant 1)$ we consider the following two cases corresponding to the definition of the $<$-relation:

3.1. $m < n$ and $\alpha_i = \beta_i$ for all $i = 1, \ldots, m$. In this case set $-\alpha + \beta := \beta_{m+1} \ldots \beta_n$.

3.2. There exists $j \leqslant \min(m, n)$ such that $\alpha_j < \beta_j$ and $\alpha_i = \beta_i$ for all $i = 1, \ldots, j-1$. In this case set $-\alpha + \beta := \beta_j \ldots \beta_n$.

In every case $-\alpha + \beta$ is again an ordinal term. The definition yields

$$\alpha \leqslant \beta \Rightarrow \alpha + (-\alpha + \beta) = \beta.$$

4. Ordinal Terms $\phi\alpha\beta$

We call the ordinal term 0 and the ordinal terms $\alpha_1 \ldots \alpha_n$ $(n \geqslant 1)$ such that $\alpha_i = (0, 0)$ $(i = 1, \ldots, n)$ *finite ordinal terms*.

Definition. $\omega := (0, (0, 0))$.

Obviously $\alpha < \omega$ holds if, and only if, α is a finite ordinal term.

Definition of $\phi\alpha\beta$ for ordinal terms α, β:

1. If $\beta = (\beta_1, \beta_2)$ where $\alpha < \beta_1$, then set $\phi\alpha\beta := \beta$.
2. If $\beta = (\beta_1, \beta_2) + \gamma'$ where $\alpha < \beta_1$ and $\gamma < \omega$, then set $\phi\alpha\beta := (\alpha, (\beta_1, \beta_2) + \gamma)$.
3. In every other case set $\phi\alpha\beta := (\alpha, \beta)$.

Corollary. *$\phi\alpha\beta$ is a principal term.*

Theorem 14.4. (1) $\beta \leqslant \phi\alpha\beta$
(2) $\alpha < \phi\alpha\beta$.

Proof of (1). If $\phi\alpha\beta = \beta$ then the assertion holds. If $\phi\alpha\beta = (\alpha, \beta)$ then $\beta < \phi\alpha\beta$ by Theorem 14.3(1). Otherwise we have $\beta = (\beta_1, \beta_2) + \gamma'$ where $\alpha < \beta_1$ and $\gamma < \omega$ and $\phi\alpha\beta = (\alpha, (\beta_1, \beta_2) + \gamma)$. In this case $(0, 0) < (\beta_1, \beta_2) + \gamma$. By Theorem 14.3(1) we obtain $(\beta_1, \beta_2) + \gamma < \phi\alpha\beta$ and $(0, 0) < \phi\alpha\beta$. Hence $\beta < \phi\alpha\beta$.

Proof of (2). If $\beta = (\beta_1, \beta_2)$ where $\alpha < \beta_1$ then $\phi\alpha\beta = \beta$. By Theorem 14.3(2) we then have $\beta_1 < \beta$. It follows that $\alpha < \phi\alpha\beta$. Otherwise we have $\phi\alpha\beta = (\alpha, \beta_0)$. Then by Theorem 14.3(2) again we have $\alpha < \phi\alpha\beta$.

Lemma 1. $\beta < \gamma \Rightarrow \phi\alpha\beta < \phi\alpha\gamma$.

Proof. Following the definition we have the following three cases to consider for $\phi\alpha\beta$ and $\phi\alpha\gamma$.

1. $\phi\alpha\beta = \beta$. Then $\beta < \gamma$ implies $\phi\alpha\beta < \phi\alpha\gamma$ since $\gamma \leqslant \phi\alpha\gamma$ by Theorem 14.4(1).

2. $\phi\alpha\beta = (\alpha, \beta_0)$ where $\beta = \beta_0$ or $\beta = \beta_0'$ and $\phi\alpha\gamma = \gamma = (\gamma_1, \gamma_2)$ where $\alpha < \gamma_1$. Now $\beta < \gamma$ implies $\beta_0 < \gamma$. Using $\alpha < \gamma_1$ it follows that $\phi\alpha\beta = (\alpha, \beta_0) < \gamma = \phi\alpha\gamma$.

3. $\phi\alpha\beta = (\alpha, \beta_0)$ where $\beta = \beta_0$ or $\beta = \beta_0'$ and $\phi\alpha\gamma = (\alpha, \gamma_0)$ where $\gamma = \gamma_0$ or $\gamma = \gamma_0'$. Now $\beta < \gamma$ implies $\beta \leqslant \gamma_0$. If $\beta = \gamma_0$ then $\gamma = \gamma_0'$. By the definition of $\phi\alpha\gamma$ we then have $\beta = \gamma_0 = (\gamma_1, \gamma_2) + \delta$ where $\alpha < \gamma_1$ and $\delta < \omega$. Since we have $\phi\alpha\beta = (\alpha, \beta_0)$ it follows that $\delta \neq 0$. Then $\beta = \beta_0'$. In any case we therefore have $\beta_0 < \gamma_0$. It follows that $\phi\alpha\beta = (\alpha, \beta_0) < (\alpha, \gamma_0) = \phi\alpha\gamma$.

Theorem 14.5. *$\phi\alpha_1\beta_1 = \phi\alpha_2\beta_2$ holds if, and only if, one of the following holds.*

1. $\alpha_1 < \alpha_2$ *and* $\beta_1 = \phi\alpha_2\beta_2$,
2. $\alpha_1 = \alpha_2$ *and* $\beta_1 = \beta_2$,
3. $\alpha_2 < \alpha_1$ *and* $\phi\alpha_1\beta_1 = \beta_2$.

Theorem 14.6. *$\phi\alpha_1\beta_1 < \phi\alpha_2\beta_2$ holds if, and only if, one of the following holds:*

1. $\alpha_1 < \alpha_2$ *and* $\beta_1 < \phi\alpha_2\beta_2$,
2. $\alpha_1 = \alpha_2$ *and* $\beta_1 < \beta_2$,
3. $\alpha_2 < \alpha_1$ *and* $\phi\alpha_1\beta_1 < \beta_2$.

Joint Proof for Theorems 14.5 and 14.6. By definition $\phi\alpha_2\beta_2$ is an ordinal term (γ_1, γ_2) with $\alpha_2 \leqslant \gamma_1$. If $\alpha_1 < \alpha_2$ then $\alpha_1 < \gamma_1$ and then again by the definition $\phi\alpha_1(\phi\alpha_2\beta_2) = \phi\alpha_2\beta_2$. The assertions of Theorems 14.5 and 14.6 for $\alpha_1 < \alpha_2$ now follow by Lemma 1. The case where $\alpha_2 < \alpha_1$ is similar. If $\alpha_1 = \alpha_2$ then the assertions follow immediately from Lemma 1. These are all the cases since $\alpha_1 < \alpha_2$ or $\alpha_1 = \alpha_2$ or $\alpha_2 < \alpha_1$.

Theorem 14.7. *For each principal term γ there exist unique α and $\beta < \gamma$ such that $\gamma = \phi\alpha\beta$. Moreover, $L\alpha < L\gamma$ and $L\beta < L\gamma$.*

Existence Proof. We have $\gamma=(\alpha, \beta_0)$ where $L\alpha<L\gamma$.

1. If $\beta_0=(\beta_1, \beta_2)+\delta$ where $\alpha<\beta_1$ and $\delta<\omega$ then $\gamma=\phi\alpha\beta'_0$. Then $\beta_0\leqslant\phi\alpha\beta_0<\phi\alpha\beta'_0$ and $(0,0)<\beta_0<\phi\alpha\beta'_0$. It follows that $\beta'_0<\phi\alpha\beta'_0$. Therefore $L\beta_0\neq 0$ and $L\beta'_0=L\beta_0+1<2\cdot L\beta_0+1\leqslant L\gamma$.
2. Otherwise $\gamma=\phi\alpha\beta_0$ and by Theorem 14.3 $\beta_0<\phi\alpha\beta_0$. Then we also have $L\beta_0<L\gamma$.

Uniqueness Proof. $\gamma=\phi\alpha_1\beta_1=\phi\alpha_2\beta_2$ where $\beta_1<\gamma$ and $\beta_2<\gamma$ implies $\alpha_1=\alpha_2$ and $\beta_1=\beta_2$ by Theorem 14.5.

Remark. A comparison of Theorems 13.9, 13.10 and 13.12 with Theorems 14.5–14.7 shows that the ordinal terms $\phi\alpha\beta$ have the properties corresponding to those of the ordinals $\phi\alpha\beta$ defined in §13.

Theorem 14.8. *For each ordinal term $\gamma\neq 0$ there are unique ordinal terms $\alpha_1,\ldots,\alpha_n$, $\beta_1,\ldots,\beta_n$ $(n\geqslant 1)$ where $\beta_i<\phi\alpha_i\beta_i$ $(i=1,\ldots,n)$, $\phi\alpha_1\beta_1\geqslant\cdots\geqslant\phi\alpha_n\beta_n$ and $\gamma=\phi\alpha_1\beta_1+\cdots+\phi\alpha_n\beta_n$.*

Proof. We have $\gamma=\gamma_1\ldots\gamma_n$ $(n\geqslant 1)$ with principal terms $\gamma_1\geqslant\cdots\geqslant\gamma_n$. By Theorem 14.7 there are unique α_i and $\beta_i<\gamma_i$ such that $\gamma_i=\phi\alpha_i\beta_i$. Hence the assertion is established.

5. Ordinal Terms ω^α, $\omega^\alpha\cdot n$ and $\omega^\alpha\cdot\beta$

Definition. $\omega^\alpha:=\phi 0\alpha$.

Corollaries. *ω^α is a principal term, $\omega^0=(0,0)$ and $\omega^{(0,0)}=\omega$.*

Lemma 2. $\alpha<\beta\Rightarrow\omega^\alpha<\omega^\beta$.

Proof. This follows from Lemma 1.

Lemma 3. *For each principal term β there is a unique ordinal term α such that $\beta=\omega^\alpha$.*

Proof. By Theorem 14.7 we have $\beta=\phi\beta_1\beta_2$. If $\beta_1=0$ then the assertion holds with $\alpha:=\beta_2$. Otherwise by Theorem 14.5 it holds for $\alpha:=\beta$. The uniqueness of α follows from Lemma 2.

Theorem 14.9. *For each ordinal term $\alpha\neq 0$ there are unique $\alpha_1\geqslant\cdots\geqslant\alpha_n$ $(n\geqslant 1)$ such that $\alpha=\omega^{\alpha_1}+\cdots+\omega^{\alpha_n}$.*

Proof. This follows from Lemma 3 and the properties of addition of ordinal terms.

Inductive Definition of $\omega^\alpha\cdot n\in\mathrm{OT}$ for $\alpha\in\mathrm{OT}$ and $n\in\mathbb{N}$.

1. $\omega^\alpha\cdot 0:=0$
2. $\omega^\alpha\cdot(n+1):=\omega^\alpha\cdot n+\omega^\alpha$.

Definition of $\nu(\alpha)\in \mathrm{OT}$ and $k(\alpha)\in \mathbb{N}$ for $\alpha\in \mathrm{OT}$.

1. If $\alpha=0$ then set $\nu(\alpha):=0$ and $k(\alpha):=0$.
2. If $\alpha\neq 0$ then by Theorem 14.9 there are unique $\alpha_1\geqslant\cdots\geqslant\alpha_n$ $(n\geqslant 1)$ where $\alpha=\omega^{\alpha_1}+\cdots+\omega^{\alpha_n}$. Then set $\nu(\alpha):=\alpha_1$ and $k(\alpha)$ to be the largest index $\leqslant n$ such that $\alpha_{k(\alpha)}=\alpha_1$.

Theorem 14.10. (1) $\omega^{\nu(\alpha)}\cdot k(\alpha)\leqslant\alpha<\omega^{\nu(\alpha)}\cdot(k(\alpha)+1)$

(2) *If* $\alpha<\omega^\beta$ *and* $\beta\neq 0$ *then* $\nu(\alpha)<\beta$ *and further* $\omega^{\nu(\alpha)}\cdot(k(\alpha)+1)<\omega^\beta$.

Proof. Immediate from the definitions.

Definition of $\omega^\alpha\cdot\beta\in \mathrm{OT}$ for $\alpha,\beta\in \mathrm{OT}$.

1. $\omega^\alpha\cdot 0:=0$
2. If $\beta\neq 0$ then by Theorem 14.9 there are unique $\beta_1\geqslant\cdots\geqslant\beta_n$ $(n\geqslant 1)$ such that $\beta=\omega^{\beta_1}+\cdots+\omega^{\beta_n}$. Then set $\omega^\alpha\cdot\beta:=\omega^{\alpha+\beta_1}+\cdots+\omega^{\alpha+\beta_n}$.

Corollaries:

$\omega^\alpha\cdot(0,0)=\omega^\alpha$

$\beta<\gamma\Rightarrow\omega^\alpha\cdot\beta<\omega^\alpha\cdot\gamma$

$\omega^\alpha\cdot\beta+\omega^\alpha\cdot\gamma=\omega^\alpha\cdot(\beta+\gamma)$

$\omega^\alpha\cdot(\omega^\beta\cdot\gamma)=\omega^{\alpha+\beta}\cdot\gamma$.

Theorem 14.11. *For each* α *and* β *there are unique* ξ *and* $\eta<\omega^\alpha$ *such that* $\beta=\omega^\alpha\cdot\xi+\eta$.

Existence Proof. 1. Suppose $\beta<\omega^\alpha$. Then the assertion holds with $\xi:=0$ and $\eta:=\beta$. 2. Suppose $\omega^\alpha\leqslant\beta$. Then there are $\beta_1\geqslant\cdots\geqslant\beta_n(n\geqslant 1)$ such that $\beta=\omega^{\beta_1}+\cdots+\omega^{\beta_n}$ and $\alpha\leqslant\beta_1$. Hence there is a largest index $m\leqslant n$ such that $\alpha\leqslant\beta_m$. We set $\xi:=\omega^{-\alpha+\beta_1}+\cdots+\omega^{-\alpha+\beta_m}$ and $\eta:=0$ if $m=n$ and $\eta:=\omega^{\beta_{m+1}}+\cdots+\omega^{\beta_n}$ if $m<n$. In any case we then have $\beta=\omega^\alpha\cdot\xi+\eta$ and $\eta<\omega^\alpha$.

Uniqueness Proof. Suppose $\beta=\omega^\alpha\cdot\xi_1+\eta_1=\omega^\alpha\cdot\xi_2+\eta_2$ where $\eta_1<\omega^\alpha$ and $\eta_2<\omega^\alpha$. Then

$$\omega^\alpha\cdot\xi_1\leqslant\beta<\omega^\alpha\cdot\xi_2+\omega^\alpha=\omega^\alpha\cdot\xi_2'.$$

It follows that $\xi_1<\xi_2'$ and therefore $\xi_1\leqslant\xi_2$. Similarly $\xi_2\leqslant\xi_1$. It follows that $\xi_1=\xi_2$. Therefore $\eta_1=\eta_2$.

Lemma 4. *If* $\gamma<\omega^\alpha\cdot\beta$ *and* $\alpha\neq 0$ *then* $\gamma'<\omega^\alpha\cdot\beta$.

Proof. $\gamma<\omega^\alpha\cdot\beta$ implies $\gamma'\leqslant\omega^\alpha\cdot\beta$. Since $\alpha\neq 0$, $\omega^\alpha\cdot\beta$ is not of the form $\gamma'=\gamma+\omega^0$. It follows that $\gamma'<\omega^\alpha\cdot\beta$.

Theorem 14.12. *If* $\gamma<\omega^\alpha\cdot\beta$ *and* $\alpha_1<\alpha$ *then there exists* β_1 *such that* $\gamma<\omega^{\alpha_1}\cdot\beta_1<\omega^\alpha\cdot\beta$.

Proof. By Theorem 14.11 we have $\gamma=\omega^{\alpha_1}\cdot\xi+\eta$ where $\eta<\omega^{\alpha_1}$. Therefore

$\omega^{\alpha_1}\cdot\xi\leqslant\gamma<\omega^{\alpha}\cdot\beta=\omega^{\alpha_1}\cdot(\omega^{-\alpha_1+\alpha}\cdot\beta)$. It follows that $\xi<\omega^{-\alpha_1+\alpha}\cdot\beta$. Now $\alpha_1<\alpha$ implies $0<-\alpha_1+\alpha$. By Lemma 4 it follows that $\xi'<\omega^{-\alpha_1+\alpha}\cdot\beta$. Putting $\beta_1:=\xi'$ we obtain $\gamma<\omega^{\alpha_1}\cdot\xi+\omega^{\alpha_1}=\omega^{\alpha_1}\cdot\beta_1<\omega^{\alpha_1}\cdot(\omega^{-\alpha_1+\alpha}\cdot\beta)=\omega^{\alpha}\cdot\beta$.

6. Ordinal Terms $n\cdot\beta$, 2^{α} and $2^{\alpha}\cdot\beta$

Since $\omega^{(0,0)}=\omega$, it follows by Theorem 14.11 that for each α there are unique ξ and $\eta<\omega$ such that

$$\alpha=\omega\cdot\xi+\eta.$$

For $\eta<\omega$, there is a unique $n\in\mathbb{N}$ such that $\eta=\omega^0\cdot n$. We define $q(\alpha):=\xi$ and $r(\alpha):=n$. In this way, we have $q(\alpha)\in\mathrm{OT}$ and $r(\alpha)\in\mathbb{N}$ for each $\alpha\in\mathrm{OT}$ such that

$$\alpha=\omega\cdot q(\alpha)+\omega^0\cdot r(\alpha)$$

holds.

Definition of $n\cdot\beta$ for $n\in\mathbb{N}$:

1. $0\cdot\beta:=0$
2. $n\cdot\beta:=\omega\cdot q(\beta)+\omega^0\cdot(n\cdot r(\beta))$ for $n\neq 0$.

Corollaries.

$1\cdot\beta=\beta$

$n\cdot(\beta+\gamma)=n\cdot\beta+n\cdot\gamma$

$n\neq 0,\qquad \beta<\gamma\Rightarrow n\cdot\beta<n\cdot\gamma.$

Definitions. $2^{\alpha}:=\omega^{q(\alpha)}\cdot 2^{r(\alpha)}$, $2^{\alpha}\cdot\beta:=\omega^{q(\alpha)}\cdot(2^{r(\alpha)}\cdot\beta)$.

Corollaries.

$2^{\alpha}\cdot(0,0)=2^{\alpha}$

$\beta<\gamma\Rightarrow 2^{\alpha}\cdot\beta<2^{\alpha}\cdot\gamma$

$2^{\alpha}\cdot(\beta+\gamma)=2^{\alpha}\cdot\beta+2^{\alpha}\cdot\gamma$

$2^{\alpha}\cdot(2^{\beta}\cdot\gamma)=2^{\alpha+\beta}\cdot\gamma.$

Theorem 14.13. If $\gamma<2^{\alpha}\cdot\beta$ and $\alpha_1+\omega\leqslant\alpha$, then there exists β_1 such that $\gamma<2^{\alpha_1}\cdot\beta_1<2^{\alpha}\cdot\beta$.

Proof. $\alpha_1+\omega\leqslant\alpha$ implies $q(\alpha_1)<q(\alpha)$. Therefore, by Theorem 14.12 there exists β_1 such that

$$\gamma<\omega^{q(\alpha_1)}\cdot\beta_1<\omega^{q(\alpha)}\cdot(2^{r(\alpha)}\cdot\beta)=2^{\alpha}\cdot\beta.$$

It follows that

$$\beta_1<\omega^{-q(\alpha_1)+q(\alpha)}\cdot(2^{r(\alpha)}\cdot\beta).$$

Since $-q(\alpha_1)+q(\alpha)>0$ and $2^{r(\alpha_1)}<\omega$, we have also

$$2^{r(\alpha_1)}\cdot\beta_1<\omega^{-q(\alpha_1)+q(\alpha)}\cdot(2^{r(\alpha)}\cdot\beta).$$

It follows that

$$\gamma<\omega^{q(\alpha_1)}\cdot\beta_1\leqslant\omega^{q(\alpha_1)}\cdot(2^{r(\alpha_1)}\cdot\beta_1)=2^{\alpha_1}\cdot\beta_1<\omega^{q(\alpha)}\cdot(2^{r(\alpha)}\cdot\beta)=2^{\alpha}\cdot\beta.$$

7. Ordinal Terms ε_α, $\omega_n(\alpha)$ and ζ_n

Definition. $\varepsilon_\alpha:=\phi(0,0)\alpha$.

These ordinal terms ε_α are called *ε-terms*. They are precisely the ordinal terms ξ such that $\omega^\xi=\xi$.

It follows from Lemma 1 that

$$\alpha<\beta\Rightarrow\varepsilon_\alpha<\varepsilon_\beta.$$

We obviously also have

$$\xi<\varepsilon_\alpha\Rightarrow\xi'<\varepsilon_\alpha.$$

Inductive Definition of $\omega_n(\alpha)\in\mathrm{OT}$ for $n\in\mathbb{N}$ and $\alpha\in\mathrm{OT}$.

1. $\omega_0(\alpha):=\alpha$
2. $\omega_{n+1}(\alpha):=\omega^{\omega_n(\alpha)}$.

Theorem 14.14. $\alpha<\varepsilon_\beta\Rightarrow\omega_n(\alpha)<\varepsilon_\beta$.

Proof by induction on n. The assertion is trivial for $n=0$. Suppose $n=m+1$. By I.H. we then have $\omega_m(\alpha)<\varepsilon_\beta$. It follows that $\omega_n(\alpha)=\omega^{\omega_m(\alpha)}<\omega^{\varepsilon_\beta}=\varepsilon_\beta$.

Theorem 4.15. *For each $\alpha<\varepsilon_\beta$ there exists $n\in\mathbb{N}$ and either $\gamma=0$ or $\gamma=\varepsilon'_\eta$ (where $\eta<\beta$) such that $\alpha<\omega_n(\gamma)$.*

Proof by induction on $L\alpha$.

1. $\alpha=0$. Then $\alpha<\omega^0=\omega_1(0)$.

2. Suppose α is a principal term. Then by Theorem 14.7 there exist ξ and $\eta<\alpha$ such that $\alpha=\phi\xi\eta$ and $L\eta<L\alpha$.

2.1. Suppose $\xi=0$. Since $\eta<\varepsilon_\beta$ and $L\eta<L\alpha$ by I.H. there exist $n\in\mathbb{N}$ and a γ of the required kind such that $\eta<\omega_n(\gamma)$. It follows that $\alpha=\omega^\eta<\omega^{\omega_n(\gamma)}=\omega_{n+1}(\gamma)$.

2.2. Suppose $\xi=(0,0)$. Then $\alpha=\varepsilon_\eta<\varepsilon_\beta$ and therefore $\eta<\beta$. It follows that $\alpha<\varepsilon'_\eta=\omega_0(\varepsilon'_\eta)$.

2.3. Suppose $(0,0)<\xi$. Then $\alpha=\varepsilon_\alpha<\varepsilon_\beta$ and therefore $\alpha<\beta$. It follows that $\alpha<\varepsilon'_\alpha=\omega_0(\varepsilon'_\alpha)$.

3. Suppose $\alpha=\alpha_1\ldots\alpha_m$ $(m\geqslant 2)$ where $\alpha_1,\ldots,\alpha_m$ are principal terms. Then

$\alpha_1 < \varepsilon_\beta$ and $L\alpha_1 < L\alpha$. By I.H. there exist therefore $n \in \mathbb{N}$ and a γ of the required kind such that $\alpha_1 < \omega_n(\gamma) \leqslant \omega_{n+1}(\gamma)$. Since $\omega_{n+1}(\gamma)$ is a principal term it follows that $\alpha < \omega_{n+1}(\gamma)$.

Inductive Definition of $\zeta_n \in \mathrm{OT}$ for $n \in \mathbb{N}$.

1. $\zeta_0 := \phi 00$,
2. $\zeta_{n+1} := \phi \zeta_n 0$,

By Theorem 14.4(2) $\zeta_n < \zeta_{n+1}$.

Theorem 14.16. *For every ordinal term α there exists $n \in \mathbb{N}$ such that $\alpha < \zeta_n$.*

Proof by induction on $L\alpha$.

1. $\alpha = 0$. Then $\alpha < \zeta_0$.

2. Suppose α is a principal term. Then by Theorem 14.7 there exist α_1, α_2 such that $\alpha = \phi\alpha_1\alpha_2$ and $L\alpha_i < L\alpha$ $(i=1, 2)$. Then by I.H. there exist $n_i \in \mathbb{N}$ such that $\alpha_i < \zeta_{n_i}$ $(i=1, 2)$. Putting $n := \max(n_1, n_2)$ it follows that $\alpha_1 < \zeta_n$ and $\alpha_2 < \zeta_n < \zeta_{n+1}$. It follows by Theorem 14.6 that $\alpha = \phi\alpha_1\alpha_2 < \phi\zeta_n 0 = \zeta_{n+1}$.

3. Suppose $\alpha = \alpha_1 \ldots \alpha_m$ $(m \geqslant 2)$ where $\alpha_1, \ldots, \alpha_m$ are principal terms. Then $L\alpha_1 < L\alpha$. Therefore by I.H. there exists $n \in \mathbb{N}$ such that $\alpha_1 < \zeta_n$. Since ζ_n is a principal term it follows that $\alpha < \zeta_n$.

8. A Mapping onto the Natural Numbers

Definitions for Natural Numbers.
If $m, n \in \mathbb{N}$ set $\pi(m, n) := \frac{1}{2}(m+n)(m+n+1)+m$, it is easy to see that $\pi: \mathbb{N} \times \mathbb{N} \to \mathbb{N}$ is a bijection of the set of ordered pairs (m, n) of natural numbers onto the set $\mathbb{N}$ of natural numbers. Let π_1 and π_2 be the inverses of π such that $\pi_1(\pi(m, n)) = m$, $\pi_2(\pi(m, n)) = n$ and $\pi(\pi_1(n), \pi_2(n)) = n$ for all $m, n \in \mathbb{N}$. Just like π the functions π_1 and π_2 are calculable arithmetic functions.

Let $p_0 := 2$. If $n > 0$ let p_n be the n-th odd prime number, thus $p_1 = 3$, $p_2 = 5$, $p_3 = 7$, $p_4 = 11$, etc.

Lemma 5. *If $k \in \mathbb{N}$ then $\pi_i(k) + 1 < p_k$ $(i=1, 2)$.*

Proof. It is easy to see that $\pi_i(k) \leqslant k$ and $k+1 < p_k$. Hence the assertion follows.

Extension of OT to $\overline{\mathrm{OT}}$. We adjoin a symbol Γ_0 to the ordinal terms of the system OT and set

$$\overline{\mathrm{OT}} := \mathrm{OT} \cup \{\Gamma_0\}.$$

If $\alpha, \beta \in \overline{\mathrm{OT}}$ then $\alpha < \beta$ if, and only if, either α, β are from OT and $\alpha < \beta$ or $\alpha \in \mathrm{OT}$ and $\beta = \Gamma_0$. Obviously $\overline{\mathrm{OT}}$ is ordered by this $<$-relation.

Let $\mathbb{N}^+$ be the set of positive natural numbers.

Inductive Definition of $\mathrm{Nr}(\alpha) \in \mathbb{N}^+$ for $\alpha \in \mathrm{OT}$ (by induction on $L\alpha$).

1. $\mathrm{Nr}(0) := 1$
2. If $\alpha = (\alpha_1, \beta_1) \ldots (\alpha_n, \beta_n) \in \mathrm{OT}$ ($n \geqslant 1$) let

$$\mathrm{Nr}(\alpha) := \prod_{i=1}^{n} p_{\pi(\mathrm{Nr}(\alpha_i)-1,\, \mathrm{Nr}(\beta_i)-1)}.$$

We also set

3. $\mathrm{Nr}(\Gamma_0) := 0$.

Then Nr is a map of $\overline{\mathrm{OT}}$ into the set $\mathbb{N}$ of natural numbers.

Inductive Definition of $\tau(m) \in \mathrm{OT}$ for $m \in \mathbb{N}^+$ (by induction on m).

1. $\tau(1) := 0$
2. If $\tau(\pi_1(k)+1) = \alpha$ and $\tau(\pi_2(k)+1) = \beta$, then let $\tau(p_k) := (\alpha, \beta)$.
3. If $m = \prod_{i=1}^{n} p_{k_i}$ ($n \geqslant 2$), $\tau(p_{k_i}) = \gamma_i$ for all $i = 1, \ldots, n$ and f is a permutation of $\{1, \ldots, n\}$ such that $\gamma_{f(1)} \geqslant \cdots \geqslant \gamma_{f(n)}$ then let $\tau(m) := \gamma_{f(1)} \ldots \gamma_{f(n)}$.

By Lemma 5 this is indeed an inductive definition. For every prime number $p_k \tau(p_k)$ is a principal term and for every natural number $m \geqslant 1$ $\tau(m)$ is an ordinal term of the system OT. We also set

4. $\tau(0) := \Gamma_0$.

Then τ is a map of $\mathbb{N}$ into $\overline{\mathrm{OT}}$.

Theorem 14.17 Nr *is a bijection of* $\overline{\mathrm{OT}}$ *onto* $\mathbb{N}$ *with inverse* $\tau: \mathbb{N} \to \overline{\mathrm{OT}}$. *That is: the following hold:*

(1) $\tau(\mathrm{Nr})(\alpha)) = \alpha$ *for all* $\alpha \in \overline{\mathrm{OT}}$

(2) $\mathrm{Nr}(\tau(m)) = m$ *for all* $m \in \mathbb{N}$.

Proof. (1) and (2) are immediate from the definition. This yields the assertion.

The $<$-relation of OT induces an ordering relation $\prec$ of the natural numbers under the bijection as follows:

$m \prec n \Leftrightarrow \tau(m) < \tau(n)$ for $m, n \in \mathbb{N}$,

$\mathrm{Nr}(\alpha) \prec \mathrm{Nr}(\beta) \Leftrightarrow \alpha < \beta$ for $\alpha, \beta \in \overline{\mathrm{OT}}$.

Theorem 14.18. $\prec$ *is a decidable arithmetic relation, that is, for any two natural numbers* m, n *it is decidable whether* $m \prec n$, $m = n$ *or* $n \prec m$.

Proof. Given natural numbers m, n the terms $\tau(m)$, $\tau(n) \in \overline{\mathrm{OT}}$ are effectively calculable. We can then decide whether $\tau(m) < \tau(n)$, $\tau(m) = \tau(n)$ or $\tau(n) < \tau(m)$.

Chapter VI

Functionals of Finite Type

In this chapter we treat the theory of functionals of finite type in which, following K Gödel [2] pure number theory can be interpreted (§19). The functionals concerned are introduced in §15 and they are shown to be calculable in §16 following W. Howard by transfinite induction up to ε_0, in fact, by using the ordinal terms of §14. In §17 we develop a formal system on the basis of the positive implicational calculus (with equality) as a logical setting for the theory of functionals of finite type.

§15. The System of Terms of Finite Type

1. Types

We now introduce *function types* in contrast to the predicate types of Chapter IV.

Inductive Definition of *types*:

1. The symbol o is a *type*. (It denotes the type of the natural numbers.)
2. If σ and τ are *types* then $(\sigma)\tau$ is also a *type*. ($(\sigma)\tau$ denotes the type of certain maps $\sigma \rightarrow \tau$ from the set of type σ into the set of type τ.)

For brevity we write $\sigma\tau$ for $(\sigma)\tau$, if the type σ is written as a single symbol.

According to the above definition every type different from o is of the form $\tau_1 \ldots \tau_n o$ $(n \geqslant 1)$.

2. Terms

We take as *primitive terms*:

1. Denumerably infinitely many *variables* of each type.
2. The *arithmetic* primitive term 0 (for the natural number zero) and $\mathscr{S}$ (for the successor function).
3. *Combinators* $K_{\sigma\tau}$ and $S_{\rho\sigma\tau}$ for all types ρ, σ, τ. (Here we use the combinators K and S which were introduced by H. B. Curry in a type-free manner but now making distinctions of type.)
4. An *iterator* J_τ for each type τ.

Inductive Definition of the *terms* and their types:

1. Every variable of type τ is a *term* of type τ.
2. The arithmetic primitive term 0 is a *term* of type o.
3. The arithmetic primitive term $\mathcal{S}$ is a *term* of type oo.
4. The combinator $K_{\sigma\tau}$ is a *term* of type $\tau\sigma\tau$.
5. The combinator $S_{\rho\sigma\tau}$ is a *term* of type $(\rho\sigma\tau)(\rho\sigma)\rho\tau$.
6. The iterator J_τ is a *term* of type $o(\tau\tau)\tau\tau$.
7. If $a^{\sigma\tau}$ and b^σ are *terms* of types $\sigma\tau$ and σ then $a^{\sigma\tau}(b^\sigma)$ is a *term* of type τ.

For brevity we write $a^{\sigma\tau}b^\sigma$ for $a^{\sigma\tau}(b^\sigma)$ if the term b^σ is not written as a composite term.

We call terms which contain no variable *functionals*.

Syntactic symbols:

x^τ, y^τ, z^τ for variables of type τ,
$a^\tau, b^\tau, c^\tau, d^\tau, e^\tau$ for terms of type τ.

We shall also use these syntactic symbols with subscripts. The superscripts always give the type.

According to the above definition of term every term which is not primitive is of the form

$$a^{\sigma_1\ldots\sigma_n\tau}\, b^{\sigma_1}\ldots b^{\sigma_n} \quad (n\geq 1)$$

where $a^{\sigma_1\ldots\sigma_n\tau}$ is a primitive term.

Inductive Definition of the *numerals* N_n $(n\in\mathbb{N})$:

1. $N_0 := 0$
2. $N_{n+1} := \mathcal{S}N_n$.

The numerals are special terms of type o.

3. Interpretation

The term $a^{\sigma\tau}b^\sigma$ denotes the image of b^σ under the map $a^{\sigma\tau}: \sigma\to\tau$. This image is of type τ.

The successor functional $\mathcal{S}$ maps each term a^o of type o onto its successor $\mathcal{S}a^0$. Thus $\mathcal{S}$ denotes a map $o\to o$ of type oo.

$K_{\sigma\tau}a^\tau$ denotes the constant function $\sigma\to\tau$ which maps every term b^σ of type σ to a^τ. Thus $K_{\sigma\tau}a^\tau b^\sigma = a^\tau$. This mapping $K_{\sigma\tau}a^\tau$ is of type $\sigma\tau$. The combinator $K_{\sigma\tau}$ denotes the function $\tau\to\sigma\tau$ which maps α^τ to the constant function $K_{\sigma\tau}\alpha_\tau$. Consequently $K_{\sigma\tau}$ is a functional of type $\tau\sigma\tau$.

$S_{\rho\sigma\tau}a^{\rho\sigma\tau}b^{\rho\sigma}$ denotes the function $\rho\to\tau$ which maps the term c^ρ to the term $a^{\rho\sigma\tau}c^\rho(b^{\rho\sigma}c^\rho)$ of type τ. Thus we have

$$S_{\rho\sigma\tau}a^{\rho\sigma\tau}b^{\rho\sigma}c^\rho = a^{\rho\sigma\tau}c^\rho(b^{\rho\sigma}c^\rho).$$

Consequently $S_{\rho\sigma\tau}a^{\rho\sigma\tau}b^{\rho\sigma}$ is of type $\rho\tau$, $S_{\rho\sigma\tau}a^{\rho\sigma\tau}$ is of type $(\rho\sigma)\rho\tau$ and $S_{\rho\sigma\tau}$ is of type $(\rho\sigma\tau)(\rho\sigma)\rho\tau$.

The iterator J_τ only has a meaning when together with a numeral N_n. Then the functional $J_\tau N_n$ maps a function $a^{\tau\tau}: \tau \to \tau$ onto its n-fold iteration. That is, the following holds:

$$J_\tau N_n a^{\tau\tau} b^\tau = a^{\tau\tau}(\ldots(a^{\tau\tau} b^\tau)\ldots)$$

where $a^{\tau\tau}$ occurs n times on the right-hand side of the equation. In particular we have

$$J_\tau 0 a^{\tau\tau} b^\tau = b^\tau$$

and

$$J_\tau N_{n+1} a^{\tau\tau} b^\tau = a^{\tau\tau}(J_\tau N_n a^{\tau\tau} b^\tau).$$

Consequently $J_\tau N_n a^{\tau\tau}$ is of type $\tau\tau$, $J_\tau N_n$ of type $(\tau\tau)\tau\tau$ and J_τ of type $o(\tau\tau)\tau\tau$.

The interpretation we have just given serves only to clarify the theory we are dealing with. In the sequel we shall not refer to this interpretation and the meaning of the terms will be fixed purely formally by an appropriate reduction procedure.

4. Reduction Procedure for Terms

Definition. A term is said to be *open* if it is of one of the following three forms:

1. $K_{\sigma\tau} a^\tau b^\sigma d_1^{\tau_1} \ldots d_m^{\tau_m}$
2. $S_{\rho\sigma\tau} a^{\rho\sigma\tau} b^{\rho\sigma} c^\rho d_1^{\tau_1} \ldots d_m^{\tau_m}$
3. $J_\tau N_n a^{\tau\tau} b^\tau d_1^{\tau_1} \ldots d_m^{\tau_m}$

where in each case $\tau = \tau_1 \ldots \tau_{m+1}$ $(m \geqslant 0)$. The term is then of type τ_{m+1}.

Inductive Definition of the *reducibility* of a term.

1. Every open term is *reducible.*
2. If $a^{\sigma\tau}$ or b^σ is *reducible* then $a^{\sigma\tau} b^\sigma$ is also *reducible.*

Inductive Definition of $a^\tau \rhd^1 b^\tau$ ("a^τ *reduces to* b^τ *in one step*").

The following hold for $\tau = \tau_1 \ldots \tau_{m+1}$ $(m \geqslant 0)$:

1. $K_{\sigma\tau} a^\tau b^\sigma d_1^{\tau_1} \ldots d_m^{\tau_m} \rhd^1 a^\tau d_1^{\tau_1} \ldots d_m^{\tau_m}$
2. $S_{\rho\sigma\tau} a^{\rho\sigma\tau} b^{\rho\sigma} c^\rho d_1^{\tau_1} \ldots d_m^{\tau_m} \rhd^1 a^{\rho\sigma\tau} c^\rho (b^{\rho\sigma} c^\rho) d_1^{\tau_1} \ldots d_m^{\tau_m}$
3. $J_\tau 0 a^{\tau\tau} b^\tau d_1^{\tau_1} \ldots d_m^{\tau_m} \rhd^1 b^\tau d_1^{\tau_1} \ldots d_m^{\tau_m}$
4. $J_\tau N_{n+1} a^{\tau\tau} b^\tau d_1^{\tau_1} \ldots d_m^{\tau_m} \rhd^1 a^{\tau\tau}(J_\tau N_n a^{\tau\tau} b^\tau) d_1^{\tau_1} \ldots d_m^{\tau_m}$
5. $b^\sigma \rhd^1 c^\sigma \Rightarrow a^{\sigma\tau} b^\sigma d_1^{\tau_1} \ldots d_m^{\tau_m} \rhd^1 a^{\sigma\tau} c^\sigma d_1^{\tau_1} \ldots d_m^{\tau_m}$,

provided $a^{\sigma\tau} b^\sigma d_1^{\tau_1} \ldots d_m^{\tau_m}$ is not open and $a^{\sigma\tau}$ is irreducible.

These reduction steps correspond to the interpretation of the terms we gave above. Namely, $a^\tau \rhd^1 b^\tau$ says that b^τ has the same meaning as a^τ in the given interpretation.

Cases 1.–4. yield reductions of open terms. Case 5. gives a reduction of a reducible term which is not open.

Corollary. *For each a^τ there exists b^τ such that $a^\tau \rhd^1 b^\tau$ if, and only if, a^τ is reducible. In this case b^τ is uniquely determined by a^τ.*

Inductive Definition of $a^\tau \rhd^n b^\tau$ ("*a^τ reduces to b^τ in n steps*") for $n \in \mathbb{N}$:

1. $a^\tau \rhd^o a^\tau$
2. $a^\tau \rhd^1 b^\tau, \qquad b^\tau \rhd^m c^\tau \Rightarrow a^\tau \rhd^{m+1} c^\tau \quad (m \geqslant 1)$.

$a^\tau \rhd b^\tau$ means that there exists $n \in \mathbb{N}$ such that $a^\tau \rhd^n b^\tau$.

Corollaries. (1) *If $a^\tau \rhd^n b^\tau$ and $\tau = \tau_1 \dots \tau_{m+1}$ $(m \geqslant 1)$ then we also have*

$$a^\tau c_1^{\tau_1} \dots c_m^{\tau_m} \rhd^n b^\tau c_1^{\tau_1} \dots c_m^{\tau_m}.$$

(2) *$a^\tau \rhd b^\tau$ and $a^\tau \rhd c^\tau$ implies $b^\tau \rhd c^\tau$ or $c^\tau \rhd b^\tau$.*

Proof. (1) is immediate from the defining rules for reduction. (2) holds because at most one (one step) reduction is possible for any given term.

5. Characterization of Numerals

Theorem 15.1. *The numerals are the only irreducible functionals of type o.*

Proof. Let a^o be an irreducible functional of type o. We prove by induction on the length of a^o that a^o is a numeral. It follows from the rules given above for types when using combinators that every term of type o which begins with a combinator is an open term. Hence the irreducible term a^o cannot begin with a combinator. So we have at most the following three cases to consider for the term a^o (which contains no variable):

1. a^o is 0
2. a^o is $\mathcal{S} b^o$
3. a^o is $J_\tau b^o c^{\tau\tau} d^\tau e_1^{\tau_1} \dots e_n^{\tau_n}$ where $\tau = \tau_1 \dots \tau_n o$ $(n \geqslant 0)$.

In cases 2 and 3 b^0 is an irreducible functional of type o. Since b^o is shorter than a^o then by I.H. b^o is, in these cases, a numeral. But then in case 3 a^o is an open, and therefore reducible, term. So this case is excluded. In the remaining first two cases a^o is a numeral.

6. Substitution Properties

We denote by $c^\sigma \binom{a^\tau}{x^\tau}$ the term which results from c^σ when every occurrence of the variable x^τ in c^σ is replaced by the term a^τ.

Lemma 1. *If c^σ is an open term and $c^\sigma \rhd^1 d^\sigma$ then $c^\sigma \binom{a^\tau}{x^\tau} \rhd^1 d^\sigma \binom{a^\tau}{x^\tau}$.*

Proof. This follows at once from the rules for reduction of open terms.

Lemma 2. *If x^τ occurs in c^σ and a^τ is reducible then $c^\sigma\binom{a^\tau}{x^\tau}$ is reducible.*

Proof. In this case the reducible term a^τ occurs in $c^\sigma\binom{a^\tau}{x^\tau}$ and hence the assertion follows by induction on the length of c^σ.

Lemma 3. *If c^σ is a term which is not open and does not begin with x^τ, and if a^τ is reducible then the term $c^\sigma\binom{a^\tau}{x^\tau}$ is not open.*

Proof. Suppose c^σ is a term $c_o^\rho c_1^{\rho_1}\dots c_n^{\rho_n}$ where $\rho=\rho_1\dots\rho_n$ ($n\geqslant 0$) and where c_o^ρ is a basic term different from x^τ. Then the term $c^\sigma\binom{a^\tau}{x^\tau}$ also begins with c_o^ρ.

1. Suppose c_o^ρ is a variable or an arithmetic basic term. Then trivially $c^\sigma\binom{a^\tau}{x^\tau}$ is not open.

2. Suppose c_o^ρ is a combinator. Then $c^\sigma\binom{a^\tau}{x^\tau}$ is not open since c^σ is not open.

3. Suppose c_o^ρ is an iterator.

3.1. Suppose $n\leqslant 2$. Then trivially $c^\sigma\binom{a^\tau}{x^\tau}$ is not open.

3.2. Suppose $n>2$. Then $\rho_1=o$ and c_1^o is not a numeral since c^σ is not open. If x^τ occurs in c_1^o then $c_1^o\binom{a^\tau}{x^\tau}$ is reducible by Lemma 2. In any case $c_1^o\binom{a^\tau}{x^\tau}$ is not a numeral. Therefore $c^\sigma\binom{a^\tau}{x^\tau}$ is not open.

Theorem 15.2. *If $a^\tau \rhd^1 b^\tau$ and $c^\sigma\binom{a^\tau}{x^\tau}\rhd^m e^\sigma$ where e^σ is irreducible then $c^\sigma\binom{b^\tau}{x^\tau}\rhd e^\sigma$.*

Proof. Without loss of generality we may assume that x^τ does not occur in a^τ, b^τ. For we can choose a variable y^τ which does not occur in a^τ, b^τ, c^σ. Then the terms $c^\sigma\binom{a^\tau}{x^\tau}$, $c^\sigma\binom{b^\tau}{x^\tau}$ are the same as the terms $c^\sigma\binom{y^\tau}{x^\tau}\binom{a^\tau}{y^\tau}$, $c^\sigma\binom{y^\tau}{x^\tau}\binom{b^\tau}{y^\tau}$ so we can replace c^σ and x^τ in the hypotheses and statement of Theorem 15.2 by $c^\sigma\binom{y^\tau}{x^\tau}$ and y^τ where y^τ does not occur in a^τ, b^τ.

Now we assume that x^τ does not occur in a^τ, b^τ and prove Theorem 15.2 by a multiple nested induction, namely, by induction

1. on the length of the term a^τ,
2. on m,
3. on the length of the term c^σ.

For brevity, in what follows we write

$$\bar{d}^\rho := d^\rho\binom{a^\tau}{x^\tau}, \qquad \tilde{d}^\rho := d^\rho\binom{b^\tau}{x^\tau}$$

for any term d^ρ.

1. Suppose c^σ is an open term.

Then we have a term d^σ such that $c^\sigma \rhd^1 d^\sigma$. It follows by Lemma 1 that $\bar{c}^\sigma \rhd^1 d^\sigma$, and $\tilde{c}^\sigma \rhd^1 \tilde{d}^\sigma$. It follows from $\bar{c}^\sigma \rhd^1 \bar{d}^\sigma$ and the hypothesis $\bar{c}^\sigma \rhd^m e^\sigma$ because of the irreducibility of e^σ that $m > 0$ and $\bar{d}^\sigma \rhd^{m-1} e^\sigma$ holds. By the second I.H. it follows that $\tilde{d}^\sigma \rhd e^\sigma$. Hence $\tilde{c}^\sigma \rhd e^\sigma$ also holds.

2. Suppose c^σ is a term which is not open and does not begin with x^τ.

If x^τ does not occur in c^σ then the assertion is trivial. Now assume that x^τ does occur in c^σ. Then by Lemma 2 $\bar{c}^\sigma$ is reducible and Lemma 3 c^σ is not open, while a^τ is reducible by hypothesis. Therefore c^σ must be of the form

$$c_o^\rho c_1^{\sigma_1} \ldots c_n^{\sigma_n} \quad \text{where } \rho = \sigma_1 \ldots \sigma_n \quad \sigma (n \geqslant 1)$$

where $\bar{c}_o^\rho$ is irreducible and $\bar{c}_1^{\sigma_1}$ is reducible. It follows by Lemma 2 that x^τ does not occur in c_o^ρ. It now follows from the hypothesis $\bar{c}^\sigma \rhd^m e^\sigma$ that there is an irreducible term $e_1^{\sigma_1}$ such that $\bar{c}_1^{\sigma_1} \rhd^{m_1} e_1^{\sigma_1}$ where $1 \leqslant m_1 \leqslant m$. Since $m_1 \leqslant m$ and $c_1^{\sigma_1}$ is shorter than c^σ it follows by the second or third I.H. that $\tilde{c}_1^{\sigma_1} \rhd e_1^{\sigma_1}$. Hence

$$\bar{c}^\sigma \rhd^{m_1} c_o^\rho e_1^{\sigma_1} \bar{c}_2^{\sigma_2} \ldots \bar{c}_n^{\sigma_n} \rhd^{m-m_1} e^\sigma$$

and

$$\tilde{c}^\sigma \rhd c_o^\rho e_1^{\sigma_1} \tilde{c}_2^{\sigma_2} \ldots \tilde{c}_n^{\sigma_n}.$$

Since $m - m_1 < m$ it follows by the second I.H. that

$$c_o^\rho e_1^{\sigma_1} \tilde{c}_2^{\sigma_2} \ldots \tilde{c}_n^{\sigma_n} \rhd e^\sigma.$$

But then $\tilde{c}^\sigma \rhd e^\sigma$ also holds.

3. Suppose c^σ is a term beginning with x^τ. Then c^σ is of the form

$$x^\tau c_1^{\sigma_1} \ldots c_n^{\sigma_n} \quad \text{where } \tau = \sigma_1 \ldots \sigma_n \sigma \quad (n \geqslant 0).$$

3.1. Suppose a^τ is open or $\bar{c}^\sigma$ is not open. Then it follows from the hypotheses of the theorem that

$$\bar{c}^\sigma \rhd^1 b^\tau \bar{c}_1^{\sigma_1} \ldots \bar{c}_n^{\sigma_n} \rhd^{m-1} e^\sigma.$$

By the second I.H. we have

$$b^\tau \tilde{c}_1^{\sigma_1} \ldots \tilde{c}_n^{\sigma_n} \rhd e^\sigma.$$

This is just the assertion $\tilde{c}^\sigma \rhd e^\sigma$.

3.2. Suppose a^τ is not open and $\tilde{c}^\sigma$ is open. Then it follows from $a^\tau \rhd^1 b^\tau$ that

$$a^\tau \quad \text{is a term} \quad a_o^\rho a_1^{\tau_1} \ldots a_r^{\tau_r}$$

and

$$b^\tau \quad \text{is a term} \quad a_o^\rho b_1^{\tau_1} a_2^{\tau_2} \ldots a_r^{\tau_r}$$

where $\rho = \tau_1 \ldots \tau_r \tau$ $(r \geqslant 1)$ and where a_o^ρ is irreducible and $a_1^{\tau_1} \rhd^1 b_1^{\tau_1}$ holds. Let y^{τ_1} be a variable which does not occur in a^τ, b^τ, c^σ.

Let d_1^σ be the term $a_o^\rho y^{\tau_1} a_2^{\tau_2} \ldots a_r^{\tau_r} c_1^{\sigma_1} \ldots c_n^{\sigma_n}$ so that $\bar{d}_1^\sigma \begin{pmatrix} a_1^{\tau_1} \\ y^{\tau_1} \end{pmatrix}$ is the term $\tilde{c}^\sigma$. Since $\tilde{c}^\sigma$ is open and a^τ, $a_1^{\tau_1}$ are reducible it follows from Lemma 3 that d_1^σ is also open. Hence we have a term d_2^σ such that $d_1^\sigma \rhd^1 d_2^\sigma$. It follows by Lemma 1'that

$$\tilde{c}^\sigma \rhd^1 \bar{d}_2^\sigma \begin{pmatrix} a_1^{\tau_1} \\ y^{\tau_1} \end{pmatrix} \rhd^{m-1} e^\sigma$$

and

$$\tilde{c}^\sigma \rhd^1 \bar{d}_2^\sigma \begin{pmatrix} b_1^{\tau_1} \\ y^{\tau_1} \end{pmatrix}.$$

It follows from the second I.H. that

$$\bar{d}_2^\sigma \begin{pmatrix} a_1^{\tau_1} \\ y^{\tau_1} \end{pmatrix} \rhd e^\sigma.$$

Since $a_1^{\tau_1} \rhd^1 b_1^{\tau_1}$ holds and $a_1^{\tau_1}$ is shorter than a^τ it follows by the first I.H. that

$$\bar{d}_2^\sigma \begin{pmatrix} b_1^{\tau_1} \\ y^{\tau_1} \end{pmatrix} \rhd e^\sigma.$$

This yields the assertion $\tilde{c}^\sigma \rhd e^\sigma$.

7. The Normal Form of a Term

Definition. b^τ is said to be a *normal form* of a term a^τ if $a^\tau \rhd b^\tau$ and b^τ is irreducible.

Corollary. *Every term has at most one normal form.*

Proof. $a^\tau \rhd b^\tau$ and $a^\tau \rhd c^\tau$ imply $b^\tau \rhd c^\tau$ or $c^\tau \rhd b^\tau$. If b^τ and c^τ are irreducible then it follows that b^τ and c^τ are identical.

In the next section we shall prove that every term has a normal form.

Remark. The existence of the normal form can be proved somewhat more simply by using a strong metamathematical predicate (see J. R. Shoenfield: Mathematical Logic 1967, pp. 225, 227). W. Howard's proof, which is given in §16, requires substantially more machinery but yields more information, since by its reference to the ordinals $<\varepsilon_0$ it gives a connection with the consistency proofs which we shall carry out in the later chapters using transfinite induction.

§16. Orders of Terms

Following a procedure of W. Howard we define the *order* of a term to be an ordinal $<\varepsilon_0$ which is decreased when the reduction process for terms is applied. It then follows by transfinite induction up to ε_0 that every term has a normal form.

We represent the ordinals by the ordinal terms we defined in §14 but in this section we restrict ourselves just to ordinal terms $<\varepsilon_0$.

By Theorem 14.9 for every ordinal term $\alpha \neq 0$ there are uniquely determined ordinal terms $\alpha_1 \geqslant \cdots \geqslant \alpha_n$ $(n \geqslant 1)$ such that

$$\alpha = \omega^{\alpha_1} + \cdots + \omega^{\alpha_n}.$$

It follows that all $\alpha_i < \alpha$ if $\alpha < \varepsilon_0$.

1. Natural Sum and Natural Product of Ordinal Terms

In order to assign ordinal terms to terms we use the *natural sum* $\alpha \# \beta$ and the *natural product* $2^\alpha \times \beta$ of the ordinal terms α, β and we first define these.

Definition of $\alpha \# \beta$.

1. $\alpha \# 0 := \alpha$ and $0 \# \beta := \beta$

2. If $\alpha = \omega^{\alpha_1} + \cdots + \omega^{\alpha_m}$ and $\beta = \omega^{\alpha_{m+1}} + \cdots + \omega^{\alpha_n}$ $(1 \leqslant m < n)$ where $\alpha_1 \geqslant \cdots \geqslant \alpha_m$ and $\alpha_{m+1} \geqslant \cdots \geqslant \alpha_n$ and f is a permutation of $\{1, \ldots, n\}$ such that $\alpha_{f(1)} \geqslant \cdots \geqslant \alpha_{f(n)}$ then

$$\alpha \# \beta := \omega^{\alpha_{f(1)}} + \cdots + \omega^{\alpha_{f(n)}}.$$

Corollaries.

(#1) $\alpha \# \beta = \beta \# \alpha$

(#2) $\alpha \# (\beta \# \gamma) = (\alpha \# \beta) \# \gamma$

(#3) $\beta < \gamma \Rightarrow \alpha \# \beta < \alpha \# \gamma$.

Definition of $2^\alpha \times \beta$.

By Theorem 14.11 for each α there are unique α_0 and $m < \omega$ such that $\alpha =$

$\omega\cdot\alpha_n+m$. We regard m as a natural number so that $\xi\cdot 2^m$ is defined as the 2^m-fold sum $\xi+\cdots+\xi$ for each ordinal ξ.

1. $2^\alpha\times 0:=0$

2. If $\beta=\omega^{\beta_1}+\cdots+\omega^{\beta_n}$ $(n\geqslant 1)$ where $\beta_1\geqslant\cdots\geqslant\beta_m$ and $\alpha=\omega\cdot\alpha_n+m$ where $m<\omega$ then set

$$2^\alpha\times\beta:=\omega^{\alpha_o\#\beta_1}.2^m+\cdots+\omega^{\alpha_o\#\beta_n}\cdot 2^m.$$

We immediately obtain from this definition:

$(\times 1)$ $\beta\leqslant 2^\alpha\times\beta$
$(\times 2)$ $\beta<\gamma\Rightarrow 2^\alpha\times\beta<2^\alpha\times\gamma$
$(\times 3)$ $2^\alpha\times(\beta\#\gamma)=(2^\alpha\times\beta)\#(2^\alpha\times\gamma)$
$(\times 4)$ $2^1\times\beta=\beta\#\beta$.

We also have

$(\times 5)$ $\alpha<\beta,\gamma\neq 0\Rightarrow 2^\alpha\times\gamma<2^\beta\times\gamma$.

Proof. If $\alpha=\omega\cdot\alpha_o+m_1<\beta=\omega\cdot\beta_o+m_2$ where $m_1<\omega$, $m_2<\omega$ then either $\alpha_o<\beta_o$ or $\alpha_o=\beta_o$, $m_1<m_2$. If $\alpha_o<\beta_o$ then $\alpha_o\#\gamma_i<\beta_o\#\gamma_i$ and $\omega^{\alpha_o\#\gamma_i}.2^{m_1}<\omega^{\beta_o\#\gamma_i}$. Hence $2^\alpha\times\gamma<2^\beta\times\gamma$. If $\alpha_o=\beta_o$, $m_1<m_2$ then the assertion follows from $2^{m_1}<2^{m_2}$.

$(\times 6)$ $2^\alpha\times(2^\beta\times\gamma)=2^{\alpha\#\beta}\times\gamma$.

Proof. If $\alpha=\omega\cdot\alpha_o+m_1$ and $\beta=\omega\cdot\beta_o+m_2$ where $m_1<\omega$, $m_2<\omega$ then $\alpha\#\beta=\omega\cdot(\alpha_o\#\beta_o)+m_1+m_2$. The assertion now follows from the definition of natural product.

$(\times 7)$ $(2^\alpha\times\beta)\#(2^\alpha\times\beta)=2^{\alpha\#1}\times\beta$.

Proof. By $(\times 3)$, $(\times 4)$ and $(\times 6)$, $(2^\alpha\times\beta)\#(2^\alpha\times\beta)=2^\alpha\times(\beta\#\beta)=2^\alpha\times(2^1\times\beta)=2^{\alpha\#1}\times\beta$.

2. Assigning Ordinal Terms to Terms

Inductive Definition of the *degree* $g\tau$ of a type τ:

1. $go:=0$
2. $g(\sigma\tau):=\max(g\sigma+1,g\tau)$.

The degree of a type is a natural number.

Inductive Definition of an ordinal term $[a^\tau]_i$ for each term a^τ and natural number i.

1. If a^τ is a variable or an arithmetic basic term 0 or $\mathcal{S}$ then set

$$[a^\tau]_i:=0\quad\text{for all } i\in\mathbb{N}.$$

2. If a^τ is a combinator of type τ, then set

$$[a^\tau]_i := \begin{cases} 1 & \text{if } i \leqslant g\tau. \\ 0 & \text{if } i > g\tau. \end{cases}$$

3. If J_ρ is an iterator of type $\tau = o(\rho\rho)\rho\rho$ then $g\tau = g\rho + 2$. Then set

$$[J_\rho]_i := \begin{cases} 1 & \text{if } i \leqslant g\rho + 1, \\ \omega & \text{if } i = g\rho + 2, \\ 0 & \text{if } i > g\rho + 2. \end{cases}$$

$$[J_\rho N_n]_i := \begin{cases} 1 & \text{if } i \leqslant g\rho + 1, \\ n & \text{if } i = g\rho + 2, \\ 0 & \text{if } i > g\rho + 2. \end{cases}$$

4. If the term $a^{\sigma\tau}b^\sigma$ is not of the form $J_\rho N_n$ then set

$$[a^{\sigma\tau}b^\sigma]_i := \begin{cases} 2^{[a^{\sigma\tau}b^\sigma]_{i+1}} \times ([a^{\sigma\tau}]_i \# [b^\sigma]_i) & \text{if } i \leqslant g\sigma, \\ [a^{\sigma\tau}]_i & \text{if } i > g\sigma. \end{cases}$$

This last rule is inductive since $[a^{\sigma\tau}b^\sigma]_{g\sigma+1} = [a^{\sigma\tau}]_{g\sigma+1}$ and therefore $[a^{\sigma\tau}b^\sigma]_i$ is determined for $i \leqslant g\sigma$ by induction on $g\sigma - i$.

Corollaries. 1. *For every term a^τ there exists $k \in \mathbb{N}$ such that $[a^\tau]_i = 0$ for all $i > k$.*
2. *If a^τ does not contain an iterator then $[a^\tau]_i < \omega$ for all $i \in \mathbb{N}$.*
3. *For all terms a^τ we have $[a^\tau]_i < \varepsilon_0$.*

Lemma 1. $[N_n]_i = 0$ *for all* $i \in \mathbb{N}$.

Proof by induction on n. By definition $[N_0]_i = [0]_i = 0$. Since $N_{n+1} = \mathscr{S} N_n$ it follows from $[\mathscr{S}]_i = 0$ and $[N_n]_i = 0$ using clause 4 of the definition that $[N_{n+1}]_i = 0$, too.

Lemma 2. (1) *If $a^{\sigma\tau}b^\sigma$ is not of the form $J_\rho N_n$ then $[a^{\sigma\tau}]_i \leqslant [a^{\sigma\tau}b^\sigma]_i$ for all $i \in \mathbb{N}$.*
(2) *For all $i \leqslant g\sigma$ we have $[b^\sigma]_i \leqslant [a^{\sigma\tau}b^\sigma]_i$.*

Proof. (1) follows immediately from the definition of $[a^{\sigma\tau}b^\sigma]_i$. (2) likewise follows from this definition if $a^{\sigma\tau}b^\sigma$ is not of the form $J_\rho N_n$. Otherwise (2) follows from Lemma 1.

Lemma 3.

$$[a^{\sigma\tau}b^\sigma]_i \leqslant \begin{cases} 2^{[a^{\sigma\tau}b^\sigma]_{i+1}} \times ([a^{\sigma\tau}]_i \# [b^\sigma]_i) & \text{if } i \leqslant g\sigma, \\ [a^{\sigma\tau}]_i & \text{if } i > g\sigma. \end{cases}$$

Proof. 1. Suppose $a^{\sigma\tau}b^\sigma$ is not of the form $J_\rho N_n$. Then the assertion follows from the definition of $[a^{\sigma\tau}b^\sigma]_i$.

2. Suppose $a^{\sigma\tau}b^{\sigma}$ is a term $J_\rho N_n$. Then $\sigma=o$ and therefore $g\sigma=0$ and

$$[J_\rho N_n]_i=\begin{cases}1<2^1\times(1\#[N_n]_o)=2^{[J_\rho N_n]_1}\times([J_\rho]_o\#[N_n]_o) & \text{if } i=0,\\ 1=[J_\rho]_i & \text{if } 1\leqslant i\leqslant g\rho+1,\\ n<\omega=[J_\rho]_i & \text{if } i=g\rho+2,\\ 0=[J_\rho]_i & \text{if } i>g\rho+2\end{cases}$$

and the assertion holds.

3. Estimates of the Order of Terms

By the *order* of a term a^τ we mean the ordinal term $[a^\tau]_o$. We shall prove that b^τ has a smaller order than a^τ if a^τ reduces to b^τ in one step.

$[a^\tau]<[b^\tau]$ means that $[a^\tau]_o<[b^\tau]_o$ and $[a^\tau]_i\leqslant[b^\tau]_i$ for all $i\leqslant g\tau$.

Lemma 4. $[a^\tau]<[K_{\sigma\tau}a^\tau b^\sigma]$.

Proof. Since $[K_{\sigma\tau}]_o=1$ we have

$$[a^\tau]_o<2^{[K_{\sigma\tau}a^\tau]_1}\times([K_{\sigma\tau}]_o\#[a^\tau]_o)=[K_{\sigma\tau}a^\tau]_o.$$

By Lemma 2(2) $[a^\tau]_i\leqslant[K_{\sigma\tau}a^\tau]_i$ for all $i\leqslant g\tau$.
By Lemma 2(1) $[K_{\sigma\tau}a^\tau]_i\leqslant[K_{\sigma\tau}a^\tau b^\sigma]_i$ for all $i\in\mathbb{N}$.
It follows that $[a^\tau]_o<[K_{\sigma\tau}a^\tau b^\sigma]_o$ and $[a^\tau]_i\leqslant[K_{\sigma\tau}a^\tau b^\sigma]_i$ for all $i\leqslant g\tau$.

Lemma 5. $[a^{\rho\sigma\tau}c^\rho(b^{\rho\sigma}c^\rho)]<[S_{\rho\sigma\tau}a^{\rho\sigma\tau}b^{\rho\sigma}c^\rho]$.

Proof. For brevity we put $a_i:=[a^{\rho\sigma\tau}]_i$, $b_i:=[b^{\rho\sigma}]_i$ and $c_i:=[c^\rho]_i$. From the definition and Lemma 3 we have

$$\xi_i:=[a^{\rho\sigma\tau}c^\rho]_i\leqslant\begin{cases}2^{\xi_{i+1}}\times(a_i\#c_i) & \text{if } i\leqslant g\rho,\\ a_i & \text{if } i>g\rho\end{cases}$$

$$\eta_i:=[b^{\rho\sigma}c^\rho]_i\leqslant\begin{cases}2^{\eta_{i+1}}\times(b_i\#c_i) & \text{if } i\leqslant g\rho,\\ b_i & \text{if } i>g\rho\end{cases}$$

$$\zeta_i:=[a^{\rho\sigma\tau}c^\rho(b^{\rho\sigma}c^\rho)]_i=\begin{cases}2^{\zeta_{i+1}}\times(\xi_i\#\eta_i) & \text{if } i\leqslant g\sigma,\\ \xi_i & \text{if } i>g\sigma.\end{cases}$$

$S_{\rho\sigma\tau}$ has type $(\rho\sigma\tau)(\rho\sigma)\rho\tau$ which has degree

$$\max(g\rho+2,g\sigma+2,g\tau+1).$$

$a^{\rho\sigma\tau}$ has type $\rho\sigma\tau$ which has degree

$$\max(g\rho+1,g\sigma+1,g\tau).$$

$b^{\rho\sigma}$ has type $\rho\sigma$ which has degree

$$\max(g\rho+1, g\sigma).$$

By definition we therefore have

$$\alpha_i := [S_{\rho\sigma\tau}a^{\rho\sigma\tau}]_i = \begin{cases} 2^{\alpha_{i+1}} \times (1 \# a_i) & \text{if } i \leqslant \max(g\rho+1, g\sigma+1, g\tau) \\ 1 & \text{if } i = \max(g\rho+2, g\sigma+2, g\tau+1) \\ 0 & \text{if } i > \max(g\rho+2, g\sigma+2, g\tau+1) \end{cases}$$

$$\beta_i := [S_{\rho\sigma\tau}a^{\rho\sigma\tau}b^{\rho\sigma}]_i = \begin{cases} 2^{\beta_{i+1}} \times (\alpha_1 \# b_i) & \text{if } i \leqslant \max(g\rho+1, g\sigma) \\ \alpha_i & \text{if } i > \max(g\rho+1, g\sigma) \end{cases}$$

$$\gamma_i := [S_{\rho\sigma\tau}a^{\rho\sigma\tau}b^{\rho\sigma}c^{\rho}]_i = \begin{cases} 2^{\gamma_{i+1}} \times (\beta_i \# c_i) & \text{if } i \leqslant g\rho \\ \beta_i & \text{if } i > g\rho. \end{cases}$$

(1) *If* $\max(g\rho+1, g\sigma) < i \leqslant \max(g\rho+1, g\sigma+1, g\tau)$ *then* $\xi_i < \gamma_i$.

Proof. Here $\zeta_i = \xi_i \leqslant a_i$ and $\gamma_i = \beta_i = \alpha_i = 2^{\alpha_{i+1}} \times (1 \# a_i) > a_i$, and therefore $\zeta_i < \gamma_i$.

(2) *If* $g\rho < i \leqslant \max(g\rho+1, g\sigma)$ *then*

$$\xi_i \# \eta_i \# \zeta_i \# 2 \leqslant 2^{\zeta_{i+1} \# 1} \times (1 \# a_i \# b_i)$$

and

$$2^{\gamma_{i+1}} \times (1 \# a_i \# b_i) \leqslant \gamma_i.$$

Proof. Here $\xi_i \leqslant a_i$, $\eta_i \leqslant b_i$ and $\zeta_i \leqslant 2^{\zeta_{i+1}} \times (\xi_i \# \eta_i)$ and so

$$\xi_i \# \eta_i \# \zeta_i \# 2 \leqslant 1 \# a_i \# b_i \# (2^{\zeta_{i+1}} \times (1 \# a_i \# b_i)) \leqslant 2^{\zeta_{i+1} \# 1} \times (1 \# a_i \# b_i).$$

Further

$$\gamma_i = \beta_i = 2^{\beta_{i+1}} \times (\alpha_i \# b_i) = 2^{\gamma_{i+1}} \times (\alpha_i \# b_i) \quad \text{and} \quad \alpha_i = 2^{\alpha_{i+1}} \times (1 \# a_i) \geqslant 1 \# a_i$$

and so $2^{\gamma_{i+1}} \times (1 \# a_i \# b_i) \leqslant \gamma_i$.

(3) *If* $i \leqslant \max(g\rho+1, g\sigma)$ *then* $\xi_i \# \eta_i \# \zeta_i \# 2 \leqslant \gamma_i$.

Proof by induction on $\max(g\rho+1, g\sigma) - i$.

1. Suppose $g\rho+1 \leqslant g\sigma = i$.

Then by (1) $\zeta_{i+1} < \gamma_{i+1}$ and so $\zeta_{i+1} \# 1 \leqslant \gamma_{i+1}$ and $2^{\zeta_{i+1} \# 1} \times (1 \# a_i \# b_i) \leqslant 2^{\gamma_{i+1}} \times (1 \# a_i \# b_i)$ where the assertion follows by (2).

2. Suppose $g\sigma < g\rho+1 = i$.

Then $\alpha_{i+1} \neq 0$, $\zeta_i = \xi_i \leqslant a_i$ and $\eta_i \leqslant b_i$ and so

$$\begin{aligned} \xi_i \# \eta_i \# \zeta_i \# 2 &\leqslant (2^1 \times (1 \# a_i)) \# b_i \leqslant (2^{\alpha_{i+1}} \times (1 \# a_i)) \# b_i \\ &= \alpha_i \# b_i \leqslant 2^{\beta_{i+1}} \times (\alpha_i \# b_i) = \beta_i = \gamma_i. \end{aligned}$$

3. Suppose $g\rho < i < \max(g\rho+1, g\sigma)$.
Then by induction hypothesis $\xi_{i+1} \# \eta_{i+1} \# \zeta_{i+1} \# 2 \leqslant \gamma_{i+1}$ and so $\zeta_{i+1} \# 1 < \gamma_{i+1}$ whence the assertion follows by (2).

4. Suppose $i \leqslant g\rho$.
Then $\xi_i \leqslant 2^{\xi_{i+1}} \times (a_i \# c_i), \eta_i \leqslant 2^{\eta_{i+1}} \times (b_i \# c_i), \zeta_i \leqslant 2^{\zeta_{i+1}} \times (\xi_i \# \eta_i), \alpha_i = 2^{\alpha_{i+1}} \times (1 \# a_i)$, $\beta_i = 2^{\beta_{i+1}} \times (\alpha_i \# b_i)$ and $\gamma_i = 2^{\gamma_{i+1}} \times (\beta_i \# c_i)$. It follows that

$$\xi_i \# \eta_i \# \zeta_i \# 2 \leqslant 2^{\zeta_{i+1} \# 1} \times (\xi_i \# \eta_i \# 1) \leqslant 2^{\xi_{i+1} \# \eta_{i+1} \# \zeta_{i+1} \# 2} \times (a_i \# b_i \# c_i)$$

and

$$2^{\gamma_{i+1}} \times (a_i \# b_i \# c_i) < 2^{\gamma_{i+1}} \times (\alpha_i \# b_i \# c_i) \leqslant 2^{\gamma_{i+1}} \times (\beta_i \# c_i) = \gamma_i.$$

The assertion now follows on using the induction hypothesis $\xi_{i+1} \# \eta_{i+1} \# \zeta_{i+1} \# 2 \leqslant \gamma_{i+1}$.

It follows from (1) and (3) that $\xi_i < \gamma_i$ for all $i \leqslant g\tau$ and therefore $[a^{\rho\sigma\tau} c^{\rho}(b^{\rho\sigma} c^{\rho})] < [S_{\rho\sigma\tau} a^{\rho\sigma\tau} b^{\rho\sigma} c^{\rho}]$.

Lemma 6. $[b^{\tau}] < [J_{\tau} 0 a^{\tau\tau} b^{\tau}]$.

Proof. By Lemma 2(1) $[J_{\tau} 0 a^{\tau\tau}]_o \geqslant [J_{\tau} O]_o = 1$. It follows that

$$[J_{\tau} 0 a^{\tau\tau} b^{\tau}]_o = 2^{[J_{\tau} K a^{\tau\tau} b^{\tau}]_1} \times ([J_{\tau} O a^{\tau\tau}]_o \# [b^{\tau}]_o) > [b^{\tau}]_o.$$

By Lemma 2(2) if $i \leqslant g\tau$ then $[b^{\tau}]_i \leqslant [J_{\tau} 0 a^{\tau\tau} b\tau]_i$. This yields the assertion.

Lemma 7. $[a^{\tau\tau}(J_{\tau} N_n a^{\tau\tau} b^{\tau})] < [J_{\tau} N_{n+1} a^{\tau\tau} b^{\tau}]$.

Proof. For brevity we write $a_i := [a^{\tau\tau}]_i$ and $b_i := [b^{\tau}]_i$. By definition and Lemma 3 we have

$$\xi_i := [J_{\tau} N_n a^{\tau\tau}]_i = \begin{cases} 2^{\xi_{i+1}} \times (1 \# a_i) & \text{if } i \leqslant g\tau + 1, \\ n & \text{if } i = g\tau + 2, \\ 0 & \text{if } i > g\tau + 2 \end{cases}$$

$$\eta_i := [J_{\tau} N_n a^{\tau\tau} b^{\tau}]_i = \begin{cases} 2^{\eta_{i+1}} \times (\xi_i \# b_i) & \text{if } i \leqslant g\tau, \\ \xi_i & \text{if } i > g\tau \end{cases}$$

$$\zeta_i := [a^{\tau\tau}(J_{\tau} N_n a^{\tau\tau} b^{\tau})]_i \leqslant \begin{cases} 2^{\zeta_{i+1}} \times (a_i \# \eta_i) & \text{if } i \leqslant g\tau \\ a_i & \text{if } i > g\tau \end{cases}$$

$$\alpha_i := [J_{\tau} N_{n+1} a^{\tau\tau}]_i = \begin{cases} 2^{\alpha_{i+1}} \times (1 \# a_i) & \text{if } i \leqslant g\tau + 1 \\ n+1 & \text{if } i = g\tau + 2 \\ 0 & \text{if } i > g\tau + 2 \end{cases}$$

$$\beta_i := [J_{\tau} N_{n+1} a^{\tau\tau} b^{\tau}]_i = \begin{cases} 2^{\beta_{i+1}} \times (\alpha_i \# b_i) & \text{if } i \leqslant g\tau, \\ \alpha_i & \text{if } i > g\tau. \end{cases}$$

We prove $\xi_i < \alpha_i$ and $\eta_i \# \zeta_i \# 1 \leqslant \beta_i$ for $i \leqslant g\tau + 1$ by induction on $g\tau + 1 - i$. From this it follows that, in particular, $\zeta_i < \beta_i$ for all $i \leqslant g\tau$ which yields the assertion.

1. Suppose $i = g\tau + 1$, then $\xi_i = 2^n \times (1 \# a_i) < 2^{n+1} \times (1 \# a_i) = \alpha_i$ and $\eta_i \# \zeta_i \# 1 = \xi_i \# a_i \# 1 = 2^n \times (1 \# a_i) \# a_i \# 1 \leqslant 2^{n+1} \times (1 \# a_i) = \alpha_i = \beta_i$.

2. Suppose $i \leqslant g\tau$. Then by the induction hypothesis $\xi_{i+1} < \alpha_{i+1}$, $\xi_{i+1} \# 1 \leqslant \alpha_{i+1}$ and $\eta_{i+1} \# \zeta_{i+1} \# 1 \leqslant \beta_{i+1}$. It follows that

$$\xi_i = 2^{\xi_{i+1}} \times (1 \# a_i) < 2^{\alpha_{i+1}} \times (1 \# a_i) = \alpha_i$$

and

$$\begin{aligned} \eta_i \# \zeta_i &\leqslant \eta_i \# (2^{\zeta_{i+1}} \times (a_i \# \eta_i)) \leqslant (2^{\zeta_{i+1}} \times a_i) \# (2^{\zeta_{i+1} \# 1} \times \eta_i) \\ &= (2^{\zeta_{i+1}} \times a_i) \# (2^{\eta_{i+1} \# \zeta_{i+1} \# 1} \times (\xi_i \# b_i)) \\ &= (2^{\zeta_{i+1}} \times a_i) \# (2^{\xi_{i+1} \# \eta_{i+1} \# \zeta_{i+1} \# 1} \times (1 \# a_i)) \# (2^{\eta_{i+1} \# \zeta_{i+1} \# 1} \times b_i). \end{aligned}$$

Therefore

$$\eta_i \# \zeta_i \# 1 \leqslant (2^{\xi_{i+1} \# \eta_{i+1} \# \zeta_{i+1} \# 2} \times (1 \# a_i)) \# (2^{\eta_{i+1} \# \zeta_{i+1} \# 1} \times b_i)$$

and by the induction hypotheses

$$\begin{aligned} \eta_i \# \zeta_i \# 1 &\leqslant (2^{\alpha_{i+1} \# \beta_{i+1}} \times (1 \# a_i)) \# (2^{\beta_{i+1}} \times b_i) \\ &= 2^{\beta_{i+1}} \times ((2^{\alpha_{i+1}} \times (1 \# a_i)) \# b_i) \\ &= 2^{\beta_{i+1}} \times (\alpha_i \# b_i) = \beta_i. \end{aligned}$$

Lemma 8. *$[b^\sigma] < [c^\sigma]$ implies $[a^{\sigma\tau} b^\sigma] < [a^{\sigma\tau} c^\sigma]$.*

Proof. By Lemma 1, $[b^\sigma] < [c^\sigma]$ implies that c^σ is not a numeral. Therefore $a^{\sigma\tau} c^\sigma$ is not of the form $J_\rho N_n$. It follows from the definition and Lemma 3 that

$$[a^{\sigma\tau} b^\sigma]_i \leqslant \begin{cases} 2^{[a^{\sigma\tau} b^\sigma]_{i+1}} \times ([a^{\sigma\tau}]_i \# [b^\sigma]_i) & \text{if } i \leqslant g\sigma \\ [a^{\sigma\tau}]_i & \text{if } i > g\sigma \end{cases}$$

$$[a^{\sigma\tau} c^\sigma]_i = \begin{cases} 2^{[a^{\sigma\tau} c^\sigma]_{i+1}} \times ([a^{\sigma\tau}]_i \# [c^\sigma]_i) & \text{if } i \leqslant g\sigma \\ [a^{\sigma\tau}]_i & \text{if } i > g\sigma. \end{cases}$$

By hypothesis $[b^\sigma]_i \leqslant [c^\sigma]_i$ for all $i \leqslant g\sigma$. It follows by induction on $g\sigma - i$ that $[a^{\sigma\tau} b^\sigma]_i \leqslant [a^{\sigma\tau} c^\sigma]_i$ for all $i \leqslant g\sigma$. Hence $[a^{\sigma\tau} b^\sigma]_i \leqslant [a^{\sigma\tau} c^\sigma]_i$ for all $i \in \mathbb{N}$. Using the hypothesis $[b^\sigma]_o < [c^\sigma]_o$ it also follows that $[a^{\sigma\tau} b^\sigma]_o < [a^{\sigma\tau} c^\sigma]_o$. This yields the assertion.

Lemma 9. *If $b^{\sigma\tau}$ is not an iterator then $[a^{\sigma\tau}] < [b^{\sigma\tau}]$ implies $[a^{\sigma\tau} c^\sigma] < [b^{\sigma\tau} c^\sigma]$.*

Proof. In this case by the definition and Lemma 3 we have

$$[a^{\sigma\tau} c^\sigma]_i \leqslant \begin{cases} 2^{[a^{\sigma\tau} c^\sigma]_{i+1}} \times ([a^{\sigma\tau}]_i \# [c^\sigma]_i) & \text{if } i \leqslant g\sigma, \\ [a^{\sigma\tau}]_i & \text{if } i > g\sigma \end{cases}$$

$$[b^{\sigma\tau} c^\sigma]_i = \begin{cases} 2^{[b^{\sigma\tau} c^\sigma]_{i+1}} \times ([b^{\sigma\tau}]_i \# [c^\sigma]_i) & \text{if } i \leqslant g\sigma \\ [b^{\sigma\tau}]_i & \text{if } i > g\sigma. \end{cases}$$

By hypothesis $[a^{\sigma\tau}]_i \leqslant [b^{\sigma\tau}]_i$ for all $i \leqslant \max(g\sigma+1, g\tau)$. By induction on i it follows that $[a^{\sigma\tau}c^\sigma]_i \leqslant [b^{\sigma\tau}c^\sigma]_i$ for all $i \leqslant g\sigma$. Hence $[a^{\sigma\tau}c^\sigma]_i \leqslant [b^{\sigma\tau}c^\sigma]_i$ for all $i \leqslant g\tau$. Using the hypothesis $[a^{\sigma\tau}]_o < [b^{\sigma\tau}]_o$ it also follows that $[a^{\sigma\tau}c^\sigma]_o < [b^{\sigma\tau}c^\sigma]_o$.

Theorem 16.1. *If* $a_1^{\tau_o} \rhd^1 a_2^{\tau_o}$ *then* $a_2^{\tau_o}$ *has a smaller order than* $a_1^{\tau_o}$.

Proof. Suppose $a_1^{\tau_o} \rhd^1 a_2^{\tau_o}$. We prove $[a_2^{\tau_o}] < [a_1^{\tau_o}]$ by induction on the length of the term $a_1^{\tau_o}$. This will yield the assertion.

There are two cases to consider when $a_1^{\tau_o} \rhd^1 a_2^{\tau_o}$:

1. For $i=1, 2$, $a_i^{\tau_o}$ is a term $b_i^\tau d_1^{\tau_1}, \ldots, d_m^{\tau_m}$ where $\tau = \tau_1, \ldots, \tau_m \tau_o$ $(m \geqslant 0)$. This gives the following four subcases for b_1^τ, b_2^τ:

1.1. b_1^τ is $K_{\sigma\tau} a^\tau b^\sigma$ and b_2^τ is a^τ,

1.2. b_1^τ is $S_{\rho\sigma\tau} a^{\rho\sigma\tau} b^{\rho\sigma} c^\rho$, and b_2^τ is $a^{\rho\sigma\tau} c^\rho (b^{\rho\sigma} c^\rho)$,

1.3. b_1^τ is $J_\tau 0 a^{\tau\tau} b^\tau$ and b_2^τ is b^τ,

1.4. b_1^τ is $J_\tau N_{n+1} a^{\tau\tau} b^\tau$ and b_2^τ is $a^{\tau\tau}(J_\tau N_n a^{\tau\tau} b^\tau)$.

2. For $i=1, 2$, $a_i^{\tau_o}$ is a term $a^{\sigma\tau} b_i^\sigma d_1^{\tau_1}, \ldots, d_m^{\tau_m}$ where $\tau = \tau_1 \ldots \tau_n \tau_o$ $(n \geqslant 0)$ where $a^{\sigma\tau}$ is irreducible and $b_1^\sigma \not\rhd^1 b_2^\sigma$.

In the first case $[b_2^\tau] < [b_1^\tau]$ holds by Lemmata 4–7. Then by Lemma 9 it follows that $[a_2^{\tau_o}] < [a_1^{\tau_o}]$. In the second case $[b_2^\sigma] < [b_1^\sigma]$ by the induction hypothesis. Then $[\alpha^{\sigma\tau} b_2^\sigma] < [a^{\sigma\tau} b_1^\sigma]$ by Lemma 8 and $[a_2^{\tau_o}] < [a_1^{\tau_o}]$ by Lemma 9.

Theorem 16.2. *Every term has a normal form.*

Proof. This follows at once from Theorem 16.1 by induction on the ordering relation of the ordinal terms $< \varepsilon_o$, in other words by transfinite induction up to ε_o.

4. Equality of Terms

Definition. If a^σ, b^τ are terms then $a^\sigma = b^\tau$ means that a^σ and b^τ have the same normal form.

Obviously this equality relation is an equivalence relation such that

$$a^\sigma = b^\tau \Rightarrow \sigma = \tau$$
$$a^\tau \rhd b^\tau \Rightarrow a^\tau = b^\tau.$$

Theorem 16.3. $a^\tau = b^\tau$ *implies* $c^\sigma \begin{pmatrix} a^\tau \\ x^\tau \end{pmatrix} = c^\sigma \begin{pmatrix} b^\tau \\ x^\tau \end{pmatrix}$.

Proof. By hypothesis a^τ and b^τ have the same normal form d^τ. It is therefore sufficient to prove:

If $a^\tau \rhd^n d^\tau$ *then* $c^\sigma \begin{pmatrix} a^\tau \\ x^\tau \end{pmatrix} = c^\sigma \begin{pmatrix} d^\tau \\ x^\tau \end{pmatrix}$.

We prove this assertion by induction on n.

1. Suppose $n=0$. Then a^τ and d^τ are identical and the assertion is trivial.
2. Suppose $n>0$. Then there exists a_1^τ such that

$$a^\tau \rhd^1 a_1^\tau \rhd^{n-1} d^\tau.$$

By Theorem 16.2 $c^\sigma\binom{a^\tau}{x^\tau}$ has a normal form e^σ. Therefore there exists $m\in\mathbb{N}$ such that

$$c^\sigma\binom{a^\tau}{x^\tau}\rhd^m e^\sigma.$$

By Theorem 15.2 it follows that

$$c\binom{a_1^\tau}{x^\tau}\rhd e^\sigma.$$

Therefore $c^\sigma\binom{a^\tau}{x^\tau}=c^\sigma\binom{a_1^\tau}{x^\tau}$. By the induction hypothesis $c^\sigma\binom{a_1^\tau}{x^\tau}=c^\sigma\binom{d^\tau}{x^\tau}$. Hence we have the assertion $c^\sigma\binom{a^\tau}{x^\tau}=c^\sigma\binom{d^\tau}{x^\tau}$.

Remark. It follows from 16.3 that equality for terms is compatible with arbitrary reductions. That is: In order to reduce a term we do not need to use the uniquely determined reduction procedure given above but can carry out the reduction of a term a^τ at any place in a^τ without altering the normal form.

§17. The Formal System FT of Functionals of Finite Type

1. The Formal Language

Inductive Definition of the *formulas* of the system FT:

1. If a^τ and b^τ are terms of the same type then $a^\tau=b^\tau$ is a *formula*.
2. If A and B are *formulas*, then $(A\rightarrow B)$ is also a *formula*.

We use $A, B, C, D, E, F, G, \mathscr{F}[a^\tau], \mathscr{G}[a^\tau]$ (possibly with subscripts) as *syntactic variables* for formulas where $\mathscr{F}$ and $\mathscr{G}$ are the corresponding nominal forms.

We use Γ, Γ_i, Δ, Δ_i to denote finite (possibly empty) sequences of formulas. As in §6 we use the abbreviation:

$$A_1\rightarrow\cdots\rightarrow A_n\rightarrow B$$

for

$$(A_1\rightarrow(\cdots\rightarrow(A_n\rightarrow B)\ldots)).$$

If Γ is the sequence $A_1, \ldots, A_n$ $(n \geqslant 1)$ then $\Gamma \to B$ denotes the formula

$$A_1 \to \cdots \to A_n \to B.$$

If Γ is empty then $\Gamma \to B$ denotes the formula B.

$\Gamma \subset \Delta$ means that every formula in the sequence Γ is also a formula in the sequence Δ.

Inductive Definition of the *zero functional* 0^τ of type τ.

1. $0^o := 0$
2. $0^{\sigma\tau} := K_{\sigma\tau} 0^\tau$

2. Deduction Procedures

The *axioms* of the system FT are

(Ax. K) $K_{\sigma\tau} a^\tau b^\sigma = a^\tau$
(Ax. S) $S_{\rho\sigma\tau} a^{\rho\sigma\tau} b^{\rho\sigma} c^\rho = a^{\rho\sigma\tau} c^\rho (b^{\rho\sigma} c^\rho)$
(Ax. J0) $J_\tau 0 b^{\tau\tau} c^\tau = c^\tau$
(Ax. $J\mathscr{S}$) $J_\tau(\mathscr{S} a^o) b^{\tau\tau} c^\tau = b^{\tau\tau}(J_\tau a^o b^{\tau\tau} c^\tau)$
(Ax. =) $a^\tau = b^\tau \to \mathscr{F}[a^\tau] \to \mathscr{F}[b^\tau]$.

The *basic inferences* of the system FT are:

(Str) $\Gamma \to B \vdash \Delta \to B$, if $\Gamma \subset \Delta$.
(Cut) $\Gamma \to A$, $A \to B \vdash \Gamma \to B$
(CI) $\mathscr{F}[0]$, $\mathscr{F}[x^o] \to \mathscr{F}[\mathscr{S} x^o] \vdash \mathscr{F}[a^o]$,
provided the variable x^o does not occur in the nominal form $\mathscr{F}$.

The basic inferences (Str) and (Cut) are the *structural inference* and the *cut* corresponding to those in §6. (CI) is the inference rule of *complete induction.*

Inductive Definition of $\models^{\underline{n}} B$ ("*B is deducible with order n*") for $n \in \mathbb{N}$:

1. If B is an axiom then $\models^{\underline{0}} B$.
2. If $A \vdash B$ is a basic inference (Str) and $\models^{\underline{n}} A$ then $\models^{\underline{n+1}} B$.
3. If $A_1, A_2 \vdash B$ is a basic inference (Cut) or (CI) and $\models^{\underline{n_i}} A_i$ $(i=1,2)$ then $\models^{\underline{n}} B$ where $n := \max(n_1, n_2) + 1$.

$\vdash B$ ("*B is deducible in the system* FT") means that there exists $n \in \mathbb{N}$ such that $\models^{\underline{n}} B$.

Theorem 17.1 (Substitution rule). *If* $\models^{\underline{n}} \mathscr{F}[x^\tau]$, *where the variable* x^τ *does not occur in the nominal form* $\mathscr{F}$, *then* $\models^{\underline{n}} \mathscr{F}[a^\tau]$ *for every term* a^τ *of type* τ.

Proof by induction on n.

1. Suppose $\mathscr{F}[x^\tau]$ is an axiom. Then $\mathscr{F}[a^\tau]$ is also an axiom.
2. Suppose $\models^{\underline{n}} \mathscr{F}[x^\tau]$ was obtained by a basic inference (Str) or (Cut). Then by

the induction hypothesis the formulas obtained from the premisses by substituting a^τ for x^τ are deducible with the same orders as the premisses. Using the appropriate basic inference (Str) or (Cut) we obtain $\vdash^{n} \mathscr{F}[a^\tau]$.

3. Suppose $\vdash^{n} \mathscr{F}[x^\tau]$ was obtained by a basic inference (CI). Then the formula $\mathscr{F}[x^\tau]$ is a formula $\mathscr{G}[b^o]$ and we have

$$\vdash^{n_1} \mathscr{G}[0] \quad \text{and} \quad \vdash^{n_2} \mathscr{G}[y^o] \rightarrow \mathscr{G}[\mathscr{S}y^o]$$

where $n = \max(n_1, n_2) + 1$ and where y^o does not occur in $\mathscr{G}$. We choose a variable z^o different from x^τ which occurs in neither $\mathscr{G}$ nor in a^τ. By I.H. (induction hypothesis) it follows that

$$\vdash^{n_2} \mathscr{G}[z^o] \rightarrow \mathscr{G}[\mathscr{S}z^o].$$

Let $\mathscr{G}^*$ be the nominal form obtained from $\mathscr{G}$ by substituting a^τ for x^τ. By I.H. it follows that

$$\vdash^{n_1} \mathscr{G}^*[0] \qquad \text{and} \quad \vdash^{n_2} \mathscr{G}*[z^o] \rightarrow \mathscr{G}*[\mathscr{S}z^o].$$

Using a (CI)-inference it follows that

$$\vdash^{n} \mathscr{G}^*\left[b^o\begin{pmatrix} a^\tau \\ x^\tau \end{pmatrix}\right],$$

but $\mathscr{G}^*\left[b^o\begin{pmatrix} a^\tau \\ x^\tau \end{pmatrix}\right]$ is the formula $\mathscr{F}[a^\tau]$.

3. The Consistency of the System FT

Definition. A formula is said to be *closed* if it contains no variable.

Inductive Definition of *truth* and *falsity* for a closed formula.

1. A closed equation $a^\tau = b^\tau$ is *true* if the functionals a^τ and b^τ have the same normal form. Otherwise the equation is *false*.

2. A closed formula $A \rightarrow B$ is *true* if A is false or B is true. Otherwise the formula is *false*.

Lemma 1. *If two functionals a^τ and b^τ have the same normal form then a closed formula $\mathscr{F}[a^\tau]$ is true if, and only if, the formula $\mathscr{F}[b^\tau]$ is true.*

Proof by induction on the number of symbols $\rightarrow$ occurring in $\mathscr{F}[a^\tau]$.

1. Suppose $\mathscr{F}[a^\tau]$ is an equation. Then there are terms c^σ, d^σ and a variable x^τ such that $\mathscr{F}[a^\tau]$ is the equation

$$(1) \quad c^\sigma\begin{pmatrix} a^\tau \\ x^\tau \end{pmatrix} = d^\sigma\begin{pmatrix} a^\tau \\ x^\tau \end{pmatrix}$$

and $\mathscr{F}[b^\tau]$ is the equation

$$(2)\quad c^\sigma\begin{pmatrix}b^\tau\\x^\tau\end{pmatrix}=d^\sigma\begin{pmatrix}b^\tau\\x^\tau\end{pmatrix}.$$

By Theorem 16.3 the functionals

$$c^\sigma\begin{pmatrix}a^\tau\\x^\tau\end{pmatrix},\quad d^\sigma\begin{pmatrix}a^\tau\\x^\tau\end{pmatrix}$$

have the same normal forms as the functionals

$$c^\sigma\begin{pmatrix}b^\tau\\x^\tau\end{pmatrix},\quad d^\sigma\begin{pmatrix}b^\tau\\x^\tau\end{pmatrix}.$$

Therefore (1) is true if, and only if, (2) is true.

2. Suppose $\mathscr{F}[a^\tau]$ contains the symbol $\rightarrow$. Then the assertion follows from the induction hypothesis and the truth definition.

Theorem 17.2. *Every deducible closed formula is true.*

Proof. Suppose C is a closed formula and $\models^n C$. We prove by induction on n that C is true.

1. Suppose C is an axiom (Ax. K), (Ax. S) or (Ax. J0). Then C is an equation $c^\tau=d^\tau$ for which $c^\tau \rhd^1 d^\tau$ holds. It follows that C is true.

2. Suppose C is an axiom (Ax. J$\mathscr{S}$):

$$J_\tau(\mathscr{S}a^0)b^{\tau\tau}c^\tau=b^{\tau\tau}(J_\tau a^0 b^{\tau\tau}c^\tau).$$

By Theorem 15.1 the normal form of a^0 is a numeral N_m. Then by the reduction procedures the equation

$$J_\tau(\mathscr{S}N_m)b^{\tau\tau}c^\tau=b^{\tau\tau}(J_\tau N_m b^{\tau\tau}c^\tau)$$

is true. Hence by Lemma 1 C is true.

3. Suppose C is an axiom (Ax. =):

$$a^\tau=b^\tau\rightarrow\mathscr{F}[a^\tau]\rightarrow\mathscr{F}[b^\tau].$$

If $a^\tau=b^\tau$ is false, then C is true. If $a^\tau=b^\tau$ is true then by Lemma 1 it follows that $\mathscr{F}[a^\tau]$ is true if, and only if, $\mathscr{F}[b^\tau]$ is true. Thus in every case C is true.

4. Suppose $\models^n C$ was obtained by a basic inference (Str). The premise of this inference is a closed formula which is true by I.H. It follows that C is true by 2. of the definition of truth.

5. Suppose $\vdash^{n} C$ was obtained by a basic inference (Cut). Then C is a formula $\Gamma \to B$ and we have

$$\vdash^{n_1} \Gamma \to A \quad \text{and} \quad \vdash^{n_2} A \to B$$

where $n = \max(n_1, n_2) + 1$. Let A^* be the formula obtained from A by replacing each variable x^τ occurring in A by the zero functional of the same type. Then by Theorem 17.1 it follows that

$$\vdash^{n_1} \Gamma \to A^* \quad \text{and} \quad \vdash^{n_2} A^* \to B.$$

$\Gamma \to A^*$ and $A^* \to B$ are closed formulas and therefore are true by I.H. It follows that $\Gamma \to B$ is true.

6. Suppose $\vdash^{n} C$ was obtained by a basic inference (CI). Then C is a formula $\mathscr{F}[a^o]$ and we have

$$\vdash^{n_1} \mathscr{F}[0] \quad \text{and} \quad \vdash^{n_2} \mathscr{F}[x^o] \to \mathscr{F}[\mathscr{S} x^o]$$

where $n = \max(n_1, n_2) + 1$. By Theorem 17.1 it follows that for all $m \in \mathbb{N}$

$$\vdash^{n_2} \mathscr{F}[N_m] \to \mathscr{F}[N_{m+1}].$$

By I.H. the closed formulas $\mathscr{F}[0]$ and $\mathscr{F}[N_m] \to \mathscr{F}[N_{m+1}]$ are true. It follows by induction on m that $\mathscr{F}[N_m]$ is true for every numeral N_m. Now the normal form of a^o is a numeral by Theorem 15.1. So by Lemma 1 it follows that $\mathscr{F}[a^o]$ is true.

Remark. The converse of Theorem 17.2 does not hold in general. Thus, for example,

$$J_o(\mathscr{S}0)\mathscr{S} = \mathscr{S} \to \mathscr{S}0 = 0$$

is a true closed formula which is not deducible in FT. However, every true closed equation is deducible in FT. This follows from the fact that every equation $a^\tau = b^\tau$ such that $a^\tau \rhd^1 b^\tau$ holds is deducible in FT.

Corollary to Theorem 17.2. *The formal system* FT *is syntactically consistent.* That is: Not every formula of the system FT is deducible in FT.

Proof. The false closed equation $\mathscr{S}0 = 0$ is not deducible in FT by Theorem 17.2.

Remark. Besides elementary methods our consistency proof for the system FT uses only Theorem 16.1 which we proved by transfinite induction up to ε_o. (Theorem 16.1 was used to prove Theorem 16.3 and hence also Lemma 1 which we used to prove Theorem 17.2.)

4. Fundamental Deduction Rules

Theorem 17.3. $\vdash A \to A$.

Proof. $K_{oo}00=0$
and

$$K_{oo}00=0 \rightarrow A \rightarrow A$$

are (Ax. K) and (Ax. =) axioms (where in the (Ax. =) axiom $\mathscr{F}$ is a 1-place nominal form A in which the nominal sign $*_1$ does not occur). Using a cut we have $\vdash A \rightarrow A$.

The next theorem shows that equality is an equivalence relation which is compatible with the mapping of terms and formulas.

Theorem 17.4 (Rules for equality).

(=1) $\vdash \mathscr{F}[a^\tau]$ *implies* $\vdash a^\tau = b^\tau \rightarrow \mathscr{F}[b^\tau]$
(=2) $\vdash a^\tau = a^\tau$
(=3) $\vdash a^\tau = b^\tau \rightarrow b^\tau = a^\tau$
(=4) $\vdash a^\tau = b^\tau \rightarrow b^\tau = c^\tau \rightarrow a^\tau = c^\tau$
(=5) $\vdash a^\tau = b^\tau \rightarrow c^\sigma\begin{pmatrix} a^\tau \\ x^\tau \end{pmatrix} = c^\sigma\begin{pmatrix} b^\tau \\ x^\tau \end{pmatrix}$.

Proof. 1. Using a structural inference on the (Ax. =)

$$a^\tau = b^\tau \rightarrow \mathscr{F}[a^\tau] \rightarrow \mathscr{F}[b^\tau]$$

we obtain

$$\mathscr{F}[a^\tau] \rightarrow a^\tau = b^\tau \rightarrow \mathscr{F}[b^\tau].$$

Then using $\vdash \mathscr{F}[a^\tau]$ by a cut we have

$$\vdash a^\tau = b^\tau \rightarrow \mathscr{F}[b^\tau].$$

2. By (=1) from the (Ax. K)

$$K_{o\tau}a^\tau 0 = a^\tau$$

we obtain

$$K_{o\tau}a^\tau 0 = a^\tau \rightarrow a^\tau = a^\tau.$$

Using a cut we have $\vdash a^\tau = a^\tau$.
3. (=3) follows from (=2) by (=1).
4. Using a structural inference on the (Ax. =)

$$b^\tau = c^\tau \rightarrow a^\tau = b^\tau \rightarrow a^\tau = c^\tau$$

we obtain (=4).

5. By (=2) we have

$$\vdash c^{\sigma}\binom{a^{\tau}}{x^{\tau}} = c^{\sigma}\binom{a^{\tau}}{x^{\tau}}.$$

Hence by (=1) we obtain (=5).

Theorem 17.5. *If* $\vdash \mathscr{F}[0]$ *and* $\vdash \mathscr{F}[\mathscr{S}x^o]$ *where the variable* x^o *does not occur in the nominal form* $\mathscr{F}$ *then* $\vdash \mathscr{F}[a^o]$ *for every term* a^o *of type o.*

Proof. By a structural inference from $\vdash \mathscr{F}[\mathscr{S}x^o]$ we obtain

$$\vdash \mathscr{F}[x^o] \to \mathscr{F}[\mathscr{S}x^o].$$

Now a (CI)-inference using $\vdash \mathscr{F}[0]$ yields $\vdash \mathscr{F}[a^o]$.

Corollary to Theorem 17.5. $\vdash a^o = 0 \to A$ *and* $\vdash a^o = \mathscr{S}x^o \to A$ *yield* $\vdash A$ *provided* x^o *occurs in neither* a^o *nor* A.

Proof. By Theorem 17.5 from the hypothesis we have $\vdash a^o = a^o \to A$. Hence by (=2) we have $\vdash A$.

Definition of *terms for definition by cases.*

$$D_\tau[a^o, b^\tau, c^\tau] := J_\tau a^o(K_{\tau\tau}c^\tau)b^\tau$$
$$\nu[a^o] := J_o a^o(K_{o\alpha}0)(\mathscr{S}0)$$

Corollaries. *The following equations are deducible*

(D0) $D_\tau[0, b^\tau, c^\tau] = b^\tau$
(D$\mathscr{S}$) $D_\tau[\mathscr{S}a^o, b^\tau, c^\tau] = c^\tau$
(ν0) $\nu[0] = \mathscr{S}0$
($\nu\mathscr{S}$) $\nu[\mathscr{S}a^o] = 0$.

Proof. By (Ax. J0), (Ax. J$\mathscr{S}$) and (Ax. K) we have

$$D_\tau[0, b^\tau, c^\tau] = J_\tau 0(K_{\tau\tau}c^\tau)b^\tau = b^\tau$$
$$D_\tau[\mathscr{S}a^o, b^\tau, c^\tau] = J_\tau(\mathscr{S}a^o)(K_{\tau\tau}c^\tau)b^\tau = K_{\tau\tau}c^\tau(J_\tau a^o(K_{\tau\tau}c^\tau)b^\tau) = c^\tau.$$

(ν0) and ($\nu\mathscr{S}$) are special cases of (D0) and (D$\mathscr{S}$).

Theorem 17.6. $\vdash \mathscr{S}a^o = 0 \to A$.

Proof by induction on the number of arrows which occur in the formula A.

1. Suppose A is an equation $b^\tau = c^\tau$. By (=5) we have

$$\vdash \mathscr{S}a^o = 0 \to D_\tau[\mathscr{S}a^o, c^\tau, b^\tau] = D_\tau[0, c^\tau, b^\tau].$$

Using (D$\mathscr{S}$) and (D0) we obtain

$$\vdash \mathscr{S}a^o = 0 \to b^\tau = c^\tau.$$

2. Suppose A is a formula $B \to C$. Then by I.H. we have

$$\vdash \mathscr{S}a^o = 0 \to C.$$

Using a structural inference we obtain

$$\vdash \mathscr{S}a^o = 0 \to B \to C.$$

5. Addition and Multiplication

Definitions.

$$a^o + b^o := J_o b^o \mathscr{S} a^o$$
$$a^o \cdot b^o := J_o b^o (J_o a^o \mathscr{S}) 0.$$

Corollaries. *The following equations are deducible*

$$(+0) \quad a^o + 0 = a^o$$
$$(+\mathscr{S}) \quad a^o + \mathscr{S}b^o = \mathscr{S}(a^o + b^o)$$
$$(\cdot 0) \quad a^o \cdot 0 = 0$$
$$(\cdot \mathscr{S}) \quad a^o \cdot \mathscr{S}b^o = a^o \cdot b^o + a^o.$$

Proof. By (Ax. J0) and (Ax. J$\mathscr{S}$) we have

$$a^o + 0 = J_o 0 \mathscr{S} a^o = a^o$$
$$a^o + \mathscr{S}b^o = J_o(\mathscr{S}b^o)\mathscr{S}a^o = \mathscr{S}(J_o b^o \mathscr{S} a^o) = \mathscr{S}(a^o + b^o)$$
$$a^o \cdot 0 = J_o 0 (J_o a^o \mathscr{S}) 0 = 0$$
$$a^o \cdot \mathscr{S}b^o = J_o(\mathscr{S}b^o)(J_o a^o \mathscr{S})0 = J_o a^o \mathscr{S}(J_o b^o (J_o a^o \mathscr{S})0) = a^o \cdot b^o + a^o$$

Remark. All the equations for the basic properties of addition and multiplication follow from the recursion equations $(+0)$, $(+\mathscr{S})$, $(\cdot 0)$ and $(\cdot \mathscr{S})$ by means of (CI)-inferences:

$$a^o + (b^o + c^o) = (a^o + b^o) + c^o$$
$$a^o \cdot (b^o + c^o) = a^o \cdot b^o + a^o \cdot c^o$$
$$a^o \cdot (b^o \cdot c^o) = (a^o \cdot b^o) \cdot c^o$$
$$a^o + b^o = b^o + a^o$$
$$a^o \cdot b^o = b^o \cdot a^o.$$

6. The Identity Functional I_τ and λ-Abstraction

Definition.

$$I_\tau := S_{\tau(\tau\tau)\tau} K_{(\tau\tau)\tau} K_{\tau\tau}$$

Corollary. *I_τ is a functional of type $\tau\tau$ and $\vdash I_\tau a^\tau = a^\tau$.*

Proof. By (Ax. S) and (Ax. K) we have

$$I_\tau a^\tau = S_{\tau(\tau\tau)\tau} K_{(\tau\tau)\tau} K_{\tau\tau} a^\tau = K_{(\tau\tau)\tau} a^\tau (K_{\tau\tau} a^\tau) = a^\tau.$$

Inductive Definition of $\lambda x^\rho(c^\tau)$.

1. $\lambda x^\rho(x^\rho) := I_\rho$.
2. If x^ρ does not occur in c^τ then set

$$\lambda x^\rho(c^\tau) := K_{\rho\tau} c^\tau.$$

3. If x^ρ in $a^{\sigma\tau} b^\sigma$ then set

$$\lambda x^\rho(a^{\sigma\tau} b^\sigma) := S_{\rho\sigma\tau}(\lambda x^\rho(a^{\sigma\tau}))(\lambda x^\rho(b^\sigma)).$$

Corollary. *$\lambda x^\rho(c^\tau)$ is a term of type $\rho\tau$ in which the variable x^ρ does not occur.*

Theorem 17.7. $\vdash \lambda x^\rho(c^\tau) d^\rho = c^\tau \begin{pmatrix} d^\rho \\ x^\rho \end{pmatrix}$.

Proof by induction on the length of the term c^τ.

1. $\lambda x^\rho(x^\rho) d^\rho = I_\rho d^\rho = d^\rho = x^\rho \begin{pmatrix} d^\rho \\ x^\rho \end{pmatrix}$.
2. If x^ρ does not occur in c^τ, then we have

$$\lambda x^\rho(c^\tau) d^\rho = K_{\rho\tau} c^\tau d^\rho = c^\tau = c^\tau \begin{pmatrix} d^\rho \\ x^\rho \end{pmatrix}.$$

3. If x^ρ occurs in $a^{\sigma\tau} b^\sigma$ then we have

$$\lambda x^\rho(a^{\sigma\tau} b^\sigma) d^\rho = S_{\rho\sigma\tau}(\lambda x^\rho(a^{\sigma\tau}))(\lambda x^\rho(b^\sigma)) d^\rho = \lambda x^\rho(a^{\sigma\tau}) d^\rho (\lambda x^\rho(b^\sigma) d^\rho).$$

Using the induction hypothesis we obtain

$$\lambda x^\rho(a^{\sigma\tau} b^\sigma) d^\rho = a^{\sigma\tau} \begin{pmatrix} d^\rho \\ x^\rho \end{pmatrix} \left[b^\sigma \begin{pmatrix} d^\rho \\ x^\rho \end{pmatrix} \right] = a^{\sigma\tau} b^\sigma \begin{pmatrix} d^\rho \\ x^\rho \end{pmatrix}.$$

Notation. In the sequel we write

$$\lambda x_1^{\sigma_1} \ldots x_n^{\sigma_n}(c^\tau)$$

for

$$\lambda x_1^{\sigma_1}(\ldots(\lambda x_n^{\sigma_n}(c^\tau))\ldots).$$

$x_1^{\sigma_1}, \ldots, x_n^{\sigma_n}$ must always be distinct variables. If these variables do not occur in $a_1^{\sigma_1}, \ldots, a_n^{\sigma_n}$ then by Theorem 17.7 we have

$$\vdash \lambda x_1^{\sigma_1} \ldots x_n^{\sigma_n}(c^\tau) a_1^{\sigma_1} \ldots a_n^{\sigma_n} = c^\tau \begin{pmatrix} a_1^{\sigma_1} \\ x_1^{\sigma_1} \end{pmatrix} \ldots \begin{pmatrix} a_n^{\sigma_n} \\ x_n^{\sigma_n} \end{pmatrix}.$$

7. The Predecessor Functional and the Arithmetic Difference

Definition of an auxiliary functional.

$$U := \lambda x^{oo} y^o (J_o y^o \mathscr{S}(x^{oo}(\mathscr{S}0))).$$

U is a functional of type $(oo)oo$. Applying Theorem 17.7 we have

$$\vdash U a^{oo} b^o = J_o b^o \mathscr{S}(a^{oo}(\mathscr{S}0)).$$

Using (Ax. J0) and (Ax. J$\mathscr{S}$) we obtain

(1) $\vdash U a^{oo} 0 = a^{oo}(\mathscr{S}0)$
(2) $\vdash U a^{oo}(\mathscr{S}0) = \mathscr{S}(a^{oo}(\mathscr{S}0)).$

Definition of the *predecessor functional*

$$P := \lambda z^o (J_{oo} z^o U(K_{oo}0)0).$$

P is a functional of type oo. By Theorem 17.7 we have

$$\vdash P c^o = J_{oo} c^o U(K_{oo}0)0.$$

Using (Ax. J0) and (Ax. J$\mathscr{S}$) we obtain

(3) $\vdash P0 = K_{oo}00 = 0$
$\vdash P(\mathscr{S}c^o) = U(J_{oo}c^o U(K_{oo}0))0.$

Using (1) we obtain

(4) $\vdash P(\mathscr{S}c^o) = J_{oo} c^o U(K_{oo}0)(\mathscr{S}0).$

Theorem 17.8. *For the predecessor functional P we have:*

(P0) $\vdash P0=0$
(P$\mathscr{S}$) $\vdash P(\mathscr{S}a^o)=a^o$.

Proof. (P0) holds by (3). By (4) we have $P(\mathscr{S}0)=J_{oo}0U(K_{oo}0)(\mathscr{S}0)=K_{oo}0(\mathscr{S}0)=0$ and therefore
(5) $\vdash P(\mathscr{S}0)=0$.

By (4) we also have

$$P(\mathscr{S}(\mathscr{S}x^o))=J_{oo}(\mathscr{S}x^o)U(K_{oo}0)(\mathscr{S}0)=U(J_{oo}x^oU(K_{oo}0))(\mathscr{S}0).$$

Using (2) and (4) we obtain

$$P(\mathscr{S}(\mathscr{S}x^o))=\mathscr{S}(J_{oo}x^oU(K_{oo}0)(\mathscr{S}0))=\mathscr{S}(P(\mathscr{S}x^o)).$$

Hence

(6) $\vdash P(\mathscr{S}x^o)=x^o\rightarrow P(\mathscr{S}(\mathscr{S}x^o))=\mathscr{S}x^o$.

By a (CI)-inference from (5) and (6) we obtain

$(P\mathscr{S})$ $\vdash P(\mathscr{S}a^o)=a^o$.

Theorem 17.9. $\vdash\mathscr{S}a^o=\mathscr{S}b^o\rightarrow a^o=b^o$.

Proof. By (=5) we have

$$\vdash\mathscr{S}a^o=\mathscr{S}b^o\rightarrow P(\mathscr{S}a^o)=P(\mathscr{S}b^o).$$

The assertion now follows using (P$\mathscr{S}$).

Definition of the *arithmetic difference* $a^o\dot{-}b^o$.

$$a^o\dot{-}b^o:=J_ob^oPa^o.$$

Corollaries. *The following equations are deducible*

$(\dot{-}0)$ $a^o\dot{-}0=a^o$
$(\dot{-}\mathscr{S})$ $a^o\dot{-}\mathscr{S}b^o=P(a^o\dot{-}b^o)$.

Proof. By (Ax. J0) and (Ax. J$\mathscr{S}$) we have

$$a^o\dot{-}0=J_o0Pa^o=a^o$$
$$a^o\dot{-}\mathscr{S}b^o=J_o(\mathscr{S}b^o)Pa^o=P(J_ob^oPa^o)=P(a^o\dot{-}b^o).$$

Theorem 17.10. *The following equations are deducible*

(1) $0 \dot{-} b^o = 0$
(2) $\mathscr{S}a^o \dot{-} \mathscr{S}b^o = a^o \dot{-} b^o$
(3) $\mathscr{S}a^o \dot{-} a^o = \mathscr{S}0$
(4) $a^o \dot{-} a^o = 0$.

Proof. 1. We have

(1.1) $\vdash 0 \dot{-} 0 = 0$

and $0 \dot{-} \mathscr{S}x^o = P(0 \dot{-} x^o)$. Using $P0 = 0$ we obtain

(1.2) $\vdash 0 \dot{-} x^o = 0 \to 0 \dot{-} \mathscr{S}x^o = 0$.

(1) follows from (1.1) and (1.2) by a (CI)-inference.

2. Now $\mathscr{S}a^o \dot{-} \mathscr{S}0 = P(\mathscr{S}a^o \dot{-} 0) = P(\mathscr{S}a^o) = a^o = a^o \dot{-} 0$, therefore

(2.1) $\vdash \mathscr{S}a^o \dot{-} \mathscr{S}0 = a^o \dot{-} 0$.

Further $\mathscr{S}a^o \dot{-} \mathscr{S}(\mathscr{S}x^o) = P(\mathscr{S}a^o \dot{-} \mathscr{S}x^o)$ and $P(a^o \dot{-} x^o) = a^o \dot{-} \mathscr{S}x^o$. Hence

(2.2) $\vdash \mathscr{S}a^o \dot{-} \mathscr{S}x^o = a^o \dot{-} x^o \to \mathscr{S}a^o \dot{-} \mathscr{S}(\mathscr{S}x^o) = a^o \dot{-} \mathscr{S}x^o$.

(2) follows from (2.1) and (2.2) by a (CI)-inference.

3. We have

(3.1) $\vdash \mathscr{S}0 \dot{-} 0 = \mathscr{S}0$

and by (2) $\vdash \mathscr{S}(\mathscr{S}x^o) \dot{-} \mathscr{S}x^o = \mathscr{S}x^o \dot{-} x^o$. Hence

(3.2) $\vdash \mathscr{S}x^o \dot{-} x^o = \mathscr{S}0 \to \mathscr{S}(\mathscr{S}x^o) \dot{-} \mathscr{S}x^o = \mathscr{S}0$.

(3) follows from (3.1) and (3.2) by a (CI)-inference.

4. From (2) and (3) we obtain $a^o \dot{-} a^o = \mathscr{S}a^o \dot{-} \mathscr{S}a^o = P(\mathscr{S}a^o \dot{-} a^o) = P(\mathscr{S}0) = 0$, and therefore

(4) $\vdash a^o \dot{-} a^o = 0$.

8. The Recursor

Definition of an auxiliary functional.

$$Q_\tau := \lambda x^{o\tau\tau} y^{o\tau} z^o (x^{o\tau\tau}(Pz^o)(y^{o\tau}(Pz^o))).$$

Q_τ is a functional of type $(o\tau\tau)(o\tau)o\tau$. By Theorem 17.7 we have

$$\vdash Q_\tau b^{o\tau\tau} d^{o\tau} c^o = b^{o\tau\tau}(Pc^o)(d^{o\tau}(Pc^o)).$$

Using (P$\mathscr{S}$), by Theorem 17.8 we have

(1) $\vdash Q_\tau b^{o\tau\tau} d^{o\tau}(\mathscr{S} c^o) = b^{o\tau\tau} c^o (d^{o\tau} c^o)$.

Definition of the *recursor* R_τ.

$$R_\tau := \lambda x^\tau y^{o\tau\tau} z^o (J_{o\tau} z^o (Q_\tau y^{o\tau\tau})(K_{o\tau} x^\tau) z^o).$$

R_τ is a functional of type $\tau(o\tau\tau)o\tau$. By Theorem 17.7 we have

(2) $\vdash R_\tau a^\tau b^{o\tau\tau} c^o = J_{o\tau} c^o (Q_\tau b^{o\tau\tau})(K_{o\tau} a^\tau) c^o$.

Theorem 17.11. (Primitive recursion for type τ).

(R0) $\vdash R_\tau a^\tau b^{o\tau\tau} 0 = a^\tau$
(R$\mathscr{S}$) $\vdash R_\tau a^\tau b^{o\tau\tau}(\mathscr{S} c^o) = b^{o\tau\tau} c^o (R_\tau a^\tau b^{o\tau\tau} c^o)$.

Proof. 1. From (2) using (Ax. J0) and (Ax. K) we have

$$R_\tau a^\tau b^{o\tau\tau} 0 = J_{o\tau} 0 (Q_\tau b^{o\tau\tau})(K_{o\tau} a^\tau) 0 = K_{o\tau} a^\tau 0 = a^\tau.$$

Hence (R0) holds.

2. Using (Ax. J$\mathscr{S}$) we obtain from (2)

$$\begin{aligned} R_\tau a^\tau b^{o\tau\tau}(\mathscr{S} c^o) &= J_{o\tau}(\mathscr{S} c^o)(Q_\tau b^{o\tau\tau})(K_{o\tau} a)(\mathscr{S} c^o) \\ &= Q_\tau b^{o\tau\tau}(J_{o\tau} c^o (Q_\tau b^{o\tau\tau})(K_{o\tau} a^\tau))(\mathscr{S} c^o). \end{aligned}$$

From (1) and (2) we obtain

$$R_\tau a^\tau b^{o\tau\tau}(\mathscr{S} c^o) = b^{o\tau\tau} c^o (J_{o\tau} c^o (Q_\tau b^{o\tau\tau})(K_{o\tau} a^\tau) c^o) = b^{o\tau\tau} c^o (R_\tau a^\tau b^{o\tau\tau} c^o).$$

Hence (R$\mathscr{S}$) also holds.

This theorem shows that $R_\tau a^\tau b^{o\tau\tau}$ is a term $f^{o\tau}$ depending on a^τ and $b^{o\tau\tau}$ for which the *primitive recursion equations*

$$\begin{aligned} &f^{o\tau} 0 = a^\tau \\ &f^{o\tau}(\mathscr{S} c^o) = b^{o\tau\tau} c^o (f^{o\tau} c^o) \end{aligned}$$

of type τ are deducible.

9. Simultaneous Recursion

Theorem 17.12 (Simultaneous n-fold primitive recursion).
Given terms $a_i^{\tau_i}$ *and* $b_i^{o\tau_1\ldots\tau_n\tau_i}$ $(i=1,\ldots,n)$ *there are terms* $f_i^{o\tau_i}$ *such that the equations*

$$f_i^{o\tau_i}0=a_i^{\tau_i}$$
$$f_i^{o\tau_i}(\mathscr{S}c^o)=b_i^{o\tau_1\ldots\tau_n\tau_i}c^o(f_1^{o\tau_1}c^o)\ldots(f_n^{o\tau_n}c^o)$$

are deducible for all $i=1,\ldots,n$ *and all terms* c^o *of type* o.

Proof by induction on n. If $n=1$ the assertion holds by Theorem 17.11 if we set

$$f_1^{o\tau_1}:=R_{\tau_1}a_1^{\tau_n}b_1^{o\tau_1\tau_1}.$$

Now suppose $n>1$. For brevity we write b_i, c, d for $b_i^{o\tau_1\ldots\tau_n\tau_i}$, c^o, d^o. Let

$$\sigma:=(o\tau_1)\ldots(o\tau_{n-1})\tau_n$$
$$s^\sigma:=\lambda x_1^{o\tau_1}\ldots x_{n-1}^{o\tau_{n-1}}(a_n^{\tau_n})$$
$$t^{o\sigma\sigma}:=\lambda x^o y^\sigma z_1^{o\tau_1}\ldots z_{n-1}^{o\tau_{n-1}}(b_n x^o(z_1^{o\tau_1}x^o)\ldots(z_{n-1}^{o\tau_{n-1}}x^o)(y^\sigma z_1^{o\tau_1}\ldots z_{n-1}^{o\tau_{n-1}}))$$
$$g:=R_\sigma s^\sigma t^{o\sigma\sigma},$$

where the variables $x_1^{o\tau_1},\ldots,x_{n-1}^{o\tau_{n-1}}$ do not occur in $a_n^{\tau_n}$ and the variables x^o, y^σ, $z_1^{o\tau_1},\ldots,z_{n-1}^{o\tau_{n-1}}$ do not occur in b_n. Then g is a term of type $o(o\tau_1)\ldots(o\tau_{n-1})\tau_n$ such that the following hold by Theorem 17.11:

$$\vdash g0=s^\sigma$$
$$\vdash g(\mathscr{S}c)=t^{o\sigma\sigma}c(gc).$$

By the definitions of s^σ, $t^{o\sigma\sigma}$ and Theorem 17.7 we have

(1) $\vdash g0e_1^{o\tau_1}\ldots e_{n-1}^{o\tau_{n-1}}=a_n^{\tau_n}$
(2) $\vdash g(\mathscr{S}c)e_1^{o\tau_1}\ldots e_{n-1}^{o\tau_{n-1}}=b_nc(e_1^{o\tau_1}c)\ldots(e_{n-1}^{o\tau_{n-1}}c)(gce_1^{o\tau_1}\ldots e_{n-1}^{o\tau_{n-1}})$.

For $i=1,\ldots,n-1$ let

$$u_i^{o\tau_i}:=K_{o\tau_i}a_i^{\tau_i}$$
$$v_i:=\begin{cases}\lambda x^o y_1^{o\tau_1}\ldots y_{n-1}^{o\tau_{n-1}}z^o(D_{\tau_i}[z^o\dot{-}x^o, y_i^{o\tau_i}z^o,\\ b_ix^o(y_1^{o\tau_1}x^o)\ldots(y_{n-1}^{o\tau_{n-1}}x^o)(gx^oy_1^{o\tau_1}\ldots y_{n-1}^{o\tau_{n-1}})])\end{cases}$$

v_i is a term of type $o(o\tau_1)\ldots(o\tau_{n-1})o\tau_i$. By the induction hypothesis there are terms h_i of type $oo\tau_i$ $(i=1,\ldots,n-1)$ such that the equations

$$h_i0=u_i^{o\tau_i}$$
$$h_i(\mathscr{S}c)=v_ic(h_1c)\ldots(h_{n-1}c)$$

are deducible for all $i=1,\ldots,n-1$ and all terms c of type o. From the definitions of $u_i^{o\tau_i}$, v_i and Theorem 17.7 we have

(3) $\vdash h_i 0d = a_i^{\tau_i}$.
(4) $\vdash h_i(\mathcal{S}c)d = D_{\tau_i}[d \dot{-} c, h_i cd, b_i c(h_1 cc)\ldots(h_{n-1}cc)(gc(h_1 c)\ldots(h_{n-1}c))]$.

By (D0) and (D$\mathcal{S}$) from (4) we obtain

(4.1) $\vdash d \dot{-} c = 0 \rightarrow h_i(\mathcal{S}c)d = h_i cd$.
(4.2) $\vdash d \dot{-} c = \mathcal{S}0 \rightarrow h_i(\mathcal{S}c)d = b_i c(h_1 cc)\ldots(h_{n-1}cc)(gc(h_1 c)\ldots(h_{n-1}c))$.

Now we define

$$f_i := \lambda x^o(h_i x^o x^o) \quad (i=1,\ldots,n-1)$$
$$f_n := \lambda x^o(gx^o(h_1 x^o)\ldots(h_{n-1}x^o)).$$

Then f_i is a term of type $o\tau_i$. By Theorem 17.7 we have

$$(5) \quad \begin{cases} \vdash f_i c = h_i cc \quad (i=1,\ldots,n-1) \\ \vdash f_n c = gc(h_1 c)\ldots(h_{n-1}c) \end{cases}$$

From (3), (4.2) and (5) we have, for $i=1,\ldots,n-1$,

$$(6) \quad \begin{cases} \vdash f_i 0 = a_i^{\tau_i} \\ \vdash f_i(\mathcal{S}c) = b_i c(f_1 c)\ldots(f_n c), \end{cases}$$

since by Theorem 17.10 (3) $\vdash \mathcal{S}c \dot{-} c = \mathcal{S}0$ holds.
From (1) and (5) we obtain

(7) $\vdash f_n 0 = a_n^{\tau_n}$.

By (4.1) we have

$$\vdash x^o \dot{-} c = 0 \rightarrow h_i(\mathcal{S}c)x^o = h_i cx^o.$$

Using (2) we obtain

$$(8) \quad \vdash \begin{cases} x^o \dot{-} c = 0 \rightarrow g(\mathcal{S}x^o)(h_1(\mathcal{S}c))\ldots(h_{n-1}(\mathcal{S}c)) \\ = b_n x^o(h_1 cx^o)\ldots(h_{n-1}cx^o)(gx^o(h_1(\mathcal{S}c))\ldots(h_{n-1}(\mathcal{S}c))) \end{cases}$$

From $\mathcal{S}x^o \dot{-} c = 0$ by Theorem 17.10 we obtain

$$x^o \dot{-} c = \mathcal{S}x^o \dot{-} \mathcal{S}c = P(\mathcal{S}x^o \dot{-} c) = P0 = 0.$$

Hence from (8) and (2) we obtain

$$\vdash \begin{cases} (x^o \dot{-} c = 0 \rightarrow gx^o(h_1(\mathcal{S}c))\ldots(h_{n-1}(\mathcal{S}c)) = gx^o(h_1 c)\ldots(h_{n-1}c)) \\ \quad \rightarrow (\mathcal{S}x^o \dot{-} c = 0 \rightarrow g(\mathcal{S}x^o)(h_1(\mathcal{S}c))\ldots(h_{n-1}(\mathcal{S}c) = g(\mathcal{S}x^o)(h_1 c)\ldots(h_{n-1}c)) \end{cases}$$

From (1) we also have

$$\vdash 0 \dot{-} c = 0 \rightarrow g0(h_1(\mathscr{S}c))\ldots(h_{n-1}(\mathscr{S}c)) = g0(h_1c)\ldots(h_{n-1}c).$$

Using a (CI)-inference we have

$$\vdash c \dot{-} c = 0 \rightarrow gc(h_1(\mathscr{S}c))\ldots(h_{n-1}(\mathscr{S}c)) = gc(h_1c)\ldots(h_{n-1}c).$$

Using $\vdash c \dot{-} c = 0$ (by Theorem 17.10) and (8) we obtain

$$\vdash g(\mathscr{S}c)(h_1(\mathscr{S}c))\ldots(h_{n-1}(\mathscr{S}c)) = b_nc(h_1cc)\ldots(h_{n-1}cc)(gc(h_1c)\ldots(h_{n-1}c)).$$

Hence by (5) we have

(9) $\vdash f_n(\mathscr{S}c) = b_nc(f_1c)\ldots(f_nc).$

With (6), (7) and (9) the proof of the theorem is complete.

Theorem 17.13 (Induction Theorem). *For $i=1,\ldots,n$ let b_i be a term of type $o\tau_1\ldots\tau_n\tau_i$. Let $\mathscr{F}[x^o, y_1^{\tau_1},\ldots,y_n^{\tau_n}]$ be a formula with distinct variables $x^o, y_1^{\tau_1},\ldots,y_n^{\tau_n}$ which do not occur in the nominal form $\mathscr{F}$ nor in the terms $b_1,\ldots,b_n$. Suppose*

(I) $\vdash \mathscr{F}[0, y_1^{\tau_1},\ldots,y_n^{\tau_n}]$
(II) $\vdash \mathscr{F}[x^o, b_1x^oy_1^{\tau_1}\ldots y_n^{\tau_n},\ldots,b_nx^oy_1^{\tau_1}\ldots y_n^{\tau_n}] \rightarrow \mathscr{F}[\mathscr{S}x^o, y_1^{\tau_1},\ldots,y_n^{\tau_n}]$

hold. Then $\vdash \mathscr{F}[a^o, a_1^{\tau_1},\ldots,a_n^{\tau_n}]$ holds for all terms $a^o, a_1^{\tau_1},\ldots,a_n^{\tau_n}$ of types $o, \tau_1,\ldots,\tau_n$.

Proof. By Theorem 17.12 for the terms a^o and $a_1^{\tau_1},\ldots,a_n^{\tau_n}$ there are terms f_i of type $o\tau_i$ such that for all $i=1,\ldots,n$ and all terms c^o of type o the equations

(1) $f_i0 = a_i^{\tau_i}$
(2) $f_i(\mathscr{S}c^o) = b_i(a^o \dot{-} \mathscr{S}c^o)(f_1c^o)\ldots(f_nc^o)$

are deducible. It follows from $a^o \dot{-} z^o = \mathscr{S}x^o$ that

$$a^o \dot{-} \mathscr{S}z^o = P(a^o \dot{-} z^o) = P(\mathscr{S}x^o) = x^o.$$

Therefore, substituting f_iz^o for $y_i^{\tau_i}$, hypothesis (II) and the equations (2) yield

(3) $\vdash a^o \dot{-} z^o = \mathscr{S}x^o \rightarrow \mathscr{F}[a^o \dot{-} \mathscr{S}z^o, f_1(\mathscr{S}z^o),\ldots,f_n(\mathscr{S}z^o)]$
$\rightarrow \mathscr{F}[a^o \dot{-} z^o, f_1z^o,\ldots,f_nz^o].$

Also by hypothesis (I) we have

(4) $\vdash a^o \dot{-} z^o = 0 \rightarrow \mathscr{F}[a^o \dot{-} \mathscr{S}z^o, f_1(\mathscr{S}z^o),\ldots,f_n(\mathscr{S}z^o)]$
$\rightarrow \mathscr{F}[a^o \dot{-} z^o, f_1z^o,\ldots,f_nz^o].$

By Theorem 17.5 from (3) and (4) we obtain

$$\vdash \mathscr{F}[a^o \dot{-} \mathscr{S}z^o, f_1(\mathscr{S}z^o), \ldots, f_n(\mathscr{S}z^o)] \to \mathscr{F}[a^o \dot{-} z^o, f_1 z^o, \ldots, f_n z^o].$$

Hence

$$(5)\ \vdash \begin{cases} (\mathscr{F}[a^o \dot{-} z^o, f_1 z^o, \ldots, f_n z^o] \to \mathscr{F}[a^o, a_1^{\tau_1}, \ldots, a_n^{\tau_n}]) \\ \quad \to (\mathscr{F}[a^o \dot{-} \mathscr{S}z^o, f_1(\mathscr{S}z^o), \ldots, f_n(\mathscr{S}z^o)] \to \mathscr{F}[a^o, a_1^{\tau_1}, \ldots, a_n^{\tau_n}]). \end{cases}$$

Also from the equations (1) we have

$$(6)\ \vdash \mathscr{F}[a^o \dot{-} 0, f_1 0, \ldots, f_n 0] \to \mathscr{F}[a^o, a_1^{\tau_1}, \ldots, a_n^{\tau_n}].$$

Using a (CI)-inference (5) and (6) yield

$$\vdash \mathscr{F}[a^o \dot{-} a^o, f_1 a^o, \ldots, f_n a^o] \to \mathscr{F}[a^o, a_1^{\tau_1}, \ldots, a_n^{\tau_n}].$$

Using $\vdash a^o \dot{-} a^o = 0$ (by Theorem 17.10) and hypothesis (I) the assertion $\vdash \mathscr{F}[a^o, a_1^{\tau_1}, \ldots, a_n^{\tau_n}]$ follows.

Theorem 17.14. *Let $\mathscr{F}[a^o, b^o]$ be a formula. Let x^o, y^o be distinct variables of type o which do not occur in the nominal form $\mathscr{F}$. Suppose*

(I) $\vdash \mathscr{F}[x^o, 0]$
(II) $\vdash \mathscr{F}[0, y^o]$
(III) $\vdash \mathscr{F}[x^o, y^o] \to \mathscr{F}[\mathscr{S}x^o, \mathscr{S}y^o]$

hold. Then $\vdash \mathscr{F}[a^o, b^o]$ holds for all terms a^o, b^o of type o.

Proof. $z^o = P(\mathscr{S}z^o)$ and hypothesis (III) yield

$$(1)\ \vdash y^o = \mathscr{S}z^o \to \mathscr{F}[x^o, Py^o] \to \mathscr{F}[\mathscr{S}x^o, y^o].$$

Also by hypothesis (I) we have

$$(2)\ \vdash y^o = 0 \to \mathscr{F}[x^o, Py^o] \to \mathscr{F}[\mathscr{S}x^o, y^o].$$

By Theorem 17.5 from (1) and (2) we obtain

$$(3)\ \vdash \mathscr{F}[x^o, Py^o] \to \mathscr{F}[\mathscr{S}x^o, y^o].$$

The assertion now follows from (3) and hypothesis (II) by Theorem 17.13.

10. The Characteristic Term of a Basic Formula

Definition. Formulas which only contain equations between terms of type *o* are said to be *basic formulas*.

Inductive Definition of the *characteristic term* $\chi[C]$ of a basic formula C.

1. $\chi[a^o = b^o] := (a^o \dot{-} b^o) + (b^o \dot{-} a^o)$
2. $\chi[A \to B] := \nu[\chi[A]] \cdot \chi[B]$.

By this definition $\chi[C]$ is a term of type o for every basic formula C.

In order to prove the next Theorem 17.15, we need the following Lemmata:

Lemma 2. *The following formulas are deducible*

(1) $0 + b^o = b^o$
(2) $0 \cdot b^o = 0$
(3) $\mathcal{S}0 \cdot b^o = b^o$
(4) $a^o + b^o = 0 \to b^o = 0$
(5) $a^o + b^o = 0 \to a^o = 0$.

Proof. 1. We have

(1.1) $\vdash 0 + 0 = 0$

and $\vdash 0 + \mathcal{S}x^o = \mathcal{S}(0 + x^o)$. Hence

(1.2) $\vdash 0 + x^o = x^o \to 0 + \mathcal{S}x^o = \mathcal{S}x^o$.

(1) follows from (1.1) and (1.2) by a (CI)-inference.

2. We have

(2.1) $\vdash 0 \cdot 0 = 0$

and $0 \cdot \mathcal{S}x^o = 0 \cdot x^o + 0 = 0 \cdot x^o$. Hence

(2.2) $\vdash 0 \cdot x^o = 0 \to 0 \cdot \mathcal{S}x^o = 0$.

(2) follows from (2.1) and (2.2) by a (CI)-inference.

3. We have

(3.1) $\vdash \mathcal{S}0 \cdot 0 = 0$

and $\mathcal{S}0 \cdot \mathcal{S}x^o = \mathcal{S}0 \cdot x^o + \mathcal{S}0 = \mathcal{S}(\mathcal{S}0 \cdot x^o + 0) = \mathcal{S}(\mathcal{S}0 \cdot x^o)$.
Hence

(3.2) $\vdash \mathcal{S}0 \cdot x^o = x^o \to \mathcal{S}0 \cdot \mathcal{S}x^o = \mathcal{S}x^o$.

(3) follows from (3.1) and (3.2) by a (CI)-inference.

4. Trivially

(4.1) $\vdash a^o + 0 = 0 \to 0 = 0$.

By Theorem 17.6 $a^o + \mathcal{S}x^o = \mathcal{S}(a^o + x^o)$ yields

(4.2) $\vdash a^o + \mathcal{S}x^o = 0 \to \mathcal{S}x^o = 0.$

(4) follows from (4.1) and (4.2) by Theorem 17.5.
5. (5) follows from (4).

Lemma 3. $\vdash (b^o \dot{-} a^o) + a^o = (a^o \dot{-} b^o) + b^o.$

Proof. By Theorem 17.10 and Lemma 2 we have

$$(0 \dot{-} x^o) + x^o = 0 + x^o = x^o \quad \text{and} \quad (x^o \dot{-} 0) + 0 = x^o \dot{-} 0 = x^o,$$

therefore

(1) $\vdash (0 \dot{-} x^o) + x^o = (x^o \dot{-} 0) + 0.$

Similarly we obtain

(2) $\vdash (y^o \dot{-} 0) + 0 = (0 \dot{-} y^o) + y^o.$

By Theorem 17.10(2) we also have

$$(\mathcal{S}y^o \dot{-} \mathcal{S}x^o) + \mathcal{S}x^o = (y^o \dot{-} x^o) + \mathcal{S}x^o = \mathcal{S}((y^o \dot{-} x^o) + x^o)$$

and similarly $(\mathcal{S}x^o \dot{-} \mathcal{S}y^o) + \mathcal{S}y^o = \mathcal{S}((x^o \dot{-} y^o) + y^o)$. Hence

(3) $\vdash (y^o \dot{-} x^o) + x^o = (x^o \dot{-} y^o) + y^o \to (\mathcal{S}y^o \dot{-} \mathcal{S}x^o) + \mathcal{S}x^o = (\mathcal{S}x^o \dot{-} \mathcal{S}y^o) + \mathcal{S}y^o.$

The assertion follows from (1)–(3) by Theorem 17.14.

Theorem 17.15. *The following hold for every basic formula C.*

(1) $\vdash \chi[C] = 0 \to C$
(2) $\vdash C \to \chi[C] = 0.$

Proof by induction on the number of $\to$ symbols, occurring in the basic formula C.
1. Suppose C is an equation $a^o = b^o$.
By Lemma 2 from the definition of $\chi[a^o = b^o]$ we have

$$\vdash \chi[a^o = b^o] = 0 \to a^o \dot{-} b^o = 0$$
$$\vdash \chi[a^o = b^o] = 0 \to b^o \dot{-} a^o = 0.$$

By Lemma 3 and Lemma 2(1) we have

$$\vdash a^o \dot{-} b^o = 0 \to b^o \dot{-} a^o = 0 \to a^o = b^o.$$

Hence we obtain

(1.1) $\vdash \chi[a^o = b^o] = 0 \rightarrow a^o = b^o$.

By Theorem 17.10 $\chi[a^o = a^o] = (a^o \dot{-} a^o) + (a^o \dot{-} a^o) = 0 + 0 = 0$. Hence

(1.2) $\vdash a^o = b^o \rightarrow \chi[a^o = b^o] = 0$.

2. Suppose C is a basic formula $A \rightarrow B$.
Since $\chi[A \rightarrow B] = \nu[\chi[A]] \cdot \chi[B]$, $\nu[0] = \mathscr{S}0$ and $\mathscr{S}0 \cdot \chi[B] = \chi[B]$ we have

(2.1) $\vdash \chi[A] = 0 \rightarrow \chi[A \rightarrow B] = \chi[B]$.

By the induction hypotheses

$$\vdash A \rightarrow \chi[A] = 0 \quad \text{and} \quad \vdash \chi[B] = 0 \rightarrow B$$

we obtain

(2.2) $\vdash \chi[A \rightarrow B] = 0 \rightarrow A \rightarrow B$.

The induction hypotheses $\vdash \chi[A] = 0 \rightarrow A$ and $\vdash B \rightarrow \chi[B] = 0$ together with the deducible formula

$$A \rightarrow (A \rightarrow B) \rightarrow B$$

yield

$$\vdash \chi[A] = 0 \rightarrow (A \rightarrow B) \rightarrow \chi[B] = 0$$

by cuts.
Hence using (2.1) we have

(2.3) $\vdash \chi[A] = 0 \rightarrow (A \rightarrow B) \rightarrow \chi[A \rightarrow B] = 0$.

It follows from $\chi[A \rightarrow B] = \nu[\chi[A]] \cdot \chi[B]$, $\nu[\mathscr{S}x^o] = 0$ and $0 \cdot \chi[B] = 0$ that

$$\vdash \chi[A] = \mathscr{S}x^o \rightarrow \chi[A \rightarrow B] = 0.$$

Hence we also have

(2.4) $\vdash \chi[A] = \mathscr{S}x^o \rightarrow (A \rightarrow B) \rightarrow \chi[A \rightarrow B] = 0$.

By Theorem 17.5 from (2.3) and (2.4) we have

(2.5) $\vdash (A \rightarrow B) \rightarrow \chi[A \rightarrow B] = 0$.

(1.1), (1.2), (2.2) and (2.5) complete the proof of Theorem 17.15.

Theorem 17.16.

$$\vdash \chi[A] = \mathscr{S} a^0 \to A \to B$$

holds for every basic formula A and every formula B.

Proof. By Theorem 17.15 we have $\vdash A \to \chi[A] = 0$. Hence $\vdash \chi[A] = \mathscr{S} a^0 \to A \to \mathscr{S} a^0 = 0$.
The assertion follows by Theorem 17.6.

Chapter VII

Pure Number Theory

In §18 we develop a formal system of pure number theory which can, in particular, be easily interpreted in the system FT of §17. The interpretation by K. Gödel [2] is carried out in §19.

§18. The Formal System PN for Pure Number Theory

1. The Formal Language

We use the following *primitive symbols*:

1. Denumerably infinitely many *free* and *bound number variables.*
2. The symbols 0, ′, +, ·, =, → and ∀.
3. Round brackets.

Nominal forms are defined in the usual way and denoted by capital script letters.

Inductive Definition of *number terms*:

1. The symbol 0 is a *number term.*
2. Every free number variable is a *number term.*
3. If t is a *number term* so is t'.
4. If s and t are *number terms* then so too are $(s+t)$ and $(s \cdot t)$.

For brevity we omit the outer round brackets. We regard · as stronger than + and thus abbreviate, for example, $((t_1 \cdot t_2)+t_3)$ by $t_1 \cdot t_2 + t_3$.

Syntactic symbols:

u, u_i, v, v_i	for free number variables
x, x_i, y, y_i, z, z_i	for bound number variables
s, s_i, t, t_i	for number terms.

Inductive Definition of *formulas*

1. If s and t are number terms then $s=t$ is a *formula.*
2. If A and B are *formulas* so too is $(A \rightarrow B)$.

3. If $\mathscr{F}[u]$ is a *formula* in which the bound variable x does not occur then $\forall x \mathscr{F}[x]$ is also a *formula*.

We use A, B, C, D, E, F, G, $\forall x \mathscr{F}[x]$, $\forall x \mathscr{G}[x]$ (possibly with indices) as syntactic symbols for formulas.

For brevity we write

$$A_1 \to \cdots \to A_n \to B$$

for

$$(A_1 \to (\cdots \to (A_n \to B)\ldots)).$$

2. The Deduction Procedure

We choose the axioms and basic inferences of the system PN so that every formula of our language which is deducible in classical predicate calculus is also deducible in PN. In doing this we do not work from positive and negative parts of formulas as we did in the system CP for classical predicate calculus, but we choose the axioms and basic rules of inference in another way to facilitate the interpretation of PN in FT (§19).

Axioms for the system PN:

(A1) $A \to A$
(A2) $\forall x \mathscr{F}[x] \to \mathscr{F}[t]$
(A3) $u' = 0 \to A$
(A4) $u' = v' \to u = v$

(A5) $u + 0 = u$
(A6) $u + v' = (u + v)'$
(A7) $u \cdot 0 = 0$
(A8) $u \cdot v' = u \cdot v + u$.

Basic Inferences for the system PN:

(B1) $B \vdash A \to B$
(B2) $A \to B \to C \to D \vdash A \to C \to B \to D$
(B3) $A \to B \to C, \quad C \to D \vdash A \to B \to D$
(B4) $A \to A \to B \vdash A \to B$
(B5) $u = u \to A \vdash A$
(B6) $\mathscr{F}[s] \vdash s = t \to \mathscr{F}[t]$
(B7) $A \to \mathscr{F}[u] \vdash A \to \forall x \mathscr{F}[x]$ provided the free number variable u does not occur in the conclusion.
(B8) $\mathscr{F}[0], \mathscr{F}[u] \to \mathscr{F}[u'] \vdash \mathscr{F}[t]$ provided the free number variable u does not occur in the nominal form $\mathscr{F}$.

(A1), (B1)–(B4) are the rules for positive implicational logic, (B5), (B6) the rules for equality, (A2), (B7) the rules for ∀-quantification. The Peano axioms for arithmetic are given by (A3), (A4), (B8) and the recursion equations for addition and multiplication are given by (A5)–(A8). Hence from a logical point of view arithmetic is axiomatised as a fragment of intuitionistic predicate calculus. In this system PN we shall also obtain all the rules of classical predicate calculus for the language of PN.

As previously we shall use

$$A_1, \ldots, A_n \vdash B$$

to denote a *permissible inference* of the system PN and $\vdash F$ to denote that F is deducible in the system PN.

3. Basic Properties of Deducibility

Lemma 1. $A \to B, B \to C \vdash A \to C$.

Proof. By (B1) from $A \to B$ we have $A \to A \to B$. By (B3) from $B \to C$ we have $A \to A \to C$. Hence $A \to C$ by (B4).

Lemma 2. $A, A \to B \vdash B$.

Proof. By (B1) from A we have $u = u \to A$. By Lemma 1 from $A \to B$ we have $u = u \to B$.
Hence B by (B5).

Lemma 3. $\mathscr{F}[v] \vdash \forall x \mathscr{F}[x]$ *provided the free number variable* v *does not occur in the nominal form* $\mathscr{F}$.

Proof. By (B1) from $\mathscr{F}[v]$ we have $u = u \to \mathscr{F}[v]$. By (B7) we have $u = u \to \forall x \mathscr{F}[x]$. Hence by (B5) we have $\forall x \mathscr{F}[x]$.

Theorem 18.1 (Substitution rule). $\mathscr{F}[u] \vdash \mathscr{F}[t]$ *provided the free number variable* u *does not occur in the nominal form* $\mathscr{F}$.

Proof. By Lemma 3 from $\mathscr{F}[u]$ we have $\forall x \mathscr{F}[x]$. Using the (A2)-axiom $\forall x \mathscr{F}[x] \to \mathscr{F}[t]$ by Lemma 2 we obtain $\mathscr{F}[t]$.

Theorem 18.2 (Rules for equality).

(1) $\vdash t = t$
(2) $\vdash s = t \to t = s$
(3) $\vdash s = t \to \mathscr{F}[s] \to \mathscr{F}[t]$
(4) $\vdash s = t \to s' = t'$.

Proof. 1. By (B5), from the (A1)-axiom,

$$u = u \rightarrow u = u$$

we obtain $\vdash u = u$. By Theorem 18.1 we have $\vdash t = t$.

2. By (1) $\vdash s = s$ and then by (B6) $\vdash s = t \rightarrow t = s$.
3. By (B6) from the (A1)-axiom $\mathscr{F}[s] \rightarrow \mathscr{F}[s]$ we have

$$\vdash s = t \rightarrow \mathscr{F}[s] \rightarrow \mathscr{F}[t].$$

4. By (1) $\vdash s' = s'$ and then by (B6) $\vdash s = t \rightarrow s' = t'$.

Lemma 4. $\vdash (A \rightarrow B \rightarrow C) \rightarrow B \rightarrow A \rightarrow C$

Proof. This follows from the (A1)-axiom

$$(A \rightarrow B \rightarrow C) \rightarrow A \rightarrow B \rightarrow C$$

by (B2).

Lemma 5. $A \rightarrow B \rightarrow C \vdash B \rightarrow A \rightarrow C$.

Proof. This follows from Lemma 4 by Lemma 2.

Lemma 6. $\vdash A \rightarrow B \rightarrow A$.

Proof. By (B1) from the (A1)-axiom $A \rightarrow A$ we have

$$\vdash B \rightarrow A \rightarrow A.$$

The assertion now follows by Lemma 5.

Lemma 7. $\vdash A \rightarrow (A \rightarrow B) \rightarrow B$.

Proof. This follows from the (A1)-axiom

$$(A \rightarrow B) \rightarrow A \rightarrow B$$

by Lemma 5.

Lemma 8. $A \rightarrow B \vdash (B \rightarrow C) \rightarrow A \rightarrow C$.

Proof. By Lemma 7

$$B \rightarrow (B \rightarrow C) \rightarrow C$$

is deducible. From this and $A \to B$ by Lemma 1 we obtain

$$A \to (B \to C) \to C.$$

By Lemma 5 we have

$$(B \to C) \to A \to C.$$

Lemma 9. *$\mathscr{F}[0], \mathscr{F}[u'] \vdash \mathscr{F}[t]$ provided the free number variable u does not occur in the nominal form $\mathscr{F}$.*

Proof. By (B1) from $\mathscr{F}[u']$ we have

$$\mathscr{F}[u] \to \mathscr{F}[u'].$$

From $\mathscr{F}[0]$ by (B8) we obtain $\mathscr{F}[t]$.

4. Properties of Negation

Definition: The *negation* of A, $\neg A$, is the formula $A \to 0'=0$.

Lemma 10. $\vdash A \to \neg\neg A$.

Proof. This follows from Lemma 7 by the definition of negation.

Lemma 11. $\vdash \neg A \to A \to B$.

Proof. By the definition of negation

$$\neg A \to A \to 0'=0$$

is an (A1)-axiom. By Theorem 18.1 from the (A3)-axiom $u'=0 \to B$ we obtain

$$\vdash 0'=0 \to B.$$

By (B3) we have

$$\vdash \neg A \to A \to B.$$

Lemma 12. $A \to B \vdash \neg B \to \neg A$.

Proof. This is a special case of Lemma 8 because of the definition of negation.

Lemma 13. $A \to B \to C \vdash A \to \neg C \to \neg B$.

Proof. By Lemma 10 we have

$$\vdash C \to \neg C \to 0' = 0.$$

By (B3) from $A \to B \to C$ we obtain

$$A \to B \to \neg C \to 0' = 0.$$

by (B2) we have

$$A \to \neg C \to \neg B.$$

Lemma 14. $\vdash \neg\neg s = t \to s = t.$

Proof. By Theorem 18.2 (1) we have $\vdash 0 = 0$. Hence by (B1) we have

(1) $\vdash \neg\neg 0 = 0 \to 0 = 0.$

By Lemmata 11 and 5 we have

$$\vdash \neg u' = 0 \to \neg\neg u' = 0 \to u' = 0.$$

By Lemma 2 using the (A3)-axiom $\neg u' = 0$ we have

(2) $\vdash \neg\neg u' = 0 \to u' = 0.$

From (1) and (2) by Lemma 9 we obtain

$$\vdash \neg\neg u = 0 \to u = 0.$$

Hence by Lemma 3

(3) $\vdash \forall x(\neg\neg x = 0 \to x = 0).$

By Theorem 18.2 (2) we have

$$\vdash 0 = v' \to v' = 0.$$

Using the (A3)-axiom

$$v' = 0 \to 0' = 0$$

by Lemma 1 we obtain

$$\vdash \neg 0 = v'.$$

By Lemmata 11 and 5 we have

$$\vdash \neg 0 = v' \to \neg\neg 0 = v' \to 0 = v'.$$

By Lemma 2 we have

$$\vdash \neg\neg 0 = v' \to 0 = v'.$$

By (B1) we obtain

(4) $\vdash \forall x(\neg\neg x = v \to x = v) \to \neg\neg 0 = v' \to 0 = v'.$

By two applications of Lemma 12 we obtain

$$\vdash \neg\neg u' = v' \to \neg\neg u = v$$

from the (A4)-axiom

$$u' = v' \to u = v.$$

By Lemma 8 we obtain

$$\vdash (\neg\neg u = v \to u = v) \to \neg\neg u' = v' \to u = v.$$

By Theorem 18.2 (4) we have

$$\vdash u = v \to u' = v'.$$

By (B3) we obtain

$$\vdash (\neg\neg u = v \to u = v) \to \neg\neg u' = v' \to u' = v'.$$

Using an (A2)-axiom by Lemma 1 we obtain

(5) $| \forall x(\neg\neg x = v \to x = v) \to \neg\neg u' = v' \to u' = v'.$

From (4) and (5) by Lemma 9 we have

$$\vdash \forall x(\neg\neg x = v \to x = v) \to \neg\neg u = v' \to u = v'.$$

Hence by (B7) we have

(6) $\vdash \forall x(\neg\neg x = v \to x = v) \to \forall x(\neg\neg x = v' \to x = v').$

By (B8) from (3) and (6) we obtain

$$\vdash \forall x(\neg\neg x = t \to x = t).$$

The assertion

$$\vdash \neg\neg s=t \to s=t$$

now follows by Lemma 2 using an (A2)-axiom.

Lemma 15. $\vdash \neg\neg C \to C$.

Proof by induction on the number of $\to$ and $\forall$ symbols occurring in the formula C.

1. Suppose C is an equation $s=t$. Then the assertion holds by Lemma 14.
2. Suppose C is a formula $A \to B$. By Lemma 7 we have

$$\vdash A \to (A \to B) \to B.$$

By two applications of Lemma 13 we obtain

$$\vdash A \to \neg\neg(A \to B) \to \neg\neg B.$$

From the induction hypothesis

$$\vdash \neg\neg B \to B$$

by (B3) we have

$$\vdash A \to \neg\neg(A \to B) \to B.$$

Hence by Lemma 5 we have

$$\vdash \neg\neg(A \to B) \to A \to B.$$

3. Suppose C is a formula $\forall x \mathscr{F}[x]$. Let u be a free number variable which does not occur in C. By two applications of Lemma 12 we obtain

$$\vdash \neg\neg\forall x \mathscr{F}[x] \to \neg\neg \mathscr{F}[u]$$

from the axiom

$$\forall x \mathscr{F}[x] \to \mathscr{F}[u].$$

Using the induction hypothesis

$$\vdash \neg\neg \mathscr{F}\{u] \to \mathscr{F}[u]$$

we obtain by Lemma 1

$$\vdash \neg\neg\forall x \mathscr{F}[x] \to \mathscr{F}[u].$$

Hence by (B7) we have

$$\vdash \neg\neg\forall x\mathscr{F}[x] \to \forall x\mathscr{F}[x].$$

Lemma 16. $\neg A \to B \vdash \neg B \to A$.

Proof. By Lemma 12 from $\neg A \to B$ we obtain $\neg B \to \neg\neg A$. By Lemma 15 the formula $\neg\neg A \to A$ is deducible. Hence by Lemma 1 we have $\neg B \to A$.

Lemma 17. $A \to B, \neg A \to B \vdash B$.

Proof. By Lemma 16 from $\neg A \to B$ we have $\neg B \to A$. By Lemma 1 using $A \to B$ we obtain $\neg B \to B$. By Lemma 10 the formula $B \to \neg\neg B$ is deducible. Hence by Lemma 1 we have $\neg B \to \neg\neg B$. Hence by (B4) we obtain $\neg\neg B$. By Lemma 15 $\neg\neg B \to B$ is a deducible formula. Hence by Lemma 2 we obtain B.

Lemma 18. $\neg A \to C, B \to C \vdash (A \to B) \to C$.

Proof. By Lemma 7 we have

$$\vdash A \to (A \to B) \to B.$$

By Lemma (B3) using $B \to C$ we obtain

$$A \to (A \to B) \to C.$$

From $\neg A \to C$ by (B1) and Lemma 5 we obtain

$$\neg A \to (A \to B) \to C.$$

Hence by Lemma 17 we obtain

$$(A \to B) \to C.$$

5. Positive and Negative Parts of Formulas

We define *positive* and *negative parts* of formulas, *P-forms*, *N-forms* and *NP-forms* as in §3 (p. 20) except that the formula $0'=0$ now plays the rôle of $\perp$. As before we use the following *syntactic symbols:*

$\mathscr{P}, \mathscr{P}_i$ for P-forms,
$\mathscr{N}, \mathscr{N}_i$ for N-forms,
$\mathscr{Q}, \mathscr{Q}_i$ for NP-forms.

Every P-form is a nominal form $*_1$ or $\mathscr{P}_0[C \to *_1]$ or $\mathscr{N}_0[\neg *_1]$ and every N-form is a nominal form $\mathscr{P}_0[*_1 \to C]$ by the inductive definition given above on p. 20. Now $\mathscr{P}_0$ and $\mathscr{N}_0$ are shorter nominal forms than $\mathscr{P}$ and $\mathscr{N}$ respectively.

Consequently we prove our theorems about P-forms and N-forms by induction on the length of these nominal forms.

Lemma 19.
(1) $\vdash A \to \mathscr{P}[A]$
(2) $\vdash \neg A \to \mathscr{N}[A]$.

Proof by induction on the lengths of $\mathscr{P}$ and $\mathscr{N}$.

1. Suppose $\mathscr{P}$ is $*_1$. Then $A \to \mathscr{P}[A]$ is the (A1)-axiom $A \to A$.
2. Suppose $\mathscr{P}$ is $\mathscr{P}_0[C \to *_1]$. By the I.H. (induction hypothesis) we then have

$$\vdash (C \to A) \to \mathscr{P}[A].$$

By Lemma 6 the formula

$$A \to C \to A$$

is deducible, hence by Lemma 1 we have

$$\vdash A \to \mathscr{P}[A].$$

3. Suppose $\mathscr{P}$ is $\mathscr{N}_0[\neg *_1]$. By I.H. we then have

$$\vdash \neg\neg A \to \mathscr{P}[A].$$

By Lemma 10 the formula $A \to \neg\neg A$ is deducible, hence by Lemma 1 $\vdash A \to \mathscr{P}[A]$.

4. Suppose $\mathscr{N}$ is $\mathscr{P}_o[*_1 \to C]$. By I.H. we then have

$$\vdash (A \to C) \to \mathscr{N}[A].$$

By Lemma 1 the formula

$$\neg A \to A \to C$$

is deducible, hence by Lemma 1 we obtain

$$\vdash \neg A \to \mathscr{N}[A].$$

Lemma 20.
(1) $\mathscr{P}[A] \vdash \neg \mathscr{P}[B] \to A$
(2) $\mathscr{N}[A] \vdash A \to \mathscr{N}[B]$.

Proof by induction on the lengths of $\mathscr{P}$ and $\mathscr{N}$.

1. Suppose $\mathscr{P}$ is $*_1$. Then the assertion holds by (B1).
2. Suppose $\mathscr{P}$ is $\mathscr{P}_o[C \to *_1]$. Then C is a negative part of $\mathscr{P}[B]$. By Lemma 19 we therefore have

$$\vdash \neg C \to \mathscr{P}[B].$$

By Lemma 16 we obtain

$$\vdash \neg \mathscr{P}[B] \to C.$$

By the I.H. from $\mathscr{P}[A]$ we obtain

$$\neg \mathscr{P}[B] \to C \to A.$$

By Lemmata 5 and 1 we obtain

$$\neg \mathscr{P}[B] \to \neg \mathscr{P}[B] \to A.$$

By (B4) we obtain $\neg \mathscr{P}[B] \to A$.

3. Suppose $\mathscr{P}$ is $\mathscr{N}_o[\neg *_1]$. By the I.H. we then obtain from $\mathscr{P}[A]$

$$\neg A \to \mathscr{P}[B].$$

By Lemma 16 we obtain

$$\neg \mathscr{P}[B] \to A.$$

4. Suppose $\mathscr{N}$ is $\mathscr{P}_o[*_1 \to C]$. Then C is a positive part of $\mathscr{N}[B]$. By Lemma 19 we therefore have

$$\vdash C \to \mathscr{N}[B].$$

By I.H. from $\mathscr{N}[A]$ we obtain

$$\neg \mathscr{N}[B] \to A \to C.$$

By (B3) we obtain

$$\neg \mathscr{N}[B] \to A \to \mathscr{N}[B].$$

By Lemma 13 and (B4) we obtain

$$\neg \mathscr{N}[B] \to \neg A.$$

By Lemma 16 we obtain

$$\neg \neg A \to \mathscr{N}[B].$$

By Lemma 10 the formula $A \to \neg \neg A$ is deducible so by Lemma 1 we obtain $A \to \mathscr{N}[B]$.

Theorem 18.3. $\vdash \mathscr{Q}[A, A]$.

Proof. By Lemma 19 we have

$$\vdash A \rightarrow \mathscr{Q}[A, A] \quad \text{and} \quad \vdash \neg A \rightarrow \mathscr{Q}[A, A].$$

By Lemma 17 we obtain $\vdash \mathscr{Q}[A, A]$.

Theorem 18.4. $\vdash \mathscr{N}[0' = 0]$.

Proof. As an (A1)-axiom we have

$$\vdash \neg 0' = 0.$$

By Lemma 19 we have

$$\vdash \neg 0' = 0 \rightarrow \mathscr{N}[0' = 0].$$

By Lemma 2 we obtain

$$\vdash \mathscr{N}[0' = 0].$$

Theorem 18.5. $\mathscr{N}[\neg A], \mathscr{N}[B] \vdash \mathscr{N}[A \rightarrow B]$.

Proof. By Lemma 20 from $\mathscr{N}[\neg A]$ and $\mathscr{N}[B]$ we obtain

$$\neg A \rightarrow \mathscr{N}[A \rightarrow B] \quad \text{and} \quad B \rightarrow \mathscr{N}[A \rightarrow B].$$

By Lemma 18 we obtain

$$(A \rightarrow B) \rightarrow \mathscr{N}[A \rightarrow B].$$

By Lemma 19 we have

$$\vdash \neg (A \rightarrow B) \rightarrow \mathscr{N}[A \rightarrow B].$$

By Lemma 17 we obtain

$$\mathscr{N}[A \rightarrow B].$$

Theorem 18.6. $\mathscr{P}[\mathscr{F}[u]] \vdash \mathscr{P}[\forall x \mathscr{F}[x]]$, *provided the free number variable u does not occur in the conclusion.*

Proof. By Lemma 20 from $\mathscr{P}[\mathscr{F}[u]]$ we obtain

$$\neg \mathscr{P}[\forall x \mathscr{F}[x]] \rightarrow \mathscr{F}[u].$$

By (B7) we obtain

$$\neg\mathscr{P}[\forall x\mathscr{F}[x]] \to \forall x\mathscr{F}[x].$$

By Lemma 16 we obtain

$$\neg\forall x\mathscr{F}[x] \to \mathscr{P}[\forall x\mathscr{F}[x]].$$

By Lemma 19 we have

$$\vdash \forall x\mathscr{F}[x] \to \mathscr{P}[\forall x\mathscr{F}[x]].$$

By Lemma 17 we obtain

$$\mathscr{P}[\forall x\mathscr{F}[x]].$$

Theorem 18.7. $\forall x\mathscr{F}[x] \to \mathscr{N}[\mathscr{F}[t]] \vdash \mathscr{N}[\forall x\mathscr{F}[\chi]]$.

Proof. By Lemma 20 from the hypothesis

$$\forall x\mathscr{F}[x] \to \mathscr{N}[\mathscr{F}[t]]$$

we obtain

$$\mathscr{F}[t] \to \forall x\mathscr{F}[x] \to \mathscr{N}[\forall x\mathscr{F}[x]].$$

Using the (A2)-axiom

$$\forall x\mathscr{F}[x] \to \mathscr{F}[t]$$

by Lemma 1 we obtain

$$\forall x\mathscr{F}[x] \to \forall x\mathscr{F}[x] \to \mathscr{N}[\forall x\mathscr{F}[x]].$$

By (B4) we obtain

$$\forall x\mathscr{F}[x] \to \mathscr{N}[\forall x\mathscr{F}[x]].$$

By Lemma 19 we have

$$\neg\forall x\mathscr{F}[x] \to \mathscr{N}[\forall x\mathscr{F}[x]].$$

By Lemma 17 we obtain

$$\mathscr{N}[\forall x\mathscr{F}[x]].$$

Theorem 18.8 (Cut rule). $\mathscr{P}[A], A \to B \vdash \mathscr{P}[B]$.

Proof. By Lemma 20 from $\mathscr{P}[A]$ we obtain

$$\neg \mathscr{P}[B] \to A.$$

By Lemma 1 using $A \to B$ we obtain

$$\neg \mathscr{P}[B] \to B.$$

By Lemma 16 we obtain

$$\neg B \to \mathscr{P}[B].$$

By Lemma 19 we have

$$\vdash B \to \mathscr{P}[B].$$

By Lemma 17 we obtain

$$\mathscr{P}[B].$$

Theorems 18.3–18.8 show that the system PN has all the deduction properties which were laid down in §3 (p. 20) for the formal system CP of classical predicate calculus. For by Theorems 18.3 and 18.4 all the axioms of the system CP are deducible in PN (with $0' = 0$ for $\bot$) and by Theorems 18.5–18.7 the basic inferences (S1)–(S3) of the system CP are permissible inferences in the system PN. By Theorem 18.8 cuts, which were shown to be permissible inferences in the system CP, are also permissible in PN. By Theorems 18.1 and 18.2, axioms (A3)–(A8) and the inference rule (B8) of complete induction the system PN is also sufficient for all the requirements of arithmetic which lie in the domain of classical predicate calculus.

6. The Consistency of the System PN

It was G. Gentzen who in 1936 first proved the consistency of a formal system for pure number theory which is equivalent to the system PN. We shall give two different kinds of consistency proof for PN. The first is in §19 by an interpretation in the system FT of functionals of finite type following K. Gödel in 1958 and the second in §23.2 by an embedding in a formal system EN which will be proved consistent by cut-elimination. Like the original proof of G. Gentzen, besides elementary methods both proofs only require transfinite induction up to ε_0.

§19. Interpretation of PN in FT

1. Sequences of Terms of the System FT

In this section we use w, w_i, x, x_i, y, y_i, z, z_i to denote finite (possibly empty) sequences of distinct variables of the system FT and we use a, a_i, b, b_i for finite (possibly empty) sequences of terms of the system FT.

$a_1{}^\frown a_2{}^\frown\ldots{}^\frown a_n$ denotes the sequence of terms obtained by concatenating the sequences of terms $a_1,\ldots,a_n$.

If a is a sequence of terms $a_1^{\sigma_1\ldots\rho_m\tau_1},\ldots,a_n^{\sigma_1\ldots\sigma_m\tau_n}$ and b is a sequence of terms $b_1^{\sigma_1},\ldots,b_m^{\sigma_m}$ then $a(b)$ denotes the sequence of terms

$$a_1^{\sigma_1\ldots\sigma_m\tau_1}b_1^{\sigma_1}\ldots b_m^{\sigma_m},\ldots,a_n^{\sigma_1\ldots\sigma_m\tau_n}b_1^{\sigma_1}\ldots b_m^{\sigma_m}.$$

The sequence of terms $a(b)$ is empty if, and only if, a is empty. If b is empty then $a(b)$ is the same as a. For brevity we write ab for $a(b)$ provided the sequence of terms b is denoted by a single symbol.

If $a(b_1{}^\frown b_2{}^\frown\ldots{}^\frown b_k)$ is defined then by the definition of sequences of terms this sequence is the same as $ab_1\ldots b_k$.

When $a:=a_1^{\sigma_1},\ldots,a_m^{\sigma_m}$ and $b:=b_1^{\tau_1},\ldots,b_n^{\tau_n}$ are sequences of terms we use the following conventions:

1. $a\sim b$ ("*a is of equal type with b*") means that $m=n$ and $\sigma_i=\tau_i$ for all $i=1,\ldots,n$. (In particular this holds if a and b are empty.)
2. $\vdash a=b$ means that $a\sim b$ and $a_i^{\tau_i}=b_i^{\tau_i}$ for all $i=1,\ldots,n$ is deducible in FT.

$$\vdash A\to a=b$$

means that $a\sim b$ and

$$A\to a_i^{\tau_i}=b_i^{\tau_i}$$

is deducible in FT for all $i=1,\ldots,n$. ($\vdash a=b$ and $\vdash A\to a=b$ are also to hold if a and b are empty.)

Given sequences of terms $a:=a_1^{\tau_1},\ldots,a_n^{\tau_n}$ and $b:=b_1^{\tau_1},\ldots,b_n^{\tau_n}$ let

$$D[c^o,a,b]$$

be the sequence of terms

$$D_{\tau_1}[c^o,a_1^{\tau_1},b_1^{\tau_1}],\ldots,D_{\tau_n}[c^o,a_n^{\tau_n},b_n^{\tau_n}].$$

Using $(D0)$ and $(D\mathscr{S})$ (§17, p. 119) we obtain

$$\vdash c^o=0\to D[c^o,a,b]=a$$
$$\vdash c^o=\mathscr{S}u^o\to D[c^o,a,b]=b.$$

Lemma 1. *Given two sequences of terms a and b, then up to equality of type, there is a uniquely determined sequence of variables x such that*

$$xa \sim b.$$

Proof. If a is a sequence of terms $a_1^{\sigma_1}, \ldots, a_m^{\sigma_m}$ and b is a sequence of terms $b_1^{\tau_1}, \ldots, b_n^{\tau_n}$, then $xa \sim b$ holds if, and only if, x is a sequence of variables

$$x_1^{\sigma_1 \ldots \sigma_m \tau_1}, \ldots, x_n^{\sigma_1 \ldots \sigma_m \tau_n}.$$

Lemma 2. *Given sequences $x_1, \ldots, x_k$ of distinct variables and a sequence of terms a there is a sequence of terms b containing no variable occurring in the sequence $x_1, \ldots, x_k$, such that*

$$\vdash b x_1 \ldots x_k = a.$$

Proof. If $x_1 \frown \ldots \frown x_k$ is the sequence of variables $y_1^{\sigma_1}, \ldots, y_m^{\sigma_m}$ and a is the sequence of terms $a_1^{\tau_1}, \ldots, a_n^{\tau_n}$ then the assertion holds for the sequence of terms

$$b := \lambda y_1^{\sigma_1} \ldots y_m^{\sigma_m} a_1^{\tau_1}, \ldots, \lambda y_1^{\sigma_1} \ldots y_m^{\sigma_m} a_n^{\tau_n}.$$

2. The Formal System QFT

We obtain the interpretation of PN in FT by using formulas in a language QFT which is a quantificational extension of the language of FT.

Inductive Definition of the *formulas* of QFT and the free occurrences of variables in such formulas.

1. Every equation $a^o = b^o$ between terms a^o, b^o of type o is a *formula* of QFT. A variable is *free* in this formula if, and only if, it occurs in a^o or b^o.

2. If A and B are *formulas* of QFT then so too is $(A \rightarrow B)$. A variable is *free* in this formula if, and only if, it is *free* in either A or B.

3. If F is a *formula* of QFT and x^τ is a variable not bound by a quantifier $\forall x^\tau$ or $\exists x^\tau$ then $\forall x^\tau F$ and $\exists x^\tau F$ are also *formulas* of QFT. A variable is *free* in such a formula if, and only if, it is different from x^τ and is *free* in F.

A formula of QFT is said to be *quantifier-free* if it contains no $\forall$ or $\exists$ quantifier. The quantifier-free formulas of QFT are exactly the *basic formulas* of the system FT (see p. 129).

We use the usual arrangements for omitting brackets and the notation $\Gamma \rightarrow B$, where Γ denotes a sequence of formulas, for the formulas of QFT as we did for FT.

A sequence of terms a is said to be *free for a formula F* of QFT if no variable y^τ occurring in a occurs in a quantifier $\forall y^\tau$ or $\exists y^\tau$ in the formula F.

We abbreviate the formulas

$$\forall x_1^{\tau_1} \ldots \forall x_n^{\tau_n} \mathscr{F}[x_1^{\tau_1}, \ldots, x_n^{\tau_n}], \qquad \exists x_1^{\tau_1} \ldots \exists x_n^{\tau_n} \mathscr{F}[x_1^{\tau_1}, \ldots, x_n^{\tau_n}], \qquad \mathscr{F}[a_1^{\tau_1}, \ldots, a_n^{\tau_n}]$$

of QFT to

$$\forall x \mathscr{F}[x], \quad \exists x \mathscr{F}[x], \quad \mathscr{F}[a],$$

where x denotes the sequence of variables $x_1^{\tau_1}, \ldots, x_n^{\tau_n}$, a the sequence of terms $a_1^{\tau_1}, \ldots, a_n^{\tau_n}$ and $\mathscr{F}$ an n-place nominal form. We allow the case $n=0$ when x and a are empty. The notation $\mathscr{F}[a]$ already assumes that the sequence of terms a is free for the formula $\mathscr{F}[x]$.

Definition of *negation* in QFT:

Let $\neg A$ be the formula $A \rightarrow \mathscr{S}0=0$.

It is not necessary to give a deduction procedure for formulas of QFT in order to give the interpretation of PN in FT. However, we shall give a deduction procedure for QFT since it will motivate the choice of the interpreting formulas.

Axioms of the system QFT:

1. All basic formulas which are deducible in the system FT.
2. The $\forall$-*axioms*

$$\forall x \mathscr{F}[x] \rightarrow \mathscr{F}[a]$$

and $\exists$-axioms

$$\mathscr{F}[a] \rightarrow \exists x \mathscr{F}[x]$$

for arbitrary sequences of variables x and sequences of terms $a \sim x$.

3. The IP_0'-*axioms* (*Independence of premise*)

$$(\forall x \mathscr{A}[x] \rightarrow \exists y \mathscr{B}[y]) \rightarrow \exists y (\forall x \mathscr{A}[x] \rightarrow \mathscr{L}[y])$$

for quantifier-free formulas $\mathscr{A}[x]$.

4. The M′-*axioms* (*Markov's principle*)

$$\neg\neg \exists x \mathscr{A}[x] \rightarrow \exists x \mathscr{A}[x]$$

for quantifier-free formulas $\mathscr{A}[x]$.

5. The AC-*axioms* (*Axioms of choice*)

$$\forall x \exists y \mathscr{F}[x, y] \rightarrow \exists z \forall x \mathscr{F}[x, zx]$$

for the sequences of variables x, y, z such that $y \sim zx$.

Basic Inferences of the system QFT

1. The basic inferences (Str), (Cut) and (CI) of the system FT for formulas of QFT (see p. 114).

2. The $\forall$-*inferences*

$$\Gamma \to \mathscr{F}[x] \vdash \Gamma \to \forall x \mathscr{F}[x]$$

provided that the variables in the sequence x do not occur in $\mathscr{F}$ and do not occur free in Γ.

3. The $\exists$-*inferences*

$$\mathscr{F}[x] \to B \vdash \exists x \mathscr{F}[x] \to B$$

provided that the variables in the sequence x do not occur in $\mathscr{F}$ and do not occur free in B.

Corollary. *Every formula*

$$A \to A$$

is deducible in QFT.

Proof. $A \to A$ is a special case of an $\forall$-axiom or $\exists$-axiom with an empty sequence of variables.

Remark. The formal system QFT corresponds to a fragment of the formal system obtained by adding the axioms IP'_o, M′ and AC to Heyting arithmetic HA^{ω}. (See A.S. Troelstra: Metamathematical Investigation of Intuitionistic Arithmetic and Analysis. Lecture Notes in Mathematics 344. Springer-Verlag 1973, p. 46 and 238.) QFT lacks the connectives $\wedge$ and $\vee$ of HA^{ω}.

Lemma 3. *The following are deducible in* QFT:
(1) $(A \to B) \to (B \to C) \to A \to C$
(2) $(A \to B) \to \neg B \to \neg A$.

Proof. The formulas

$$(A \to B) \to A \to B \quad \text{and} \quad B \to (B \to C) \to C$$

are deducible. Hence by a cut

$$(A \to B) \to A \to (B \to C) \to C.$$

By a structural inference we have
(1) $(A \to B) \to (B \to C) \to A \to C$
(2) is a special case of (1) with $\mathscr{S}0=0$ for C.

Lemma 4. *The following are permissible inferences of the system* QFT:
(1) $A \to B \vdash (B \to C) \to A \to C$

(2) $A \to B \vdash \neg B \to \neg A$
(3) $B \to C \vdash (A \to B) \to A \to C$.

Proof. This follows from Lemma 3 by cuts and a structural inference.

Lemma 5. *The following are deducible in* QFT:
(1) $\mathscr{S}0=0 \to B$
(2) $\neg A \to A \to B$.

Proof of (1) by induction on the length of the formula B.

1. Suppose B is an equation $a^o = b^o$. Then the assertion holds by Theorem 17.6.
2. Suppose B is a formula $A \to C$. Then by I.H.

$$\mathscr{S}0=0 \to C$$

is deducible. The assertion follows by a structural inference.

3. Suppose B is a formula $\forall x \mathscr{F}[x]$ or $\exists x \mathscr{F}[x]$ where the nominal form $\mathscr{F}$ contains no variable in the sequence x. Then by I.H.

$$\mathscr{S}0=0 \to \mathscr{F}[x]$$

is deducible. The assertion follows by an $\forall$-inference or cut using an $\exists$-axiom.

(2) follows from (1) by Lemma 4(3).

Lemma 6. *If A is a quantifier-free formula, then*

$$A \to B, \qquad \neg A \to B \vdash B$$

is a permissible inference of the system QFT.

Proof. By Theorems 17.5 and 17.6 the formulas

$$\chi[A]=0 \to A \quad \text{and} \quad \chi[A]=\mathscr{S}x^o \to \neg A$$

are deducible in FT and therefore also in QFT. Using cuts with

$$A \to B \quad \text{and} \quad \neg A \to B$$

we obtain

$$\chi[A]=0 \to B \quad \text{and} \quad \chi[A]=\mathscr{S}x^o \to B.$$

Using a structural inference and a basic inference (CI) we obtain

$$\chi[A]=\chi[A] \to B.$$

Hence we obtain B.

Lemma 7. *If A is a quantifier-free formula then the formula*

$$\neg\neg A \to A$$

is deducible in QFT.

Proof. The formulas

$$A \to \neg\neg A \to A \quad \text{and} \quad \neg A \to \neg\neg A \to A$$

are deducible in QFT. Hence by Lemma 6 we obtain $\neg\neg A \to A$.

We say that two formulas F and G are *equivalent* in QFT if the formulas $F \to G$ and $G \to F$ are deducible in QFT.

Theorem 19.1. *Let $x_1, y_1, z_1, x_2, y_2, z_2$ be sequences of distinct variables which do not occur in the nominal forms $\mathscr{A}$ and $\mathscr{B}$ and which are such that $y_1 \sim z_1 x_1 y_2$ and $x_2 \sim z_2 x_1$. Let $\mathscr{A}[x_1, y_1]$ and $\mathscr{B}[x_2, y_2]$ be quantifier-free formulas, then the formulas*

$$\exists x_1 \forall y_1 \mathscr{A}[x_1, y_1] \to \exists x_2 \forall y_2 \mathscr{B}[x_2, y_2]$$

and

$$\exists z_2{}^\frown z_1 \forall x_1{}^\frown y_2 (\mathscr{A}[x_1, z_1 x_1 y_2] \to \mathscr{B}[z_2 x_1, y_2])$$

are equivalent in QFT.

Proof. 1. The formula

$$\neg\mathscr{A}[x_1, y_1] \to \mathscr{A}[x_1, y_1] \to \mathscr{B}[x_2, y_2]$$

is deducible in QFT by Lemma 5(2). Using a cut with an $\exists$-axiom we obtain

$$\neg\mathscr{A}[x_1, y_1] \to \exists y_1(\mathscr{A}[x_1, y_1] \to \mathscr{B}[x_2, y_2]).$$

By Lemma 4(2) and a cut using the formula

$$\neg\neg\mathscr{A}[x_1, y_1] \to \mathscr{A}[x_1, y_1]$$

(which is deducible by Lemma 2) we obtain

$$\neg\exists y_1(\mathscr{A}[x_1, y_1] \to \mathscr{B}[x_2, y_2]) \to \mathscr{A}[x_1, y_1].$$

By an $\forall$-inference and Lemma 4(2) we obtain

$$\neg\forall y_1 \mathscr{A}[x_1, y_1] \to \neg\neg\exists y_1(\mathscr{A}[x_1, y_1] \to \mathscr{B}[x_2, y_2]).$$

Using a cut with an M′-axiom we obtain

$$\neg\forall y_1\mathscr{A}[x_1, y_1] \to \exists y_1(\mathscr{A}[x_1, y_1] \to \mathscr{B}[x_2, y_2]).$$

By Lemma 3(2) the formula

$$(\forall y_1\mathscr{A}[x_1, y_1] \to \mathscr{B}[x_2, y_2]) \to \neg\mathscr{B}[x_2, y_2] \to \neg\forall y_1\mathscr{A}[x_1, y_1]$$

is deducible and hence by a cut and a structural inference we obtain

$$\neg\mathscr{B}[x_2, y_2] \to (\forall y_1\mathscr{A}[x_1, y_1] \to \mathscr{B}[x_2, y_2]) \to \exists y_1(\mathscr{A}[x_1, y_1] \to \mathscr{B}[x_2, y_2]).$$

Obviously the formula

$$\mathscr{B}[x_2, y_2] \to (\forall y_1\mathscr{A}[x_1, y_1] \to \mathscr{B}[x_2, y_2]) \to \exists y_1(\mathscr{A}[x_1, y_1] \to \mathscr{B}[x_2, y_2])$$

is also deducible in QFT. Hence by Lemma 6 we obtain

$$(\forall y_1\mathscr{A}[x_1, y_1] \to \mathscr{B}[x_2, y_2]) \to \exists y_1(\mathscr{A}[x_1, y_1] \to \mathscr{B}[x_2, y_2]).$$

By Lemma 4(3) from an $\forall$-axiom we obtain

$$(\forall y_1\mathscr{A}[x_1, y_1] \to \forall y_2\mathscr{B}[x_2, y_2]) \to \forall y_1\mathscr{A}[x_1, y_1] \to \mathscr{B}[x_2, y_2].$$

By a cut and an $\forall$-inference we obtain

$$(\forall y_1\mathscr{A}[x_1, y_1] \to \forall y_2\mathscr{B}[x_2, y_2]) \to \forall y_2\exists y_1(\mathscr{A}[x_1, y_1] \to \mathscr{B}[x_2, y_2]).$$

By a cut using an AC-axiom we obtain

$$(\forall y_1\mathscr{A}[x_1, y_1] \to \forall y_2\mathscr{B}[x_2, y_2]) \to \exists z\forall y_2(\mathscr{A}[x_1, zy_2] \to \mathscr{B}[x_2, y_2]).$$

By a cut using an $\exists$-axiom and an $\exists$-inference we obtain

$$\exists x_2(\forall y_1\mathscr{A}[x_1, y_1] \to \forall y_2\mathscr{B}[x_2, y_2]) \to \exists x_2{\frown}z\forall y_2(\mathscr{A}[x_1, zy_2] \to \mathscr{B}[x_2, y_2]).$$

By a cut using an IP_o'-axiom we obtain

$$(\forall y_1\mathscr{A}[x_1, y_1] \to \exists x_2\forall y_2\mathscr{B}[x_2, y_2]) \to \exists x_2{\frown}z\forall y_2(\mathscr{A}[x_1, zy_2] \to \mathscr{B}[x_2, y_2].$$

From an $\exists$-axiom by Lemma 4(1) we obtain

$$(\exists x_1\forall y_1\mathscr{A}[x_1, y_1] \to \exists x_2\forall y_2\mathscr{B}[x_2, y_2]) \to \forall y_1[x_1, y_1] \to \exists x_2\forall y_2\mathscr{B}[x_2, y_2].$$

By a cut and an $\forall$-inference we obtain

$$(\exists x_1\forall y_1\mathscr{A}[x_1, y_1] \to \exists x_2\forall y_2\mathscr{B}[x_2, y_2]) \to \forall x_1\exists x_2{\frown}z\forall y_2(\mathscr{A}[x_1, zy_2] \to$$
$$\to \mathscr{B}[x_2, y_2]).$$

By a cut using an AC-axiom we obtain

$$(\exists x_1 \forall y_1 \mathscr{A}[x_1, y_1] \to \exists x_2 \forall y_2 \mathscr{B}[x_2, y_2]) \to \exists z_2{}^\frown z_1 \forall x_1{}^\frown y_2 (\mathscr{A}[x_1, z_1 x_1 y_2] \\ \to \mathscr{B}[z_2 x_1, y_2]).$$

2. By structural inferences and a cut from the $\forall$-axioms

$$\forall x_1{}^\frown y_2 (\mathscr{A}[x_1, z_1 x_1 y_2] \to \mathscr{B}[z_2 x_1, y_2]) \to \mathscr{A}[x_1, z_1 x_1 y_2] \to \mathscr{B}[z_2 x_1, y_2]$$

and

$$\forall y_1 \mathscr{A}[x_1, y_1] \to \mathscr{A}[x_1, z_1 x_1 y_2]$$

we obtain

$$\forall x_1{}^\frown y_2 (\mathscr{A}[x_1, z_1 x_1 y_2] \to \mathscr{B}[z_2 x_1, y_2]) \to \forall y_1 \mathscr{A}[x_1, y_1] \to \mathscr{B}[z_2 x_1, y_2].$$

By an $\forall$-inference and a cut using an $\exists$-axiom we obtain

$$\forall x_1{}^\frown y_2 (\mathscr{A}[x_1, z_1 x_1 y_2] \to \mathscr{B}[z_2 x_1, y_2]) \to \forall y_1 \mathscr{A}[x_1, y_1] \to \exists x_2 \forall y_2 \mathscr{B}[x_2, y_2].$$

By structural inferences and $\exists$-inferences we obtain

$$\exists z_2{}^\frown z_1 \forall x_1{}^\frown y_2 (\mathscr{A}[x_1, z_1 x_1 y_2] \to \mathscr{B}[z_2 x_1, y_2]) \to \exists x_1 \forall y_1 \mathscr{A}[x_1, y_1] \\ \to \exists x_2 \forall y_2 \mathscr{B}[x_2, y_2].$$

This completes the proof of Theorem 19.1.

3. Interpreting Formulas

We use formulas

$$\exists x \forall y \mathscr{A}[x, y]$$

of the system QFT as *interpreting formulas* (for formulas of the system PN) where x, y are sequences of distinct variables which do not occur in the nominal form $\mathscr{A}$ and $\mathscr{A}[x, y]$ is a quantifier-free formula. x and y may be empty. The choice of bound variables in the interpreting formulas is not significant. That is to say: two interpreting formulas

$$\exists x_1 \forall y_1 \mathscr{A}[x_1, y_1] \quad \text{and} \quad \exists x_2 \forall y_2 \mathscr{A}[x_2, y_2]$$

where $x_1 \sim x_2$ and $y_1 \sim y_2$ are identified with each other. We regard every free number variable as a variable of type o and every number term t' as an abbreviation for $\mathscr{S}t$. The number terms of the system PN are then to be understood as terms of type o in QFT where $+$ and $\cdot$ are to be understood according to the definition in §17 (p. 120).

Inductive Definition of an *interpreting formula* F^* for each formula F of the system PN.

1. If F is an equation $s=t$, then F^* is the same equation where s, t are understood as terms of type o.
2. If F is a formula $A \to B$ and A^*, B^* are the interpreting formulas

$$\exists x_1 \forall y_1 \mathscr{A}[x_1, y_1], \qquad \exists x_2 \forall y_2 \mathscr{B}[x_2, y_2],$$

where the variables in the sequences x_1, y_1, x_2, y_2 are chosen to be distinct from one another and not to occur in the nominal forms $\mathscr{A}$, $\mathscr{B}$ then F^* is the interpreting formula

$$\exists z_2 \frown z_1 \forall x_1 \frown y_2 (\mathscr{A}[x_1, z_1 x_1 y_2] \to \mathscr{B}[z_2 x_1, y_2])$$

where z_1, z_2 are chosen by Lemma 1 so that $y_1 \sim z_1 x_1 y_2$ and $x_2 \sim z_2 x_1$. The variables in the sequences z_1, z_2 are to be distinct from each other and from the variables in the sequences x_1, y_2 and are not to occur in the nominal forms $\mathscr{A}$, $\mathscr{B}$. (This choice of F^* is motivated by Theorem 19.1.)

3. If F is a formula $\forall x \mathscr{F}[x]$ where the bound number variable x does not occur in the nominal form $\mathscr{F}$, u is a free number variable which also does not occur in $\mathscr{F}$ and $\mathscr{F}[u]^*$ is the interpreting formula

$$\exists x_1 \forall y_1 \mathscr{A}[x_1, y_1, u],$$

where u is distinct from the variables in the sequences x_1, y_1 and does not occur in the nominal form $\mathscr{A}$, then F^* is the interpreting formula

$$\exists z \forall u \frown y_1 \mathscr{A}[zu, y_1, u],$$

where z is chosen by Lemma 1 so that $x_1 \sim zu$. The variables in the sequence z are to be distinct and also distinct from u and the variables in the sequence y_1 and not to occur in the nominal form $\mathscr{A}$. $u \frown y_1$ denotes the sequence of variables obtained by concatenating the variable u (of type o) with the sequence y_1. (This choice of F^* is motivated by an AC-axiom.)

Definition. An interpreting formula

$$\exists x \forall y \mathscr{A}[x, y]$$

is said to be *valid* if there is a sequence of terms a, in which no variable in the sequence y occurs, such that $a \sim x$ holds and the quantifier-free formula $\mathscr{A}[a, y]$ is deducible in the system FT.

Theorem 19.2 (Interpretation Theorem). *If F is a formula deducible in the system* PN *then its interpreting formula F^* is valid.*

Corollary. *The formal system* PN *is sententially consistent.* That is to say: There is no formula A such that A and $\neg A$ are both deducible in PN.

Proof. $\neg A$ is defined to be the formula $A \to 0'=0$. If A and $A \to 0'=0$ were deducible in PN then by §18 (Lemma 2) the formula $0'=0$ would also be deducible in PN. But by Theorem 17.2 its interpreting formula $\mathscr{S}0=0$ is not deducible in FT and therefore is not valid. It follows by Theorem 19.2 that $0'=0$ is not deducible in PN.

We prove Theorem 19.2 by induction on the deduction, viz. by induction on the length of an assumed deduction of F in PN. We write $\vdash A$ to indicate that A is deducible in the system FT.

4. Interpretations of the Axioms of the System PN

We first prove that the interpreting formula F^* of each axiom F of the system PN is valid.

1. Suppose F is an axiom

(A1) $A \to A$.

Then F^* is an interpreting formula

$$\exists z_2 \frown z_1 \forall x_1 \frown y_2 (\mathscr{A}[x_1, z_1 x_1 y_2] \to \mathscr{A}[z_2 x_1, y_2]).$$

By Lemma 2 there are sequences of terms a_1, a_2 in which no variable from the sequences x_1, y_2 occurs such that

$$\vdash a_1 x_1 y_2 = y_2 \quad \text{and} \quad \vdash a_2 x_1 = x_1.$$

Hence we have

$$\vdash \mathscr{A}[x_1, a_1 x_1 y_2] \to \mathscr{A}[a_2 x_1, y_2].$$

Thus F^* is valid.

2. Suppose F is an axiom

(A2) $\forall x \mathscr{F}[x] \to \mathscr{F}[t]$.

Then the formulas $\forall x \mathscr{F}[x]$ and $\mathscr{F}[t]$ have interpreting formulas

$$\exists x_1 \forall u \frown y_1 \mathscr{A}[x_1 u, y_1, u] \quad \text{and} \quad \exists x_2 \forall y_2 \mathscr{A}[x_2, y_2, t].$$

Hence F^* is an interpretating formula

$$\exists x_2 \frown z_0 \frown z_1 \forall x_1 \frown y_2 (\mathscr{A}[x_1(z_0 x_1 y_2), z_1 x_1 y_2, z_0 x_1 y_2] \to \mathscr{A}[z_2 x_1, y_2, t]).$$

By Lemma 2 there are sequences of terms a_0, a_1, a_2 in which no variable from the sequences x_1, y_2 occurs such that

$$\vdash a_0 x_1 y_2 = t, \qquad \vdash a_1 x_1 y_2 = y_2, \qquad \vdash a_2 x_1 = x_1 t.$$

It follows that $\vdash x_1(a_0 x_1 y_2) = a_2 x_1$ and

$$\vdash \mathscr{A}[x_1(a_0 x_1 y_2), a_1 x_1 y_2, a_0 x_1 y_2] \to \mathscr{A}[a_2 x_1, y_2, t].$$

Thus F^* is valid.

3. Suppose F is an axiom

(A3) $u' = 0 \to A$.

Then F^* is an interpreting formula

$$\exists x \forall y (\mathscr{S} u = 0 \to \mathscr{A}[x, y]).$$

By Theorem 17.6 we have

$$\vdash \mathscr{S} u = 0 \to \mathscr{A}[x, y].$$

Thus F^* is valid.

4. Suppose F is an axiom

(A4) $u' = v' \to u = v$.

Then the interpreting formula F^* is the quantifier-free formula

$$\mathscr{S} u = \mathscr{S} v \to u = v,$$

which is deducible in FT by Theorem 17.9 and therefore valid.

5. Suppose F is one of the axioms (A5)–(A8). Then F^* is an equation which is deducible in FT by §17 (p. 120) and therefore valid.

5. Interpretations of the Basic Inferences in the System PN

We now prove that assuming the interpreting formulas of all premises of a basic inference of the system PN are valid, then the interpreting formula of the conclusion is valid.

1. Suppose G is the conclusion of a basic inference

(B1) $B \vdash A \rightarrow B.$

Then B^* is an interpreting formula

$$\exists x_2 \forall y_2 \mathscr{B}[x_2, y_2]$$

and G^* is an interpreting formula

$$\exists z_2 \frown z_1 \forall x_1 \frown y_2 (\mathscr{A}[x_1, z_1 x_1 y_2] \rightarrow \mathscr{B}[z_2 x_1, y_2]).$$

By hypothesis there is a sequence of terms a_2 in which no variable from the sequence y_2 occurs such that

$$\vdash \mathscr{B}[a_2, y_2].$$

Then by Lemma 2 there is a sequence of terms b_2 in which no variable from the sequences x_1, y_2 occurs such that

$$\vdash b_2 x_1 = a_2.$$

It follows that

$$\vdash \mathscr{A}[x_1, z_1 x_1 y_2] \rightarrow \mathscr{B}[b_2 x_1, y_2].$$

Thus G^* is valid.

2. Suppose F is the premise and G the conclusion of a basic inference

(B2) $A \rightarrow B \rightarrow C \rightarrow D \vdash A \rightarrow C \rightarrow B \rightarrow D.$

Then F^* is an interpreting formula

$$\begin{cases} \exists z_4 \frown z_3 \frown z_2 \frown z_1 \forall x_1 \frown x_2 \frown x_3 \frown y_4 (\mathscr{A}[x_1, z_1 x_1 x_2 x_3 y_4] \rightarrow \mathscr{B}[x_2, z_2 x_1 x_2 x_3 y_4] \\ \quad \rightarrow \mathscr{C}[x_3, z_3 x_1 x_2 x_3 y_4] \rightarrow \mathscr{D}[z_4 x_1 x_2 x_3, y_4]) \end{cases}$$

and G^* is an interpreting formula

$$\begin{cases} \exists w_4 \frown w_2 \frown w_3 \frown w_1 \forall x_1 \frown x_1 \frown x_3 \frown x_2 \frown y_4 (\mathscr{A}[x_1, w_1 x_1 x_3 x_2 y_4] \rightarrow \mathscr{C}[x_3, w_3 x_1 x_3 x_2 y_4] \\ \quad \rightarrow \mathscr{B}[x_2, w_2 x_1 x_3 x_2 y_4] \rightarrow \mathscr{D}[w_4 x_1 x_3 x_2, y_4]). \end{cases}$$

By hypothesis there are sequences of terms a_1, a_2, a_3, a_4 in which no variable from the sequences x_1, x_2, x_3, y_4 occurs such that

$$\vdash \begin{cases} \mathscr{A}[x_1, a_1 x_1 x_2 x_3 y_4] \rightarrow \mathscr{B}[x_2, a_2 x_1 x_2 x_3 y_4] \\ \quad \rightarrow \mathscr{C}[x_3, a_3 x_1 x_2 x_3 y_4] \rightarrow \mathscr{D}[a_4 x_1 x_2 x_3, y_4]. \end{cases}$$

By Lemma 2 there are sequences of terms b_1, b_2, b_3, b_4 in which no variable from the sequences x_1, x_2, x_3, y_4 occurs such that

$$\vdash b_i x_1 x_3 x_2 y_4 = a_i x_1 x_2 x_3 y_4 \quad (i=1,2,3)$$

and

$$\vdash b_4 x_1 x_3 x_2 = a_4 x_1 x_2 x_3 .$$

Using the equality rules and a structural inference we obtain

$$\vdash \begin{cases} \mathscr{A}[x_1, b_1 x_1 x_3 x_2 y_4] \to \mathscr{C}[x_3, b_3 x_1 x_3 x_2 y_4] \\ \quad \to \mathscr{B}[x_2, b_2 x_1 x_3 x_2 y_4] \to \mathscr{D}[b_4 x_1 x_3 x_2, y_4]. \end{cases}$$

Thus G^* is valid.

3. Suppose F_1, F_2 are the premises and G the conclusion of a basic inference

(B3) $A \to B \to C, \qquad C \to D \vdash A \to B \to D.$

Then F_1^*, F_2^* are interpreting formulas

$$\exists z_3 \frown z_2 \frown z_1 \forall x_1 \frown z_2 \frown y_3 (\mathscr{A}[x_1, z_1 x_1 x_2 y_3] \to \mathscr{B}[x_2, z_2 x_1 x_2 y_3] \to \mathscr{C}[z_3 x_1 x_2, y_3]),$$
$$\exists z_4 \frown z_5 \forall x_3 \frown y_4 (\mathscr{C}[x_3, z_5 x_3 y_4] \to \mathscr{D}[z_4 x_3, y_4]),$$

and G^* is an interpreting formula

$$\exists w_4 \frown w_2 \frown w_1 \forall x_1 \frown x_2 \frown y_4 (\mathscr{A}[x_1, w_1 x_1 x_2 y_4] \to \mathscr{B}[x_2, w_2 x_1 x_2 y_4]$$
$$\to \mathscr{D}[w_4 x_1 x_2, y_4]).$$

By hypothesis there are sequences of terms, $a_1, \ldots, a_5$ in which no variable from the sequences x_1, x_2, x_3, y_3, y_4 occurs such that

$$\vdash \mathscr{A}[x_1, a_1 x_1 x_2 y_3] \to \mathscr{B}[x_2, a_2 x_1 x_2 y_3] \to \mathscr{C}[a_3 x_1 x_2, y_3]$$

and

$$\vdash \mathscr{C}[x_3, a_5 x_3 y_4] \to \mathscr{D}[a_4 x_3, y_4].$$

Substituting $a_5(a_3 x_1 x_2) y_4$ for y_3 and $a_3 x_1 x_2$ for x_3 we obtain by Theorem 17.1

$$\vdash \begin{cases} \mathscr{A}[x_1, a_1 x_1 x_2 (a_5(a_3 x_1 x_2) y_4) \to \mathscr{B}[x_2, a_2 x_1 x_2 (a_5(a_3 x_1 x_2) y_4)] \\ \quad \to \mathscr{C}[a_3 x_1 x_2, a_5(a_3 x_1 x_2) y_4] \end{cases}$$

and

$$\vdash \mathscr{C}[a_3 x_1 x_2, a_5(a_3 x_1 x_2) y_4] \to \mathscr{D}[a_4(a_3 x_1 x_2), y_4].$$

By a cut we obtain

$$\vdash\begin{cases}\mathscr{A}[x_1, a_1x_1x_2(a_5(a_3x_1x_2)y_4)] \to \mathscr{B}[x_2, a_2x_1x_2(a_5(a_3x_1x_2)y_4)] \\ \quad \to \mathscr{D}[a_4(a_3x_1x_2), y_4].\end{cases}$$

By Lemma 2 there are sequences of terms b_1, b_2, b_4 in which no variable from the sequences x_1, x_2, y_4 occurs such that

$$\vdash b_1x_1x_2y_4 = a_1x_1x_2(a_5(a_3x_1x_2)y_4),$$
$$\vdash b_2x_1x_2y_4 = a_2x_1x_2(a_5(a_3x_1x_2)y_4),$$
$$\vdash b_4x_1x_2 = a_4(a_3x_1x_2).$$

Hence we have

$$\vdash \mathscr{A}[x_1, b_1x_1x_2y_4] \to \mathscr{B}[x_2, b_2x_1x_2y_4] \to \mathscr{D}[b_4x_1x_2, y_4].$$

Thus G^* is valid.

4. Suppose F is the premise and G the conclusion of a basic inference

(B4) $A \to A \to B \vdash A \to B.$

Then F^* is an interpreting formula

$$\exists z_2 \frown z_1 \frown z_0 \forall x_0 \frown x_1 \frown y_2(\mathscr{A}[x_0, z_0x_0x_1y_2] \to \mathscr{A}[x_1, z_1x_0x_1y_2] \to \mathscr{B}[z_2x_0x_1, y_2])$$

and G^* is an interpreting formula

$$\exists w_2 \frown w_1 \forall x_1 \frown y_2(\mathscr{A}[x_1, w_1x_1y_2] \to \mathscr{B}[w_2x_1, y_2]).$$

By hypothesis there are sequences of terms a_0, a_1, a_2 in which no variable from the sequences x_0, x_1, y_2 occurs such that

$$\vdash \mathscr{A}[x_0, a_0x_0x_1y_2] \to \mathscr{A}[x_1, a_1x_0x_1y_2] \to \mathscr{B}[a_2x_0x_1, y_2].$$

Substituting x_1 for x_0 by Theorem 17.1 we obtain

(1) $\vdash \mathscr{A}[x_1, a_0x_1x_1y_2] \to \mathscr{A}[x_1, a_1x_1x_1y_2] \to \mathscr{B}[a_2x_1x_1, y_2].$

By Lemma 2 there are sequences of terms b_1, b_2 in which no variable from the sequences x_1, y_2 occurs such that

(2) $\vdash b_2x_1 = a_2x_1x_1$
$\vdash b_1x_1y_2 = D[\chi[\mathscr{A}[x_1, a_0x_1x_1y_2]], a_1x_1x_1y_2, a_0x_1x_1y_2].$

Hence we have (see p. 119)

(3) $\vdash \chi[\mathscr{A}[x_1, a_0x_1x_1y_2]] = 0 \to b_1x_1y_2 = a_1x_1x_1y_2$
(4) $\vdash \chi[\mathscr{A}[x_1, a_0x_1x_1y_2]] = \mathscr{S}u^0 \to b_1x_1y_2 = a_0x_1x_1y_2].$

By Theorem 17.5 we have

$$\vdash \chi[\mathscr{A}[x_1, a_0x_1x_1y_2]] = 0 \rightarrow \mathscr{A}[x_1, a_0x_1x_1y_2].$$

By a cut with (1) we obtain

$$\vdash \chi[\mathscr{A}[x_1, a_0x_1x_1y_2]] = 0 \rightarrow \mathscr{A}[x_1, a_1x_1x_1y_2] \rightarrow \mathscr{B}[a_2x_1x_1, y_2].$$

Using (2) and (3) we obtain

(5) $\vdash \chi[\mathscr{A}[x_1, a_0x_1x_1y_2]] = 0 \rightarrow \mathscr{A}[x_1, b_1x_1y_2] \rightarrow \mathscr{B}[b_2x_1, y_2].$

By Theorem 17.16 we have

$$\vdash \chi[\mathscr{A}[x_1, a_0x_1x_1y_2]] = \mathscr{S}u^0 \rightarrow \mathscr{A}[x_1, a_0x_1x_1y_2] \rightarrow \mathscr{B}[b_2x_1, y_2].$$

Using (4) we obtain

(6) $\vdash \chi[\mathscr{A}[x_1, a_0x_1x_1y_2]] = \mathscr{S}u^0 \rightarrow \mathscr{A}[x_1, b_1x_1y_2] \rightarrow \mathscr{B}[b_2x_1, y_2].$

By the Corollary to Theorem 17.5 from (5) and (6) we obtain

$$\vdash \chi[x_1, b_1x_1y_2] \rightarrow \mathscr{B}[b_2x_1, y_2].$$

Thus G^* is valid.

5. and 6. Suppose F is the premise and G the conclusion of a basic inference

(B5) $u = u \rightarrow A \vdash A$

or

(B6) $\mathscr{F}[s] \vdash s = t \rightarrow \mathscr{F}[t].$

Then trivially F^* is valid implies G^* is valid.

7. Suppose F is the premise and G the conclusion of a basic inference

(B7) $A \rightarrow \mathscr{F}[u] \vdash A \rightarrow \forall x \mathscr{F}[x].$

Then F^* is an interpreting formula

$$\exists z_2 \frown z_1 \forall x_1 \frown y_2 (\mathscr{A}[x_1, z_1x_1y_2] \rightarrow \mathscr{B}[z_2x_1, y_2, u])$$

and G^* is an interpreting formula

$$\exists w_2 \frown w_1 \forall x_1 \frown u \frown y_2 (\mathscr{A}[x_1, w_1x_1uy_2] \rightarrow \mathscr{B}[w_2x_1u, y_2, u]).$$

By hypothesis there are sequences of terms a_1, a_2 in which no variable from the sequences x_1, y_2 occurs such that

$$\vdash \mathscr{A}[x_1, a_1 x_1 y_2] \rightarrow \mathscr{B}[a_2 x_1, y_2, u].$$

By Lemma 2 there are sequences of terms b_1, b_2 in which no variable from the sequence $x_1 \frown u \frown y_2$ occurs such that

$$\vdash b_1 x_1 u y_2 = a_1 x_1 y_2 \quad \text{and} \quad \vdash b_2 x_1 u = a_2 x_1.$$

Hence

$$\vdash \mathscr{A}(x_1, b_1 x_1 u y_2] \rightarrow \mathscr{B}[b_2 x_1 u, y_2, u].$$

Thus G^* is valid.

8. Suppose F_1, F_2 are the premises and G the conclusion of a basic inference

(B8) $\mathscr{F}[0], F[u] \rightarrow \mathscr{F}[u'] \vdash \mathscr{F}[t]$.

Then F_1^*, F_2^* are interpreting formulas

$$\exists x \forall y \mathscr{A}[x, y, 0],$$
$$\exists z_2 \frown z_1 \forall x \frown y (\mathscr{A}[x, z_1 x y, u] \rightarrow \mathscr{A}[z_2 x, y, \mathscr{S} u]),$$

where u does not occur in the nominal form $\mathscr{A}$ and G^* is an interpreting formula

$$\exists x \forall y \mathscr{A}[x, y, t].$$

By hypothesis there are sequences of terms a_0, a_1, a_2 in which no variable from the sequences x, y occurs such that

(1) $\vdash \mathscr{A}[a_0, y, 0]$
$\vdash \mathscr{A}[x, a_1 x y, u] \rightarrow \mathscr{A}[a_2 x, y, \mathscr{S} u]$.

We may assume that u does not occur in a_0. By Lemma 2 there are sequences of terms b_1, b_2 in which neither u nor any variable from the sequences x, y occurs such that

$$\vdash b_i u = a_i \quad (i = 1, 2).$$

Hence

(2) $\vdash \mathscr{A}[x, b_1 u x y, u] \rightarrow \mathscr{A}[b_2 u x, y, \mathscr{S} u]$.

By Theorem 17.12 there is a sequence of terms f in which neither u nor any variable from the sequences x, y occurs such that

(3) $\vdash f0 = a_0$
(4) $\vdash f(\mathscr{S}u) = b_2u(fu)$.

From (1) and (3) we obtain

(5) $\vdash \mathscr{A}[f0, y, 0]$.

From (2) and (4) substituting fu for x we obtain

(6) $\vdash \mathscr{A}[fu, b_1u(fu)y, u] \rightarrow \mathscr{A}[f(\mathscr{S}u), y, \mathscr{S}u]$.

From (5) and (6) by Theorem 17.13 we obtain

$\vdash \mathscr{A}[ft, y, t]$.

Thus G^* is valid.

This completes the proof of Theorem 19.2.

Part C

Subsystems of Analysis

Chapter VIII

Predicative Analysis

In this chapter we consider formal and semi-formal systems in which the Peano axioms hold for number variables and there is quantification over predicate variables. In these systems the real numbers are definable by predicators as certain sets of rational numbers and universal and existential statements about real numbers are formalizable. Here certain subsystems of classical analysis will be delimited. We restrict ourselves in this chapter to *predicative* subsystems of analysis, that is, to systems which are interpreted in such a way that a concept which is defined under the assumption of a collection of concepts does not itself belong to this collection.

§20. Systems of Δ_1^1-Analysis

The systems in this section have the same language as for classical analysis but are restricted in the definition of predicators (sets) by Δ_1^1-comprehension for which a predicative interpretation will be made possible (in §22).

1. The Formal Language of Second Order Arithmetic

As *primitive symbols* we use

1. Denumerably infinitely many free and bound number variables and predicate variables. (All predicate variables are to be 1-place.)
2. The symbols 0, $'$, $\perp$, $\rightarrow$, $\forall$ and λ.
3. Symbols for n-place calculable arithmetic functions and n-place decidable arithmetic predicates ($n \geqslant 1$). (The conditions governing the use of these symbols will be given on page 168.)
4. Round brackets and comma.

We use *nominal forms* in the usual way and denote them by capital script letters.

Inductive definition of *terms*:

1. The symbol 0 is a term.
2. Every free number variable is a term.

3. If t is a term then so too is t .

4. If $\mathfrak{f}$ is a symbol for an n-place calculable arithmetic function ($n \geqslant 1$) and $t_1, \ldots, t_n$ are terms, then $\mathfrak{f}(t_1, \ldots, t_n)$ is also a term.

Terms built up according to 1. and 3. only are called *numerals*. A term is said to be *numerical* if it contains no free number variables.

The *prime formulae* are:

1. The symbol $\perp$,

2. $\mathfrak{P}(t_1, \ldots, t_n)$ where $\mathfrak{P}$ is a symbol for an n-place decidable arithmetic predicate ($n \geqslant 1$) and $t_1, \ldots, t_n$ are terms.

A prime formula is said to be *constant* if it contains no free number variable.

Inductive definition of *formulas* and *predicators*:

1. Every prime formula is a formula.
2. Every free predicate variable is a predicator.
3. If P is a predicator and t a term, then $P(t)$ is a formula.
4. If A and B are formulas then so too is $(A \rightarrow B)$.
5. If $\mathscr{F}[0]$ is a formula and x is a bound number variable which does not occur in $\mathscr{F}$, then $\forall x \mathscr{F}[x]$ is a formula and $\lambda x \mathscr{F}[x]$ is a predicator.
6. If U is a free predicate variable, $\mathscr{F}[U]$ is a formula and X is a bound predicate variable which does not occur in $\mathscr{F}$ then $\forall X \mathscr{F}[X]$ is a formula.

A formula is said to be *simple* if it is not of the form $(A \rightarrow B)$. A predicator is said to be *elementary* if it contains no bound predicate variables. By the *length* of a formula we mean the number of symbols $\rightarrow$, $\forall$ and λ which occur in the formula.

Syntactic symbols:

a, b, c, d	for free number variables,
x, y, z	for bound number variables,
U, V, W	for free predicate variables,
X, Y, Z	for bound predicate variables,
s, t	for terms,
P, Q	for predicators,
A, B, C, D, E, F, G	for formulas,
i, j, k, m, n	for natural numbers and their corresponding numerals.

(A natural number n corresponds to the numeral in which the symbol occurs precisely n times.)
We shall also use these syntactic symbols with indices.

Conditions for the use of *arithmetic functions and predicates symbols*:

1. A symbol $\mathfrak{f}$ for an n-place calculable arithmetic function ($n \geqslant 1$) may only be used as a primitive symbol provided there is a general procedure by which a numeral may be calculated as the *value* of $\mathfrak{f}(m_1, \ldots, m_n)$ for each n numerals $m_1, \ldots, m_n$.

2. A symbol $\mathfrak{P}$ for an n-place decidable arithmetic predicate ($n \geqslant 1$) may only be used as a primitive symbol provided there is a general procedure which decides, for each n numerals $m_1, \ldots, m_n$, whether $\mathfrak{P}(m_1, \ldots, m_n)$ is *true* or *false*.

Inductive definition of the *value* of a *numerical term*:

1. The term 0 has value 0.
2. If t is a numerical term of value m, then t' has value m'.
3. If $\mathfrak{f}$ is a symbol for an n-place calculable arithmetic function ($n \geqslant 1$) and $t_1, \ldots, t_n$ are numerical terms with values $m_1, \ldots, m_n$, then $\mathfrak{f}(t_1, \ldots, t_n)$ has the value given for $\mathfrak{f}(m_1, \ldots, m_n)$ by the procedure for $\mathfrak{f}$ which we required to exist.

By this definition every numerical term has a calculable value which is a numeral.

Definition of the *truth-value* of a *constant prime formula*.

1. The formula $\perp$ is *false*.
2. If $\mathfrak{P}$ is a symbol for an n-place decidable arithmetic predicate ($n \geqslant 1$) and $t_1, \ldots, t_n$ are numerical terms with values $m_1, \ldots, m_n$ then $\mathfrak{P}(t_1, \ldots, t_n)$ is *true* or *false* according as $\mathfrak{P}(m_1, \ldots, m_n)$ is decided to be true or false by the procedure which we required to exist for $\mathfrak{P}$.

By this definition every constant prime formula is, decidably, either true or false.

Two formulas are said to be *equivalent* if they are formulas $\mathscr{F}[s_1, \ldots, s_n]$ and $\mathscr{F}[r_1, \ldots, t_n]$ where s_i and $t_i (i = 1, \ldots, n)$ are numerical terms of equal values.

Theorem 20.1. *Equivalence of formulae is an equivalence relation.*

Proof. By definition equivalence is reflexive and symmetric. A numerical term t is said to be a *maximal-numerical* component of a formula F, if t occurs in F but not as a component of a longer numerical term. By a substitution which transforms F into an equivalent formula every maximal-numerical component of F is transformed into a numerical term of equal value while every other component of F remains unaltered. Hence it follows that equivalence of formulas is also transitive.

We define *positive* and *negative parts* of formulas, *P-forms*, *N-forms* and *NP-forms* in the usual way (see p. 20). As before we define:

The *minimal positive parts* of a formula F are the simple positive parts of F. The *minimal negative parts* of a formula F are those negative parts of F which are not of the form $(A \to \perp)$. The *minimal parts* of F are the minimal positive and negative parts of F.

$F \models^{s} G$ (*G follows structurally from F*) denotes that to every minimal positive (negative) part of F there is an identical positive (negative) part of G.

We use $\mathscr{P}$ for P-forms, $\mathscr{N}$ for N-forms, $\mathscr{Q}$ for NP-forms and $\mathscr{E}$ for P-forms and N-forms as syntactic symbols.

As in §4.4 we define

$$\neg A := (A \to \perp)$$
$$(A \wedge B) := \neg(A \to \neg B)$$
$$(A \vee B) := (\neg A \to B)$$
$$(A \leftrightarrow B) := \neg((A \to B) \to \neg(B \to A))$$
$$\exists x \mathscr{F}[x] := \neg \forall x \neg \mathscr{F}[x]$$
$$\exists X \mathscr{F}[X] := \neg \forall X \neg \mathscr{F}[X].$$

For brevity we omit the outer round brackets in formulas. We also write $A_1 \to \cdots \to A_n \to B$ for $(A_1 \to (\cdots \to (A_n \to B)\ldots))$.

2. The Formal System DA

Axioms:

(Ax1) $\mathscr{P}[A]$, if A is a true constant prime formula,

(Ax2) $\mathscr{N}[A]$, if A is a false constant prime formula.

(Ax3) $\mathscr{Q}[A, B)$ if A, B are equivalent formulas of length 0.

(Ax4) $\mathscr{F}[a_1, \ldots, a_n]$ $(n \geq 1)$, if for every n numerals $m_1, \ldots, m_n$ $\mathscr{F}[m_1, \ldots, m_n]$ is one of the axioms (Ax1)–(Ax3).

The minimal parts indicated in axioms (Ax1)–(Ax3) are called the *principal parts* of these axioms.

Remark. It is not in general decidable whether a formula is an (Ax4) axiom. A formula, therefore, may only be used as an (Ax4) axiom if there is a general procedure which shows that the formula does satisfy the conditions for such an axiom.

Basic inferences:

(S1) $\mathscr{N}[\neg A], \mathscr{N}[B] \vdash \mathscr{N}[(A \to B)]$, if B is not the formula $\perp$.

(S2.0) $\mathscr{P}[\mathscr{F}[a]] \vdash \mathscr{P}[\forall x \mathscr{F}[x]]$, if a does not occur in the conclusion.

(S2.1) $\mathscr{P}[\mathscr{F}[U]] \vdash \mathscr{P}[\forall X \mathscr{F}[X]]$, if U does not occur in the conclusion.

(S3.0) $\mathscr{F}[t] \to \mathscr{N}[\forall x \mathscr{F}[x]] \vdash \mathscr{N}[\forall x \mathscr{F}[x]]$

(S3.E) $\mathscr{F}[P] \to \mathscr{N}[\forall X \mathscr{F}[X]] \vdash \mathscr{N}[\forall X \mathscr{F}[X]]$, if P is an elementary predicator.

(S3.Δ) $\begin{cases} \forall x(\forall Y \mathscr{A}[x, Y] \leftrightarrow \exists Y \mathscr{B}[x, Y]), \\ \mathscr{F}[\lambda x \forall Y \mathscr{A}[x, Y]] \to \mathscr{N}[\forall X \mathscr{F}[X]] \vdash \mathscr{N}[\forall X \mathscr{F}[X]], \end{cases}$

if no bound predicate variables occur in $\mathscr{A}$ and $\mathscr{B}$.

(S4) $\mathscr{E}[\mathscr{F}[t]] \vdash \mathscr{E}[\lambda x \mathscr{F}[x](t)]$

(cut) $\mathscr{P}[A], A \to B \vdash C$, if $\mathscr{P}[B] \models^{s} C$ holds.

(ind) $\mathscr{F}[0], \mathscr{F}[a] \to \mathscr{F}[a'] \vdash \mathscr{F}[t]$, if a does not occur in $\mathscr{F}$.

(*Complete induction* inference)

Remark on (S3.E) and (S3.Δ). If $\mathscr{F}[U]$ is a formula containing no bound predicate variable then $\forall Y \mathscr{F}[Y]$ is said to be a *Π_1^1-formula* and $\exists Y \mathscr{F}[Y]$ is said to be a *Σ_1^1-formula*. If the formula

$$\forall Y \mathscr{A}[Y] \leftrightarrow \exists Y \mathscr{B}[Y]$$

is deducible for a Π_1^1-formula $\forall Y \mathscr{A}[Y]$ and a Σ_1^1-formula $\exists Y \mathscr{B}[Y]$ then $\forall Y \mathscr{A}[Y]$ and $\exists Y \mathscr{B}[Y]$ are said to be *Δ_1^1-formulas*. Similarly we call a predicator $\lambda x \forall Y \mathscr{A}[x, Y]$ a *Δ_1^1-predicator* if there is a deducible formula

$$\forall x(\forall Y \mathscr{A}[x, Y] \leftrightarrow \exists Y \mathscr{B}[x, Y])$$

where no bound predicate variables occur in $\mathscr{A}$ and $\mathscr{B}$. So by (S3.E) and (S3.Δ) basic inferences of the form

(S3.1) $\mathscr{F}[P] \to \mathscr{N}[\forall X \mathscr{F}[X]] \vdash \mathscr{N}[\forall X \mathscr{F}[X]]$

in the system DA are restricted to those where P is an elementary predicator or a Δ_1^1-predicator. This formation of such predicators is called *Δ_1^1-comprehension*. If one allows (S3.1) as a basic inference for every predicator P then with the axioms and the remaining basic inferences of the system DA one obtains full classical arithmetic of second order which is not, however, predicatively interpretable. The restriction in (S3.1) to (S3.E) and (S3.Δ) which makes a predicative interpretation possible (in §22) means that the domain of the quantifiers $\forall X$ is restricted to those predicators which satisfy Δ_1^1-comprehension.

Inductive definition of DA $\models\!\frac{m,n}{}\, F$:

1. If F is an axiom of the system DA, then DA $\models\!\frac{0,0}{}\, F$ holds.
2. If DA $\models\!\frac{m,n}{}\, F_1$ holds for the premise F_1 of a basic inference (S2.0), (S2.1), (S3.0), (S3.E) or (S4) of the system DA, then DA $\models\!\frac{m,n+1}{}\, F$ holds for the conclusion F of that basic inference.
3. If DA$\models\!\frac{m_i,n_i}{}\, F_i$ $(i=1,2)$ holds for the premises F_1 and F_2 of a basic inference (S1), or (cut) of the system DA, then DA $\models\!\frac{m,n}{}\, F$ holds, where $m:=\max(m_1, m_2)$ and $n:=\max(n_1,n_2)+1$, for the conclusion F of that basic inference.
4. If DA$\models\!\frac{m_i,n_i}{}\, F_i$ $(i=1,2)$ holds for the premises F_1 and F_2 of a basic inference (S3.Δ) or (ind), then DA $\models\!\frac{m,0}{}\, F$ holds, where $m:=\max(m_1,m_2)+1$, for the conclusion F of that basic inference.

A formula F is said to be *deducible* in the system DA if there exist natural numbers m, n such that DA $\models\!\frac{m,n}{}\, F$ holds. If DA $\models\!\frac{0,n}{}\, F$ holds then F is deducible in DA without use of the inference rule (S3.Δ) and (ind). We also abbreviate DA $\models\!\frac{m,n}{}\, F$ to $\vdash F$.

3. Deducible Formulas and Permissible Inference of the System DA

Let $=$ be the symbol for the 2-place decidable arithmetic predicate such that $=(m_1, m_2)$ is *true* if, and only if, the numerals m_1, m_2 are identical. We use $=$ as a primitive symbol of our formal language and write $s=t$ for $=(s,t)$.

Theorem 20.2. $\vdash t=t$.

Proof. $t=t$ is an (Ax1) if t is numerical, otherwise an (Ax4).

Theorem 20.3. $\vdash s=t \rightarrow \mathcal{Q}[\mathcal{F}[s], \mathcal{F}[t]]$.

Proof by induction on the length of $\mathcal{F}[s]$.

1. Let $\mathcal{F}[s]$ have length 0. Then the formula

$$s=t \rightarrow \mathcal{Q}[\mathcal{F}[s], \mathcal{F}[t]]$$

is an (Ax2) or an (Ax3) if s and t are numerical, otherwise it is an (Ax4).

2. Let $\mathcal{F}[s]$ have length >0. Then the assertion follows from the I.H. using basic inferences (S1), (S2.0), (S3.0), (S2.1), (S3.E) and (S4) as in the proof of Theorem 4.1.

By Theorems 20.2 and 20.3 all the laws of an identity relation hold for $s=t$.

Theorem 20.4. $\vdash \mathcal{Q}[C, C]$.

Proof. By Theorems 20.2 and 20.3 we have $\vdash t=t$ and $\vdash t=t \rightarrow \mathcal{Q}[C, C]$. By a cut we obtain $\vdash \mathcal{Q}[C, C]$.

Corollary to Theorem 20.4. *The formal system* DA *is s-complete, that is, every s-valid formula of our formal language is deducible in* DA. (*This follows just like the Corollary to Theorem* 4.1.)

By virtue of this corollary it follows in particular that all the inferences given in Theorem 4.8 for $\wedge$ and $\vee$ are also permissible in the system DA and also

$$\mathcal{P}[(A \rightarrow B)], \mathcal{P}[(B \rightarrow A) \vdash \mathcal{P}[(A \leftrightarrow B)],$$
$$\mathcal{P}[(A \leftrightarrow B] \vdash \mathcal{P}[(A \rightarrow B)] \quad \text{and} \quad \mathcal{P}[(A \leftrightarrow B)] \vdash \mathcal{P}[(B \rightarrow A)].$$

Theorem 20.5 (Structural inference rule). *If* $F \overset{s}{\models} G$ *holds, then* $F \vdash G$ *is a permissible inference.*

Proof. If $F \overset{s}{\models} G$ holds, then

$$F \rightarrow F, F \vdash G$$

is a cut. Using $\vdash F \rightarrow F$ (by Theorem 20.4) the assertion follows.

Theorem 20.6 (Basic properties of quantifiers).

a) $\vdash \forall x \mathcal{F}[x] \rightarrow \mathcal{F}[t]$ *and* $\vdash \mathcal{F}[t] \rightarrow \exists x \mathcal{F}[x]$

b) $\vdash \forall X \mathcal{F}[X] \rightarrow \mathcal{F}[P]$ *and* $\vdash \mathcal{F}[P] \rightarrow \exists X \mathcal{F}[X]$, *if* P *is an elementary or* Δ_1^1*-predicator.*

Proof. By Theorem 20.4,

$$\vdash \mathcal{F}[t] \rightarrow \forall x \mathcal{F}[x] \rightarrow \mathcal{F}[t] \quad \text{and} \quad \vdash \neg \mathcal{F}[t] \rightarrow \mathcal{F}[t] \rightarrow \neg \forall x \neg \mathcal{F}[x]$$

hold. Assertion a) follows using basic inferences (S3.0). Assertion b) follows similarly using basic inferences (S3.E) and (S3.Δ).

Theorem 20.7 (Substitution rules). *The following are permissible inferences*:

a) $\mathcal{F}[a] \vdash \mathcal{F}[t]$, *if* a *does not occur in* $\mathcal{F}$.

b) $\mathcal{F}[U] \vdash \mathcal{F}[P]$, *if* U *does not occur in* $\mathcal{F}$ *and* P *is an elementary or* Δ_1^1*-predicator.*

Proof. This follows from Theorem 20.6 using basic inferences (S2.0) and (S2.1).

Theorem 20.8 (Introduction of existential quantifiers). *The following are permissible inferences*:

a) $\mathcal{N}[\mathcal{F}[a]] \vdash \mathcal{N}[\exists x \mathcal{F}[x]]$, *if a does not occur in the conclusion.*
b) $\mathcal{N}[\mathcal{F}[U]] \vdash \mathcal{N}[\exists X \mathcal{F}[X]]$, *if U does not occur in the conclusion.*

Proof. By a structural rule we obtain $\mathcal{N}[\neg\neg\mathcal{F}[a]]$ from $\mathcal{N}[\mathcal{F}[a]]$ and then by a basic inference (S2.0) $\mathcal{N}[\neg\forall x \neg\mathcal{F}[x]]$. b) follows similarly using a basic inference (S2.1).

Theorem 20.9 (Extensionality Theorem).

$$\vdash \forall x(\mathcal{A}[x] \leftrightarrow \mathcal{B}[x]) \rightarrow \mathcal{F}[\lambda x \mathcal{A}[x]] \rightarrow \mathcal{F}[\lambda x \mathcal{B}[x]].$$

Proof by induction on the length of the formula $\mathcal{F}[U]$. If U does not occur in $\mathcal{F}$ then the assertion holds by Theorem 20.4. If $\mathcal{F}[U]$ is a formula $U(t)$, then the assertion follows by basic inferences (S4) from

$$\vdash \forall x(\mathcal{A}[x] \leftrightarrow \mathcal{B}[x]) \rightarrow \mathcal{A}[t] \rightarrow \mathcal{B}[t].$$

In all the remaining cases the assertion follows from the IH.

Corollary. $\forall x(\mathcal{A}[x] \leftrightarrow \mathcal{B}[x]), \mathcal{F}[\lambda x \mathcal{A}[x]] \vdash \mathcal{F}[\lambda x \mathcal{B}[x]]$ *is a permissible inference.*

Let $<$ be the symbol for the 2-place decidable arithmetic predicate such that $<(m_1, m_2)$ is *true* if, and only if, the numeral m_1 is shorter than the numeral m_2. We use $<$ as a primitive symbol of our formal language and write $s<t$ for $<(s, t)$.

Theorem 20.10 (Induction Theorem). *If* l *is a symbol for a* 1-*place calculable arithmetic function then*

$$\vdash \forall x(\forall y(\mathrm{l}(y)<\mathrm{l}(x) \rightarrow \mathcal{F}[y]) \rightarrow \mathcal{F}[x]) \rightarrow \mathcal{F}[t] \quad \textit{holds.}$$

Proof. Let

$$F := \forall x(\forall y(\mathrm{l}(y)<\mathrm{l}(x) \rightarrow \mathcal{F}[y]) \rightarrow \mathcal{F}[x]),$$
$$\mathcal{G}[t] := \forall x(\mathrm{l}(x)<t \rightarrow \mathcal{F}[x]).$$

Then we have

$$\vdash F \rightarrow \mathcal{G}[0]$$
$$\vdash (F \rightarrow \mathcal{G}[a]) \rightarrow (F \rightarrow \mathcal{G}[a']).$$

$\vdash F \rightarrow \mathcal{G}[\mathrm{l}(t)']$ follows by a basic inference (ind).

Hence

$$\vdash F \rightarrow \mathscr{F}[t]$$

which was to be proved.

4. The Semi-Formal System DA*

In DA* we use the same formal language of second order arithmetic as before but now exclude free number variables. Thus in DA* every term is *numerical* and every prime formula *constant*.

Axioms of the system DA*: (Ax1)–(Ax3) as in DA.

Basic inferences of the system DA*: As in DA, but replacing (S2.0) and (ind) by

(S2.0*) $\mathscr{P}[\mathscr{F}[n]]$ for every numeral $n \vdash \mathscr{P}[\forall x \mathscr{F}[x]]$.

All axioms and basic inferences of the system DA* are to be restricted to formulas in which no free number variable occurs. We call the system DA* *semi-formal* since, as opposed to formal systems, it has basic inferences (S2.0*) with infinitely many premises.

In what follows small Greek letters (possibly with indices) always denote ordinal terms of the system OT (of §14).

Inductive definition of $\mathrm{DA}^* \models^{\alpha} F$ and $\mathrm{DA}^* \models^{\alpha}_{d} F$:

1. If F is an axiom of the system DA*, then $\mathrm{DA}^* \models^{\alpha} F$ and $\mathrm{DA}^* \models^{\alpha}_{d} F$ hold for every ordinal term α of the system OT.

2. If $\mathrm{DA}^* \models^{\alpha_i} F_i$ holds where $\alpha_i < \alpha$ for every premise F_i of a basic inference of the system DA*, then $\mathrm{DA}^* \models^{\alpha} F$ holds for the conclusion F of this basic inference.

3. If $\mathrm{DA}^* \models^{\alpha_i}_{d} F_i$ holds where $\alpha_i < \alpha$ for every premise F_i of a basic inference of the system DA* which is not an inference (S3.Δ), then $\mathrm{DA}^* \models^{\alpha}_{d} F$ holds for the conclusion F of this basic inference.

4. If $\mathrm{DA}^* \models^{\alpha_i}_{d} F_i$ holds where $\alpha_i + \omega \leqslant \alpha$ $(i = 1, 2)$ for the premises F_1 and F_2 of an inference (S3.Δ) of the system DA*, then $\mathrm{DA}^* \models^{\alpha}_{d} F$ holds for the conclusion F of this inference.

Corollaries. 1. *If* $\mathrm{DA}^* \models^{\alpha} F$ *(or* $\mathrm{DA}^* \models^{\alpha}_{d} F$*) and* $\alpha < \beta$, *then* $\mathrm{DA}^* \models^{\beta} F$ *(or* $\mathrm{DA}^* \models^{\beta}_{d} F$*) also holds.*

2. *If* $\mathrm{DA}^* \models^{\alpha}_{d} F$, *then* $\mathrm{DA}^* \models^{\alpha} F$.

A formula F is said to be *deducible* (or *distinguished deducible*) in DA* with order α, if $\mathrm{DA}^* \models^{\alpha} F$ (or $\mathrm{DA}^* \models^{\alpha}_{d} F$) holds.

Theorem 20.11 (Replacement rule). *If* $\mathrm{DA}^* \models^{\alpha} F$ *(or* $\mathrm{DA}^* \models^{\alpha}_{d} F$*) holds and F, G are equivalent formulas then* $\mathrm{DA}^* \models^{\alpha} G$ *(or* $\mathrm{DA}^* \models^{\alpha}_{d} G$*) holds.*

Theorem 20.12 (Structural rule). *If* $\text{DA}^* \models^{\alpha} F$ *(or* $\text{DA}^* \models^{\alpha}_{d} F$*) and* $F \models^{s} G$ *hold for formulas* F, G *of the system* DA*, *then* $\text{DA}^* \models^{\alpha} G$ *(or* $\text{DA}^* \models^{\alpha}_{d} G$*) holds.*

Both these theorems are proved by transfinite induction on α.

5. Embedding DA in DA*

If F is a formula of the system DA then we denote by F^* a formula of the system DA* which results from F when all the free number variables occurring in F are replaced by numerals. In the following we write $\omega \cdot m + n$ for $\omega \cdot m + \omega^0 \cdot n$.

Theorem 20.13 (Embedding Theorem).

$$\text{DA} \models^{m,n} F \quad \textit{implies} \quad \text{DA}^* \models^{\omega \cdot m + n}_{d} F^*.$$

Proof by induction on m with subsidiary induction on n. Suppose $\text{DA} \models^{m,n} F$ holds.

1. $m=0$ and $n=0$. Then F is an axiom of the system DA and F^* is an axiom of the system DA*. Therefore $\text{DA}^* \models^{0}_{d} F^*$ holds.

2. $m \neq 0$ or $n \neq 0$. Then F is the conclusion of a basic inference BI of the system DA.

2.1. If BI is not a basic inference (S2.0) (S3.Δ) or (ind) then the assertion follows immediately from the I.H.

2.2 Suppose BI is a basic inference (S2.0). Then F is a formula $\mathscr{P}[\forall x \mathscr{F}[x]]$ and we have $\text{DA} \models^{m,n_0} \mathscr{P}[\mathscr{F}[a]]$ where a does not occur in F and $n=n_0+1$. F^* is a formula $\mathscr{P}^*[\forall x \mathscr{F}^*[x]]$. By I.H. $\text{DA}^* \models^{\omega \cdot m + n_0}_{h} \mathscr{P}^*[\mathscr{F}^*[k]]$ holds for every numeral k. Using a basic inference (S2.0*) we obtain $\text{DA}^* \models^{\omega \cdot m + n}_{d} F^*$.

2.3 Suppose BI is a basic inference (S3.Δ). Then we have $\text{DA} \models^{m_i, n_i} F_i$ for the premises F_i $(i=1,2)$ of this inference where $m = \max(m_1, m_2)+1$ and $n=0$. F^* is the conclusion of a basic inference (S3.Δ) of the system DA* with premises F_i^* such that by I.H. $\text{DA}^* \models^{\omega \cdot m_i + n_i}_{d} F_i^*$ $(i=1,2)$ holds. Since $\omega \cdot m_i + n_i + \omega \leqslant \omega \cdot m$, $\text{DA}^* \models^{\omega \cdot m}_{d} F^*$ follows.

2.4. Suppose BI is a basic inference (ind). Then we have $\text{DA} \models^{m_1, n_1} \mathscr{F}[0]$ and $\text{DA} \models^{m_2, n_2} \mathscr{F}[a] \rightarrow \mathscr{F}[a']$ where $m = \max(m_1, m_2)+1$ and $n=0$, a does not occur in $\mathscr{F}$, F is a formula $\mathscr{F}[t]$ and F^* is a formula $\mathscr{F}^*[t^*]$. Let $m_0 := \max(m_1, m_2)$ and $n_0 := \max(n_1, n_2)$. By I.H. $\text{DA}^* \models^{\omega \cdot m_0 + n_0}_{d} \mathscr{F}^*[0]$ and $\text{DA}^* \models^{\omega \cdot m_0 + n_0}_{d} \mathscr{F}^*[k] \rightarrow \mathscr{F}^*[k']$ for every numeral k hold. Using cuts we obtain $\text{DA}^* \models^{\omega \cdot m_0 + (n_0 + k)}_{d} \mathscr{F}^*[k]$ and hence also $\text{DA}^* \models^{\omega \cdot m}_{d} \mathscr{F}^*[k]$ for every numeral k. Since t^* is a numerical term $\text{DA}^* \models^{\omega \cdot m}_{d} F^*$ follows by Theorem 20.11.

Corollaries. 1. *For each formula* F *deducible in* DA *there exists* $\alpha < \omega^2$ *such that* $\text{DA}^* \models^{\alpha}_{d} F^*$.

2. *If* F *is deducible in* DA *without use of basic inferences* (S3.Δ) *and* (ind) *then there exists* $\alpha < \omega$ *such that* $\text{DA}^* \models^{\alpha}_{d} F^*$.

6. General Properties of Deduction in the System DA*

Below we consider only formulas of the system DA* and we write $\vdash^{\alpha} F$ for DA* $\vdash^{\alpha} F$.

$\vdash^{<\alpha} F$ denotes that there exists $\xi<\alpha$ such that DA* $\vdash^{\xi} F$ holds.

A formula F is said to be *finitely deducible* if $\vdash^{<\omega} F$ holds.

Theorem 20.14. *The following formulas are finitely deducible.*

a) *all s-valid formulas,*

b) $\forall x\mathscr{F}[x]\to\mathscr{F}[n]$ and $\mathscr{F}[n]\to\exists x\mathscr{F}[x]$,

c) $\forall X\mathscr{F}[X]\to\mathscr{F}[P]$ and $\mathscr{F}[P]\to\exists X\mathscr{F}[X]$, *if P is an elementary predicator.*

Proof. This follows from the second corollary to Theorem 20.13 since the formulas concerned are deducible in DA without use of a basic inference (ind).

A formula B is said to be a *logical consequence* of formulas $A_1,\ldots,A_n$ if B can be obtained by cuts from the formulas $A_1,\ldots,A_n$ and formulas which are finitely deducible by Theorem 20.14.

Theorem 20.15. *If $\vdash^{\alpha} A_i$ $(i=1,\ldots,n)$ hold and B is a logical consequence of $A_1,\ldots,A_n$* then $\vdash^{<\alpha+\omega} B$ *holds.*

Proof. Trivial.

Theorem 20.16. *If $\vdash^{\alpha}\mathscr{N}[\mathscr{F}[U]]$ holds where U occurs in neither $\mathscr{N}$ nor $\mathscr{F}$ then $\vdash^{\beta}\mathscr{N}[\exists X\mathscr{F}[X]$ for $\alpha<\beta$.*

Proof. By Theorem 20.12 from $\vdash^{\alpha}\mathscr{N}[\mathscr{F}[U]]$ we obtain $\vdash^{\alpha}\mathscr{N}[\neg\neg\mathscr{F}[U]]$ and then using a basic inference (S2.1), $\vdash^{\beta}\mathscr{N}[\neg\forall X\neg\mathscr{F}[X]]$. That is $\vdash^{\beta}\mathscr{N}[\exists X\mathscr{F}[X]]$.

Theorem 20.17. *If $\vdash^{\alpha}\forall x(\forall Y\mathscr{A}[x,Y]\leftrightarrow\exists Y\mathscr{B}[x,Y])$ and $\vdash^{\beta}\mathscr{F}[\lambda x\forall Y\mathscr{A}[x,Y]]$ where neither $\mathscr{A}$ nor $\mathscr{B}$ contains bound predicate variables then $\vdash^{\gamma}\exists X\mathscr{F}[X]$ if* $\max(\alpha,\beta)<\gamma$.

Proof. By Theorem 20.12 from the second hypothesis we obtain

$$\vdash^{\beta}\neg\mathscr{F}[\lambda x\forall Y\mathscr{A}[x,Y]]\to\forall X\neg\mathscr{F}[X]\to\bot.$$

Using the first hypothesis a basic inference (S3.Δ) yields $\vdash^{\gamma}\forall X\neg\mathscr{F}[X]\to\bot$. That is $\vdash^{\gamma}\exists X\mathscr{F}[X]$.

7. Subsystems of DA and DA*

We denote by EN (*elementary number theory*) the subsystem of DA in which no bound predicate variables occur. In this system the basic inferences (S2.1), (S3.E) and (S3.Δ) disappear.

We let EA (*elementary analysis*) be the subsystem of DA in which the basic inference (S3.Δ) is not allowed.

The corresponding subsystems of DA* are denoted by EN* and EA*.

§21. Deductions of Transfinite Induction

In this section we develop deductions of formalized transfinite induction (over OT) in systems and subsystems of Δ_1^1-analysis.

If Σ is one of the formal systems EN, EA, DA and Σ* is one of the semi-formal systems EN*, EA*, DA* then

$\Sigma \vdash F$ denotes that the formula F is deducible in Σ,

$\Sigma^* \models^{\alpha} F (\Sigma^* \models^{<\alpha} F)$, that the formula F is deducible in Σ* with order α (with an order $<\alpha$).

The following obviously hold:

$$\mathrm{EN} \vdash F \Rightarrow \mathrm{EA} \vdash F \quad \text{and} \quad \mathrm{EA} \vdash F \Rightarrow \mathrm{DA} \vdash F,$$
$$\mathrm{EN}^* \models^{\alpha} F \Rightarrow \mathrm{EA}^* \models^{\alpha} F \quad \text{and} \quad \mathrm{EA}^* \models^{\alpha} F \Rightarrow \mathrm{DA}^* \models^{\alpha} F.$$

By Theorem 20.13

$$\Sigma \vdash F \Rightarrow \Sigma^* \models^{\leq \omega^2} F^* \quad \text{also holds.}$$

1. Formalisation of Transfinite Induction

In §14 we defined a bijective mapping Nr from $\overline{\mathrm{OT}}$ onto the set ℕ of natural numbers with inverse τ (p. 97). We regard ℕ as the set of numerals and denote by $\bar{\alpha}$ the numeral Nr α of an ordinal term α from $\overline{\mathrm{OT}}$.

Corresponding to the relations $<$ and $\leqslant$ on $\overline{\mathrm{OT}}$ the given mapping induces 2-place decidable arithmetic predicates $\prec$ and $\preccurlyeq$ defined as follows:

$$\prec(m, n) :\Leftrightarrow \tau(m) < \tau(n)$$
$$\preccurlyeq(m, n) :\Leftrightarrow \tau(m) \leqslant \tau(n).$$

We use $\prec$ and $\preccurlyeq$ as primitive symbols of our formal language and write

$$\begin{aligned}
s \prec t \quad & \text{for} \quad \prec(s, t), \\
s \preccurlyeq t \quad & \text{for} \quad \preccurlyeq(s, t), \\
A_1 \to \cdots \to A_n \to B \quad & \text{for} \quad (A_1 \to (\cdots \to (A_n \to B)\ldots)), \\
t_1 \prec t_2 \prec t_3 \to B \quad & \text{for} \quad t_1 \prec t_2 \to t_2 \prec t_3 \to B, \\
t_1 \preccurlyeq t_2 \prec t_3 \to B \quad & \text{for} \quad t_1 \preccurlyeq t_2 \to t_2 \prec t_3 \to B, \\
\forall x \prec t \mathscr{F}[x] \quad & \text{for} \quad \forall x (x \prec t \to \mathscr{F}[x]).
\end{aligned}$$

The relation $\prec$ is easily shown in EN to be a *linear ordering relation*, that is, the following formulas are deducible in EN for arbitrary terms t_1, t_2, t_3:

$$\neg t_1 \prec t_1$$
$$t_1 \prec t_2 \to t_2 \prec t_3 \to t_1 \prec t_3$$
$$\neg t_1 \prec t_2 \to \neg t_1 = t_2 \to t_2 \prec t_1.$$

The combinations $\alpha+\beta$, $\phi\alpha\beta$, $\omega^\alpha \cdot n$, $\omega_m(\beta)$ and $\omega^m \cdot n$ of ordinal terms $\alpha, \beta \in \mathrm{OT}$ where $m, n \in \mathbb{N}$ correspond under $\prec$ to 2-place calculable arithmetic functions $\hat{+}$, $\hat{\phi}$, $\hat{\omega}$, $\tilde{\omega}$ and $\dot{\omega}$ defined as follows:

$$\hat{+}(m,n) := \begin{cases} \overline{\tau(m)+\tau(n)}, & \text{if } \tau(m) \neq \Gamma_0 \text{ and } \tau(n) \neq \Gamma_0, \\ \overline{\Gamma_0} & , \text{if } \tau(m) = \Gamma_0 \text{ or } \tau(n) = \Gamma_0. \end{cases}$$

$$\hat{\phi}(m,n) := \begin{cases} \overline{\phi(\tau(m))(\tau(n))}, & \text{if } \tau(m) \neq \Gamma_0 \text{ and } \tau(n) \neq \Gamma_0, \\ \overline{\Gamma_0}, & \text{if } \tau(m) = \Gamma_0 \text{ or } \tau(n) = \Gamma_0. \end{cases}$$

$$\hat{\omega}(m,n) := \begin{cases} \overline{\omega^{\tau(m)} \cdot n}, & \text{if } \tau(m) \neq \Gamma_0 \text{ and } n \neq 0, \\ \overline{0}, & \text{if } n = 0, \\ \overline{\Gamma_0}, & \text{if } \tau(m) = \Gamma_0 \text{ and } n \neq 0. \end{cases}$$

$$\tilde{\omega}(m,n) := \begin{cases} \overline{\omega_m(\tau(n))}, & \text{if } \tau(n) \neq \Gamma_0, \\ \overline{\Gamma_0}, & \text{if } \tau(n) = \Gamma_0. \end{cases}$$

$$\dot{\omega}(m,n) := \overline{\omega^m \cdot n}.$$

We use $\hat{+}$, $\hat{\phi}$, $\hat{\omega}$, $\tilde{\omega}$, $\dot{\omega}$ as primitive symbols of our formal language and write

$$\begin{aligned} s \hat{+} t &\quad \text{for} \quad \hat{+}(s,t), \\ \hat{\omega}^s t &\quad \text{for} \quad \hat{\omega}(s,t), \\ \tilde{\omega}_s(t) &\quad \text{for} \quad \tilde{\omega}(s,t), \\ \dot{\omega}^s t &\quad \text{for} \quad \dot{\omega}(s,t), \\ \hat{\omega}^s &\quad \text{for} \quad \hat{\omega}(s,1), \\ \dot{\omega}^s &\quad \text{for} \quad \dot{\omega}(s,1), \\ \hat{\varepsilon}_t &\quad \text{for} \quad \hat{\phi}(\bar{1},t). \end{aligned}$$

The following equations are true constant prime formulas for ordinal terms α, β from OT and numerals m, n

$$\bar{\alpha} \hat{+} \bar{\beta} = \overline{\alpha+\beta}, \qquad \hat{\phi}(\bar{\alpha}, \bar{\beta}) = \overline{\phi\alpha\beta},$$
$$\hat{\omega}^{\bar{\alpha}} n = \overline{\omega^\alpha \cdot n}, \qquad \tilde{\omega}_m(\bar{\beta}) = \overline{\omega_m(\beta)}, \qquad \omega^m n = \overline{\omega^m \cdot n},$$
$$\hat{\omega}^{\bar{\alpha}} = \overline{\omega^\alpha}, \qquad \dot{\omega}^n = \overline{\omega^n}, \qquad \varepsilon_{\bar{\alpha}} = \overline{\varepsilon_\alpha} \quad \text{and} \quad \hat{\varepsilon}_{\bar{\alpha}} = \overline{\phi 1 \alpha}.$$

From this it follows that the following equations are axioms of the system EN for arbitrary terms s and t:

$$\hat{\omega}^s 0 = \bar{0}, \qquad \hat{\omega}^s t' = \hat{\omega}^s t \mathbin{\hat{+}} \hat{\omega}^s,$$
$$\dot{\omega}^s 0 = \bar{0}, \qquad \dot{\omega}^s t' = \dot{\omega}^s t \mathbin{\hat{+}} \dot{\omega}^s,$$
$$\tilde{\omega}_0(t) = t, \qquad \tilde{\omega}_{s'}(t) = \hat{\omega}_s^{\tilde{\omega}(t)},$$
$$\hat{\omega}^t = \hat{\phi}(\bar{0}, t), \qquad \hat{\varepsilon}_t = \hat{\phi}(\bar{1}, t).$$

Also for all the properties of the ordinal terms $\alpha + \beta$, $\phi\alpha\beta$, $\omega^\alpha \cdot n$, $\omega_m(\beta)$ and $\omega^m \cdot n$ developed in §14 the corresponding formulas are all deducible in EN, for example

$$s_1 \prec s_2 \rightarrow \hat{\phi}(s_1, \hat{\phi}(s_2, t)) = \hat{\phi}(s_2, t).$$

We express the *progressiveness* $\mathcal{P}\imath[P]$ of a predicate P with respect to $\prec$ and *transfinite induction up to* t, $\mathcal{I}[t]$, with respect to $\prec$ by the following formulas:

$$\mathcal{P}\imath[P] :\!= \forall x_2(\forall x_1 \prec x_2 P(x_1) \rightarrow P(x_2))$$
$$\mathcal{I}[t] :\!= \forall X(\mathcal{P}\imath[X] \rightarrow \forall x \prec t X(x)).$$

For these the following hold:

($\mathcal{P}\imath$1) EN $\vdash \mathcal{P}\imath[P] \rightarrow \forall x \prec t P(x) \rightarrow P(t)$, if P is an elementary predicator.
($\mathcal{P}\imath$2) EA $\vdash \mathcal{P}\imath[P] \rightarrow \forall x \prec t P(x) \rightarrow P(t)$ for every predicator P.
($\mathcal{I}$1) EA $\vdash \mathcal{I}[t] \rightarrow \mathcal{P}\imath[P] \rightarrow P(t)$, if P is an elementary predicator.
($\mathcal{I}$2) DA $\vdash \mathcal{I}[t] \rightarrow \mathcal{P}\imath[P] \rightarrow P(t)$, if P is an elementary or a Δ_1^1-predicator.
($\mathcal{I}$3) EA $\vdash \mathcal{I}[t] \rightarrow s \prec t \rightarrow \mathcal{I}[s]$
($\mathcal{I}$4) EA $\vdash \forall x \prec t \mathcal{I}[x] \rightarrow \mathcal{I}[t]$.

($\mathcal{P}\imath$1), ($\mathcal{P}\imath$2) and ($\mathcal{I}$1)–($\mathcal{I}$3) are easy to prove. ($\mathcal{I}$4) is obtained as follows: From ($\mathcal{I}$1) we have

$$\text{EA} \vdash \forall x \prec t \mathcal{I}[x] \rightarrow \mathcal{P}\imath[U] \rightarrow \forall x \prec t U(x).$$

Using a basic inference (S2.1), ($\mathcal{I}$4) follows.

2. Deductions in EN

We set

$$\mathcal{S}[P, t] :\!= \forall y(\forall x \prec y P(x) \rightarrow \forall x \prec y \mathbin{\hat{+}} t P(x))$$
$$\mathfrak{s}[P] :\!= \lambda z \mathcal{S}[P, \hat{\omega}^z].$$

Lemma 1. *For every elementary predicator P* EN $\vdash \mathscr{P}\imath[P] \rightarrow \mathscr{P}\imath[\mathscr{d}[P]]$ *holds.*

Proof. We obtain

$$\mathrm{EN} \vdash \mathscr{d}[P](s) \rightarrow \forall x < b \,\hat{+}\, \hat{\omega}^s a P(x) \rightarrow \forall x < b \,\hat{+}\, \hat{\omega}^s a' P(x)$$

from the definition of $\mathscr{d}[P]$ and the properties of $\hat{\omega}^s t$. Applying a basic inference (ind) we obtain

(1) $\mathrm{EN} \vdash \mathscr{d}[P](s) \rightarrow \forall x < b P(x) \rightarrow \forall x < b \,\hat{+}\, \hat{\omega}^s t P(x)$

for every term t. For ordinal terms α, β and $\gamma \neq 0$ such that $\alpha < \beta + \omega^\gamma$ an ordinal term $\gamma_0 < \gamma$ and a numeral n can be determined such that $\alpha < \beta + \omega^{\gamma_0} \cdot n$. Hence one can define 3-place calculable arithmetic functions f and g for which the following hold:

$$\mathrm{EN} \vdash a < b \,\hat{+}\, \hat{\omega}^c \rightarrow a < b \,\hat{+}\, \hat{\omega}^{f(a,b,c)} g(a, b, c)$$
$$\mathrm{EN} \vdash \bar{0} < c \rightarrow f(a, b, c) < c.$$

By (1) with $s := f(a, b, c)$ and $t := g(a, b, c)$ we obtain

(2) $\mathrm{EN} \vdash \bar{0} < c \rightarrow \forall x < c \mathscr{d}[P](x) \rightarrow \forall x < b P(x) \rightarrow a < b \,\hat{+}\, \hat{\omega}^c \rightarrow P(a).$

Further

$$\mathrm{EN} \vdash c = \bar{0} \rightarrow a < b \,\hat{+}\, \hat{\omega}^c \rightarrow a \leqslant b$$

holds. Using ($\mathscr{P}\imath$1) we obtain

(3) $\mathrm{EN} \vdash c = \bar{0} \rightarrow \mathscr{P}\imath[P] \rightarrow \forall x < b P(x) \rightarrow a < b \,\hat{+}\, \hat{\omega}^c \rightarrow P(a).$

From (2) and (3) we obtain

$$\mathrm{EN} \vdash \mathscr{P}\imath[P] \rightarrow \forall x < c \;\; \mathscr{d}[P](x) \rightarrow \forall x < b P(x) \rightarrow \forall x < b \,\hat{+}\, \hat{\omega}^c P(x).$$

Using basic inferences (S2.0) and (S4) we obtain

$$\mathrm{EN} \vdash \mathscr{P}\imath[P] \rightarrow \forall x < c \;\; \mathscr{d}[P](x) \rightarrow \mathscr{d}[P](c).$$

The assertion follows from this by a basic inference (S2.0).

Lemma 2. *For every elementary predicator P and every numeral n* EN $\vdash \mathscr{P}\imath[P] \rightarrow \forall x < \tilde{\omega}_n(\bar{0}) P(x)$ *holds.*

Proof by induction on n. If $n=0$ the assertion is trivial for $\tilde{\omega}_0(\bar{0})=\bar{0}$. We now go from n to n'. Like P $\mathscr{J}[P]$ is also an elementary predicator. Hence by hypothesis $\mathrm{EN} \vdash \mathscr{P}\imath[\mathscr{J}[P]] \to \forall x < \tilde{\omega}_n(\bar{0})\mathscr{J}[P](x)$ holds. Using Lemma 1 and $(\mathscr{P}\imath 1)$ we obtain

(1) $\mathrm{EN} \vdash \mathscr{P}\imath[P] \to \mathscr{J}[P](\tilde{\omega}_n(\bar{0}))$.

By the definition of $\mathscr{J}[P]$ $\mathrm{EN} \vdash \mathscr{J}[P](t) \to \forall x < sP(x) \to \forall x < s \hat{+} \hat{\omega}^t P(x)$ holds. For $s:=\bar{0}$ and $t:=\tilde{\omega}_n(0)$, $\mathrm{EN} \vdash \forall x < sP(x)$ and $\mathrm{EN} \vdash s \hat{+} \hat{\omega}^t = \tilde{\omega}_{n'}(\bar{0})$ holds. Hence we have

(2) $\mathrm{EN} \vdash \mathscr{J}[P](\tilde{\omega}_n(\bar{0})) \to \forall x < \tilde{\omega}_{n'}(\bar{0})P(x)$.

The assertion $\mathrm{EN} \vdash \mathscr{P}\imath[P] \to \forall x < \tilde{\omega}_{n'}(\bar{0})P(x)$ for n' follows from (1) and (2).

Theorem 21.1. *For every elementary predicator P and every ordinal term $\alpha < \varepsilon_0$ $\mathrm{EN} \vdash \mathscr{P}\imath[P] \to \forall x < \bar{\alpha}P(x)$ holds.*

Proof By Theorem 14.14 if $\alpha < \varepsilon_0$ there exists a numeral n such that $\alpha < \omega_n(0)$ and therefore $\bar{\alpha} < \tilde{\omega}_n(\bar{0})$. Hence the assertion follows from Lemma 2.

3. Deductions in EN*

Lemma 3. *If Σ is one of the formal systems* EN, EA *or* DA *and F is a formula of the corresponding semi-formal system Σ^* such that $\Sigma \vdash F$, then $\Sigma^* \models^{\leq \omega^2} F$.*

Proof. This holds by theorem 20.13.

Lemma 4. *If Σ^* is one of the semi-formal systems* EN*, EA* *or* DA* *and Σ^* $\models^{\leq \alpha} \mathscr{P}[\mathscr{F}[n]]$ where $\alpha \neq 0$ holds for every numeral $n < \bar{\beta}$ then $\Sigma^* \models^{\underline{\alpha}} \mathscr{P}[\forall x < \bar{\beta}\mathscr{F}[x]]$.*

Proof. From the hypothesis $\Sigma^* \models^{\leq \alpha} \mathscr{P}[(n < \bar{\beta} \to \mathscr{F}[n])]$ for every numeral $n < \bar{\beta}$ holds by Theorem 20.12. For every other numeral n $\mathscr{P}[(n < \bar{\beta} \to \mathscr{F}[n])]$ is an (Ax2). Hence the assertion follows by a basic inference (S2.0*).

Lemma 5. *The following hold for every elementary predicator P:*

a) $\mathrm{EN}^* \models^{\leq \omega^2 + \omega \cdot \alpha} \mathscr{P}\imath[P] \to P(m)$ *for every numeral* $m < \hat{\varepsilon}_{\bar{\alpha}}$

b) $\mathrm{EN}^* \models^{\omega^2 + \omega \cdot \alpha + 3n + 2} \mathscr{P}\imath[P] \to \forall x < \tilde{\omega}_n(\hat{\varepsilon}_{\bar{\alpha}} + \bar{1})P(x)$.

Proof by induction on α.

a) If $\alpha = 0$ then a) holds by Theorem 21.1 and Lemma 3. Suppose now that $\alpha \neq 0$. Then for $m < \hat{\varepsilon}_{\bar{\alpha}}$ there exist, by Theorem 14.14, an ordinal term $\xi < \alpha$ and a numeral n such that $m < \tilde{\omega}_n(\hat{\varepsilon}_{\bar{\xi}} \hat{+} \bar{1})$. Hence the assertion a) follows from the I.H. for b).

b) We use a subsidiary induction on n:
b1) $n=0$. By Lemma 4 from a) we obtain

$$\mathrm{EN}^* \models^{\omega^2+\omega\cdot\alpha} \mathscr{P}\imath[P] \to \forall x < \hat{\varepsilon}_{\bar{\alpha}} P(x).$$

By Lemma 3 using ($\mathscr{P}\imath$ 1) we obtain

$$\mathrm{EN}^* \models^{\omega^2+\omega\cdot\alpha+1} \mathscr{P}\imath[\mathrm{P}] \to P(\hat{\varepsilon}_{\bar{\alpha}}).$$

Hence by Lemma 4 using a) we obtain

$$\mathrm{EN}^* \models^{\omega^2+\omega\cdot\alpha+2} \mathscr{P}\imath[P] \to \forall x < \hat{\varepsilon}_{\bar{\alpha}} \hat{+} \bar{1} P(x).$$

This yields the assertion b) for $n=0$ since $\tilde{\omega}_0(\hat{\varepsilon}_{\bar{\alpha}} \hat{+} \bar{1}) = \hat{\varepsilon}_{\bar{\alpha}} \hat{+} \bar{1}$.
b2) From n to n'. Like P, $\jmath[P]$ is also an elementary predicator. Therefore by hypothesis

$$\mathrm{EN}^* \models^{\omega^2+\omega\cdot\alpha+3\cdot n+2} \mathscr{P}\imath[\jmath[P]] \to \forall x < \tilde{\omega}_n(\hat{\varepsilon}_{\bar{\alpha}} \hat{+} \bar{1}) \jmath[P](x) \text{ holds.}$$

By Lemma 1, ($\mathscr{P}\imath$1), the definition of $\jmath[P]$ and Lemma 3 the following hold

$$\mathrm{EN}^* \models^{\leq\omega^2} \mathscr{P}\imath[P] \to \mathscr{P}\imath[\jmath P]]$$

$$\mathrm{EN}^* \models^{\leq\omega^2} \mathscr{P}\imath[\jmath[P]] \to \forall x < \tilde{\omega}_n(\hat{\varepsilon}_{\bar{\alpha}} \hat{+} \bar{1}) \jmath[P](x) \to \jmath[P](\tilde{\omega}_n(\hat{\varepsilon}_{\bar{\alpha}} \hat{+} \bar{1}))$$

$$\mathrm{EN}^* \models^{\leq\omega^2} \jmath[P](\tilde{\omega}_n(\hat{\varepsilon}_{\bar{\alpha}} \hat{+} \bar{1})) \to \forall x < \tilde{\omega}_{n'}(\hat{\varepsilon}_{\bar{\alpha}} \hat{+} \bar{1}) P(x).$$

Using cuts the assertion

$$\mathrm{EN}^* \models^{\omega^2+\omega\cdot\alpha+3n+5} \mathscr{P}\imath[P] \to \forall x < \tilde{\omega}_{n'}(\hat{\varepsilon}_{\bar{\alpha}} \hat{+} \bar{1}) P(x) \quad \text{holds for } n'.$$

Theorem 21.2 $\mathrm{EN}^* \models^{\omega^2+\omega\cdot\alpha} \mathscr{P}\imath[P] \to \forall x < \hat{\varepsilon}_{\bar{\alpha}} P(x)$ *holds for every elementary predicator P.*

Proof. This follows from Lemmata 5a) and 4.

4. Deductions in EA and EA*

Lemma 6. $\mathrm{EA} \vdash \forall y(\forall x < y \mathscr{F}[x] \to \mathscr{F}[y]) \to \forall x < \bar{\alpha}[x]$ *holds for* $\alpha < \varepsilon_0$.

Proof. As for Theorem 21.1 we have

$$\mathrm{EA} \vdash \mathscr{P}\imath[\lambda z \mathscr{F}[z]] \to \forall x < \bar{\alpha} \lambda z \mathscr{F}[z](x)$$

for every predicate $\lambda z \mathscr{F}[z]$, since in EA there is no restriction to elementary predicators. Hence the assertion.

Lemma 7.

a) $EA \vdash \mathscr{I}[s] \to \mathscr{I}[t] \to \mathscr{I}[s \hat{+} \hat{\omega}^t]$
b) $EA \vdash \mathscr{I}[s] \to \mathscr{I}[t] \to \mathscr{I}[s \hat{+} t]$
c) $EA \vdash \mathscr{I}[t] \to \mathscr{I}[\tilde{\omega}_s(t)]$
d) $EA \vdash \forall x < t \mathscr{I}[\hat{\varepsilon}_x] \to \mathscr{I}[\hat{\varepsilon}_t]$ *for* $t < \overline{\Gamma}_0$.

Proof. a) The following hold by Lemma 1, ($\mathscr{I}$1) and the definition of $\mathscr{J}[U]$

$$EA \vdash \mathscr{P}\mathfrak{r}[U] \to \mathscr{P}\mathfrak{r}[\mathscr{J}[U]]$$
$$EA \vdash \mathscr{I}[t] \to \mathscr{P}\mathfrak{r}[\mathscr{J}[U]] \to \mathscr{J}[U](t)$$
$$EA \vdash \mathscr{J}[U](t) \to \forall x < s U(x) \to \forall x < s \hat{+} \hat{\omega}^t U(x).$$

Using cuts

$$EA \vdash \mathscr{I}[t] \to \mathscr{P}\mathfrak{r}[U] \to \forall x < s U(x) \to \forall x < s \hat{+} \hat{\omega}^t U(x) \quad \text{follows.}$$

Further

$$EA \vdash \mathscr{I}[s] \to \mathscr{P}\mathfrak{r}[U] \to \forall x < s U(x) \quad \text{holds.}$$

Hence we have

$$EA \vdash \mathscr{I}[s] \to \mathscr{I}[t] \to \mathscr{P}\mathfrak{r}[U] \to \forall x < s \hat{+} \hat{\omega}^t U(x).$$

Using (S2.1) assertion a) follows.
b) follows from a) by ($\mathscr{I}$3) since $s \hat{+} t \preccurlyeq s \hat{+} \hat{\omega}^t$.
c) Trivially $EA \vdash \mathscr{I}[\overline{0}]$ holds. Using a) for $s := \overline{0}$,

$$EA \vdash \mathscr{I}[t] \to \mathscr{I}[\hat{\omega}^t].$$

By the definition of $\tilde{\omega}_s(t)$ we have

$$EA \vdash \mathscr{I}[\tilde{\omega}_a(t)] \to \mathscr{I}[\tilde{\omega}_{a'}(t)].$$

Using a basic inference (ind) we have

$$EA \vdash \mathscr{I}[t] \to \mathscr{I}[\tilde{\omega}_s(t)].$$

d) By Theorem 14.14 there are calculable arithmetic functions f, g and h such that

(1) $EA \vdash a < \hat{\varepsilon}_{\overline{0}} \to a < \tilde{\omega}_{f(a)}(\overline{0})$
(2) $EA \vdash a < \hat{\varepsilon}_t \to a < \tilde{\omega}_{g(a,t)}(\hat{\varepsilon}_{h(a,t)} \hat{+} \overline{1})$
(3) $EA \vdash \overline{0} < t \to h(a, t) < t$

From $EA \vdash \mathscr{I}[\bar{0}]$ and c) we have $EA \vdash \mathscr{I}[\tilde{\omega}_{f(a)}(\bar{0})]$. Using (1) and ($\mathscr{I}4$) we have

(4) $EA \vdash \mathscr{I}[\hat{\varepsilon}_{\bar{0}}]$.

By Theorem 21.1 $EA \vdash \mathscr{I}[\bar{1}]$ holds. Using b) and c) we have

$$EA \vdash \mathscr{I}[\hat{\varepsilon}_{h(a,t)}] \rightarrow \mathscr{I}[\tilde{\omega}_{g(a,t)}(\hat{\varepsilon}_{h(a,t)} \hat{+} \bar{1})].$$

Using (2), (3) and ($\mathscr{I}4$) we have

(5) $EA \vdash \bar{0} \prec t \rightarrow \forall x \prec t \mathscr{I}[\hat{\varepsilon}_x] \rightarrow \mathscr{I}[\hat{\varepsilon}_t]$.

The assertion d) follows from (4) and (5).

Theorem 21.3. *If* $\alpha < \phi 1 \varepsilon_0$ *then* $EA \vdash \mathscr{I}[\bar{\alpha}]$.

Proof. Letting $\mathscr{F}[t] := t \prec \Gamma_0 \rightarrow \mathscr{I}[\hat{\varepsilon}_t]$ we obtain from Lemma 7d)

$$EA \vdash \forall y(\forall x \prec y \mathscr{F}[x] \rightarrow \mathscr{F}[y]).$$

By Lemma 6 $EA \vdash \mathscr{F}[\bar{\beta}]$ follows for all $\beta < \varepsilon_0$. Since $\varepsilon_\beta = \phi 1 \beta$, $EA \vdash \mathscr{I}[\overline{\phi 1 \beta}]$ follows for all $\beta < \varepsilon_0$. If $\alpha < \phi 1 \varepsilon_0$ there exists $\beta < \varepsilon_0$ such that $\alpha < \phi 1 \beta$. Hence by ($\mathscr{I}3$) we obtain $EA \vdash \mathscr{I}[\bar{\alpha}]$ for $\alpha < \phi 1 \varepsilon_0$.

Theorem 21.4. $EA^* \models^{\omega^2 + 2 \cdot \alpha + 1} \mathscr{I}[\hat{\varepsilon}_{\bar{\alpha}}]$ holds for $\alpha < \Gamma_0$.

Proof by induction on α. From Lemmata 7d) and 4 we have

(1) $EA^* \models^{\leq \omega^2} \forall x \prec \bar{\alpha} \mathscr{I}[\hat{\varepsilon}_x] \rightarrow \mathscr{I}[\hat{\varepsilon}_{\bar{\alpha}}]$ for $\alpha < \Gamma_0$.

By I.H. $EA^* \models^{\leq \omega^2 + 2 \cdot \alpha} \mathscr{I}[\hat{\varepsilon}_n]$ holds for every numeral $n \prec \bar{\alpha}$. By Lemma 5 we obtain

(2) $EA^* \models^{\omega^2 + 2 \cdot \alpha} \forall x \prec \bar{\alpha} \mathscr{I}[\hat{\varepsilon}_x]$.

The assertion follows from (1) and (2).

5. The Formula $\mathscr{R}[P, Q, t]$

So far the deductions of formalized transfinite induction have essentially depended on the passage from a predicator P to the predicator $\mathscr{J}[P]$ which yields a step up in the transfinite induction. In order to deduce the formula $\mathscr{I}[\bar{\alpha}]$ in the systems DA and DA* for higher ordinal terms α we need to go from a predicator P to essentially stronger predicators. To this end following a method suggested by S. Feferman [1] we use a formula $\mathscr{R}[P, Q, t]$ by which we recursively define a sequence of suitable predicators $Q_n (n \prec t)$ depending on a predicator P. Hence by using basic inferences (S3.Δ) we can deduce the existence of sufficiently strong predicators.

In §14 (p. 96) a bijective mapping π of $\mathbb{N}\times\mathbb{N}$ onto $\mathbb{N}$ was defined with inverses π_1 and π_2. These definitions give calculable arithmetic functions such

$$\pi(\pi_1(n), \pi_2(n))=n,$$
$$\pi_i(\pi(n_1, n_2))=n_i \quad (i=1,2).$$

We use π, π_1, π_2 as primitive symbols of our formal language and write

$$P(s,t) \quad \text{for } P(\pi(s,t)),$$
$$P_s \quad \text{for } \lambda y P(\pi(s,y)).$$

In addition we use h and e as symbols for the 1-place calculable functions which are defined as follows:

1. If $\tau(m)=0$ or $\tau(m)=\Gamma_0$ then let $h(m):=\bar{0}$ and $e(m):=m$.
2. If $0<\tau(m)<\Gamma_0$ then by Theorem 14.9 there exist unique ordinal terms $\alpha_1\geqslant\cdots\geqslant\alpha_n$ $(n\geqslant 1)$ of the system OT such that

$$\tau(m)=\omega^{\alpha_1}+\cdots+\omega^{\alpha_n}.$$

Then let

$$h(m):=\begin{cases}\bar{0} & \text{, if } n=1,\\ \overline{\omega^{\alpha_1}+\cdots+\omega^{\alpha_{n-1}}}, & \text{if } n>1.\end{cases}$$

$$e(m):=\overline{\alpha_n}$$

Let $\mathscr{R}[P,Q,t]$ be the formula

$$\begin{cases}\forall y(P(y)\leftrightarrow Q(\bar{0},y))\wedge\forall x(\bar{0}\prec x\prec t\rightarrow\forall y(Q(x,y)\leftrightarrow\forall z(h(x)\preccurlyeq z\prec x\rightarrow\\ \mathscr{S}[Q_z,\hat{\phi}(e(x),y)]))).\end{cases}$$

Here the restriction of the universal quantifier $\forall z$ to $z\prec x$ is sufficient to obtain the desired deduction in DA*. The additional restriction of $\forall z$ to $h(x)\preccurlyeq z$ allows us, following H. Schwichtenberg, to also obtain the desired deduction in DA.

Lemma 8. $\mathrm{DA}\vdash\forall y(\forall x\prec y\mathscr{G}[U,V,W,x,t]\rightarrow\mathscr{G}[U,V,W,y,t])$ *where*

$$\mathscr{G}[U,V,W,s,t]:=s\prec t\rightarrow\mathscr{R}[U,V,t]\rightarrow\mathscr{R}[U,W,t]\rightarrow\forall y(V_s(y)\leftrightarrow W_s(y)).$$

Proof. $\mathscr{R}[U,V,t]$ implies that for $s\prec t$ V_s is uniquely determined by U and the $V_x(x\prec s)$. Hence the assertion follows by the Extensionality Theorem 20.9.

Let the one-place calculable arithmetic function l be defined as follows:

1. If $\tau(m)=0$ or $\tau(m)=\Gamma_0$, let $\mathrm{l}(m):=0$.
2. If $0<\tau(m)<\Gamma_0$, then by Theorem 14.8 there are uniquely ordinal terms $\alpha_1,\ldots,\alpha_n,\beta_1,\ldots,\beta_n$ $(n\geqslant 1)$ of the system OT such that $\beta_i<\phi\alpha_i\beta_i(i=1,\ldots,n)$, $\phi\alpha_1\beta_1\geqslant\cdots\geqslant\phi\alpha_n\beta_n$ and

$$\tau(m)=\phi\alpha_1\beta_1+\cdots+\phi\alpha_n\beta_n.$$

Then let

$$\mathrm{l}(m):=L\alpha_1+\cdots+L\alpha_n+L\beta_1+\cdots+L\beta_n+n,$$

where L is the *length* of the ordinal term defined in §14 (p. 86).

Lemma 9. $\mathrm{DA}\vdash\mathscr{R}[U,V,t]\to\bar{0}<a<t\to\forall x<a\mathscr{P}\imath[V_x]\to\mathscr{P}\imath[V_a]$.

Proof. We first prove that $\mathscr{S}[V_c,d]$ is implied by the following formulas (1)–(7).
(1) $\mathscr{R}[U,V,t]$
(2) $a<t$
(3) $\forall x<a\mathscr{P}\imath[V_x]$
(4) $\forall y<bV_a(y)$
(5) $\forall y(l(y)<l(d)\to y<\hat{\phi}(e(a),b)\to\forall z(h(a)\leqslant z<a\to\mathscr{S}[V_z,y]))$
(6) $d<\hat{\phi}(e(a),b)$
(7) $h(a)\leqslant c<a$.

Now using calculable arithmetic functions we can distinguish for d the three cases $\bar{0}$, $d_1\hat{+}d_2$ and $\hat{\phi}(d_1,d_2)$ where in the last two cases $l(d_i)<l(d)$. The case $d=\bar{\Gamma}_0$ is excluded by (6).

1. $d=\bar{0}$. Then trivially $\mathscr{S}[V_c,d]$ holds.
2. $d=d_1\hat{+}d_2$ where $l(d_i)<l(d)$ $(i=1,2)$. Then $d_i<\hat{\phi}(e(a),b)$ follows from (6). Using (7) and (5) we obtain $\mathscr{S}[V_c,d_i]$. Hence we obtain

$$\forall x<a_0V_c(x)\to\forall x<a_0\hat{+}d_1V_c(x)$$

and

$$\forall x<a_0\hat{+}d_1V_c(x)\to\forall x<a_0\hat{+}dV_c(x),$$

hence $\mathscr{S}[V_c,d]$.

3. $d=\hat{\phi}(d_1,d_2)$ where $l(d_i)<l(d)$ $(i=1,2)$.

3.1. $d_1<e(a)$. If $c_1:=c\hat{+}\hat{\omega}^{d_1}$ then $h(a)\leqslant c_1<a$ follows from (7). From (6) we obtain $d_2<\hat{\phi}(e(a),b)$. Using (7) and (5) we obtain $\mathscr{S}[V_{c_1},d_2]$. Hence $\forall x<d_2V_{c_1}(x)$. From (3) we obtain $\mathscr{P}\imath[V_{c_1}]$. By $(\mathscr{P}\imath 1)$ we have $V_{c_1}(d_2)$. Since $\bar{0}<c_1<a$ using (2) and (1) by the definition of $\mathscr{R}$ we obtain

$$\forall z(h(c_1)\leqslant z<c_1\to\mathscr{S}[V_z,\hat{\phi}(e(c_1),d_2)]).$$

Since $h(c_1) \leqslant c < c_1$ and $e(c_1) = d_1$ the assertion $\mathscr{S}[V_c, \hat{\phi}(d_1, d_2)]$ follows.

3.2. $e(a) = d_1$. Then from (6) we have $d_2 < b$. Using (4) we obtain $V_a(d_2)$. By (1), (2) and (7) from the definition of $\mathscr{R}$ we obtain $\mathscr{S}[V_c, \hat{\phi}(e(a), d_2)]$.

3.3. $e(a) < d_1$. Then from (6) we have $d < b$. As in 3.2 we obtain $\mathscr{S}[V_c, \hat{\phi}(e(a), d)]$. Using $d \leqslant \hat{\phi}(e(a), d)$ we obtain $\mathscr{S}[V_c, d]$. Hence we have shown

$$\mathrm{DA} \vdash \begin{cases} \mathscr{R}[U, V, t] \to a < t \to \forall x < a \mathscr{P}\imath[V_x] \to \forall y < b V_a(y) \to \forall y(l(y) < l(d) \\ \quad \to y < \hat{\phi}(e(a), b) \to \forall z(h(a) \leqslant z < a \to \mathscr{S}[V_z, y])) \to d < \hat{\phi}(e(a), b) \\ \quad \to h(a) \leqslant c < a \to \mathscr{S}[V_c, d]. \end{cases}$$

Using a basic inference (S2.0)

$$\mathrm{DA} \vdash \begin{cases} \mathscr{R}[U, V, t] \to a < t \to \forall x < a \mathscr{P}\imath[V_x] \to \forall y < b V_a(y) \to d < \hat{\phi}(e(a), b) \\ \quad \to \forall z(h(a) \leqslant z < a \to \mathscr{S}[V_z, d]) \end{cases}$$

follows by the Induction Theorem 20.10. With the aid of ($\mathscr{P}\imath 1$) we get

$$\mathrm{DA} \vdash \mathscr{P}\imath[V_c] \to \forall x < \hat{\phi}(e(a), b) \mathscr{S}[V_c, x] \to \mathscr{S}[V_c, \hat{\phi}(e(a), b)].$$

From the last two formulas we have

$$\mathrm{DA} \vdash \begin{cases} \mathscr{R}[U, V, t] \to a < t \to \forall x < a \mathscr{P}\imath[V_x] \to \forall y < b V_a(y) \\ \quad \to \forall z(h(a) \leqslant z < a \to \mathscr{S}[V_z, \hat{\phi}(e(a), b)]). \end{cases}$$

From the definition of $\mathscr{R}$ we have

$$\mathrm{DA} \vdash \mathscr{R}[U, V, t] \to \bar{0} < a < t \to \forall x < a \mathscr{P}\imath[V_x] \to \forall y < b V_a(y) \to V_a(b).$$

The assertion follows by a basic inference (S2.0).

Lemma 10. $\mathrm{DA} \vdash \forall X \exists Y \mathscr{R}[X, Y, t] \to \hat{\omega}^s < t \to \mathscr{I}[\hat{\omega}^s] \to \mathscr{I}[\hat{\phi}(s, \bar{0})]$.

Proof. By the definition of $\mathscr{R}$ we have

(1) $\mathrm{DA} \vdash \mathscr{R}[U, V, t] \to \forall y(U(y) \leftrightarrow V(\bar{0}, y))$.

Hence $\mathrm{DA} \vdash \mathscr{R}[U, V, t] \to \mathscr{P}\imath[U] \to \mathscr{P}\imath[V_{\bar{0}}]$. Using Lemma 9 we obtain.

$$\mathrm{DA} \vdash \mathscr{R}[U, V, t] \to \mathscr{P}\imath[U] \to a < t \to \forall x < a \mathscr{P}\imath[V_x] \to \mathscr{P}\imath[V_a].$$

For the predicator $P := \lambda x(x < t \to \mathscr{P}\imath[V_x])$ we obtain

(2) $\mathrm{DA} \vdash \mathscr{R}[U, V, t] \to \mathscr{P}\imath[U] \to \mathscr{P}\imath[P]$.

By ($\mathscr{I}1$) and the definition of P

(3) $\mathrm{DA} \vdash \mathscr{I}[\hat{\omega}^s] \to \mathscr{P}\imath[P] \to \hat{\omega}^s < t \to \mathscr{P}\imath[V_{\hat{\omega}^s}]$ holds.

Trivially

(4) $\mathrm{DA} \vdash \mathscr{P}\imath[V_{\hat{\omega}^s}] \to V(\hat{\omega}^s, \bar{0})$ holds.

Obviously $h(\hat{\omega}^s) = \bar{0}$ and $e(\hat{\omega}^s) = s$. Therefore from the definition of $\mathscr{R}$ we have

(5) $\mathrm{DA} \vdash \mathscr{R}[U, V, t] \to \hat{\omega}^s < t \to V(\hat{\omega}^s, \bar{0}) \to \mathscr{S}[V_{\bar{0}}, \hat{\phi}(s, \bar{0})]$.

Further

(6) $\mathrm{DA} \vdash \mathscr{S}[V_{\bar{0}}, \hat{\phi}(s, \bar{0})] \to \forall x < \hat{\phi}(s, \bar{0}) V(\bar{0}, x)$ holds.

From (1)–(6) we obtain

$$\mathrm{DA} \vdash \mathscr{R}[U, V, t] \to \hat{\omega}^s < t \to \mathscr{I}[\hat{\omega}^s] \to \mathscr{P}\imath[U] \to \forall x < \hat{\phi}(s, \bar{0}) U(x).$$

By Theorems 20.8 and 20.6 we obtain

$$\mathrm{DA} \vdash \forall X \exists Y \mathscr{R}[X, Y, t] \to \hat{\omega}^s < t \to \mathscr{I}[\hat{\omega}^s] \to \mathscr{P}\imath[U] \to \forall x < \hat{\phi}(s, \bar{0}) U(x).$$

The assertion follows by a basic inference (S2.1).

6. Deductions in DA

Lemma 11. *For every numeral n*

$$\mathrm{DA} \vdash \pi_1(a) < \dot{\omega}^n \to \mathscr{R}[U, V, \pi_1(a) \hat{+} \bar{1}] \to \mathscr{R}[U, W, \pi_1(a) \hat{+} \bar{1}] \to V(a) \to W(a).$$

Proof. Let $\mathscr{G}[U, V, W, s, t]$ be the formula given in Lemma 8. Since $\omega^n < \varepsilon_0$ it follows from Lemmata 6 and 8 that

$$\mathrm{DA} \vdash \pi_1(a) < \dot{\omega}^n \to \mathscr{G}[U, V, W, \pi_1(a), t].$$

That is

$$\mathrm{DA} \vdash \pi_1(a) < \dot{\omega}^n \to \pi_1(a) < t \to \mathscr{R}[U, V, t] \to \mathscr{R}[U, W, t] \to \forall y (V_{\pi_1(a)}(y) \leftrightarrow W_{\pi_1(a)}(y)).$$

The assertion follows with $t := \pi_1(a) \hat{+} \bar{1}$.

Lemma 12. $\mathrm{DA} \vdash \forall X \exists Y \mathscr{R}[X, Y, \dot{\omega}^n]$ *holds for every numeral n.*

Proof by induction on n. Obviously $\mathrm{DA} \vdash \mathscr{R}[U, \lambda z U(\pi_2(z)), \dot{\omega}^0]$ holds. So the assertion holds for $n = 0$.

We prove the assertion for n' under the assumption

(1) $\mathrm{DA} \vdash \forall X \exists Y \mathscr{R}[X, Y, \dot{\omega}^n]$.

Elementary predicators P and Q depending on the free predicate variables U, V, W and the free number variable a can be defined so that the following hold:

(2.1) $\mathrm{DA} \vdash \forall x < \dot{\omega}^n a \forall y (P(x, y) \leftrightarrow V(x, y))$
(2.2) $\mathrm{DA} \vdash \forall x_0 \forall y (P(\dot{\omega}^n a \hat{+} x_0, y) \leftrightarrow W(x_0, y))$
(2.3) $\mathrm{DA} \vdash \forall y (Q(y) \leftrightarrow \forall z (h(\dot{\omega}^n a) \leqslant z < \dot{\omega}^n a \rightarrow \mathscr{S}[V_z, \hat{\phi}(e(\dot{\omega}^n a), y)]))$.

Here W does not occur in Q. Using the Extensionality Theorem 20.9 one obtains the following properties of P and Q from the definition of $\mathscr{R}$: From (2.1) we have

(3.1) $\mathrm{DA} \vdash \mathscr{R}[U, V, \dot{\omega}^n a] \rightarrow \mathscr{R}[U, P, \dot{\omega}^n a]$.

Using the formula

$$\mathscr{R}[Q, W, \dot{\omega}^n] \rightarrow \forall y (W(\bar{0}, y) \leftrightarrow Q(y))$$

which is deducible in DA, from (2.1)–(2.3) we obtain

(3.2) $\mathrm{DA} \vdash \begin{cases} \mathscr{R}[Q, W, \dot{\omega}^n a] \rightarrow \forall y (P(\dot{\omega}^n a, y) \leftrightarrow \\ \quad \forall z (h(\dot{\omega}^n a) \leqslant z < \dot{\omega}^n a \rightarrow \mathscr{S}[P_z, \hat{\phi}(e(\dot{\omega}^n a), y)])). \end{cases}$

From (2.2) we obtain

(3.3) $\mathrm{DA} \vdash \begin{cases} \mathscr{R}[Q, W, \dot{\omega}^n] \rightarrow \forall x_0 (\bar{0} < x_0 < \dot{\omega}^n \rightarrow \forall y (P(\dot{\omega}^n a \hat{+} x_0, y) \leftrightarrow \\ \quad \forall z_0 (h(x_0) \leqslant z_0 < x_0 \rightarrow \mathscr{S}[P_{\dot{\omega}^n a + z_0}, \hat{\phi}(e(x_0), y)]))). \end{cases}$

If $\dot{\omega}^n a < x < \dot{\omega}^n a'$ then $\bar{0} < x_0 < \dot{\omega}^n$ where $x = \dot{\omega}^n a \hat{+} x_0$, $h(x) = \dot{\omega}^n a \hat{+} h(x_0)$ and $e(x) = e(x_0)$. Therefore from (3.3) we have

(3.4) $\mathrm{DA} \vdash \begin{cases} \mathscr{R}[Q, W, \dot{\omega}^n] \rightarrow \forall x (\dot{\omega}^n a < x < \dot{\omega}^n a' \rightarrow \forall y (P(x, y) \leftrightarrow \\ \quad \forall z (h(x) \leqslant z < x \rightarrow \mathscr{S}[P_z, \hat{\phi}(e(x), y)]))). \end{cases}$

From (3.1), (3.2) and (3.4) we obtain

$$\mathrm{DA} \vdash \mathscr{R}[Q, W, \dot{\omega}^n] \rightarrow \mathscr{R}[U, V, \dot{\omega}^n a] \rightarrow \mathscr{R}[U, P, \dot{\omega}^n a'].$$

By Theorem 20.6 we obtain

$$\mathrm{DA} \vdash \mathscr{R}[Q, W, \dot{\omega}^n] \rightarrow \mathscr{R}[U, V, \dot{\omega}^n a] \rightarrow \exists Y \mathscr{R}[U, Y, \dot{\omega}^n a'].$$

Since W does not occur in Q by Theorem 20.8 we have

$$\mathrm{DA} \vdash \exists Y \mathscr{R}[Q, Y, \dot{\omega}^n] \rightarrow \mathscr{R}[U, V, \dot{\omega}^n a] \rightarrow \exists Y \mathscr{R}[U, Y, \dot{\omega}^n a'].$$

By Theorem 20.6 from (1) we obtain $\mathrm{DA} \vdash \exists Y \mathscr{R}[Q, Y, \dot{\omega}^n]$. There follows

$$\mathrm{DA} \vdash \mathscr{R}[U, V, \dot{\omega}^n a] \rightarrow \exists Y \mathscr{R}[U, Y, \dot{\omega}^n a'].$$

By Theorems 20.8, 20.6 and using (S2.1) we successively obtain

$$\mathrm{DA} \vdash \exists Y \mathscr{R}[U, Y, \dot{\omega}^n a] \rightarrow \exists Y \mathscr{R}[U, Y, \dot{\omega}^n a']$$
$$\mathrm{DA} \vdash \forall X \exists Y \mathscr{R}[X, Y, \dot{\omega}^n a] \rightarrow \exists Y \mathscr{R}[U, Y, \dot{\omega}^n a']$$
$$\mathrm{DA} \vdash \forall X \exists Y \mathscr{R}[X, Y, \dot{\omega}^n a] \rightarrow \forall X \exists Y \mathscr{R}[X, Y, \dot{\omega}^n a'].$$

As for $n=0$ we also have $\mathrm{DA} \vdash \forall X \exists Y \mathscr{R}[X, Y, \dot{\omega}^n 0]$.
Using a basic inference (ind) we obtain

(4.1) $\mathrm{DA} \vdash \forall X \exists Y \mathscr{R}[X, Y, \dot{\omega}^n t]$ for every term t.

There is a one-place calculable arithmetic function f such that

(4.2) $\mathrm{DA} \vdash a < \dot{\omega}^{n'} \rightarrow a < \dot{\omega}^n f(a)$.

From (4.1) and (4.2) we obtain

(5) $\mathrm{DA} \vdash \forall x_0 < \dot{\omega}^{n'} \exists Y \mathscr{R}[U, Y, x_0]$.

With the aid of (5) we now prove that there is a Δ_1^1-predicator $\not{p}[U]$ such that $\mathrm{DA} \vdash \mathscr{R}[U, \not{p}[U], \dot{\omega}^{n'}]$. We set

$$\mathscr{A}[U, V, s, t] := \pi_1(s) < t \rightarrow \mathscr{R}[U, V, \pi_1(s) \hat{+} \bar{1}] \rightarrow V(s)$$
$$\mathscr{B}[U, V, s, t] := \pi_1(s) < t \rightarrow (\mathscr{R}[U, V, \pi_1(s) \hat{+} \bar{1}] \wedge V(s)).$$

Then

$$\mathrm{DA} \vdash \mathscr{R}[U, V, \pi_1(a) \hat{+} \bar{1}] \rightarrow \mathscr{A}[U, V, a, \dot{\omega}^{n'}] \rightarrow \mathscr{B}[U, V, a, \dot{\omega}^{n'}] \quad \text{holds.}$$

By Theorems 20.6 and 20.8 we obtain

$$\mathrm{DA} \vdash \exists Y \mathscr{R}[U, Y, \pi_1(a) \hat{+} \bar{1}] \rightarrow \forall Y \mathscr{A}[U, Y, a, \dot{\omega}^{n'}] \rightarrow \exists Y \mathscr{B}[U, Y, a, \dot{\omega}^{n'}].$$

Since $\pi_1(a) < \dot{\omega}^{n'}$ we also have $\pi_1(a) \hat{+} 1 < \dot{\omega}^{n'}$. Therefore by (5) we obtain

$$\mathrm{DA} \vdash \pi_1(a) < \dot{\omega}^{n'} \rightarrow \forall Y \mathscr{A}[U, Y, a, \dot{\omega}^{n'}] \rightarrow \exists Y \mathscr{B}[U, Y, a, \dot{\omega}^{n'}].$$

Since $\pi_1(a) \prec \dot{\omega}^{n'}$ occurs as a premise of an implication in $\mathscr{B}[U, Y, a, \dot{\omega}^{n'}]$ we obtain

(6.1) $\mathrm{DA} \vdash \forall Y \mathscr{A}[U, Y, a, \dot{\omega}^{n'}] \rightarrow \exists Y \mathscr{B}[U, Y, a, \dot{\omega}^{n'}]$.

By Lemma 11 we obtain

$$\mathrm{DA} \vdash \pi_1(a) \prec \dot{\omega}^{n'} \rightarrow \mathscr{R}[U, V, \pi_1(a) \hat{+} \bar{1}] \rightarrow \mathscr{R}[U, W, \pi_1(a) \hat{+} \bar{1}] \rightarrow V(a) \rightarrow W(a).$$

By the definitions of $\mathscr{A}$ and $\mathscr{B}$ we obtain

$$\mathrm{DA} \vdash \mathscr{B}[U, V, a, \dot{\omega}^{n'}] \rightarrow \mathscr{A}[U, W, a, \dot{\omega}^{n'}].$$

By Theorem 20.8 using (S2.1) we obtain

(6.2) $\mathrm{DA} \vdash \exists Y \mathscr{B}[U, Y, a, \dot{\omega}^{n'}] \rightarrow \forall Y \mathscr{A}[U, Y, a, \dot{\omega}^{n'}]$.

From (6.1) and (6.2) we obtain

(7) $\mathrm{DA} \vdash \forall x(\forall Y \mathscr{A}[U, Y, x, \dot{\omega}^{n'}] \leftrightarrow \exists Y \mathscr{B}[U, Y, x, \dot{\omega}^{n'}])$.

Thus

$$\not{p}[U] := \lambda x \forall Y \mathscr{A}[U, Y, x, \dot{\omega}^{n'}]$$

is a Δ_1^1-predicator.
By Theorem 20.6 using (6.2) we get

$$\mathrm{DA} \vdash \forall Y \mathscr{A}[U, Y, \pi(a, b), \dot{\omega}^{n'}] \rightarrow \mathscr{A}[U, V, \pi(a, b), \dot{\omega}^{n'}]$$

$$\mathrm{DA} \vdash \mathscr{B}[U, V, \pi(a, b), \dot{\omega}^{n'}] \rightarrow \forall Y \mathscr{A}[U, Y, \pi(a, b), \dot{\omega}^{n'}].$$

Using basic inferences (S4) we obtain

$$\mathrm{DA} \vdash \not{p}[U](a, b) \rightarrow \mathscr{A}[U, V, \pi(a, b), \dot{\omega}^{n'}]$$
$$\mathrm{DA} \vdash \mathscr{B}[U, V, \pi(a, b), \dot{\omega}^{n'}] \rightarrow \not{p}[U](a, b).$$

From the definitions of $\mathscr{A}$ and $\mathscr{B}$ we obtain

(8.1) $\mathrm{DA} \vdash a \prec \dot{\omega}^{n'} \rightarrow \mathscr{R}[U, V, a \hat{+} \bar{1}] \rightarrow \not{p}[U](a, b) \rightarrow V(a, b)$
(8.2) $\mathrm{DA} \vdash a \prec \dot{\omega}^{n'} \rightarrow \mathscr{R}[U, V, a \hat{+} \bar{1}] \rightarrow V(a, b) \rightarrow \not{p}[U](a, b)$.

Further

(8.3) $\mathrm{DA} \vdash c \prec \dot{\omega}^{n'} \rightarrow a \prec c \rightarrow a \prec \dot{\omega}^{n'}$
(8.4) $\mathrm{DA} \vdash \mathscr{R}[U, V, c] \rightarrow a \prec c \rightarrow \mathscr{R}[U, V, a \hat{+} \bar{1}]$ hold.

From (8.1)–(8.4) we obtain

$$\mathrm{DA} \vdash c \prec \dot{\omega}^{n'} \to \mathscr{R}[U, V, c] \to \forall x \prec c \forall y (V(x, y) \leftrightarrow \not{p}[U](x, y)).$$

Using the Extensionality Theorem 20.9 we obtain

$$\mathrm{DA} \vdash c \prec \dot{\omega}^{n'} \to \mathscr{R}[U, V, c] \to \mathscr{R}[U, \not{p}[U], c].$$

By Theorem 20.8 we obtain

$$\mathrm{DA} \vdash c \prec \dot{\omega}^{n'} \to \exists Y \mathscr{R}[U, Y, c] \to \mathscr{R}[U, \not{p}[U], c].$$

Using (5) we obtain

$\mathrm{DA} \vdash \forall x_0 \prec \dot{\omega}^{n'} \mathscr{R}[U, \not{p}[U], x_0]$. Then
$\mathrm{DA} \vdash \mathscr{R}[U, \not{p}[U], \dot{\omega}^{n'}]$

also holds since $\dot{\omega}^{n'}$ is a limit term. Since $\not{p}[U]$ is a Δ_1^1-predicator,

$\mathrm{DA} \vdash \exists Y \mathscr{R}[U, Y, \dot{\omega}^{n'}]$ follows by Theorem 20.6.

Using a basic inference (S2.1) the assertion

$\mathrm{DA} \vdash \forall X \exists Y \mathscr{R}[X, Y, \dot{\omega}^{n'}]$ follows for n'.

Theorem 21.5. $\mathrm{DA} \vdash \mathscr{I}[\bar{\alpha}]$ *holds for every ordinal term* $\alpha < \phi\omega 0$.

Proof. Suppose $\alpha < \phi\omega 0$. Then there is a numeral n such that $\bar{\alpha} \prec \hat{\phi}(\bar{n}, \bar{0})$. By Lemma 10 we have

$$\mathrm{DA} \vdash \forall X \exists Y \mathscr{R}[X, Y, \dot{\omega}^{n'}] \to \hat{\omega}^{\bar{n}} \prec \dot{\omega}^{n'} \to \mathscr{I}[\hat{\omega}^{\bar{n}}] \to \mathscr{I}[\hat{\phi}(\bar{n}, \bar{0})].$$

By Lemma 12 $\mathrm{DA} \vdash \forall X \exists Y \mathscr{R}[X, Y, \dot{\omega}^{n'}]$ holds. Further $\hat{\omega}^{\bar{n}} = \dot{\omega}^{n} \prec \dot{\omega}^{n'}$ and by Theorem 21.3 $\mathrm{DA} \vdash \mathscr{I}[\dot{\omega}^{n}]$ also holds. Using cuts $\mathrm{DA} \vdash \mathscr{I}[\hat{\phi}(\bar{n}, \bar{0})]$ holds. Since $\bar{\alpha} \prec \hat{\phi}(\bar{n}, \bar{0})$, $\mathrm{DA} \vdash \mathscr{I}[\bar{\alpha}]$ follows by ($\mathscr{I}3$).

7. Deductions in DA*

As previously, we put

$$\mathscr{A}[U, V, n, m] := \pi_1(n) \prec m \to \mathscr{R}[U, V, \pi_1(n) \hat{+} \bar{1}] \to V(n)$$
$$\mathscr{B}[U, V, n, m] := \pi_1(n) \prec m \to (\mathscr{R}[U, V, \pi_1(n) \hat{+} \bar{1}] \wedge V(n)).$$

Lemma 13. *For every ordinal term α of* OT *and every numeral n*

$$\mathrm{DA}^* \vdash^{\leq \omega^2+\omega\cdot\alpha} \exists Y \mathscr{B}[U, Y, n, \overline{\omega\cdot\alpha}] \rightarrow \forall Y \mathscr{A}[U, Y, n, \overline{\omega\cdot\alpha}].$$

Proof. Let $\mathscr{G}[U, V, W, s, t]$ be the formula given in Lemma 8. From Lemmata 8 and 3 we have

(1) $\mathrm{DA}^* \vdash^{\leq \omega^2} \forall x \prec \bar{\beta} \mathscr{G}[U, V, W, x, t] \rightarrow \mathscr{G}[U, V, W, \bar{\beta}, t]$

for every numerical term t. We prove

$$\mathrm{DA}^* \vdash^{\omega^2+2\cdot\beta+1} \mathscr{G}[U, V, W, \bar{\beta}, t]$$

by induction on β. For $\beta_0 < \beta$, $2\cdot\beta_0+1 < 2\cdot\beta$. Therefore by I.H.

$$\mathrm{DA}^* \vdash^{\leq \omega^2+2\cdot\beta} \mathscr{G}[U, V, W, n, t]$$

for every numeral $n \prec \bar{\beta}$. By Lemma 4 we have

$$\mathrm{DA}^* \vdash^{\omega^2+2\cdot\beta} \forall x \prec \bar{\beta} \mathscr{G}[U, V, W, x, t].$$

Using a cut with (1) the assertion

$$\mathrm{DA}^* \vdash^{\omega^2+2\cdot\beta+1} \mathscr{G}[U, V, W, \bar{\beta}, t]$$

follows, that is

(2) $\mathrm{DA}^* \vdash^{\omega^2+2\cdot\beta+1} \bar{\beta} \prec t \rightarrow \mathscr{R}[U, V, t] \rightarrow \mathscr{R}[U, W, t] \rightarrow \forall y (V_{\bar{\beta}}(y) \leftrightarrow W_{\bar{\beta}}(y))$.

$\bar{\beta} \prec \overline{\beta+1}$ and $\forall y (V_{\bar{\beta}}(y) \leftrightarrow W_{\bar{\beta}}(y)) \rightarrow V(\bar{\beta}, \pi_2(n)) \rightarrow W(\bar{\beta}, \pi_2(n))$

are finitely deducible formulas.

Using (2) for $t := \overline{\beta+1}$ by two cuts we obtain

$$\mathrm{DA}^* \vdash^{\omega^2+2\cdot\beta+3} \mathscr{R}[U, V, \overline{\beta+1}] \rightarrow \mathscr{R}[U, W, \overline{\beta+1}] \rightarrow V(\bar{\beta}, \pi_2(n)) \rightarrow W(\bar{\beta}, \pi_2(n)).$$

For $\pi_1(n) = \bar{\beta} \prec \overline{\omega\cdot\alpha}$ we have $2\cdot\beta+3 < \omega\cdot\alpha$. It follows that

(3) $\mathrm{DA}^* \vdash^{\leq \omega^2+\omega\cdot\alpha} \mathscr{R}[U, V, \pi_1(n) \hat{+} \bar{1}] \rightarrow \mathscr{R}[U, W, \pi_1(n) \hat{+} \bar{1}] \rightarrow V(n) \rightarrow W(n)$

for $\pi_1(n) \prec \overline{\omega\cdot\alpha}$. We obtain

(4) $\mathrm{DA}^* \vdash^{\leq \omega^2+\omega\cdot\alpha} \mathscr{B}[U, V, n, \overline{\omega\cdot\alpha}] \rightarrow \mathscr{A}[U, W, n, \overline{\omega\cdot\alpha}]$

for $\pi_1(n) < \overline{\omega\cdot\alpha}$ as a logical consequence of (3). (4) also holds for $\overline{\omega\cdot\alpha} \leqslant \pi_1(n)$, since the formula concerned is in this case an (Ax2). By Theorem 20.16 and a basic inference (S2.1) we obtain

$$\mathrm{DA}^* \vdash^{<\omega^2+\omega\cdot\alpha} \exists Y \mathscr{B}[U, Y, n, \overline{\omega\cdot\alpha}] \to \forall Y \mathscr{A}[U, Y, n, \overline{\omega\cdot\alpha}]$$

for every numeral n.

Lemma 14. *For every ordinal term $\omega\cdot\alpha+n$ of* OT *where $n<\omega$*

$$\mathrm{DA}^* \vdash^{\omega^2+\omega\cdot\alpha+n+9} \exists Y \mathscr{R}[U, Y, \overline{\omega\cdot\alpha+n}] \quad \textit{holds.}$$

Proof by induction on $\omega\cdot\alpha+n$. For $\alpha=0$ the assertion follows from Lemmata 12 and 13. We may therefore assume $\alpha\neq 0$.

Case 1. $n=n_0+1$. An elementary predicator $\mathfrak{p}[V]$ may be defined such that the following hold:

$$\mathrm{DA} \vdash \forall x < \overline{\omega\cdot\alpha+n_0}\, \forall y(\mathfrak{p}[V](x, y) \leftrightarrow V(x, y))$$

$$\mathrm{DA} \vdash \forall y(\mathfrak{p}[V](\overline{\omega\cdot\alpha+n_0}, y) \leftrightarrow \forall z(h(\overline{\omega\cdot\alpha+n_0}) \leqslant z < \overline{\omega\cdot\alpha+n_0}$$
$$\to \mathscr{S}[V_z, \hat{\phi}(e(\overline{\omega\cdot\alpha+n_0}), y)])).$$

With the aid of the Extensionality Theorem 20.9 we then obtain

$$\mathrm{DA} \vdash \mathscr{R}[U, V, \overline{\omega\cdot\alpha+n_0}] \to \mathscr{R}[U, \mathfrak{p}[V], \overline{\omega\cdot\alpha+n}].$$

By Theorems 20.6, 20.8 and Lemma 3 we obtain

$$\mathrm{DA}^* \vdash^{<\omega^2} \exists Y \mathscr{R}[U, Y, \overline{\omega\cdot\alpha+n_0}] \to \exists Y \mathscr{R}[U, Y, \overline{\omega\cdot\alpha+n}].$$

By I.H.

$$\mathrm{DA}^* \vdash^{\omega^2+\omega\cdot\alpha+n+8} \exists Y \mathscr{R}[U, Y, \overline{\omega\cdot\alpha+n_0}].$$

The assertion follows by a cut.

Case 2. $n=0$. From the definitions of $\mathscr{A}$ and $\mathscr{B}$ we have

$$\mathrm{DA} \vdash \mathscr{R}[U, V, \pi_1(n) \hat{+} \bar{1}] \to \mathscr{A}[U, V, n, \overline{\omega\cdot\alpha}] \to \mathscr{B}[U, V, n, \overline{\omega\cdot\alpha}].$$

By Theorems 20.6, 20.8 and Lemma 3 we obtain

$$\mathrm{DA}^* \vdash^{\leq \omega^2} \exists Y \mathscr{R}[U, Y, \pi_1(n) \hat{+} \bar{1}] \to \forall Y \mathscr{A}[U, Y, n, \overline{\omega \cdot \alpha}] \\ \to \exists Y \mathscr{B}[U, Y, n, \overline{\omega \cdot \alpha}].$$

For $\pi_1(n) = \bar{\beta} < \overline{\omega \cdot \alpha}$ we have $\pi_1(n) \hat{+} \bar{1} = \overline{\beta + 1}$ and $\beta + 9 < \omega \cdot \alpha$. Hence by the I.H. we have

$$\mathrm{DA}^* \vdash^{\leq \omega^2 + \omega \cdot \alpha} \exists Y \mathscr{R}[U, Y, \pi_1(n) \hat{+} \bar{1}],$$

and consequently also

(1) $\mathrm{DA}^* \vdash^{\leq \omega^2 + \omega \cdot \alpha} \forall Y \mathscr{A}[U, Y, n, \overline{\omega \cdot \alpha}] \to \exists Y \mathscr{B}[U, Y, n, \overline{\omega \cdot \alpha}]$

for $\pi_1(n) < \overline{\omega \cdot \alpha}$. This also holds for $\overline{\omega \cdot \alpha} \leqslant \pi_1(n)$, since the formula

$$\forall Y \mathscr{A}[U, Y, n, \overline{\omega \cdot \alpha}] \to \mathscr{B}[U, V, n, \overline{\omega \cdot \alpha}]$$

is in this case an (Ax2). Hence (1) holds for every numeral n. Using Lemma 13 we obtain

$$\mathrm{DA}^* \vdash^{\leq \omega^2 + \omega \cdot \alpha} \forall Y \mathscr{A}[U, Y, n, \overline{\omega \cdot \alpha}] \leftrightarrow \exists Y \mathscr{B}[U, Y, n, \overline{\omega \cdot \alpha}].$$

Using a basic inference (S2.0*) we obtain

(2) $\mathrm{DA}^* \vdash^{\omega^2 + \omega \cdot \alpha} \forall x (\forall Y \mathscr{A}[U, Y, x, \overline{\omega \cdot \alpha}] \leftrightarrow \exists Y \mathscr{B}[U, Y, x, \overline{\omega \cdot \alpha}])$.

By the definitions of $\mathscr{A}$ and $\mathscr{B}$ we obtain from the formulas

$$\forall Y \mathscr{A}[U, Y, \pi(m, n), \overline{\omega \cdot \alpha}] \to \mathscr{A}[U, V, \pi(m, n), \overline{\omega \cdot \alpha}]$$
$$\mathscr{B}[U, V, \pi(m, n), \overline{\omega \cdot \alpha}] \to \exists Y \mathscr{B}[U, Y, \pi(m, n), \overline{\omega \cdot \alpha}]$$

which are deducible in DA, that

$$\mathrm{DA} \vdash m < \overline{\omega \cdot \alpha} \to \mathscr{R}[U, V, m \hat{+} \bar{1}] \to \forall Y \mathscr{A}[U, Y, \pi(m, n), \overline{\omega \cdot \alpha}] \to V(m, n)$$
$$\mathrm{DA} \vdash m < \overline{\omega \cdot \alpha} \to \mathscr{R}[U, V, m \hat{+} \bar{1}] \to V(m, n) \to \exists Y \mathscr{B}[U, Y, \pi(m, n), \overline{\omega \cdot \alpha}].$$

Using Lemmata 3 and 13 we obtain

$$\mathrm{DA}^* \vdash^{\leq \omega^2} m < \overline{\omega \cdot \alpha} \to \mathscr{R}[U, V, m \hat{+} \bar{1}] \to \forall Y \mathscr{A}[U, Y, \pi(m, n), \overline{\omega \cdot \alpha}] \to V(m, n)$$
$$\mathrm{DA}^* \vdash^{\leq \omega^2 + \omega \cdot \alpha} m < \overline{\omega \cdot \alpha} \to \mathscr{R}[U, V, m \hat{+} 1] \to V(m, n) \\ \to \forall Y \mathscr{A}[U, Y, \pi(m, n), \overline{\omega \cdot \alpha}].$$

Using a basic inference (S4) for the predicator

$$\mathscr{p}[U] := \lambda x \forall Y \mathscr{A}[U, Y, x, \overline{\omega \cdot \alpha}]$$

we obtain as a logical consequence

$$\mathrm{DA}^* \vdash^{\leq \omega^2 + \omega \cdot \alpha} m < \overline{\omega \cdot \alpha} \to \mathscr{R}[U, V, m \hat{+} \bar{1}] \to (V(m, n) \leftrightarrow \mathscr{p}[U](m, n)).$$

Using a basic inference (S2.0*) we obtain

(3) $\mathrm{DA}^* \vdash^{\omega^2 + \omega \cdot \alpha} m < \overline{\omega \cdot \alpha} \to \mathscr{R}[U, V, m \hat{+} \bar{1}] \to \forall y (V(m, y) \leftrightarrow \mathscr{p}[U](m, y)).$

For $k < \overline{\omega \cdot \alpha}$ the formulas $m < k \to m < \overline{\omega \cdot \alpha}$ and

$$\mathscr{R}[U, V, k] \to m < k \to \mathscr{R}[U, V, m \hat{+} \bar{1}]$$

are deducible in DA and therefore in DA* with orders $< \omega^2$. Therefore using two cuts with (3) we obtain

$$\mathrm{DA}^* \vdash^{\omega^2 + \omega \cdot \alpha + 2} \mathscr{R}[U, V, k] \to m < k \to \forall y (V(m, y) \leftrightarrow \mathscr{p}[U](m, y))$$

and by a basic inference (S2.0*)

(4) $\mathrm{DA}^* \vdash^{\omega^2 + \omega \cdot \alpha + 3} \mathscr{R}[U, V, k] \to \forall x < k \forall y (V(x, y) \leftrightarrow \mathscr{p}[U](x, y))$

for $k < \overline{\omega \cdot \alpha}$. With the aid of the Extensionality Theorem 20.9 the formula

$$\mathscr{R}[U, V, k] \to \forall x < k \forall y (V(x, y) \leftrightarrow \mathscr{p}[U](x, y)) \to \mathscr{R}[U, \mathscr{p}[U], k]$$

is deducible in DA and therefore in DA* with order $< \omega^2$.

Therefore by a cut using (4) we obtain

$$\mathrm{DA}^* \vdash^{\omega^2 + \omega \cdot \alpha + 4} \mathscr{R}[U, V, k] \to \mathscr{R}[U, \mathscr{p}[U], k] \quad \text{for } k < \overline{\omega \cdot \alpha}.$$

By Theorem 20.16 we have

$$\mathrm{DA}^* \vdash^{\omega^2 + \omega \cdot \alpha + 5} \exists Y \mathscr{R}[U, Y, k] \to \mathscr{R}[U, \mathscr{p}[U], k].$$

By I.H. we have

$\mathrm{DA}^* \vdash^{\leq \omega^2 + \omega \cdot \alpha} \exists Y \mathscr{R}[U, Y, k]$ for $k < \overline{\omega \cdot \alpha}$. It follows that
$\mathrm{DA}^* \vdash^{\omega^2 + \omega \cdot \alpha + 6} \mathscr{R}[U, \mathscr{p}[U], k]$ and by Lemma 4
$\mathrm{DA}^* \vdash^{\omega^2 + \omega \cdot \alpha + 7} \forall x < \overline{\omega \cdot \alpha}\, \mathscr{R}[U, \mathscr{p}[U], x].$

The formula

$$\forall x \prec \overline{\omega\cdot\alpha}\, \mathscr{R}[U, p[U], x] \to \mathscr{R}[U, p[U], \overline{\omega\cdot\alpha}]$$

is deducible in DA and therefore in DA* with order $<\omega^2$. Hence we obtain

(5) $\mathrm{DA}^* \mid\!\underline{\omega^2+\omega\cdot\alpha+8}\ \mathscr{R}[U, p[U], \overline{\omega\cdot\alpha}]$.

By Theorem 20.17 the assertion

$\mathrm{DA}^* \mid\!\underline{\omega^2+\omega\cdot\alpha+9}\ \exists Y \mathscr{R}[U, Y, \overline{\omega\cdot\alpha}]$ follows from (2) and (5).

As in §14 (p. 96) let $\zeta_0 = \phi 00 = 1$ and $\zeta_{n+1} = \phi\zeta_n 0$.

Theorem 21.6. $\mathrm{DA}^* \mid\!\underline{\zeta_n+13}\ \mathscr{I}[\overline{\zeta_{n+1}}]$ *for* $n \geq 1$.

Proof by induction on n. By Lemmata 10 and 3 we have

$$\mathrm{DA}^* \mid\!\underline{{}^{<\omega^2}}\ \forall X \exists Y \mathscr{R}[X, Y, \overline{\zeta_n+1}] \to \hat{\omega}^{\bar{\zeta}_n} \prec \overline{\zeta_n+1} \to \mathscr{I}[\hat{\omega}^{\bar{\zeta}_n}] \to \mathscr{I}[\hat{\phi}(\bar{\zeta}_n, \bar{0})].$$

Since $\hat{\phi}(\bar{\zeta}_n, \bar{0}) = \overline{\zeta_{n+1}}$ and $\omega^{\zeta_n} = \zeta_n$, therefore $\hat{\omega}^{\bar{\zeta}_n} = \bar{\zeta}_n \prec \overline{\zeta_n+1}$ for $n \geq 1$. It follows that

(1) $\mathrm{DA}^* \mid\!\underline{{}^{<\omega^2}}\ \forall X \exists Y \mathscr{R}[X, Y, \overline{\zeta_n+1}] \to \mathscr{I}[\bar{\zeta}_n] \to \mathscr{I}[\overline{\zeta_{n+1}}]$.

$\omega^2 + \zeta_n = \zeta_n$ for $n \geq 1$. Therefore by Lemma 14 we obtain

$\mathrm{DA}^* \mid\!\underline{\zeta_n+10}\ \exists Y \mathscr{R}[U, Y, \overline{\zeta_n+1}]$. Using (S2.1) we obtain
$\mathrm{DA}^* \mid\!\underline{\zeta_n+11}\ \forall X \exists Y \mathscr{R}[U, Y, \overline{\zeta_n+1}]$. By a cut with (1) we obtain

(2) $\mathrm{DA}^* \mid\!\underline{\zeta_n+12}\ \mathscr{I}[\bar{\zeta}_n] \to \mathscr{I}[\overline{\zeta_{n+1}}]$.

If $n=1$, then $\zeta_n = \varepsilon_0 < \phi\omega 0$. Then by Theorem 21.5 and Lemma 3 $\mathrm{DA}^* \mid\!\underline{{}^{<\omega^2}}\ \mathscr{I}[\bar{\zeta}_n]$. By the I.H. in every case we have $\mathrm{DA}^* \mid\!\underline{{}^{<\zeta_n}}\ \mathscr{I}[\bar{\zeta}_n]$. Using a cut with (2) we obtain $\mathrm{DA}^* \mid\!\underline{\zeta_n+13}\ \mathscr{I}[\overline{\zeta_{n+1}}]$.

§22. The Semi-Formal System RA* for Ramified Analysis

1. The Formal Language

We use the following as *primitive symbols*:

1. Denumerably infinitely many bound number variables.
2. Denumerably infinitely many free (1-place) predicate variables of level α for each ordinal term α of the system OT.

3, Denumerably infinitely many bound (1-place) predicate variables of level β for each ordinal term $\beta \neq 0$ of the system OT.

4. The symbols 0, ′, $\perp$, $\rightarrow$, $\forall$ and λ.

5. Symbols for n-place calculable arithmetic functions and n-place decidable arithmetic predicates ($n \geqslant 1$). (The conditions specified on p. 168 for the use of these symbols have to be satisfied.)

6. Round brackets and comma.

In this section small Greek letters (possibly with indices) always denote ordinal terms of the system OT.

We use *nominal forms* in the usual way and denote them by capital script letters. *Terms*, *numerals* and *prime formulas* are defined as in §20.1. In the formal language of the system RA* every term is *numerical* and every prime formula *constant*. Therefore by §20.1 we have: Every term has a calculable *value* and every prime formula is decidably *true* or *false*.

Inductive definition of *formulas* and *predicators*:

1. Every prime formula is a formula of level 0.
2. Every free predicate variable of level α is a predicator of level α.
3. If P^α is a predicator of level α and t is a term, then $P^\alpha(t)$ is a formula of level α.
4. If A and B are formulas of levels α and β then $(A \rightarrow B)$ is a formula of level $\gamma := \max(\alpha, \beta)$.
5. If $\mathscr{F}[0]$ is a formula of level α and x is a bound number variable which does not occur in $\mathscr{F}$, then $\forall x \mathscr{F}[x]$ is a formula of level α and $\lambda x \mathscr{F}[x]$ is a predicator of level α.
6. If U^β is a free predicate variable of level $\beta \neq 0$, $\mathscr{F}[U^\beta]$ is a formula of level α and X^β a bound predicate variable of level β which does not occur in $\mathscr{F}$, then $\forall X^\beta \mathscr{F}[X^\beta]$ is a formula of level $\gamma = \max(\alpha, \beta)$.

Corollaries. 1. *The level of a formula is* 0 *if it contains no predicate variables, otherwise it is the maximum of the levels of the predicate variables which occur in it.*

2. *If s, t are terms then: If $\mathscr{F}[s]$ is a formula, $\mathscr{F}[t]$ is also a formula.*

3. *If P^α is a predicator of level α, $\mathscr{F}[P^\alpha]$ a formula and Q^β a predicator of level β which contains no bound variable occurring in $\mathscr{F}$, then $\mathscr{F}[Q^\beta]$ is also a formula.*

Syntactic symbols:

x, y, z	for bound number variables,
$U^\alpha, V^\alpha, W^\alpha$	for free predicate variables of level α,
$X^\beta, Y^\beta, Z^\beta$	(where $\beta \neq 0$) for bound predicate variables of level β,
P^α, Q^α	for predicators of level α,
s, t	for terms,
A, B, C, D, E, F, G	for formulas,
i, j, k, m, n	for natural numbers and the corresponding numerals.

We shall also use these syntactic symbols with subscripts.

In the same way as before we write $A \rightarrow B$ for $(A \rightarrow B)$ and $\neg A$ for $(A \rightarrow \perp)$. We also denote an ordinal term $\omega^0 \cdot n$ by n for short. As before, the *length* of a formula is the number of $\rightarrow$, $\forall$ and λ symbols, which occur in it.

A predicator P^{α} is said to be $\mathcal{F}$*-permissible* if $\mathcal{F}[P^{\alpha}]$ is a formula. In particular this is the case when $\mathcal{F}[U^0]$ is a formula and P^{α} contains no bound variable occurring in $\mathcal{F}$.

We now assign an ordinal term of the system OT as a *rank* to each formula.

Inductive definition of the *rank* of a formula:

1. Every prime formula has rank 0.
2. Every formula $U^{\alpha}(t)$ has rank $\rho := \omega \cdot \alpha$.
3. If A and B are formulas of ranks ρ_1 and ρ_2 then the formula $(A \to B)$ has rank $\rho := \max(\rho_1, \rho_2) + 1$.
4. If $\mathcal{F}[0]$ is a formula of rank ρ_0 in which x does not occur then the formulas $\forall x \mathcal{F}[x]$ and $\lambda x \mathcal{F}[x](t)$ have rank $\rho := \rho_0 + 1$.
5. If $\mathcal{F}[U^0]$ is a formula of rank ρ_0 and $\forall X^{\beta} \mathcal{F}[X^{\beta}]$ is a formula of level γ then $[U^0]$ has rank $\rho := \max(\omega \cdot \gamma, \rho_0 + 1)$.

Corollaries. 1. Two formulas $\mathcal{F}[s]$ and $\mathcal{F}[t]$ have the same rank.
2. Two formulas $\mathcal{F}[U^{\alpha}]$ and $\mathcal{F}[V^{\alpha}]$ have the same rank.

Theorem 22.1. *Every formula of level* γ *has rank* $\rho = \omega \cdot \gamma + n$ (*where* n *is an abbreviation for* $\omega^0 \cdot n$ *and therefore* $\rho < \omega \cdot \gamma + \omega$).

Proof. By induction on the length of the formula.

Theorem 22.2. *If* P^{α} *is an* $\mathcal{F}$*-permissible predicator of level* $\alpha < \gamma$ *and* $\mathcal{F}[U^0]$ *is a formula of level* γ *then* $\mathcal{F}[P^{\alpha}]$ *has the same rank as* $\mathcal{F}[U^0]$.

Proof by induction on the length of the formula $\mathcal{F}[U^0]$. We may assume that $\mathcal{F}[P^{\alpha}]$ and $\mathcal{F}[U^0]$ are different formulas since otherwise the assertion is trivial. It then follows since $\gamma \neq 0$ that $\mathcal{F}[U^0]$ has length >0.

1. Suppose $\mathcal{F}[U^0]$ is a formula $\mathcal{F}_1[U^0] \to \mathcal{F}_2[U^0]$ or $\forall x \mathcal{G}[U^0, x]$ or $\lambda x \mathcal{G}[U^0, x](t)$. Then the assertion follows from Theorem 22.1 and the I.H.
2. Suppose $\mathcal{F}[U^0]$ is a formula $\forall X^{\beta} \mathcal{G}[U^0, X^{\beta}]$. Then $\mathcal{G}[U^0, V^9]$ is a formula of level $\gamma_0 \leqslant \gamma$. If $\gamma_0 = \gamma$ then by I.H. $\mathcal{G}[U^0, V^0]$ and $\mathcal{G}[P^{\alpha}, V^0]$ have the same rank ρ_0. In this case both formulas $\mathcal{F}[U^0]$ and $\mathcal{F}[P^{\alpha}]$ have rank $\rho_0 + 1$. If $\gamma_0 < \gamma$ then $\mathcal{G}[P^{\alpha}, V^0]$ is also a formula of level $< \gamma$. In this case both formulas $\mathcal{F}[U^0]$ and $\mathcal{F}[P^{\alpha}]$ have rank $\omega \cdot \gamma$.

Theorem 22.3. a) A and B have smaller ranks than $A \to B$.
b) $\mathcal{F}[t]$ has smaller rank than $\forall x \mathcal{F}[x]$ and $\lambda x \mathcal{F}[x](t)$.
c) Every formula $\mathcal{F}[P^{\alpha}]$ where $\alpha < \beta$ has smaller rank than $\forall X^{\beta} \mathcal{F}[X^{\beta}]$.

Proof. a) and b) follow immediately from the definition of the rank of a formula. c) is obtained as follows: If γ is the level of the formula $\forall X^{\beta} \mathcal{F}[X^{\beta}]$ then $\mathcal{F}[U^0]$ is a formula of a level $\gamma_0 \leqslant \gamma$. If $\gamma_0 = \gamma$, then by Theorem 22.2 $\mathcal{F}[U^0]$ and $\mathcal{F}[P^{\alpha}]$ have the same rank ρ_0 while the formula $\forall X^{\beta} \mathcal{F}[X^{\beta}]$ has rank $\rho_0 + 1$. If $\gamma_0 < \gamma$ then $\mathcal{F}[P^{\alpha}]$ is also a formula of a level $< \gamma$. By Theorem 22.1 it has a rank $< \omega \cdot \gamma$ while the formula $\forall X^{\beta} \mathcal{F}[X^{\beta}]$ has rank $\omega \cdot \gamma$ in this case.

2. The Deduction Procedures

In the same way as before we define *positive* and *negative parts* of formulas, $F \models^{s} G$ and the *equivalence* of formulas. We also use the same syntactic symbols as before for P-forms, N-forms and NP-forms.

Axioms of the system RA* (corresponding to those for DA*):
(Ax1) $\mathscr{P}[A]$, if A is a true prime formula.
(Ax2) $\mathscr{N}[A]$, if A is a false prime formula
(Ax3) $\mathscr{Q}[A, B]$, if A and B are equivalent formulas of length 0.

We call the minimal parts indicated in the axioms the *principal parts* of these axioms.

Basic inferences of the system RA*:
(S1) $\mathscr{N}[\neg A], \mathscr{N}[B] \vdash \mathscr{N}[(A \to B)]$, if B is not the formula $\perp$.
(S2.0*) $\mathscr{P}[\mathscr{F}[n]]$ for every numeral $n \vdash \mathscr{P}[\forall x \mathscr{F}[x]]$.
(S2.1*) $\mathscr{P}[\mathscr{F}[P^{\alpha}]]$ for every $\mathscr{F}$-permissible predicator P^{α} of a level $\alpha < \beta \vdash \mathscr{P}[\forall X^{\beta} \mathscr{F}[X^{\beta}]]$.
(S3.0) $\mathscr{F}[t] \to \mathscr{N}[\forall x \mathscr{F}[x]] \vdash \mathscr{N}[\forall x \mathscr{F}[x]]$.
(S3.1) $\mathscr{F}[P^{\alpha}] \to \mathscr{N}[\forall X^{\beta} \mathscr{F}[X^{\beta}]] \vdash \mathscr{N}[\forall X^{\beta}] \mathscr{F}[X^{\beta}]]$, if $\alpha < \beta$.
(S4) $\mathscr{E}[\mathscr{F}[t]] \vdash \mathscr{E}[\lambda x \mathscr{F}[x](t)]$.
(Cut) $\mathscr{P}[A], A \to B \vdash C$, if $\mathscr{P}[B] \models^{s} C$ holds.

All premises and all conclusions of these basic inferences must be formulas. The basic inferences (S2.0*) and (S2.1*) have infinitely many premises. Every other basic inference has either one or two premises. We call the basic inferences (S1)–(S4) *principal inferences*. We call the minimal part indicated in the conclusion of a principal inference the *principal part* of that inference. The formula denoted by A in the premises of a cut is called the *cut formula*. The *rank of a cut* is the rank of its cut formula.

Inductive definition of RA* $\models^{\alpha}_{\rho} F$
1. If F is an axiom of the system RA* then RA* $\models^{\alpha}_{\rho} F$ holds for all ordinal terms α and ρ of the system OT.
2. If RA* $\models^{\alpha_i}_{\rho} F_i$ and $\alpha_i < \alpha$ for every premise F_i of a principal inference or a cut of rank $< \rho$ then RA* $\models^{\alpha}_{\rho} F$ holds for the conclusion F of that inference.

Corollary. *If* RA* $\models^{\alpha}_{\rho} F$, $\alpha \leqslant \beta$ *and* $\rho \leqslant \sigma$, *then* RA* $\models^{\beta}_{\sigma} F$ *also holds.*

A formula F is said to be *deducible* in RA* with *order* α and *rank* ρ if RA* $\models^{\alpha}_{\rho} F$ holds. Thus α is an upper bound for the orders of the basic inferences which occur in the deduction of F while ρ says that every cut which occurs in the deduction of F has a rank $< \rho$. If RA* $\models^{\alpha}_{0} F$ holds then the formula F has a cut-free deduction with order α.

Predicativity of the system RA*. On the basis of the intended interpretation of the formulas and predicators every formula and every predicator which contains a universal quantifier $\forall X^\beta$ depends on the totality of those predicators which are to form the domain of this universal quantifier. The basic inferences (S2.1*) and (S3.1) of the system RA* are so formulated that the domain of a universal quantifier $\forall X^\beta$ consists only of predicators of level $<\beta$ while every formula and every predicator which contains the universal quantifier $\forall X^\beta$ belongs, according to the definition, to a level $\geqslant \beta$. Therefore by means of the levels we have given a predicative construction of our formal language. We shall also show on pp. 217–220 that the transfinite induction over ordinal terms of the system OT which we use here is predicatively provable. Thus not only the formal language but also the deduction procedures of the system RA* are seen to be predicative.

3. Weak Inferences

Below we write $\vdash^{\alpha}_{\rho} F$ for RA* $\vdash^{\alpha}_{\rho} F$. $F \vdash G$ is said to be a *weak inference* if the following always holds: If $\vdash^{\alpha}_{\rho} F$ then $\vdash^{\alpha}_{\rho} G$.

Theorem 22.4 (Replacement rule). *If F and G are equivalent formulas $F \vdash G$ is a weak inference.*

Proof. In this case $\vdash^{\alpha}_{\rho} G$ follows from $\vdash^{\alpha}_{\rho} F$ by induction on α.

Theorem 22.5 (Inversion rules). *The following are weak inferences*:

a) $\mathscr{N}[(A \to B)] \vdash \mathscr{N}[\neg A]$
b) $\mathscr{N}[(A \to B)] \vdash \mathscr{N}[B]$
c) $\mathscr{P}[\forall x \mathscr{F}[x]] \vdash \mathscr{P}[\mathscr{F}[t]]$
d) $\mathscr{P}[\forall X^\beta \mathscr{F}[X^\beta]] \vdash \mathscr{P}[\mathscr{F}[P^\alpha]]$, *if P^α is $\mathscr{F}$-permissible and $\alpha < \beta$.*
e) $\mathscr{E}[\lambda x \mathscr{F}[x](t)] \vdash \mathscr{E}[\mathscr{F}[t]]$.

The inductive proofs proceed as for Theorem 4.3.

Theorem 22.6 (Structural rule). *If $F \vdash^{s} G$ holds then $F \vdash^{s} G$ is a weak inference.*

Proof. Using Theorem 22.5, as for Theorem 4.4.

4. Elimination of Cuts

Lemma 1. *If A is a true prime formula and $\vdash^{\alpha}_{0} A \to B$ holds, then $\vdash^{\alpha}_{0} \mathscr{P}[B]$ for every P-form $\mathscr{P}$.*

Proof by induction on α.

1. Suppose $A \to B$ is an axiom. If A is not a principal part of this axiom, then $\mathscr{P}[B]$ is an axiom of the same sort. Otherwise B is a formula $\mathscr{P}_1[A_1]$, where A_1

is also a true prime formula. Then $\mathscr{P}[B]$ is the (Ax1) $\mathscr{P}[\mathscr{P}_1[A_1]]$. In each case $\vdash_0^\alpha \mathscr{P}[B]$ follows.

2. Suppose $A \to B$ is the conclusion of a principal inference. Then the assertion follows from the I.H. using the structural rule.

Lemma 2. *If A is a false prime formula and $\vdash_0^\alpha \mathscr{P}[A]$ holds, then $\vdash_0^\alpha \mathscr{P}[B]$ for every formula B.*

Proof by induction on α.

1. Suppose $\mathscr{P}[A]$ is an axiom. If A is not a principal part of this axiom then $\mathscr{P}[B]$ is an axiom of the same sort. Otherwise $\mathscr{P}[A]$ is an (Ax3) $\mathscr{Q}[A_1, A]$ where A_1 is also a false prime formula and $\mathscr{P}[B]$ is the (Ax2) $\mathscr{Q}[A_1, B]$. In each case $\vdash_0^\alpha \mathscr{P}[B]$ follows.

2. Suppose $\mathscr{P}[A]$ is the conclusion of a principal inference. Then the assertion follows immediately from the I.H.

Lemma 3. *If A is a formula $U^\alpha(t)$ or $\forall x \mathscr{F}[x]$ or $\forall X^\beta \mathscr{F}[X^\beta]$ of rank ρ and $\vdash_\rho^\gamma \mathscr{P}[A]$ and $\vdash_\rho^\delta A \to B$ hold, then $\vdash_\rho^{\gamma+\delta} \mathscr{P}[B]$.*

Proof by induction on δ.

1. Suppose $A \to B$ is an axiom. If $\mathscr{P}[B]$ is also an axiom then the assertion is trivial. Otherwise A is a formula $U^\alpha(t)$ and B is a formula $\mathscr{P}_1[U^\alpha(s)]$ where s and t are terms of equal value. Then $\vdash_\rho^{\gamma+\delta} \mathscr{P}[\mathscr{P}_1[U^\alpha(s)]]$ follows from $\vdash_\rho^\gamma \mathscr{P}[U^\alpha(t)]$ by the replacement rule and the structural rule.

2. Suppose $A \to B$ is the conclusion of a principal inference with principal part A. Then A is a formula $\forall x \mathscr{F}[x]$ or $\forall X^\beta \mathscr{F}[X^\beta]$ and we have $\delta_0 < \delta$ such that

$$\vdash_\rho^{\delta_0} \mathscr{F}[t] \to (\forall x \mathscr{F}[x] \to B) \quad \text{or} \quad \vdash_\rho^{\delta_0} \mathscr{F}[P^\alpha] \to (\forall X^\beta \mathscr{F}[X^\beta] \to B),$$

where, in the second case, $\alpha < \beta$. By the structural rule and the I.H. we obtain

$$\vdash_\rho^{\gamma+\delta_0} \mathscr{F}[t] \to \mathscr{P}[B] \quad \text{or} \quad \vdash_\rho^{\gamma+\delta_0} \mathscr{F}[P^\alpha] \to \mathscr{P}[B].$$

In this case by the inversion rules, from $\vdash_\rho^\gamma \mathscr{P}[A]$ we obtain

$$\vdash_\rho^\gamma \mathscr{P}[\mathscr{F}[t]] \quad \text{or} \quad \vdash_\rho^\gamma \mathscr{P}[\mathscr{F}[P^\alpha]].$$

Then $\mathscr{F}[t]$ or $\mathscr{F}[P^\alpha]$, respectively, has a rank $<\rho$ by Theorem 22.3. Therefore $\vdash_\rho^{\gamma+\delta} \mathscr{P}[B]$ follows by a cut with cut-formula $\mathscr{F}[t]$ or $\mathscr{F}[P^\alpha]$.

3. Suppose $A \to B$ is the conclusion of a principal inference whose principal part is in B. Then the assertion follows from the I.H. by the structural rule.

4. Suppose $A \to B$ is the conclusion of a cut $\mathscr{P}_1[C]$, $C \to D \vdash A \to B$ such that $\mathscr{P}_1[D] \models^{s} A \to B$, whose cut-formula C has a rank $<\rho$. From the formulas $\mathscr{P}_1[C]$ and D we obtain formulas $\mathscr{P}_0[C]$ and D_0 such that

$$\mathscr{P}_1[C] \models^{s} A \to \mathscr{P}_0[C] \quad \text{and} \quad C \to D \models^{s} A \to (C \to D_0)$$

by either removing or introducing negative parts A in such a way that $\mathscr{P}_0[D_0] \models^{s} B$ holds. Now there exist $\delta_i < \delta$ ($i = 1, 2$) such that

$$\vdash_{\rho}^{\delta_1} \mathscr{P}_1[C] \quad \text{and} \quad \vdash_{\rho}^{\delta_2} C \to D.$$

By the structural rule we obtain

$$\vdash_{\rho}^{\delta_1} A \to \mathscr{P}_0[C] \quad \text{and} \quad \vdash_{\rho}^{\delta_2} A \to (C \to D_0).$$

By I.H. and the structural rule we obtain

$$\vdash_{\rho}^{\gamma+\delta_1} \mathscr{P}[\mathscr{P}_0[C]] \quad \text{and} \quad \vdash_{\rho}^{\gamma+\delta_2} C \to \mathscr{P}[D_0].$$

By a cut with cut-formula C of rank $<\rho$ we obtain $\vdash_{\rho}^{\gamma+\delta} \mathscr{P}[B]$, since $\mathscr{P}[\mathscr{P}_0[\mathscr{P}[D_0]]] \models^{s} \mathscr{P}[B]$ holds.

Lemma 4. $\vdash_{\rho+1}^{\alpha} F$ *implies* $\vdash_{\rho}^{\phi 0\alpha} F$.

Proof by induction on α. If F is an axiom then the assertion is trivial. Suppose now that F is the conclusion of a basic inference BI. Then there exist $\alpha_i < \alpha$ such that $\vdash_{\rho+1}^{\alpha_i} F_i$ for every premise F_i of BI. Hence by I.H. $\vdash_{\rho}^{\phi 0\alpha_i} F_i$.

1. Suppose BI is a principal inference or a cut of rank $<\rho$. By BI we obtain $\vdash_{\rho}^{\phi 0\alpha}$ since $\phi 0\alpha_i < \phi 0\alpha$ for all $\alpha_i < \alpha$.

2. Suppose BI is a cut $\mathscr{P}[A], A \to B \vdash F$ such that $\mathscr{P}[B] \models^{s} F$ where the cut-formula A has rank ρ. Then $\vdash_{\rho}^{\phi 0\alpha_1} \mathscr{P}[A]$ and $\vdash_{\rho}^{\phi 0\alpha_2} A \to B$ hold where $\alpha_i < \alpha (i = 1, 2)$. It suffices to prove $\vdash_{\rho}^{\phi 0\alpha} \mathscr{P}[B]$ since then $\vdash_{\rho}^{\phi 0\alpha} F$ follows by the structural rule.

2.1. Suppose A is a prime formula. Then $\rho = 0$ and A is either true or false. In either case $\vdash_{\rho}^{\phi 0\alpha} \mathscr{P}[B]$ follows by Lemmata 1 and 2.

2.2. Suppose A is a formula $U^{\alpha_0}(t)$ or $\forall x \mathscr{F}[x]$ or $\forall X^{\beta} \mathscr{F}[X^{\beta}]$. In these cases $\vdash_{\rho}^{\phi 0\alpha} \mathscr{P}[B]$ follows by Lemma 3 since $\phi 0\alpha_1 + \phi 0\alpha_2 < \phi 0\alpha$.

2.3. Suppose A is a formula $\lambda x \mathscr{F}[x](t)$. Then by the inversion rule we have $\vdash_{\rho}^{\phi 0\alpha_1} \mathscr{P}[\mathscr{F}[t]]$ and $\vdash_{\rho}^{\phi 0\alpha_2} \mathscr{F}[t] \to B$. By Theorem 22.3 $\mathscr{F}[t]$ has a rank $<\rho$. Therefore by a cut with cut-formula $\mathscr{F}[t]$ we have $\vdash_{\rho}^{\phi 0\alpha} \mathscr{P}[B]$.

2.4. Suppose A is a formula $A_1 \to A_2$. Then by Theorem 22.3 A_1 and A_2 have ranks $<\rho$. From $\vdash_{\rho}^{\phi 0\alpha_2} (A_1 \to A_2) \to B$ by the inversion rule we have $\vdash_{\rho}^{\phi 0\alpha_2} \neg A_1 \to B$ and $\vdash_{\rho}^{\phi 0\alpha_2} A_2 \to B$. Using $\vdash_{\rho}^{\phi 0\alpha_1} \mathscr{P}[(A_1 \to A_2)]$ we obtain $\vdash_{\rho}^{\phi 0\alpha_0+1} A_1 \to \mathscr{P}[B]$ by a cut with cut-formula A_2 and by the structural rule where $\alpha_0 := \max(\alpha_1, \alpha_2)$. Using $\vdash_{\rho}^{\phi 0\alpha_2} \neg A_1 \to B$ we obtain $\vdash_{\rho}^{\phi 0\alpha} \mathscr{P}[B]$ by a cut with cut-formula A_1, where $\phi 0\alpha_2 < \phi 0\alpha$, $\phi 0\alpha_0 + 1 < \phi 0\alpha$ and $\neg \mathscr{P}[B] \to B \models^{s} \mathscr{P}[B]$.

Theorem 22.7 (First Cut Elimination Theorem). $\vdash_{\rho}^{\alpha} F$ *implies* $\vdash_{0}^{\phi\rho\alpha} F$.

Proof by induction on ρ with a subsidiary induction on α. If F is an axiom the assertion is trivial. Suppose now that F is the conclusion of a basic inference BI. Then there exist $\alpha_i < \alpha$ such that $\vdash_{\rho}^{\alpha_i} F_i$ for each premise F_i of BI. By the subsidiary I.H. we have $\vdash_{0}^{\phi\rho\alpha_i} F_i$.

1. Suppose BI is a principal inference. Then by BI we have $\vdash_0^{\phi\rho\alpha} F$ since $\phi\rho\alpha_i < \phi\rho\alpha$ for all $\alpha_i < \alpha$.

2. Suppose BI is a cut of rank $\rho_0 < \rho$. Using this cut we obtain $\vdash_{\rho_0+1}^{\phi\rho\alpha} F$. Since $\rho \neq 0$, $\phi 0(\phi\rho\alpha) = \phi\rho\alpha$. Therefore by Lemma 4 $\vdash_{\rho_0}^{\phi\rho\alpha} F$. Since $\rho_0 < \rho$ we also have $\phi\rho_0(\phi\rho\alpha) = \phi\rho\alpha$. Therefore by the main I.H. we obtain $\vdash_0^{\phi\rho\alpha} F$.

Theorem 22.7 can be sharpened following W. Tait [3] thus:

Theorem 22.8 (Second Cut Elimination Theorem). $\vdash_{\beta+\omega^\nu}^{\alpha} F$ *implies* $\vdash_{\beta}^{\phi\nu\alpha} F$.

Proof by induction on ν with a subsidiary induction on α. The assertion holds for $\nu = 0$ by Lemma 4. If F is an axiom the assertion is trivial. Now suppose $\nu \neq 0$ and F is the conclusion of a basic inference BI. Then there exist $\alpha_i < \alpha$ such that $\vdash_{\beta+\omega^\nu}^{\alpha_i} F_i$ for each premise F_i of BI. By the subsidiary I.H. we obtain $\vdash_{\beta}^{\phi\nu\alpha_i} F_i$.

1. Suppose BI is a principal inference or a cut of rank $< \beta$. Then using BI we have $\vdash_{\beta}^{\phi\nu\alpha} F$.

2. Suppose BI is a cut of rank ρ where $\beta \leqslant \rho < \beta + \omega^\nu$. Then there exist $\nu_0 < \nu$ and $n \in \mathbb{N}$ such that $\rho < \beta + \omega^{\nu_0} \cdot n$. Using this cut we obtain $\vdash_{\beta+\omega^{\nu_0}\cdot n}^{\phi\nu\alpha} F$. Since $\nu_0 < \nu$, $\phi\nu_0(\phi\nu\alpha) = \phi\nu\alpha$. Therefore by n-fold application of the main I.H. we obtain $\vdash_{\beta}^{\phi\nu\alpha} F$.

5. Further Properties of Deductions

Theorem 22.9. *If C_1 and C_2 are equivalent formulas of rank ρ, then* $\vdash_0^{2\cdot\rho} \mathcal{Q}[C_1, C_2]$.

Proof by induction on ρ as in the proof of Theorem 4.1. For example, if C_k is a formula $\forall X^\beta \mathcal{F}_k[X^\beta]$ and P_i^α is an $\mathcal{F}_2$-permissible predicator, then P_i^α is also $\mathcal{F}_1$-permissible and by Theorem 22.3 $\mathcal{F}_k[P_i^\alpha]$ is a formula of rank $\rho_i < \rho$. Then by I.H.

$$\vdash_0^{2\cdot\rho_i} \mathcal{F}_1[P_i^\alpha] \to \mathcal{Q}[\forall X^\beta \mathcal{F}_1[X^\beta], \mathcal{F}_2[P_i^\alpha]] \quad \text{holds.}$$

Hence by basic inferences (S3.1) we obtain

$$\vdash_2^{2\cdot\rho_i+1} \mathcal{Q}[\forall X^\beta \mathcal{F}_1[X^\beta], \mathcal{F}_2[P_i^\alpha]]$$

and using a basic inference (S2.1*)

$$\vdash_0^{2\cdot\rho} \mathcal{Q}[\forall X^\beta \mathcal{F}_1[X^\beta], \forall X^\beta \mathcal{F}_2[X^\beta]],$$

since $2\cdot\rho_i + 1 < 2\cdot\rho$ holds for all $\rho_i < \rho$.

Theorem 22.10. *If $0 < \beta < \gamma$ and $\forall X^\gamma \mathcal{F}[X^\gamma]$ is a formula of rank ρ, then*

$$\vdash_0^{2\cdot\rho+1} \mathcal{Q}[\forall X^\gamma \mathcal{F}[X^\gamma], \forall X^\beta \mathcal{F}[X^\beta]].$$

Proof. By Theorem 22.9

$$\vdash_0^{2\cdot\rho} \mathcal{Q}[\forall X^\gamma \mathcal{F}[X^\gamma], \forall X^\gamma \mathcal{F}[X^\gamma]]$$

holds. By an inversion rule we obtain

$$\vdash_0^{2\cdot\rho} \mathcal{Q}[\forall X^\gamma \mathcal{F}[X^\gamma], \mathcal{F}[P^\alpha]]$$

for every $\mathcal{F}$-permissible predicator P^α of level $\alpha<\beta$. Hence the assertion follows by a basic inference (S2.1*).

Definitions. As before we set

$$(A \leftrightarrow B) := \neg((A \to B) \to \neg(B \to A))$$
$$\exists X^\beta \mathcal{F}[X^\beta] := \neg \forall X^\beta \neg \mathcal{F}[X^\beta].$$

Theorem 22.11 (Extensionality Theorem). *If* $\mathcal{F}[\lambda x \mathcal{A}_1[x]]$ *and* $\mathcal{F}[\lambda x \mathcal{A}_2[x]]$ *are formulas of maximal rank* ρ *then*

$$\vdash_0^{2\cdot\rho+2} \forall x(\mathcal{A}_1[x] \leftrightarrow \mathcal{A}_2[x]) \to \mathcal{Q}[\mathcal{F}[\lambda x \mathcal{A}_i[x]], \mathcal{F}[\lambda x \mathcal{A}_j[x]]]$$

holds for $(i,j)=(1,2)$ *and* $(i,j)=(2,1)$.

Proof by induction on ρ. For brevity we set

$$A := \forall x(\mathcal{A}_1[x] \leftrightarrow \mathcal{A}_2[x]),$$
$$P_k := \lambda x \mathcal{A}_k[x] \quad (k=1,2).$$

The assertion then reads

$$\vdash_0^{2\cdot\rho+2} A \to \mathcal{Q}[\mathcal{F}[P_i], \mathcal{F}[P_j]].$$

1. Suppose $\mathcal{F}[P_i]$ has length 0. Then P_i does not occur in $\mathcal{F}[P_i]$ so $\mathcal{F}[P_1]$ and $\mathcal{F}[P_2]$ are the same. In this case the assertion holds by Theorem 22.9.

2. Suppose $\mathcal{F}[P_i]$ is a formula $P_i(t)$. Then by Theorem 22.3 the formulas $\mathcal{A}_1(t)$ and $\mathcal{A}_2(t)$ have a maximal rank $\rho_0<\rho$. We set

$$\mathcal{N} := \begin{cases} (*_1 \to \neg(\mathcal{A}_2[t] \to \mathcal{A}_1[t])) & \text{if } (i,j)=(1,2), \\ ((\mathcal{A}_1[t] \to \mathcal{A}_2[t]) \to \neg *_1), & \text{if } (i,j)=(2,1). \end{cases}$$

Then $\mathcal{N}$ is an N-form such that $\neg\mathcal{N}[(\mathcal{A}_i[t] \to \mathcal{A}_j[t])]$ is the formula $(\mathcal{A}_1[t] \leftrightarrow \mathcal{A}_2[t])$. By Theorem 22.9 we have

$$\vdash_0^{2\cdot\rho_0} \neg\mathcal{N}[\neg\mathcal{A}_i[t]] \to (A \to \mathcal{Q}[\mathcal{A}_i[t], \mathcal{A}_j[t]]),$$
$$\vdash_0^{2\cdot\rho_0} \neg\mathcal{N}[\mathcal{A}_j[t]] \to (A \to \mathcal{Q}[\mathcal{A}_i[t], \mathcal{A}_j[t]]).$$

Hence by a basic inference (S1)

$$\vdash_{0}^{2\cdot\rho_0+1} (\mathscr{A}_1[t] \leftrightarrow \mathscr{A}_2[t]) \to (A \to \mathscr{Q}[\mathscr{A}_i[t], \mathscr{A}_j[t]])$$

and by a basic inference (S3.0)

$$\vdash_{0}^{2\cdot\rho} A \to \mathscr{Q}[\mathscr{A}_i[t], \mathscr{A}_j[t]],$$

since $2\cdot\rho_0+1<2\cdot\rho$. The assertion follows by two basic inferences (S4).

3. Suppose $\mathscr{F}[P_i]$ is a formula $\mathscr{F}_1[P_i] \to \mathscr{F}_2[P_i]$ or $\lambda x\mathscr{G}[P_i, x](t)$ or $\forall x\mathscr{G}[P_i, x]$ or $\forall X^\beta\mathscr{G}[P_i, X^\beta]$. Then the assertion follows from the I.H. by appropriate basic inferences.

Theorem 22.12. *If* $\forall Y^{\sigma_i}\mathscr{A}[n, Y^{\sigma_i}]$ *and* $\exists Y^{\sigma_i}\mathscr{B}[n, Y^{\sigma_i}]$ *are formulas of level* $\sigma_i (i=1,2)$ *where* $0<\sigma_1<\sigma_2$ *and*

$$\vdash_{\rho}^{\alpha} \forall x(\forall Y^{\sigma_i}\ \mathscr{A}[x, Y^{\sigma_i}] \leftrightarrow \exists Y^{\sigma_i}\mathscr{B}[x, Y^{\sigma_i}]) \quad (i=1,2)$$

holds where $\omega\cdot\sigma_2+\omega\leqslant\alpha$ *and* $\omega\cdot\sigma_2+\omega\leqslant\rho$ *then*

$$\vdash_{\rho}^{\alpha+5} \forall x(\forall Y^{\sigma_1}\mathscr{A}[x, Y^{\sigma_1}] \leftrightarrow \forall Y^{\sigma_2}\mathscr{A}[x, Y^{\sigma_2}]).$$

Proof. For brevity we set

$$\mathscr{A}_i[n] := \forall Y^{\sigma_i}\mathscr{A}[n, Y^{\sigma_i}] \quad \text{and} \quad \mathscr{B}_i[n] := \exists Y^{\sigma_i}\mathscr{B}[n, Y^{\sigma_i}] \quad (i=1,2).$$

Let the maximal rank of the formulas $\mathscr{A}_i[n]$, $\mathscr{B}_i[n] (i=1,2)$ be ρ_0. By Theorem 22.1 $\rho_0<\omega\cdot\sigma_2+\omega$. Then $2\cdot\rho_0<\omega\cdot\sigma_2+\omega$ also and therefore $2\cdot\rho_0<\alpha$ and $\rho_0<\rho$. From the hypotheses $\vdash_{\rho}^{\alpha} \forall x(\mathscr{A}_i[x] \leftrightarrow \mathscr{B}_i[x]) (i=1,2)$ by the inversion rule we obtain

$$\vdash_{\rho}^{\alpha} (\mathscr{A}_i[n] \leftrightarrow \mathscr{B}_i[n]),$$

that is,

$$\vdash_{\rho}^{\alpha} \neg((\mathscr{A}_i[n] \to \mathscr{B}_i[n]) \to \neg(\mathscr{B}_i[n] \to \mathscr{A}_i[n])).$$

Hence by the inversion rules we obtain

$$\vdash_{\rho}^{\alpha} \neg\neg(\mathscr{A}_1[n] \to \mathscr{B}_1[n]) \quad \text{and} \quad \vdash_{\rho}^{\alpha} \neg\neg(\mathscr{B}_2[n] \to \mathscr{A}_2[n]).$$

Using the axioms $\bot \to \mathscr{B}_1[n]$ and $\bot \to \mathscr{A}_2[n]$ we obtain by cuts with cut-formula $\bot$ and rank $0<\rho$

$$\vdash_{\rho}^{\alpha+1} \neg(\mathscr{A}_1[n] \to \mathscr{B}_1[n]) \to \mathscr{B}_1[n] \quad \text{and} \quad \vdash_{\rho}^{\alpha+1} \neg(\mathscr{B}_2[n] \to \mathscr{A}_2[n]) \to \mathscr{A}_2[n].$$

By the structural rule we obtain

(1) $\vdash_{\rho}^{\alpha+1} \mathscr{A}_1[n] \to \mathscr{B}_1[n]$ and $\vdash_{\rho}^{\alpha+1} \mathscr{B}_2[n] \to \mathscr{A}_2[n]$.

The formula $\forall Y^{\sigma_2} \neg \mathscr{B}[n, Y^{\sigma_2}]$ has a rank $<\rho_0$. Since $2 \cdot \rho_0 < \alpha$ it follows by Theorem 22.10 that

$$\vdash_0^{\alpha} \neg \forall Y^{\sigma_1} \neg \mathscr{B}[n, Y^{\sigma_1}] \to \neg \forall Y^{\sigma_2} \neg \mathscr{B}[n, Y^{\sigma_2}],$$

that is

(2) $\vdash_0^{\alpha} \mathscr{B}_1[n] \to \mathscr{B}_2[n]$.

The formulas $\mathscr{B}_1[n]$ and $\mathscr{B}_2[n]$ have ranks $<\rho$. Therefore by two cuts from (1) and (2) we obtain

(3) $\vdash_{\rho}^{\alpha+3} \mathscr{A}_1[n] \to \mathscr{A}_2[n]$.

By Theorem 22.10 we also have

(4) $\vdash_0^{\alpha} \mathscr{A}_2[n] \to \mathscr{A}_1[n]$.

From (3) and (4) by the structural rule we obtain

$\vdash_{\rho}^{\alpha+3} \neg\neg(\mathscr{A}_1[n] \to \mathscr{A}_2[n])$ and $\vdash_0^{\alpha} \neg\neg(\mathscr{A}_2[n] \to \mathscr{A}_1[n])$.

Using a basic inference (S1) we obtain

$$\vdash_{\rho}^{\alpha+4} (\mathscr{A}_1[n] \leftrightarrow \mathscr{A}_2[n])$$

for every numeral n. The assertion $\vdash_{\rho}^{\alpha+5} \forall x(\mathscr{A}_1[x] \leftrightarrow \mathscr{A}_2[x])$ follows by a basic inference (S2.0*).

6. Interpretations of EA* and DA* in RA*

If F is a formula of the system DA* and $\sigma \neq 0$ then we denote by F^{σ} a formula of RA* which results from F by replacing every bound predicate variable X by X^{σ} and every free predicate variable by a predicator of a level $<\sigma$. Then F^{σ} is a formula of a level $\leqslant \sigma$. It has a rank $<\omega \cdot \sigma + \omega$.

For the subsystem EA* of DA* which lacks the basic inference (S3.Δ) we have:

Theorem 22.13 (First Interpretation Theorem). EA* $\vdash^{\alpha} F$ *implies* RA* $\vdash_{\omega+\omega}^{\omega+\alpha} F^1$.

Proof by induction on α. Suppose $\mathrm{EA}^* \vdash^{\alpha} F$ holds.

1. Suppose F is an axiom of the system EA*. If F is an (Ax1) or (Ax2) then the assertion holds since then F^1 is an axiom of the system RA*. If F is an (Ax3) then F^1 is a formula $\mathscr{Q}[C_1, C_2]$ where C_1 and C_2 are equivalent formulas of level 0. They have a rank $<\omega$. Thus the assertion holds by Theorem 22.9.

2. Suppose F is the conclusion of a basic inference BI of the system EA*.

2.1. Suppose BI is a basic inference (S2.1). Then F is a formula $\mathscr{P}[\forall X \mathscr{F}[X]]$ and there exists $\alpha_0 < \alpha$ such that

$$\mathrm{EA}^* \vdash^{\alpha_0} \mathscr{P}[\mathscr{F}[U]],$$

where U does not occur in F. F^1 is a formula $\mathscr{P}^1[\forall X^1 \mathscr{F}^1[X^1]]$. By I.H.

$$\mathrm{RA}^* \vdash^{\omega+\alpha_0}_{\omega+\omega} \mathscr{P}^1[\mathscr{F}^1[P_0]]$$

holds for every $\mathscr{F}^1$-permissible predicator P_0 of level 0. The assertion follows using a basic inference (S2.1*).

2.2. Suppose BI is not a basic inference (S2.1). If BI is a cut with cut-formula A then BI is replaced by a cut with a cut-formula A_1 of level $\leqslant 1$ whose rank is $<\omega+\omega$. In every case the assertion follows immediately from the I.H.

Theorem 22.14 (Second Interpretation Theorem). $\mathrm{DA}^* \vdash^{\alpha} F$ *(or* $\mathrm{DA}^* \vdash^{\alpha}_{d} F$*) implies*

$$\mathrm{RA}^* \vdash^{\omega\cdot\sigma+\omega+6\cdot\alpha}_{\omega\cdot\sigma+\omega} F^{\sigma} \textit{ for any } \sigma=\omega^{\alpha}\cdot\beta \textit{ (or } \sigma=2^{\alpha}\cdot\beta\textit{) such that } \beta\neq 0.$$

Proof by induction on α. Suppose $\mathrm{DA}^* \vdash^{\alpha} F$ (or $\mathrm{DA}^* \vdash^{\alpha}_{d} F$) holds and $\sigma=\omega^{\alpha}\cdot\beta$ (or $\sigma=2^{\alpha}\cdot\beta$) where $\beta\neq 0$.

1. Suppose F is an axiom of the system DA*. If F is an (Ax1) or (Ax2) then the assertion holds since then F^{σ} is an axiom of the system RA*. If F is an (Ax3) then F^{σ} is a formula $\mathscr{Q}[C_1, C_2]$ where C_1, C_2 are equivalent formulas of a level $<\sigma$. They have a rank $\rho<\omega\cdot\sigma$. Then $2\cdot\rho<\omega\cdot\sigma$ also. Therefore the assertion holds by Theorem 22.9.

2. Suppose F is the conclusion of a basic inference BI of the system DA* which is not an inference (S3.Δ). Then there exist $\alpha_i<\alpha$ such that $\mathrm{DA}^* \vdash^{\alpha_i} F_i$ (or $\mathrm{DA}^* \vdash^{\alpha_i}_{d} F_i$) for each premise F_i of BI. Putting $\beta_i := \omega^{-\alpha_i+\alpha}\cdot\beta$ (or $\beta_i := 2^{-\alpha_i+\alpha}\cdot\beta$) we have $\beta_i\neq 0$ and $\omega^{\alpha_i}\cdot\beta_i=\omega^{\alpha}\cdot\beta=\sigma$ (or $2^{\alpha_i}\cdot\beta_i=2^{\alpha}\cdot\beta=\sigma$). Therefore by the I.H.

$$\mathrm{RA}^* \vdash^{\omega\cdot\sigma+\omega+6\cdot\alpha_i}_{\omega\cdot\sigma+\omega} F_i^{\sigma}$$

holds for appropriate formulas F_i^{σ}. The assertion follows as in the proof of Theorem 22.13.

3. Suppose F is the conclusion of a basic inference (S3.Δ) of the system DA*. Then F^{σ} is a formula $\mathscr{N}[\forall X^{\sigma} \mathscr{F}[X^{\sigma}]]$ and by I.H. there exist $\alpha_i<\alpha$ (or $\alpha_i+\omega\leqslant\alpha$) $(i=1, 2)$ such that

(1) $\mathrm{RA}^* \vdash^{\omega\cdot\sigma+\omega+6\cdot\alpha_1}_{\omega\cdot\sigma+\omega} \forall x(\forall Y^{\sigma} \mathscr{A}[x, Y^{\sigma}] \leftrightarrow \exists Y^{\sigma} \mathscr{B}[x, Y^{\sigma}])$,

(2) $\mathrm{RA}^* \vdash^{\omega\cdot\sigma+\omega+6\cdot\alpha_2}_{\omega\cdot\sigma+\omega} \mathscr{F}[\lambda x \forall Y^{\sigma} \mathscr{A}[x, Y^{\sigma}]] \rightarrow \mathscr{N}[\forall X^{\sigma} \mathscr{F}[X^{\sigma}]]$,

where $\mathscr{A}[n, U^0]$ and $\mathscr{B}[u, U^0]$ are formulas of a maximal level $\sigma_0<\sigma$. Since $\alpha_1<\alpha$ and $\sigma=\omega^\sigma\cdot\beta$ (or $\alpha_1+\omega\leqslant\alpha$ and $\sigma=2^\alpha\cdot\beta$), by Theorem 14.12 (or 14.13) there is an ordinal term $\sigma_1=\omega^{\alpha_1}\cdot\beta_1$ (or $\sigma_1=2^{\alpha_1}\cdot\beta_1$) such that $\sigma_0<\sigma_1<\sigma$. By Theorem 22.11

$$\text{(3)}\quad \mathrm{RA}^* \vdash_{0}^{\omega\cdot\sigma+\omega} \begin{cases} \forall x(\forall Y^{\sigma_1}\mathscr{A}[x, Y^{\sigma_1}] \leftrightarrow \forall Y^\sigma\mathscr{A}[x, Y^\sigma]) \\ \quad \to (\mathscr{F}[\lambda x\forall Y^{\sigma_1}\mathscr{A}[x, Y^{\sigma_1}]] \to \mathscr{F}[\lambda x\forall Y^\sigma\mathscr{A}[x, Y^\sigma]] \end{cases} \text{ holds.}$$

By a cut from (2) and (3) we obtain

$$\mathrm{RA}^* \vdash_{\omega\cdot\sigma+\omega}^{\omega\cdot\sigma+\omega+6\cdot\alpha_2+1} \begin{cases} \mathscr{F}[\lambda x\forall Y^{\sigma_1}\mathscr{A}[x, Y^{\sigma_1}]] \to (\forall x(\forall Y^{\sigma_1}\mathscr{A}[x, Y^{\sigma_1}] \\ \quad \leftrightarrow \forall Y^\sigma\mathscr{A}[x, Y^\sigma]) \to \mathscr{N}[\forall X^\sigma\mathscr{F}[X^\sigma]). \end{cases}$$

Here $\lambda x\forall Y^{\sigma_1}\mathscr{A}[x, Y^{\sigma_1}]$ is an $\mathscr{F}$-permissible predicator of level $\sigma_1<\sigma$. Therefore using a basic inference (S3.1) we obtain

$$\text{(4)}\quad \mathrm{RA}^* \vdash_{\omega\cdot\sigma+\omega}^{\omega\cdot\sigma+\omega+6\cdot\alpha_2+2} \begin{cases} \forall x(\forall Y^{\sigma_1}\mathscr{A}[x, Y^{\sigma_1}] \leftrightarrow \forall Y^\sigma\mathscr{A}[x, Y^\sigma]) \\ \quad \to \mathscr{N}[\forall X^\sigma\mathscr{F}[X^\sigma]]. \end{cases}$$

By I.H. we also have

$$\text{(5)}\quad \mathrm{RA}^* \vdash_{\omega\cdot\sigma_1+\omega}^{\omega\cdot\sigma_1+\omega+6\cdot\alpha_1} \forall x(\forall Y^{\sigma_1}\mathscr{A}[x, Y^{\sigma_1}] \leftrightarrow \exists Y^{\sigma_1}\mathscr{B}[x, Y^{\sigma_1}])$$

From (1) and (5) by Theorem 22.12 we obtain

$$\text{(6)}\quad \mathrm{RA}^* \vdash_{\omega\cdot\sigma+\omega}^{\omega\cdot\sigma+\omega+6\cdot\alpha_1+5} \forall x(\forall Y^{\sigma_1}\mathscr{A}[x, Y^{\sigma_1}] \leftrightarrow \forall Y^\sigma\mathscr{A}[x, Y^\sigma]),$$

and we have $6\cdot\alpha_1+5<6\cdot\alpha$ and $6\cdot\alpha_2+2<6\cdot\alpha$. Therefore by a cut from (4) and (6) we obtain the assertion

$$\mathrm{RA}^* \vdash_{\omega\cdot\sigma+\omega}^{\omega\cdot\sigma+\omega+6\cdot\alpha} \mathscr{N}[\forall X^\sigma\mathscr{F}[X^\sigma]].$$

§23. The Limits of the Deducibility of Transfinite Induction

1. Orders of Deductions of Induction in RA*

As in §21 we define in RA*:

$$\mathscr{P}\!\imath[P^\alpha] := \forall x_2(\forall x_1 \prec x_2 P^\alpha(x_1) \to P^\alpha(x_2))$$
$$\mathscr{I}^\beta[t] := \forall X^\beta(\mathscr{P}\!\imath[X^\beta] \to \forall x \prec t X^\beta(x)).$$

Lemma 1. *Assume:*

I. *M_+ and M_- are disjoint finite sets of numerals, M_+ is non-empty and* $\gamma := \min\{\tau(n) \mid n \in M_+\}$.

II. *F is a formula with the following properties: The minimal negative parts of F are* $\mathscr{P}\imath[U^\alpha]$, *all* $U^\alpha(n)$ *such that* $n \in M_-$ *and otherwise only true prime formulas (if any). The minimal positive parts of F are only the* $U^\alpha(n)$ *such that* $n \in M_+$.

III. $\mathrm{RA}^* \vdash_0^\delta F$.

Then: $\gamma < \omega \cdot \delta$.

Proof by induction on δ. By hypotheses I and II F can only have been obtained by a basic inference (S3.0). Therefore it follows from hypothesis III that there exist $\delta_0 < \delta$ and a numerical term t such that

(1) $\mathrm{RA}^* \vdash_0^{\delta_0} (\forall x_1 < t\, U^\alpha(x_1) \rightarrow U^\alpha(t)) \rightarrow F$.

Suppose the numerical term t has value m. Then by Theorems 22.4 and 22.5a) and b) from (1) we obtain

(2) $\mathrm{RA}^* \vdash_0^{\delta_0} \neg \forall x_1 < m\, U^\alpha(x_1) \rightarrow F$

(3) $\mathrm{RA}^* \vdash_0^{\delta_0} U^\alpha(m) \rightarrow F$.

1. Suppose $m \notin M_+$. Then the formula $U^\alpha(m) \rightarrow F$ satisfies hypotheses I and II of our lemma with the same set M_+ as for F. Therefore from (3) by I.H. $\gamma < \omega \cdot \delta_0$ and therefore $\gamma < \omega \cdot \delta$.

2. Suppose $m \in M_+$. Then $\gamma \leqslant \tau(m)$. Since M_- is a finite set and $\delta_0 < \delta$ there exists a numeral $k \notin M_-$ such that $\omega \cdot \delta_0 \leqslant \tau(k) < \omega \cdot \delta$. By Theorem 22.5c) from (2) we obtain

(4) $\mathrm{RA}^* \vdash_0^{\delta_0} \neg(k < m \rightarrow U^\alpha(k)) \rightarrow F$.

Assume that $\tau(k) < \gamma$. Then $k < m$ is a true prime formula and the formula $\neg(k < m \rightarrow U^\alpha(k)) \rightarrow F$ satisfies hypotheses I and II of our lemma with $\{k\} \cup M_+$ instead of M_+ and $\tau(k) = \min\{\tau(n) \mid n \in \{k\} \cup M_+\}$. By I.H. from (4) it follows that $\tau(k) < \omega \cdot \delta_0$ contradicting the choice of k. Thus the assumption $\tau(k) < \gamma$ is refuted. Hence $\gamma \leqslant \tau(k)$ whence $\gamma < \omega \cdot \delta$.

Theorem 23.1. $\mathrm{RA}^* \vdash_0^\delta \mathscr{P}\imath[U^\alpha] \rightarrow \forall x < \bar{\gamma}\, U^\alpha(x)$ *implies* $\gamma \leqslant \omega \cdot \delta$.

Proof. By Theorem 22.5c) from the hypothesis we obtain

$$\mathrm{RA}^* \vdash_0^\delta \mathscr{P}\imath[U^\alpha] \rightarrow (\xi < \bar{\gamma} \rightarrow U^\alpha(\xi)).$$

By Lemma 1 we have $\xi < \omega \cdot \delta$ for all $\xi < \gamma$, therefore $\gamma \leqslant \omega \cdot \delta$.

Theorem 23.2. $\mathrm{RA}^* \vdash_0^\delta \mathscr{I}^\beta[\bar{\gamma}]$ (*where* $\beta \neq 0$) *implies* $\gamma \leqslant \omega \cdot \delta$.

Proof. By Theorem 22.5d) from the hypothesis we obtain

$$\mathrm{RA}^* \vdash_0^\delta \mathscr{P}\imath[U^\alpha] \rightarrow \forall x < \bar{\gamma}\, U^\alpha(x)$$

for $\alpha < \beta$. By Theorem 23.1 we obtain $\gamma \leqslant \omega \cdot \delta$.

2. The Limiting Numbers of the Systems EN, EA and DA

By the *limiting number* of the system EN we mean the least ordinal term γ such that the formula

$$\mathscr{P}\imath[U] \to \forall x < \bar{\gamma} U(x)$$

is not deducible in EN.

Theorem 23.3. *The formal system* EN *has limiting number* ε_0.

Proof. Suppose

(1) $\mathrm{EN} \vdash \mathscr{P}\imath[U] \to \forall x < \bar{\gamma} U(x)$.

By Lemma 3 of §21 we obtain

(2) $\mathrm{EN}^* \vdash^{\leq \omega^2} \mathscr{P}\imath[U] \to \forall x < \bar{\gamma} U(x)$.

EN* can be regarded as a subsystem of RA* in which every free predicate variable belongs to level 0 and every formula has a rank $< \omega$. It follows from (1) that the formula

$$\mathscr{P}\imath[U^0] \to \forall x < \bar{\gamma} U^0(x)$$

has a deduction in this subsystem with cut-rank restricted to $< \omega$. Therefore there exists a numeral n such that

(3) $\mathrm{RA}^* \vdash^{\omega^2}_{\omega^0 \cdot n} \mathscr{P}\imath[U^0] \to \forall x < \bar{\gamma} U^0(x)$.

Now $\phi 0 \omega^2 < \varepsilon_0$ and $\phi 0 \alpha < \varepsilon_0$ for $\alpha < \varepsilon_0$. Therefore it follows from (3) by n-fold application of Theorem 22.8 that there exists $\delta < \varepsilon_0$ such that

$$\mathrm{RA}^* \vdash^{\delta}_{0} \mathscr{P}'\imath[U^0] \to \forall x < \bar{\gamma} U^0(x).$$

By Theorem 23.1 it follows that $\gamma \leqslant \omega \cdot \delta < \varepsilon_0$. Consequently EN has a limiting number $\leqslant \varepsilon_0$. It follows from Theorem 21.1 that ε_0 is the limiting number of EN.

By the *limiting number* of the system EA (or DA) we mean the least ordinal term γ such that the formula $\mathscr{I}[\bar{\gamma}]$ is not deducible in EA (or in DA).

Theorem 23.4. *The formal system* EA *has limiting number* $\phi 1 \varepsilon_0$.

Proof. Suppose

(1) $\mathrm{EA} \vdash \mathscr{I}[\bar{\gamma}]$ holds.

By Lemma 3 of §21 we obtain

(2) $\mathrm{EA}^* \vdash^{\leq \omega^2} \mathscr{I}[\bar{\gamma}]$.

By Theorem 22.13 EA* can be interpreted in RA* where every formula has a rank $<\omega+\omega$. It follows from (1) that the formula $\mathscr{I}^1[\bar{\gamma}]$ has a deduction in RA* with cut ranks restricted to $<\omega+\omega$. Therefore there is a numeral n such that, by Theorem 22.13,

(3) $\mathrm{RA}^* \vdash^{\omega^2}_{\omega^1+\omega^0\cdot n} \mathscr{I}^1[\bar{\gamma}]$ holds.

It follows from (3) by n-fold application of Theorem 22.8 that there exists $\delta<\varepsilon_0$ such that

$$\mathrm{RA}^* \vdash^{\delta}_{\omega^1} \mathscr{I}^1[\bar{\gamma}].$$

Hence by Theorem 22.8

$$\mathrm{RA}^* \vdash^{\phi 1\delta}_{0} \mathscr{I}^1[\bar{\gamma}].$$

By Theorem 23.2 we have $\gamma \leqslant \phi 1\delta < \phi 1\varepsilon_0$. Consequently EA has a limiting number $\leqslant \phi 1\varepsilon_0$. It follows from Theorem 21.3 that $\phi 1\varepsilon_0$ is the limiting number of EA.

Theorem 23.5. *The formal system* DA *has limiting number* $\phi\omega 0$.

Proof. Suppose

$$\mathrm{DA} \vdash \mathscr{I}[\bar{\gamma}]$$

holds. By Theorem 20.13 there exists an ordinal term $\omega\cdot m+n<\omega^2$ such that

$$\mathrm{DA}^* \vdash^{\omega\cdot m+n}_{d} \mathscr{I}[\bar{\gamma}].$$

Putting $\sigma:=2^{\omega\cdot m+n}=\omega^m\cdot 2^n$ from Theorem 22.14 we obtain

$$\mathrm{RA}^* \vdash^{\omega\cdot\sigma+\omega+6\cdot(\omega\cdot m+n)}_{\omega\cdot\sigma+\omega} \mathscr{I}^\sigma[\bar{\gamma}].$$

Here $\omega\cdot\sigma+\omega\leqslant\omega\cdot\sigma+\omega+6\cdot(\omega\cdot m+n)<\omega^{m+2}$. Therefore we also have

$$\mathrm{RA}^* \vdash^{\omega^{m+2}}_{\omega^{m+2}} \mathscr{I}^\sigma[\bar{\gamma}].$$

By Theorem 22.8 we obtain

$$\mathrm{RA}^* \vdash^{\phi(m+2)(\omega^{m+2})}_{0} \mathscr{I}^\sigma[\bar{\gamma}].$$

By Theorem 23.2 we obtain $\gamma \leqslant \omega \cdot \phi(m+2)(\omega^{m+2}) < \phi\omega 0$. Consequently DA has a limiting number $\leqslant \phi\omega 0$. By Theorem 21.5 it follows that $\phi\omega 0$ is the limiting number of DA.

3. The Autonomous Ordinal Terms of the Systems EN*, EA* and DA*

Inductive definition of the *autonomous ordinal terms* of the semi-formal system EN*:

1. If γ is less than the limiting number of EN then γ is autonomous in EN*.
2. If δ is autonomous in EN* and

$$\mathrm{EN}^* \mid\!\frac{\delta}{\quad} \mathscr{P}\!\imath[U] \to \forall x \prec \bar{\gamma}\, U(x)$$

holds where $\delta < \gamma$, then γ is also autonomous in EN*.

Lemma 2. *If γ is autonomous in* EN*, *then $\gamma+1$ is also autonomous in* EN*.

Proof by induction on γ. If γ is less than the limiting number of EN then $\gamma+1$ is also less than the limiting number of EN and therefore autonomous in EN*. Otherwise there is an autonomous ordinal term $\delta < \gamma$ such that

(1) $\mathrm{EN}^* \mid\!\frac{\delta}{\quad} \mathscr{P}\!\imath[U] \to \forall x \prec \bar{\gamma}\, U(x)$.

We may assume that $\delta \geqslant \omega^2$. The formula

$$\mathscr{P}\!\imath[U] \to \forall x \prec \bar{\gamma}\, U(x) \to \forall x \prec \overline{\gamma+1}\, U(x)$$

is deducible in EN. By Lemma 3 of §21 it follows that

(2) $\mathrm{EN}^* \mid\!\frac{<\omega^2}{\quad} \mathscr{P}\!\imath[U] \to \forall x \prec \bar{\gamma}\, U(x) \to \forall x \prec \overline{\gamma+1}\, U(x)$.

From (1) and (2) by a cut we obtain

$$\mathrm{EN}^* \mid\!\frac{\delta+1}{\quad} \mathscr{P}\!\imath[U] \to \forall x \prec \overline{\gamma+1}\, U(x).$$

Here $\delta+1 < \gamma+1$ and by I.H. $\delta+1$ is autonomous in EN*. Consequently $\gamma+1$ is also autonomous in EN*.

Lemma 3. *The autonomous ordinal terms of* EN* *form a segment. That is: If γ is autonomous in* EN* *and $\gamma_0 < \gamma$ then γ_0 is also autonomous in* EN*.

Proof by induction on γ. If γ is smaller than the limiting number of EN, then γ_0 is autonomous in EN*. Otherwise there is an autonomous ordinal term $\delta < \gamma$ such that

(1) $\mathrm{EN}^* \mid\!\frac{\delta}{\quad} \mathscr{P}\!\imath[U] \to \forall x \prec \bar{\gamma}\, U(x)$.

We may assume that $\delta \geqslant \omega^2$. The formula

$$\forall x \prec \bar{\gamma} U(x) \rightarrow \forall x \prec \bar{\gamma}_0 U(x)$$

is deducible in EN for $\gamma_0 < \gamma$. It follows by Lemma 3 of §21 that

(2) $\mathrm{EN}^* \vdash^{\leq \omega^2} \forall x \prec \bar{\gamma} U(x) \rightarrow \forall x \prec \bar{\gamma}_0 U(x)$.

By a cut from (1) and (2) we obtain

(3) $\mathrm{EN}^* \vdash^{\delta+1} \mathscr{P}\!r[U] \rightarrow \forall x \prec \bar{\gamma}_0 U(x)$.

By Lemma 2 $\delta+1$ is autonomous in EN*. Consequently γ_0 is autonomous in EN* by I.H. if $\gamma_0 \leqslant \delta_0 + 1$ and otherwise by (3).

Definition. Following S. Feferman [1] we denote by Aut(EN*) the least ordinal term which is not autonomous in EN*.

Theorem 23.6. $\mathrm{Aut}(\mathrm{EN}^*) = \phi 20$.

1. *Proof* of $\phi 20 \leqslant \mathrm{Aut}(\mathrm{EN}^*)$. Set $\alpha_0 := \varepsilon_0$ and $\alpha_{n+1} := \varepsilon_{\alpha_n} = \phi 1 \alpha_n$. Then $\alpha_n < \alpha_{n+1}$ and $\omega^2 + \omega \cdot \alpha_n = \alpha_n$. Therefore from Theorem 21.2 we have

(1) $\mathrm{EN}^* \vdash^{\omega^2} \mathscr{P}\!r[U] \rightarrow \forall x \prec \bar{\alpha}_0 U(x)$

(2) $\mathrm{EN}^* \vdash^{\alpha_n} \mathscr{P}\!r[U] \rightarrow \forall x \prec \overline{\alpha_{n+1}} U(x)$.

By Theorem 21.1 ω^2 is autonomous in EN*. It follows by induction on n using (1) and (2) that every ordinal term α_n is autonomous in EN*. For each ordinal term $\gamma < \phi 20$ there is a natural number n such that $\gamma < \alpha_n$. Therefore by Lemma 3 every ordinal term $< \phi 20$ is autonomous in EN* and therefore $\phi 20 \leqslant \mathrm{Aut}(\mathrm{EN}^*)$.

2. *Proof* of $\mathrm{Aut}(\mathrm{EN}^*) \leqslant \phi 20$. By Theorem 23.3 $\phi 20$ is greater than the limiting number of EN. Suppose

(1) $\mathrm{EN}^* \vdash^{\delta} \mathscr{P}\!r[U] \rightarrow \forall x \prec \bar{\gamma} U(x)$

holds, where $\delta < \phi 20$. We have to prove that γ is also less than $\phi 20$. We regard EN* as a subsystem of RA* in which every free predicate variable belongs to level 0 and every formula has a rank $< \omega$. Then it follows from (1) that

(2) $\mathrm{RA}^* \vdash^{\delta}_{\omega 1} \mathscr{P}\!r[U^0] \rightarrow \forall x \prec \bar{\gamma} U^0(x)$.

By Theorem 22.8 we obtain

$$\mathrm{RA}^* \vdash^{\phi 1 \delta}_{0} \mathscr{P}\!r[U^0] \rightarrow \forall x \prec \bar{\gamma} U^0(x).$$

By Theorem 23.1 we have $\gamma \leqslant \omega \cdot \phi 1 \delta < \phi 20$. Hence $\mathrm{Aut}(\mathrm{EN}^*) \leqslant \phi 20$.

Below we denote by Σ one of the formal systems EA, DA and by Σ^* the corresponding semi-formal system.

Inductive definition of the *autonomous ordinal terms* of the semi-formal system Σ^*.

1. If γ is less than the limiting number of Σ then γ is autonomous in Σ^*.
2. If δ is autonomous in Σ^* and $\Sigma^* \overset{\delta}{\vdash} \mathscr{I}[\bar{\gamma}]$ holds where $\delta < \gamma$, then γ is also autonomous in Σ^*.

Lemma 4. *If γ is autonomous in Σ^* then, for all $n < \omega$, $\gamma + n$ is autonomous in Σ^*.*

Proof by induction on γ. If γ is less than the limiting number of Σ then $\gamma + n$ is also less than the limiting number of Σ and therefore $\gamma + n$ is autonomous in Σ^*. Otherwise there is an autonomous ordinal term $\delta < \gamma$ such that

(1) $\Sigma^* \overset{\delta}{\vdash} \mathscr{I}[\bar{\gamma}]$.

We may assume that $\omega^2 \leqslant \delta$ and $1 \leqslant n$. By Lemma 7b) of §21 the formula

$$\mathscr{I}[\bar{\gamma}] \to \mathscr{I}[\bar{n}] \to \mathscr{I}[\overline{\gamma + n}]$$

is deducible in Σ. By Theorem 21.3 $\mathscr{I}[\bar{n}]$ is also deducible in Σ. Using a cut and Lemma 3 of §21 we obtain

(2) $\Sigma^* \overset{<\omega^2}{\vdash} \mathscr{I}[\bar{\gamma}] \to \mathscr{I}[\overline{\gamma + n}]$.

From (1) and (2) we obtain

$$\Sigma^* \overset{\delta + 1}{\vdash} \mathscr{I}[\overline{\gamma + n}].$$

Here $\delta + 1 < \gamma + n$ and by I.H. $\delta + 1$ is autonomous in Σ^*. Consequently $\gamma + n$ is also autonomous in Σ^*.

Lemma 5. The autonomous ordinal terms of Σ^* form a segment.

Proof. This result follows in the same way as Lemma 3 since the formula

$$\mathscr{I}[\bar{\gamma}] \to \mathscr{I}[\bar{\gamma}_0]$$

is deducible in Σ for $\gamma_0 < \gamma$.

Definition. Again following S. Feferman [1] we denote by Aut(Σ^*) the least ordinal term which is not autonomous in Σ^*.

Theorem 23.7. Aut(EA*) $= \phi 20$.

1. *Proof* of $\phi 20 \leqslant$ Aut(EA*). Let α_n be defined as in the proof of Theorem 23.6.

Then $\omega^2+2\cdot\alpha_n=\alpha_n$. Therefore from Theorem 21.4 we have

$$\mathrm{EA}^* \vdash^{\alpha_n+1} \mathscr{I}[\overline{\alpha_{n+1}}].$$

By Theorem 23.4 α_0+1 is autonomous in EA*. Using Lemma 4 it follows by induction on n that every ordinal term α_n is autonomous in EA*. As in the proof of Theorem 23.6 it follows that $\phi 20\leqslant \mathrm{Aut}(\mathrm{EA}^*)$.

2. *Proof* of $\mathrm{Aut}(\mathrm{EA}^*)\leqslant\phi 20$. By Theorem 23.4 $\phi 20$ is greater than the limiting number of EA. Suppose

$$\mathrm{EA}^* \vdash^{\delta} \mathscr{I}[\bar{\gamma}]$$

where $\delta<\phi 20$. We have to prove that γ is then also less than $\phi 20$.
By Theorem 22.13 we have

$$\mathrm{RA}^* \vdash^{\omega+\delta}_{\omega^1+\omega^1} \mathscr{I}^1[\bar{\gamma}].$$

By two applications of Theorem 22.8 we obtain

$$\mathrm{RA}^* \vdash^{\phi 1(\phi 1(\omega+\delta))}_{0} \mathscr{I}^1[\bar{\gamma}].$$

By Theorem 23.2 we obtain $\gamma\leqslant\omega\cdot\phi 1(\phi 1(\omega+\delta))<\phi 20$. Hence $\phi 20$ is not autonomous in EA* and therefore $\mathrm{Aut}(\mathrm{EA}^*)\leqslant\phi 20$.

Theorem 23.8. $\mathrm{Aut}(\mathrm{DA}^*)=\Gamma_0$.

1. *Proof* of $\Gamma_0\leqslant\mathrm{Aut}(\mathrm{DA}^*)$. By Theorem 21.6 we have

$$\mathrm{DA}^* \vdash^{\zeta_n+13} \mathscr{I}[\overline{\zeta_{n+1}}].$$

By Theorem 23.5 ζ_0+13 is autonomous in DA*. It follows by induction on n using Lemma 4 that every ordinal term ζ_n is autonomous in DA*. By Theorem 14.16 to each ordinal term $\gamma<\Gamma_0$ there is a natural number n such that $\gamma<\zeta_n$. Therefore by Lemma 5 every ordinal term $<\Gamma_0$ is autonomous in DA* and therefore $\Gamma_0\leqslant\mathrm{Aut}(\mathrm{DA}^*)$.

2. *Proof* of $\mathrm{Aut}(\mathrm{DA}^*)\leqslant\Gamma_0$. By Theorem 23.5 Γ_0 is greater than the limiting number of DA. Suppose

$$\mathrm{DA}^* \vdash^{\delta} \mathscr{I}[\bar{\gamma}]$$

holds where $\delta<\Gamma_0$. We have to prove that γ is then also less than Γ_0. By Theorem 22.14 for $\sigma:=\omega^\delta$ we have

$$\mathrm{RA}^* \vdash^{\omega\cdot\sigma+\omega+6\cdot\delta}_{\omega\cdot\sigma+\omega} \mathscr{I}^\sigma[\bar{\gamma}].$$

It follows by Theorem 22.7 that for $\alpha := \phi(\omega \cdot \sigma + \omega)(\omega \cdot \sigma + \omega + 6 \cdot \delta) < \Gamma_0$ that

$$\mathrm{RA}^* \vdash_0^\alpha \mathscr{I}^\sigma[\bar{\gamma}].$$

By Theorem 23.2 we obtain $\gamma \leqslant \omega \cdot \alpha < \Gamma_0$. Hence Γ_0 is not autonomous in DA* and therefore $\mathrm{Aut}(\mathrm{DA}^*) \leqslant \Gamma_0$.

4. The Autonomous Ordinal Terms of the System RA*

If Σ is a formal system with the basic inference rule (ind) and Σ^* is the semi-formal system obtained from Σ by replacing (ind) by (S2.0*) then by Theorem 20.13 infinite deductions in Σ^* with orders $<\omega^2$ correspond to the finite deductions of the system Σ. It is therefore fitting to allow the following inductive definition of the autonomous terms of the system RA* to begin with the ordinal terms $<\omega^2$.

Inductive definition of the *autonomous ordinal terms* of the semi-formal system RA*:

1. Every ordinal term $<\omega^2$ is autonomous in RA*.
2. If $\mathrm{RA}^* \vdash_\rho^\alpha \mathscr{I}^1[\bar{\gamma}]$ holds where α and ρ are autonomous ordinal terms which are less than γ, then γ is also autonomous in RA*.

Lemma 6. *If γ is autonomous in* RA* *and $\omega^2 \leqslant \gamma$, then $\gamma+1$ and ω^γ are also autonomous in* RA*.

Proof by induction on γ. By the hypothesis there exist autonomous ordinal terms $\alpha < \gamma$ and $\rho < \gamma$ such that

(1) $\mathrm{RA}^* \vdash_\rho^\alpha \mathscr{I}^1[\bar{\gamma}]$.

It follows from Lemma 7 of §21 that

$$\mathrm{EA} \vdash \mathscr{I}[\bar{\gamma}] \to \mathscr{I}[\overline{\gamma+1}]$$
$$\mathrm{EA} \vdash \mathscr{I}[\bar{\gamma}] \to \mathscr{I}[\overline{\omega^\gamma}].$$

By Lemma 3 of §21 and Theorem 22.13 it follows that there exist $\delta_1 < \omega^2$ and $\delta_2 < \omega^2$ such that

(2) $\mathrm{RA}^* \vdash_{\omega+\omega}^{\omega+\delta_1} \mathscr{I}^1[\bar{\gamma}] \to \mathscr{I}^1[\overline{\gamma+1}]$

(3) $\mathrm{RA}^* \vdash_{\omega+\omega}^{\omega+\delta_2} \mathscr{I}^1[\bar{\gamma}] \to \mathscr{I}^1[\overline{\omega^\gamma}]$.

We may assume that $\omega+\delta_1 \leqslant \alpha$, $\omega+\delta_2 \leqslant \alpha$ and $\omega+\omega \leqslant \rho$. The formula $\mathscr{I}^1[\bar{\gamma}]$ has a rank $<\omega+\omega$. Therefore by cuts from (1)–(3) we obtain

(4) $\mathrm{RA}^* \vdash_\rho^{\alpha+1} \mathscr{I}^1[\overline{\gamma+1}]$

(5) $\mathrm{RA}^* \vdash_\rho^{\alpha+1} \mathscr{I}^1[\overline{\omega^\gamma}]$.

If $\alpha<\omega^2$, then $\alpha+1<\omega^2$ is autonomous in RA*. Otherwise $\alpha+1$ is autonomous by the I.H. Since $\alpha+1<\gamma+1$, it follows by (4) that also $\gamma+1$ is autonomous in RA*. If $\omega^\gamma=\gamma$, then ω^γ is autonomous by the hypotheses. Otherwise $\gamma<\omega^\gamma$ and $\alpha+1<\omega^\gamma$. Then ω^γ is autonomous in RA* by (5).

Lemma 7. *The autonomous ordinal terms of* RA* *form a segment.*

Proof. Suppose $\gamma_0<\gamma$ and γ is autonomous in RA*. We prove by induction on γ that γ_0 is also autonomous in RA*. This is trivial for $\gamma\leqslant\omega^2$. Now suppose $\omega^2<\gamma$. Then there exist autonomous ordinal terms $\alpha<\gamma$ and $\rho<\gamma$ such that

(1) $\mathrm{RA}^* \,|\!\frac{\alpha}{\rho}\, \mathscr{I}^1[\bar{\gamma}]$.

For $\gamma_0<\gamma$ we have

$$\mathrm{EA} \vdash \mathscr{I}[\bar{\gamma}] \to \mathscr{I}[\bar{\gamma}_0].$$

By Lemma 3 of §21 and Theorem 22.13 it follows that there exists $\delta<\omega^2$ such that

(2) $\mathrm{RA}^* \,|\!\frac{\omega+\delta}{\omega+\omega}\, \mathscr{I}^1[\bar{\gamma}] \to \mathscr{I}^1[\bar{\gamma}_0]$.

We may assume that $\omega+\delta\leqslant\alpha$ and $\omega+\omega\leqslant\rho$. By a cut from (1) and (2) we obtain

(3) $\mathrm{RA}^* \,|\!\frac{\alpha+1}{\rho}\, \mathscr{I}^1[\bar{\gamma}_0]$.

By Lemma 6 $\alpha+1$ is autonomous in RA*. If $\gamma_0\leqslant\alpha+1$ or $\gamma_0\leqslant\rho$ then γ_0 is autonomous in RA* by I.H., otherwise γ_0 is autonomous in RA* by (3).

Definition. As before we denote by Aut(RA*) the least ordinal term which is not autonomous in RA*.

Lemma 8. *If* $\mathrm{RA}^* \,|\!\frac{\alpha}{\rho}\, \mathscr{I}^\sigma[\bar{\gamma}]$ *where* $1<\sigma$ *then*

$$\mathrm{RA}^* \,|\!\frac{\alpha+1}{\rho}\, \mathscr{I}^1[\bar{\gamma}].$$

Proof. It follows from the hypothesis by Theorem 22.5 d) that

$$\mathrm{RA}^* \,|\!\frac{\alpha}{\rho}\, \mathscr{P}\!\imath[P^0] \to \forall x<\bar{\gamma}P^0(x)$$

holds for every permissible predicator P^0 of level 0. The assertion follows using a basic inference (S2.1*).

Theorem 23.9. $\mathrm{Aut}(\mathrm{RA}^*)=\Gamma_0$.

1. *Proof* that $\Gamma_0 \leqslant \mathrm{Aut}(\mathrm{RA}^*)$. By Theorem 21.3

$$\mathrm{EA} \vdash \mathscr{I}[\overline{\zeta_1}]$$

where $\zeta_1 = \phi 10 = \varepsilon_0$.

It follows by Lemma 3 of §21 and Theorem 22.13 that there exists $\delta < \omega^2$ such that

$$\mathrm{RA}^* \vdash^{\omega+\delta}_{\omega+\omega} \mathscr{I}^1[\overline{\zeta_1}].$$

Hence ζ_1 is autonomous in RA*. By Theorem 21.6 if $n \geqslant 1$ then

$$\mathrm{DA}^* \vdash^{\zeta_n+13} \mathscr{I}[\overline{\zeta_{n+1}}].$$

By Theorem 22.14 we have

$$\mathrm{RA}^* \vdash^{\omega\cdot\sigma+\omega+6\cdot(\zeta_n+13)}_{\omega\cdot\sigma+\omega} \mathscr{I}^\sigma[\overline{\zeta_{n+1}}],$$

where $\sigma := \omega^{\zeta_n+13}$. Now $\omega\cdot\sigma+\omega+6\cdot(\zeta_n+13) < \omega^{\zeta_n+14}$. Therefore, we have by Lemma 8

$$\mathrm{RA}^* \vdash^{\omega^{\zeta_n+14}}_{\omega^{\zeta_n+14}} \mathscr{I}^1[\overline{\zeta_{n+1}}].$$

If ζ_n is autonomous in RA*, then by Lemma 6 ζ_n+14 and ω^{ζ_n+14} are also autonomous in RA*. Since $\omega^{\zeta_n+14} < \zeta_{n+1}$, it follows that ζ_{n+1} is autonomous in RA*. By induction on n, every ordinal term $\zeta_n (n \geqslant 1)$ is autonomous in RA*. Then by Theorem 14.15 and Lemma 7 every ordinal term $< \Gamma_0$ is autonomous in RA* and therefore $\Gamma_0 \leqslant \mathrm{Aut}(\mathrm{RA}^*)$.

2. *Proof* that $\mathrm{Aut}(\mathrm{RA}^*) \leqslant \Gamma_0$. Suppose

$$\mathrm{RA}^* \vdash^{\alpha}_{\rho} \mathscr{I}^1[\bar{\gamma}]$$

holds where $\alpha < \Gamma_0$ and $\rho < \Gamma_0$. It follows by Theorem 22.7 that

$$\mathrm{RA}^* \vdash^{\phi\rho\alpha}_{0} \mathscr{I}^1[\bar{\gamma}].$$

Hence by Theorem 23.2

$$\gamma \leqslant \omega\cdot\phi\rho\alpha < \Gamma_0.$$

Consequently Γ_0 is not autonomous in RA* and therefore $\mathrm{Aut}(\mathrm{RA}^*) \leqslant \Gamma_0$.

5. The Limits of Predicativity

Predicative inferences are formalized in a natural way in a system of ramified analysis of the same sort as RA*. But such a system can only be seen to be predicative if the only ordinal terms permitted to be orders of deductions and levels of formulas are those for which transfinite induction can be proved using solely predicative means. If RA* $\vdash_{\rho}^{\alpha} \mathscr{I}^1[\bar{\gamma}]$ holds then in RA* there is a deduction of the formula $\mathscr{I}^1[\bar{\gamma}]$ of order α in which only formulas of level 1 and levels $<\rho$ occur. Therefore from the proof of Theorem 23.9 we see that: to every ordinal term $\alpha \geqslant \omega^2$ and $<\Gamma_0$ we can fix a strictly increasing sequence of ordinal terms $\alpha_0, \ldots, \alpha_n$ such that $\alpha_0=\omega^2$ and $\alpha_n=\alpha$ in such a way that a deduction of order $<\alpha_i$ can be given for formalized transfinite induction up to $\alpha_{i+1} (i<n)$ in which, besides formulas of level 1, only formulas of level $<\alpha_i$ occur. In this way transfinite induction is predicatively provable up to each ordinal term $<\Gamma_0$. On the other hand, by Theorem 23.9, a deduction of transfinite induction up to Γ_0 must have an order $\geqslant\Gamma_0$ or it must contain occurrences of formulas of levels up to Γ_0. Therefore transfinite induction up to Γ_0 is not strictly predicatively provable.

Our formalization of transfinite induction is based on a given well-ordering $\prec$ of the natural numbers. For a differently defined well-ordering W the associated ordinal number $\mathrm{Aut}_W(\mathrm{RA}^*)$ may be $<\Gamma_0$. But the proof that $\mathrm{Aut}(\mathrm{RA}^*) \leqslant \Gamma_0$ can be carried over, without any difficulty, to every well-ordering W. We also have:

1. For every ordinal term $\alpha<\Gamma_0$ there is a predicatively provable well-ordering of order type α.

2. There is no predicatively provable well-ordering of any order type $\geqslant\Gamma_0$.

Thus Γ_0 is the limiting number for the predicative provability of transfinite induction. (Cf. S. Feferman [1])

The semi-formal system of Δ_1^1-analysis with deduction of orders $<\Gamma_0$ must now also be considered predicative since by Theorem 22.14 DA* can be interpreted in RA*. In DA* by Theorem 23.8 formalized transfinite induction is even provable autonomously up to the limits of predicative provability.

The proofs of Theorem 23.3–23.5 show that every formal system Σ we have considered can be embedded in a subsystem of RA* for which eliminability of cuts is provable by means of transfinite induction up to the limiting number of Σ. Hence we have a proof of the consistency of Σ by transfinite induction up to the limiting number for Σ. (Such a proof for pure number theory was given by G. Gentzen [2] and [3].) Likewise, for every semi-formal system Σ^* (with deductions of orders $<\mathrm{Aut}(\Sigma^*)$) we have considered we have a consistency proof by transfinite induction up to $\mathrm{Aut}(\Sigma^*)$. These consistency proofs yield still stronger methods than for the demonstration of predicativity, since, for example, in order to deduce $\mathscr{I}^1[\zeta_{n+1}]$ in RA* we need a subsystem of RA* for which the elimination of cuts can be carried out only with the use of deductions of orders $>\zeta_{n+1}$.

Chapter IX

Higher Ordinals and Systems of Π_1^1-Analysis

Our aim in this chapter is the proof-theoretic treatment of Π_1^1-Analysis following G. Takeuti [4]. In §§24–26 starting from normal functions we present an extended constructive system of ordinals with the necessary properties for this and in §29 we give a constructive proof of wellordering for this system. In §27 and §28 we carry out a constructive consistency proof for a formal and for a semiformal system of Π_1^1-Analysis, using transfinite induction on the system or ordinal terms of §25.

§24. Normal Functions on a Segment $\mathbb{O}^*$ of the Ordinals

In this section we consider *ordinals* in the sense of the classical theory of ordinals based on the axioms of Zermelo–Fraenkel or Bernays–Gödel set theory. These ordinals, as defined in §13, characterize the *order types* of the *well-ordered sets*. However, we do not deal here with the class of all ordinals but only with a proper segment $\mathbb{O}^*$ which nevertheless is larger than the segment $\mathbb{O}$ dealt with in §13. We characterize $\mathbb{O}^*$ itself by an axiom system without tying this to a general axiom system for set theory. Thus we use the notions *map* (*function*) and *set* in a naive way. But these may also be regarded as being determined axiomatically (in the context of a general axiom system for set theory).

1. Axiomatic Characterization of the Segment $\mathbb{O}^*$ of the Ordinals

Ax. I. $\mathbb{O}^*$ is a non-empty set *well-ordered* by the relation $<$. (See §13.1.)

We call the elements of $\mathbb{O}^*$ *ordinals*. We denote by 0 (*zero*) the least element of $\mathbb{O}^*$ (with respect to the relation $<$). In this section the letters α, β, γ, δ, ξ, η, ζ (possibly with indices) always denote ordinals from the set $\mathbb{O}^*$. $\alpha \leqslant \beta$ means that $\alpha < \beta$ or $\alpha = \beta$. $\alpha > \beta$ ($\alpha \geqslant \beta$) stands for $\beta < \alpha$ ($\beta \leqslant \alpha$). We denote by $\mathbb{O}^*(\alpha)$ the segment of all ordinals $\xi < \alpha$.

An ordinal α is said to be *finite* (*transfinite*) if $\mathbb{O}^*(\alpha)$ is a finite (infinite) set.

If M_1 and M_2 are subsets of $\mathbb{O}^*$ then $M_1 \sim M_2$ denotes that there is a bijection of M_1 onto M_2. This relation is obviously an equivalence relation. An ordinal α is said to be a *cardinal* if there is no $\xi < \alpha$ such that $\mathbb{O}^*(\xi) \sim \mathbb{O}^*(\alpha)$.

Ax. II. There is a map $\aleph$ from $\mathbb{O}^*$ to $\mathbb{O}^*$ with the following properties:

II.1. $\aleph(\alpha)$ is a transfinite cardinal of $\mathbb{O}^*$.

II.2. For each transfinite cardinal β of $\mathbb{O}^*$ there is an α such that $\aleph(\alpha) = \beta$.

II.3. If $\alpha < \beta$ and $\mathbb{O}^*(\aleph(\alpha)) \sim \mathbb{O}^*(\xi)$, then $\xi < \aleph(\beta)$.

II.4. $\alpha < \aleph(\alpha)$.

From II.3 (for $\xi = \aleph(\alpha)$) we have: If $\alpha < \beta$, then $\aleph(\alpha) < \aleph(\beta)$. It follows by II.1 and II.2 that $\aleph$ is an order preserving map of $\mathbb{O}^*$ onto the set of transfinite cardinals in $\mathbb{O}^*$. In the classical theory of ordinals there is a map $\aleph$ with properties II.1–II.3 from the class of all ordinals onto the class of all transfinite cardinals. This map has fixed points $\alpha = \aleph(\alpha)$. By II.4 $\mathbb{O}^*$ is restricted to the segment of those ordinals which are less than the least fixed point of $\aleph$.

By II.4, for each ordinal α there is an ordinal $> \alpha$ in $\mathbb{O}^*$. Using Ax. I it follows that for each α there is a smallest ordinal $> \alpha$ in $\mathbb{O}^*$. We call this the *successor* α' of α. An ordinal β is said to be a *limit number* if $\beta \neq 0$ and there is no α such that $\alpha' = \beta$.

A set $M \subset \mathbb{O}^*$ is said to be *bounded* if there exists β such that $\alpha \leqslant \beta$ for all $\alpha \in M$. By Ax. I for every bounded set $M \subset \mathbb{O}^*$ there is a unique least ordinal β such that $\alpha \leqslant \beta$ for all $\alpha \in M$. As previously we call this ordinal β the supremum $\sup M$ of the set M.

We say that a set $M \subseteq \mathbb{O}^*$ has *cardinality* α if α is a cardinal such that $M \sim \mathbb{O}^*(\alpha)$. In this case α is uniquely determined by M.

The closure of a set $M_0 \subseteq \mathbb{O}^*$ under finitely many n_i-placed functions of ordinals f_i $(i = 1, \ldots, m)$ is the smallest set $M \subseteq \mathbb{O}^*$ such that $M_0 \subseteq M$ and $f_i(\alpha_1, \ldots, \alpha_{n_i}) \in M$ for all $\alpha_1, \ldots, \alpha_{n_i} \in M$ $(i = 1, \ldots, m)$.

Ax. III. If M is the closure of the set $\{0\} \cup \mathbb{O}^*(\beta)$ under finitely many functions of ordinals then every subset of M has cardinality $\leqslant \max(\beta, \aleph(0))$.

This property of $\mathbb{O}^*$ is provable in the classical theory of ordinals.

Ax. IV. If M is a subset of $\mathbb{O}^*(\aleph(\alpha'))$ which has cardinality $< \aleph(\alpha')$ then $\sup M < \aleph(\alpha')$.

This property of $\mathbb{O}^*$ which we call the *regularity* of the cardinal $\aleph(\alpha')$ is provable in the classical theory of ordinals using the axiom of choice.

Ax. I–Ax. IV comprise all the properties of the segment $\mathbb{O}^*$ of the ordinals which will be used in the sequel without reference to a general system of axioms for set theory.

2. Basic Properties of $\mathbb{O}^*$

By Ax. I the *induction theorem* formulated in §13 (p. 75) also holds for $\mathbb{O}^*$.

As in §13 we denote by $\mathbb{N}$ the set $\{0, 0', 0'', \ldots\}$ of *finite* ordinals. Then putting $\omega := \aleph(0)$, $\omega = \sup \mathbb{N}$ holds.

We define $\mathbb{O}^*$-*segments* and the ordering functions in the same way as in §13 (p. 77). Every *proper* $\mathbb{O}^*$-segment (that is, every $\mathbb{O}^*$-segment different from $\mathbb{O}^*$) is a set $\mathbb{O}^*(\alpha)$. Every set $M \subseteq \mathbb{O}^*$ has precisely one ordering function f which maps an $\mathbb{O}^*$-segment A onto M preserving order. Then $\alpha \leqslant f(\alpha)$ for all $\alpha \in A$. We denote the domain A of the ordering function f by $D(f)$.

We also define *continuity* of an ordering function as in §13. A *normal function* on $\mathbb{O}^*$ is a continuous ordering function with domain $\mathbb{O}^*$. A set $M \subseteq \mathbb{O}^*$ is said to be *closed* if for every nonempty bounded set $M_0 \subseteq M$ we have $\sup M_0 \in M$.

Lemma 1. *If the ordering function f of a set $M \subseteq \mathbb{O}^*$ has domain $\mathbb{O}^*$ and M is closed then f is a normal function.*

Proof like that for Theorem 13.5.

We set $\omega := \aleph(0)$, $\Omega_0 := 0$ and $\Omega_\alpha := \aleph(\alpha)$ for $\alpha \neq 0$. Let K be the set of transfinite cardinals $> \omega$, then $K := \{\Omega_\alpha \mid \alpha \neq 0\}$.

Lemma 2. *If $K \subseteq M \subseteq \mathbb{O}^*$ then the ordering function of M has domain $\mathbb{O}^*$.*

Proof. Let f be the ordering function of M. It follows from $K \subseteq M$ that for every ordinal $\alpha \in \mathbb{O}^*$ there is an ordinal $\xi \in D(f)$ such that $f(\xi) = \Omega_{\alpha'}$, where $\alpha \leqslant \xi$. Thus $\alpha \in D(f)$ too. Hence $D(f) = \mathbb{O}^*$.

Lemma 3. a) *If α is transfinite, then $\mathbb{O}^*(\alpha') \sim \mathbb{O}^*(\alpha)$.*
b) *Every transfinite cardinal is a limit number.*

Proof. a) If α is transfinite, then the function f given by $f(n) := n'$ for $n < \omega$, $f(\xi) := \xi$ for $\omega \leqslant \xi < \alpha$ and $f(\alpha) := 0$ is a bijective map of $\mathbb{O}^*(\alpha')$ onto $\mathbb{O}^*(\alpha)$.
b) follows from a).

Addition of ordinals is defined as in §13.5. Let $B_\alpha := \{\xi \mid \alpha \leqslant \xi \in \mathbb{O}^*\}$, f_α be the ordering function of B_α and $\alpha + \beta := f_\alpha(\beta)$ for $\beta \in D(f_\alpha)$.

Lemma 4. *If $\alpha < \beta$ and $\beta \in K$, then $\beta \in D(f_\alpha)$ and $\alpha + \beta = \beta$.*

Proof. There exists $\gamma \in D(f_\alpha)$ such that $0 < \gamma \leqslant \alpha + \gamma = \beta$. There are the following two cases to consider for $\beta \in K$.

1. $\beta = \aleph(\delta')$. If $M := \{\alpha + \xi \mid \xi < \gamma\}$ then $M \subseteq \mathbb{O}^*(\aleph(\delta'))$ and $M \sim \mathbb{O}^*(\gamma)$. By Lemma 3b) $\aleph(\delta')$ is a limit number. Therefore $\sup M = \aleph(\delta')$. By Ax. IV it follows that M has cardinality $\geqslant \aleph(\delta')$. Thus $\beta = \aleph(\delta') \leqslant \gamma$. It follows that $\beta = \gamma$ and therefore $\beta \in D(f_\alpha)$ and $\alpha + \beta = \beta$.

2. $\beta = \aleph(\delta)$ where δ is a limit ordinal. Assume that $\gamma < \beta$. Then there exists $\xi < \delta$ such that $\gamma < \aleph(\xi') < \beta$ and $\alpha < \aleph(\xi')$. It follows from 1. that $\aleph(\xi') \in D(f_\alpha)$ and $\alpha + \gamma < \alpha + \aleph(\xi') = \aleph(\xi') < \beta$ contradicting $\alpha + \gamma = \beta$. Hence $\beta = \gamma$ and therefore $\beta \in D(f_\alpha)$ and $\alpha + \beta = \beta$.

Theorem 24.1. *f_α is a normal function.*

Proof. For each ordinal $\xi \in \mathbb{O}^*$ there exists $\beta \in K$ such that $\alpha < \beta$ and $\xi < \beta$. Then by Lemma 4 $\beta \in D(f_\alpha)$ and therefore $\xi \in D(f_\alpha)$ too. Hence $D(f_\alpha) = \mathbb{O}^*$. Since B_α is closed it follows by Lemma 1 that the ordering function of B_α is a normal function.

By virtue of Theorem 24.1 we obtain all the properties of addition of ordinals developed in §13.5. We set $1 := 0'$, $2 := 0''$, ... Then $\alpha' = \alpha + 1$.

We define *additive principal numbers* and the ordering function $\alpha \mapsto \omega^\alpha$ as in §13. Corresponding to Theorem 13.7 there are, for each $\alpha \neq 0$, uniquely determined $\alpha_1 \geqslant \cdots \geqslant \alpha_n$ such that $\alpha = \omega^{\alpha_1} + \cdots + \omega^{\alpha_n}$. We call the additive principal numbers $\omega^{\alpha_1}, \ldots, \omega^{\alpha_n}$ involved the *components* of the ordinal α. The ordinal 0 has no components. We denote the set of additive principal numbers in $\mathbb{O}^*$ by P and the finite set of components of α by $P(\alpha)$. The set $P(0)$ is empty. $\alpha \in P$ holds if, and only if, $P(\alpha) = \{\alpha\}$. In general we have $P(\beta) \subseteq P(\alpha + \beta) \subseteq P(\alpha) \cup P(\beta)$.

Theorem 24.2. $K \subseteq P$.

Proof. If $\beta \in K$ then by Lemma 4 $\beta \in P$.

Definition of the *level* $S\alpha$ of an ordinal α. Given α there is a unique smallest ordinal β such that $\mathbb{O}^*(\beta) \sim \mathbb{O}^*(\alpha)$. This ordinal is a cardinal. If $\beta < \omega$ let $S\alpha := 0$. Otherwise there exists $\gamma \neq 0$ such that $\beta = \Omega_\gamma$. Then let $S\alpha := \gamma$.

Corollaries $\Omega_{S\alpha} \leqslant \alpha < \Omega_{S\alpha+1}$ *and* $S\Omega_\gamma = \gamma$.

Proof. If α is finite then $S\alpha = 0$ and $\Omega_{S\alpha} = 0 \leqslant \alpha < \Omega_{S\alpha+1}$. Otherwise there exists γ such that $\mathbb{O}^*(\aleph(\gamma)) \sim \mathbb{O}^*(\alpha)$. Then $S\alpha = \gamma$, $\Omega_{S\alpha} \leqslant \aleph(\gamma) \leqslant \alpha$ and by Ax. II.3

$$\alpha < \aleph(\gamma + 1) = \Omega_{S\alpha+1}.$$

By definition we also have $S\Omega_\gamma = \gamma$.

3. Definitions of the Functions θ_α

Following a definition of S. Feferman and P. H. G. Aczel [1] we define functions θ_α as follows.

Inductive definition of sets of ordinals $C_n(\alpha, \beta)$, $C(\alpha, \beta)$, In (α) and ordering functions θ_α:

(D1) $C_0(\alpha, \beta) := \mathbb{O}^*(\beta)$

(D2) $P(\xi) \subseteq C_n(\alpha, \beta) \Rightarrow \xi \in C_{n+1}(\alpha, \beta)$

(D3) $\xi < \alpha$, $\xi \in C(\xi, \theta\xi\eta)$, $\eta < \theta\xi\eta$ and $\xi, \eta \in C_n(\alpha, \beta) \Rightarrow \theta\xi\eta \in C_{n+1}(\alpha, \beta)$

(D4) $\xi \neq 0,\ \xi \in C_n(\alpha, \beta) \Rightarrow \Omega_\xi \in C_{n+1}(\alpha, \beta)$

(D5) $C(\alpha, \beta) := \bigcup_{n \in \mathbb{N}} C_n(\alpha, \beta)$

(D6) $\mathrm{In}\,(\alpha) := \{\beta \mid \beta \notin C(\alpha, \beta)\} \cup K$

(D7) Let θ_α be the ordering function of the set $\mathrm{In}\,(\alpha)$. We write $\theta\alpha\beta$ for $\theta_\alpha(\beta)$.

This definition is inductive since it only assumes knowing the functions θ_ξ and sets $C(\xi, \theta\xi\eta)$ for $\xi < \alpha$. Since $K \subseteq \mathrm{In}\,(\alpha)$ it follows by Lemma 2 that the ordering function θ_α of $\mathrm{In}\,(\alpha)$ has domain $\mathbb{O}^*$. Hence $\theta\alpha\beta$ is defined for all $\alpha, \beta \in \mathbb{O}^*$.

S. Feferman's definition refers to the closure $\mathrm{Cl}_\alpha(\beta)$ of the set $\{0\} \cup \mathbb{O}^*(\beta)$ under addition and the functions $\theta_\xi (\xi < \alpha)$ and $\xi \mapsto \Omega_\xi$ instead of $C(\alpha, \beta)$. By W. Buchholz [1], $C(\alpha, \beta) = \mathrm{Cl}_\alpha(\beta)$. Thus the functions θ_α defined here agree with the functions defined by S. Feferman. The condition $\xi \in C(\xi, \theta\xi\eta)$ in (D3) which is the essential difference from the definition of $\mathrm{Cl}_\alpha(\beta)$ is included since it makes the most important properties of the functions θ_α easier to prove than on the basis of $\mathrm{Cl}_\alpha(\beta)$. The stratification of $C(\alpha, \beta)$ into $C_n(\alpha, \beta)$ $(n \in \mathbb{N})$ introduced by W. Buchholz helps in carrying out the proofs of induction.

The property $\beta \notin C(\alpha, \beta)$ occurring in (D6) says that β is *inaccessible* from ordinals $< \beta$ by certain functions such that the only θ_ξ which occur among these functions are ones for which $\xi < \alpha$. We therefore denote by $\mathrm{In}\,(\alpha)$ the set of α-*inaccessible* ordinals. If $\alpha < \Gamma_0$ then the set $\mathrm{Cr}\,(\alpha)$ of α-*critical* numbers defined in §13 is a subset of $\mathrm{In}\,(\alpha)$ as we shall show in §25.

4. Properties of the θ Functions

Lemma 5. a) $\beta \leqslant \theta\alpha\beta$,
- b) $\beta_1 < \beta_2 \Rightarrow \theta\alpha\beta_1 < \theta\alpha\beta_2$,
- c) $\xi < \beta \Rightarrow \xi \in C(\alpha, \beta)$,
- d) $0 \in C(\alpha, \beta)$,
- e) $\theta\alpha\beta \in P$,
- f) $\gamma \in C_n(\alpha, \beta) \Rightarrow P(\gamma) \subseteq C_n(\alpha, \beta)$,
- g) $C_n(\alpha, \beta) \subseteq C_{n+1}(\alpha, \beta)$.

Proofs. a) and b) hold because θ_α is an ordering function.

c) holds by (D1) and (D5).

d) holds by (D2) and (D5) since $P(0)$ is empty.

e) Assume that $\theta\alpha\beta \notin P$. Then $\gamma < \theta\alpha\beta$ for all $\gamma \in P(\theta\alpha\beta)$ and therefore by (D1) $P(\theta\alpha\beta) \subset C_0(\alpha, \theta\alpha\beta)$ and by (D2) and (D5) $\theta\alpha\beta \in C(\alpha, \theta\alpha\beta)$. By Theorem 24.2 if we also had $\theta\alpha\beta \notin K$, we should have $\theta\alpha\beta \notin \mathrm{In}\,(\alpha)$ by (D6) contradicting (D7). Consequently $\theta\alpha\beta \in P$.

f) Suppose $\gamma \in C_n(\alpha, \beta)$. If $n=0$ or $\gamma \in P$ then trivially $P(\gamma) \subseteq C_n(\alpha, \beta)$. If $n \neq 0$ and $\gamma \notin P$ then from e) and Theorem 24.2 we have that $\gamma \in C_n(\alpha, \beta)$ can only hold by (D2). In this case $P(\gamma) \subseteq C_{n-1}(\alpha, \beta)$. It follows that for all $\gamma_i \in P(\gamma)$, $P(\gamma_i) = \{\gamma_i\} \subseteq C_{n-1}(\alpha, \beta)$ and by (D2) $\gamma_i \in C_n(\alpha, \beta)$. Thus $P(\gamma) \subseteq C_n(\alpha, \beta)$.

g) By f) from $\gamma \in C_n(\alpha, \beta)$ we have $P(\gamma) \subseteq C_n(\alpha, \beta)$ and by (D2) $\gamma \in C_{n+1}(\alpha, \beta)$. Hence $C_n(\alpha, \beta) \subseteq C_{n+1}(\alpha, \beta)$.

Theorem 24.3. a) $\gamma \in C(\alpha, \beta) \Leftrightarrow P(\gamma) \subseteq C(\alpha, \beta)$.

b) *If* $\xi < \alpha$, $\xi \in O(\xi, \theta\xi\eta)$ *and* $\xi, \eta \in C(\alpha, \beta)$ *then* $\theta\xi\eta \in C(\alpha, \beta)$.

Proofs. a) By (D5) and Lemma 5 f) we obtain $P(\gamma) \subseteq C(\alpha, \beta)$ from $\gamma \in C(\alpha, \beta)$. If $P(\gamma) \subseteq C(\alpha, \beta)$ holds then by (D5) and Lemma 5 g) there exists n such that $P(\gamma) \subseteq C_n(\alpha, \beta)$. Hence by (D2) and (D5) $\gamma \in C(\alpha, \beta)$.

b) If $\xi, \eta \in C(\alpha, \beta)$ then by (D5) and Lemma 5 g) there exists n such that $\xi, \eta \in C_n(\alpha, \beta)$. If $\eta < \theta\xi\eta$ then from $\xi < \alpha$ and $\xi \in C(\xi, \theta\xi\eta)$ by (D3) and (D5) we obtain $\theta\xi\eta \in C(\alpha, \beta)$. Otherwise $\theta\xi\eta = \eta \in C(\alpha, \beta)$.

Lemma 6. a) $\alpha_1 \leqslant \alpha_2, \beta_1 \leqslant \beta_2 \Rightarrow C_n(\alpha_1, \beta_1) \subseteq C_n(\alpha_2, \beta_2)$, $C(\alpha_1, \beta_1) \subseteq C(\alpha_2, \beta_2)$.

b) *If* α *is a limit number, then* $C(\alpha, \beta) = \bigcup_{\xi < \alpha} C(\xi, \beta)$.

c) *If* β *is a limit number, then* $C(\alpha, \beta) = \bigcup_{\eta < \beta} C(\alpha, \eta)$.

Proofs. a) If $\alpha_1 \leqslant \alpha_2$, $\beta_1 \leqslant \beta_2$ then $C_n(\alpha_1, \beta_1) \subseteq C_n(\alpha_2, \beta_2)$ by induction on n. By (D5) we obtain $C(\alpha_1, \beta_1) \subseteq C(\alpha_2, \beta_2)$.

b) Suppose α is a limit number and $\gamma \in C_n(\alpha, \beta)$. We prove, by induction on n, that there exists $\xi < \alpha$ such that $\gamma \in C_n(\xi, \beta)$.

b1). Suppose $\gamma \in C_n(\alpha, \beta)$ by (D1), then $\gamma \in C_n(0, \beta)$.

b2). Suppose $\gamma \in C_n(\alpha, \beta)$ by (D2), then $\gamma_i \in C_{n-1}(\alpha, \beta)$ for all $\gamma_i \in P(\gamma)$. By I.H. there exist $\xi_i < \alpha$ such that $\gamma_i \in C_{n-1}(\xi_i, \beta)$. Let ξ be the maximum of these ξ_i then by a) $P(\gamma) \subseteq C_{n-1}(\xi, \beta)$ and by (D2) $\gamma \in C_n(\xi, \beta)$.

b3). If $\gamma \in C_n(\alpha, \beta)$ holds by (D3) then $\gamma = \theta\gamma_1\gamma_2$ where $\gamma_1 < \alpha$ and $\gamma_1, \gamma_2 \in C_{n-1}(\alpha, \beta)$. By I.H. there exist $\xi_i < \alpha$ such that $\gamma_i \in C_{n-1}(\xi_i, \beta)$ $(i = 1, 2)$. Let $\xi := \max(\gamma_1 + 1, \xi_1, \xi_2)$ then $\xi < \alpha$, $\gamma_1 < \xi$ and by a) $\gamma_1, \gamma_2 \in C_{n-1}(\xi, \beta)$. By (D3) we obtain $\gamma \in C_n(\xi, \beta)$.

b4). Suppose $\gamma \in C_n(\alpha, \beta)$ holds by (D4), then the assertion follows immediately from I.H.

Hence $C_n(\alpha, \beta) \subseteq \bigcup_{\xi < \alpha} C_n(\xi, \beta)$ and consequently by (D5) $C(\alpha, \beta) \subseteq \bigcup_{\xi < \alpha} C(\xi, \beta)$.

By a) we also have $\bigcup_{\xi < \alpha} C(\xi, \beta) \subseteq C(\alpha, \beta)$.

c) Suppose β is a limit number and $\gamma \in C_n(\alpha, \beta)$. We prove, by induction on n, that there exists $\eta < \beta$ such that $\gamma \in C_n(\alpha, \eta)$. If $n = 0$, then this holds for $\eta = \gamma + 1$ since then $\gamma + 1 < \beta$. If $n \neq 0$ then $\gamma \in C_n(\alpha, \beta)$ holds on the basis of finitely many hypothesis $\gamma_i \in C_{n-1}(\alpha, \beta)$ $(i = 1, \ldots, m)$. By I.H. there exist $\eta_i < \beta$ such that $\gamma_i \in C_{n-1}(\alpha, \eta_i)$. Let $\eta := \max(\eta_1, \ldots, \eta_m)$ then by a) $\gamma_i \in C_{n-1}(\alpha, \eta)$ $(i = 1, \ldots, m)$. Hence $\gamma \in C_n(\alpha, \eta)$. Thus $C_n(\alpha, \beta) \subset \bigcup_{\eta < \beta} C_n(\alpha, \eta)$. By (D5) we obtain $C(\alpha, \beta) \subseteq \bigcup_{\eta < \beta} C(\alpha, \eta)$.

By a) we also have $\bigcup_{\eta < \beta} C(\alpha, \eta) \subseteq C(\alpha, \beta)$.

Theorem 24.4. a) $\mathrm{In}(0)=P$, $\theta 0\beta=\omega^\beta$.

b) $\alpha_1<\alpha_2 \Rightarrow \mathrm{In}(\alpha_2)\subseteq \mathrm{In}(\alpha_1)$, $\quad \theta\alpha_1\beta\leqslant\theta\alpha_2\beta$.

c) *If* α *is a limit number, then* $\mathrm{In}(\alpha)=\bigcap_{\xi<\alpha}\mathrm{In}(\xi)$.

Proofs. a) If $\beta\in P$ and $\beta\notin K$ then we obtain $\beta\notin C_n(0,\beta)$ by induction on n. By (D5), (D6) and Theorem 24.2 we obtain $P\subseteq\mathrm{In}(0)$. By Lemma 5 e) we also have $\mathrm{In}(0)\subseteq P$. Hence $\mathrm{In}(0)=P$. It follows that $\theta 0\beta=\omega^\beta$, since $\beta\mapsto\omega^\beta$ is the ordering function of P.

b) If $\alpha_1<\alpha_2$ then by Lemma 6 a) $C(\alpha_1,\beta)\subseteq C(\alpha_2,\beta)$. By (D6) we obtain $\mathrm{In}(\alpha_2)\subseteq\mathrm{In}(\alpha_1)$ and by (D7) $\theta\alpha_1\beta\leqslant\theta\alpha_2\beta$ by induction on β.

c) Follows from Lemma 6 b) by (D6).

Lemma 7. *For each* α,β *there exists a least ordinal* $\gamma\notin C(\alpha,\beta)$. *This ordinal* γ *has the properties:* $\beta\leqslant\gamma$, $S\beta=S\gamma$, $C(\alpha,\beta)=C(\alpha,\gamma)$ *and* $\gamma\in\mathrm{In}(\alpha)$.

Proof. Let M be the closure of the set $\{0\}\cup\mathbb{O}^*(\beta)$ under addition and the functions $(\xi,\eta)\mapsto\theta\xi\eta$ and $\xi\mapsto\Omega_\xi$. By (D1)–(D6) $C(\alpha,\beta)\subseteq M$ and therefore we also have $M_0:=C(\alpha,\beta)\cap\mathbb{O}^*(\Omega_{S\beta+1})\subseteq M$. By Ax. III M_0 has cardinality $\leqslant\max(\beta,\aleph(0))<\Omega_{S\beta+1}$. Hence M_0 is a proper subset of $\mathbb{O}^*(\Omega_{S\beta+1})$. Therefore there exists a least ordinal $\gamma<\Omega_{S\beta+1}$ such that $\gamma\notin C(\alpha,\beta)$. By Lemma 5 c) $\beta\leqslant\gamma$. Since $\gamma<\Omega_{S\beta+1}$, $S\beta=S\gamma$. By the minimality of $\gamma\notin C(\alpha,\beta)$ we have $\xi\in C(\alpha,\beta)$ for all $\xi<\gamma$. Consequently $C_0(\alpha,\gamma)\subseteq C(\alpha,\beta)$. It follows by induction on n that $C_n(\alpha,\gamma)\subseteq C(\alpha,\beta)$ for all $n\in\mathbb{N}$. Therefore by (D5) $C(\alpha,\gamma)\subseteq C(\alpha,\beta)$. By Lemma 6 a) we also have $C(\alpha,\beta)\subseteq C(\alpha,\gamma)$. Thus $\gamma\notin C(\alpha,\beta)=C(\alpha,\gamma)$ and by (D6) $\gamma\in\mathrm{In}(\alpha)$.

Lemma 8. a) $C(\alpha,0)=C(\alpha,\theta\alpha 0)$.

b) $C(\alpha,\theta\alpha\beta+1)=C(\alpha,\theta\alpha(\beta+1))$.

Proofs. a) By Lemma 7 there exists $\gamma\in\mathrm{In}(\alpha)$ such that $C(\alpha,0)=C(\alpha,\gamma)$. It follows that $\theta\alpha 0\leqslant\gamma$ and by Lemma 6 a)

$$C(\alpha,0)\subseteq C(\alpha,\theta\alpha 0)\subseteq C(\alpha,\gamma)=C(\alpha,0),$$

and therefore $C(\alpha,0)=C(\alpha,\theta\alpha 0)$.

b) By Lemma 7 there exists $\gamma\in\mathrm{In}(\alpha)$ such that $C(\alpha,\theta\alpha\beta+1)=C(\alpha,\gamma)$ and $\theta\alpha\beta+1\leqslant\gamma$. It follows that $\theta\alpha\beta+1\leqslant\theta\alpha(\beta+1)\leqslant\gamma$ and by Lemma 6 a)

$$C(\alpha,\theta\alpha\beta+1)\subseteq C(\alpha,\theta\alpha(\beta+1))\subseteq C(\alpha,\gamma)=C(\alpha,\theta\alpha\beta+1),$$

and therefore $C(\alpha,\theta\alpha\beta+1)=C(\alpha,\theta\alpha(\beta+1))$.

Theorem 24.5. a) $S\theta\alpha\beta=S\beta$.

b) $\theta\alpha\gamma=\gamma$ *for* $\gamma\in K$.

Proofs. a) There exists γ such that $\Omega_{S\beta+1}=\theta\alpha\gamma$ since $\Omega_{S\beta+1}\in K\subseteq \mathrm{In}(\alpha)$. For $\xi<\Omega_{S\beta+1}$, by Lemma 7, there exists $\eta\in\mathrm{In}(\alpha)$ such that $\xi\leqslant\eta<\Omega_{S\beta+1}$. Let $M:=\{\theta\alpha\eta\mid\eta<\gamma\}$ then $\sup M=\Omega_{S\beta+1}$. Thus $M\subseteq\mathbb{O}^*(\Omega_{S\beta+1})$ and $M\sim\mathbb{O}^*(\gamma)$. By Ax. IV we obtain $\Omega_{S\beta+1}\leqslant\gamma$. Therefore

$$\beta\leqslant\theta\alpha\beta<\theta\alpha\Omega_{S\beta+1}\leqslant\theta\alpha\gamma=\Omega_{S\beta+1}.$$

Hence $S\theta\alpha\beta=S\beta$.

b) If $\gamma\in K$ there exists β such that $\gamma=\theta\alpha\beta$ since $K\subseteq\mathrm{In}(\alpha)$. By a), $S\gamma=S\beta$. It follows that

$$\gamma=\Omega_{S\gamma}\leqslant\beta\leqslant\theta\alpha\beta=\gamma,$$

and therefore $\beta=\gamma$ and $\theta\alpha\gamma=\gamma$.

Theorem 24.6. a) $\Omega_\gamma\in C(\alpha,\beta)\Leftrightarrow\gamma\in C(\alpha,\beta)$.
b) $\gamma\in C(\alpha,\beta)\Rightarrow\Omega_{S\gamma}\in C(\alpha,\beta)$.

Proofs. a) The assertion is trivial if $\gamma=0$. Now suppose $\gamma\neq0$. If $\Omega_\gamma\in C(\alpha,\beta)$ then there is a least number n such that $\Omega_\gamma\in C_n(\alpha,\beta)$. If $n=0$ then $\gamma<\Omega_\gamma<\beta$, and therefore, also, $\gamma\in C_0(\alpha,\beta)$. If $n\neq0$, then by Theorems 24.2 and 24.5 b) we have that $\Omega_\gamma\in C(\alpha,\beta)$ can only hold because of (D4). Therefore $\gamma\in C_{n-1}(\alpha,\beta)$. Therefore from $\Omega_\gamma\in C(\alpha,\beta)$ we have, in every case, $\gamma\in C(\alpha,\beta)$. By (D4) and (D5) from $\gamma\in C(\alpha,\beta)$ we obtain $\Omega_\gamma\in C(\alpha,\beta)$.

b) Suppose $\gamma\in C(\alpha,\beta)$ holds then there exists n such that $\gamma\in C_n(\alpha,\beta)$. By Theorem 24.5 a) we obtain $\Omega_{S\gamma}\in C(\alpha,\beta)$ by induction on n.

Lemma 9. $\alpha_1<\alpha_2,\ \forall\xi(\alpha_1\leqslant\xi<\alpha_2\to\xi\notin C(\alpha_1,\beta))\Rightarrow C(\alpha_1,\beta)=C(\alpha_2,\beta)$.

Proof. Suppose $\gamma\in C_n(\alpha_2,\beta)$. We prove by induction on n that $\gamma\in C_n(\alpha_1,\beta)$. If $\gamma\in C_n(\alpha_2,\beta)$ holds by (D1), (D2) or (D4), then $\gamma\in C_n(\alpha_1,\beta)$ follows either trivially or by the I.H. Now suppose $\gamma\in C_n(\alpha_2,\beta)$ holds by (D3). Then $\gamma=\theta\xi\eta$ where $\xi<\alpha_2$ and $\xi,\eta\in C_{n-1}(\alpha_2,\beta)$. By I.H. we have $\xi,\eta\in C_{n-1}(\alpha_1,\beta)$. Using $\xi<\alpha_2$ and the hypothesis $\forall\xi(\alpha_1\leqslant\xi<\alpha_2\to\xi\notin C(\alpha_1,\beta))$ we obtain $\xi<\alpha_1$. By (D3) we obtain $\gamma\in C_n(\alpha_1,\beta)$. Hence $C_n(\alpha_2,\beta)\subseteq C_n(\alpha_1,\beta)$ holds. By (D5) we obtain $C(\alpha_2,\beta)\subseteq C(\alpha_1,\beta)$. By Lemma 6 a) $C(\alpha_1,\beta)\subseteq C(\alpha_2,\beta)$ also holds.

Theorem 24.7. a) $\alpha_1<\alpha_2,\ \forall\xi(\alpha_1\leqslant\xi<\alpha_2\to\xi\notin C(\alpha_1,\theta\alpha_2\beta))\Rightarrow\theta\alpha_1\beta=\theta\alpha_2\beta$.
b) $\exists\xi(\alpha_1\leqslant\xi<\alpha_2\wedge\xi\in C(\alpha_1,\theta\alpha_2\beta))\Rightarrow\theta\alpha_1(\theta\alpha_2\beta)=\theta\alpha_2\beta$.

Proofs. a) When $\eta\leqslant\theta\alpha_2\beta$ we obtain

$$\forall\xi(\alpha_1\leqslant\xi<\alpha_2\to\xi\notin C(\alpha_1,\eta))$$

from the hypothesis. By Lemma 6 a) and by Lemma 9 $C(\alpha_1,\eta)=C(\alpha_2,\eta)$. By (D7) we obtain $\theta\alpha_1\beta=\theta\alpha_2\beta$.

b) If $\theta\alpha_2\beta \in K$, then $\theta\alpha_1(\theta\alpha_2\beta)=\theta\alpha_2\beta$ holds by Theorem 24.5 b). Now suppose $\theta\alpha_2\beta \notin K$. By Theorem 24.4 b) $\alpha_1 \leqslant \xi < \alpha_2$ implies $\theta\alpha_2\beta \in \mathrm{In}(\xi)$. Therefore there exists $\eta \leqslant \theta\xi\eta = \theta\alpha_2\beta$. Using $\xi \in C(\alpha_1, \theta\alpha_2\beta)$, by Lemma 6 a) we obtain

$$\xi \in C(\xi, \theta\xi\eta) \quad \text{and} \quad \xi \in C(\alpha_2, \theta\alpha_2\beta).$$

If we had $\eta < \theta\alpha_2\beta$, then by Lemma 5 c) we should have $\eta \in C(\alpha_2, \theta\alpha_2\beta)$ and by Theorem 24.3 b) $\theta\xi\eta \in C(\alpha_2, \theta\alpha_2\beta)$ which contradicts (D6) since $\theta\xi\eta = \theta\alpha_2\beta \notin K$. Thus $\eta = \theta\alpha_2\beta$. By Theorem 24.4 b) we have

$$\theta\alpha_2\beta \leqslant \theta\alpha_1(\theta\alpha_2\beta) \leqslant \theta\xi\eta = \theta\alpha_2\beta,$$

and therefore $\theta\alpha_1(\theta\alpha_2\beta)=\theta\alpha_2\beta$.

Lemma 10. a) *If $\theta\alpha_1\beta_1 = \theta\alpha_2\beta_2$ where $\alpha_1 \leqslant \alpha_2$, then $\beta_1 = \beta_2$ or $\beta_1 = \theta\alpha_2\beta_2$.*
b) $\alpha_1 < \alpha_2$, $\alpha_1 \in C(\alpha_2, \theta\alpha_2\beta) \Rightarrow \theta\alpha_1(\theta\alpha_2\beta) = \theta\alpha_2\beta$.

Proofs. a) If $\alpha_1 \leqslant \alpha_2$ then by Theorem 24.7 $\theta\alpha_1\beta_2 = \theta\alpha_2\beta_2$ or $\theta\alpha_1(\theta\alpha_2\beta_2) = \theta\alpha_2\beta_2$. If $\theta\alpha_1\beta_1 = \theta\alpha_2\beta_2$, then $\beta_1 = \beta_2$ or $\beta_1 = \theta\alpha_2\beta_2$.

b) Suppose $\alpha_1 < \alpha_2$ and $\alpha_1 \in C(\alpha_2, \theta\alpha_2\beta)$. If $\alpha_1 \notin C(\alpha_1, \theta\alpha_2\beta)$ then $C(\alpha_1, \theta\alpha_2\beta) \neq C(\alpha_2, \theta\alpha_2\beta)$. Then, by Lemma 9, there exists $\xi \in C(\alpha_1, \theta\alpha_2\beta)$ such that $\alpha_1 \leqslant \xi < \alpha_2$. By Theorem 24.7 b) it also follows in the case that $\alpha_1 \in C(\alpha_1, \theta\alpha_2\beta)$ that $\theta\alpha_1(\theta\alpha_2\beta) = \theta\alpha_2\beta$.

Lemma 11. $\theta\alpha\beta \notin K$, $\theta\alpha\beta \leqslant \gamma \in C(\alpha, \theta\alpha\beta) \Rightarrow S\beta < S\gamma$.

Proof. By hypothesis there exists m such that $\gamma \in C_m(\alpha, \theta\alpha\beta)$. We prove $S\beta < S\gamma$ by induction on m. Since $\theta\alpha\beta \leqslant \gamma$, $m \neq 0$ therefore $m = n+1$.

1. Suppose $\gamma \in C_{n+1}(\alpha, \theta\alpha\beta)$ holds by (D2). Since $\theta\alpha\beta \in P$, it follows from $\theta\alpha\beta \leqslant \gamma$ that there exists $\gamma_0 \in P(\gamma)$ where $\theta\alpha\beta \leqslant \gamma_0$. Hence $\gamma_0 \in C_n(\alpha, \theta\alpha\beta)$. By I.H. we obtain $S\beta < S\gamma_0$. Therefore we also have $S\beta < S\gamma$.
2. Suppose $\gamma \in C_{n+1}(\alpha, \theta\alpha\beta)$ holds by (D3). Then $\gamma = \theta\xi\eta$ where $\xi < \alpha$ and $\xi, \eta \in C_n(\alpha, \theta\alpha\beta)$. By Lemma 10 b) we have $\theta\xi(\theta\alpha\beta) = \theta\alpha\beta$. Using $\theta\alpha\beta \leqslant \gamma = \theta\xi\eta$ we obtain $\theta\alpha\beta \leqslant \eta$. By I.H. we obtain $S\beta < S\eta$. Then by Theorem 24.5 a) we also have $S\beta < S\gamma$.
3. Suppose $\gamma \in C_{n+1}(\alpha, \theta\alpha\beta)$ holds by (D4). Then $\gamma \in K$. Since $\theta\alpha\beta \notin K$ and $\theta\alpha\beta \leqslant \gamma$, we have $\theta\alpha\beta < \gamma$ and $S\beta < S\gamma$.

Theorem 24.8. a) $\beta \notin K \Rightarrow C(\alpha, \theta\alpha\beta) \cap \mathbb{O}^*(\Omega_{S\beta+1}) = \mathbb{O}^*(\theta\alpha\beta)$.
b) $C(\alpha, 0) \cap \mathbb{O}^*(\Omega_1) = \mathbb{O}^*(\theta\alpha 0)$.

Proofs. a) By Lemma 5 c) we have

(1) $\mathbb{O}^*(\theta\alpha\beta) \subseteq C(\alpha, \theta\alpha\beta)$.

By Theorem 24.5 a) $\theta\alpha\beta < \Omega_{S\beta+1}$ and therefore

(2) $\mathbb{O}^*(\theta\alpha\beta) \subseteq \mathbb{O}^*(\Omega_{S\beta+1})$.

By Theorem 24.5 b) $\beta \notin K$ implies $\theta\alpha\beta \notin K$. If $\theta\alpha\beta \leqslant \gamma \in C(\alpha, \theta\alpha\beta)$ then by Lemma 11 $\Omega_{S\beta+1} \leqslant \gamma$. Hence we also have

(3) $C(\alpha, \theta\alpha\beta) \cap \mathbb{O}^*(\Omega_{S\beta+1}) \subseteq \mathbb{O}^*(\theta\alpha\beta)$.

Assertion a) follows from (1)–(3).

b) Follows from a) since $S0 = 0 \notin K$ and by Lemma 8 a) $C(\alpha, 0) = C(\alpha, \theta\alpha 0)$.

Theorem 24.9. *θ_α is a normal function.*

Proof. Since θ_α has domain $\mathbb{O}^*$, by Lemma 1 it remains only to prove that In (α) is closed. Let M be a non-empty bounded subset of In (α). We have to prove that $\sup M$ is then a member of In (α). If $\sup M \in K$ or $\sup M \in M$ then trivially $\sup M \in \mathrm{In}(\alpha)$. Now suppose $\sup M \notin K$ and $\sup M \notin M$. Then $\sup M$ is a limit number. Assume that $\sup M \in C(\alpha, \sup M)$. Then by Lemma 6 c) there exists $\eta < \sup M$ such that $\sup M \in C(\alpha, \eta)$. Since $M \subseteq \mathrm{In}(\alpha)$ and $\sup M \notin K$ there is then $\theta\alpha\xi \notin K$ such that $\eta \leqslant \theta\alpha\xi \leqslant \sup M$ and $S\theta\alpha\xi = S\xi = S \sup M$. By Lemma 6 a) $\sup M \in C(\alpha, \eta)$ implies $\sup M \in C(\alpha, \theta\alpha\xi)$. By Lemma 11 we have $S\xi < S \sup M$ contradicting the choice of ξ. Consequently $\sup M \notin C(\alpha, \sup M)$ and by (D6) $\sup M \in \mathrm{In}(\alpha)$.

5. The Sets $\overline{\mathrm{In}}(\alpha)$ and Functions $\overline{\theta}_\alpha$

We introduce sets $\overline{\mathrm{In}}(\alpha)$ of ordinals and functions $\overline{\theta}_\alpha$ on the ordinals following W. Buchholz [2] and use these to develop a notation system for a segment of the ordinals.

Definition. $\overline{\mathrm{In}}(\alpha) := \mathrm{In}(\alpha) \backslash \mathrm{In}(\alpha+1)$. That is to say $\gamma \in \overline{\mathrm{In}}(\alpha)$ is to hold if, and only if, $\gamma \in \mathrm{In}(\alpha)$ and $\gamma \notin \mathrm{In}(\alpha+1)$.

In this case by Theorem 24.4 b) $\gamma \in \mathrm{In}(\xi)$ for all $\xi \leqslant \alpha$ and $\gamma \notin \mathrm{In}(\eta)$ for all $\eta > \alpha$. $\overline{\mathrm{In}}(\alpha)$ is therefore the set of those ordinals which are *maximally* α-inaccessible.

Corollaries. 1. $\overline{\mathrm{In}}(\alpha) \subseteq P$.

2. *$\overline{\mathrm{In}}(\alpha)$ and K are disjoint sets.*

3. *If $\alpha_1 \neq \alpha_2$ then $\overline{\mathrm{In}}(\alpha_1)$ and $\overline{\mathrm{In}}(\alpha_2)$ are disjoint sets.*

Theorem 24.10. *$\overline{\mathrm{In}}(\alpha)$ is the set of those ordinals $\theta\alpha\beta$ such that $\alpha \in C(\alpha, \theta\alpha\beta)$ and $\beta < \theta\alpha\beta$.*

Proof. 1. Suppose $\gamma \in \overline{\mathrm{In}}(\alpha)$, then $\gamma \in \mathrm{In}(\alpha)$ and $\gamma \notin \mathrm{In}(\alpha+1)$. Then there exists β such that $\gamma = \theta\alpha\beta \notin K$, $\theta\alpha\beta \notin C(\alpha, \theta\alpha\beta)$ and $\theta\alpha\beta \in C(\alpha+1, \theta\alpha\beta)$. It follows that $C(\alpha, \theta\alpha\beta) \neq C(\alpha+1, \theta\alpha\beta)$ and by Lemma 9 $\alpha \in C(\alpha, \theta\alpha\beta)$. Since $\theta\alpha\beta \in C(\alpha+1, \theta\alpha\beta)$ there exists a least number n such that $\theta\alpha\beta \in C_n(\alpha+1, \theta\alpha\beta)$. Because $\theta\alpha\beta \in P$ and $\theta\alpha\beta \notin K$ this can only hold by (D3). Then we have $\eta < \theta\xi\eta = \theta\alpha\beta$ where $\xi \leqslant \alpha$. By Theorem 24.4 b) we have $\theta\alpha\beta \leqslant \theta\alpha\eta$. Then $\beta \leqslant \eta$, and therefore $\beta < \theta\alpha\beta$.

2. Suppose $\alpha \in C(\alpha, \theta\alpha\beta)$ and $\beta < \theta\alpha\beta$. By Theorem 24.5 b) we then have $\theta\alpha\beta \notin K$. By Lemmata 6 a) and 5 c) it follows that $\alpha \in C(\alpha+1, \theta\alpha\beta)$ and $\beta \in C(\alpha+1, \theta\alpha\beta)$. By Theorem 24.3 b) we obtain $\theta\alpha\beta \in C(\alpha+1, \theta\alpha\beta)$. By (D6) using $\theta\alpha\beta \notin K$ we have $\theta\alpha\beta \notin \mathrm{In}(\alpha+1)$. Hence we have $\theta\alpha\beta \in \overline{\mathrm{In}}(\alpha)$.

Lemma 12. *$\xi \in C(\xi, \theta\xi\eta)$, $\eta < \theta\xi\eta$ and $\beta \leqslant \theta\xi\eta \in C(\alpha, \beta)$ imply $\xi < \alpha$ and $\xi, \eta \in C(\alpha, \beta)$.*

Proof. By Theorem 24.5 b) $\eta < \theta\xi\eta$ implies $\theta\xi\eta \notin K$. Since $\theta\xi\eta \in C(\alpha, \beta)$ there is a least number n such that $\theta\xi\eta \in C_n(\alpha, \beta)$. Since $\beta \leqslant \theta\xi\eta \in P$ and $\theta\xi\eta \notin K$ this can only hold by (D3). Then we have $\theta\xi\eta = \theta\xi_0\eta_0$ where $\xi_0 < \alpha$, $\xi_0 \in C(\xi_0, \theta\xi_0\eta_0)$, $\eta_0 < \theta\xi_0\eta_0$ and $\xi_0, \eta_0 \in C(\alpha, \beta)$. By Theorem 24.10 $\theta\xi\eta \in \overline{\mathrm{In}}(\xi)$ and $\theta\xi_0\eta_0 \in \overline{\mathrm{In}}(\xi_0)$. If $\xi \neq \xi_0$, $\overline{\mathrm{In}}(\xi)$ and $\overline{\mathrm{In}}(\xi_0)$ are disjoint. Therefore from $\theta\xi\eta = \theta\xi_0\eta_0$ we have $\xi = \xi_0$. So we also have $\eta = \eta_0$ and therefore $\xi < \alpha$ and $\xi, \eta \in C(\alpha, \beta)$.

Definition. We set $\mu(\alpha)$ to be the least ordinal η such that $\alpha \in C(\alpha, \theta\alpha\eta)$. (Such an ordinal $\mu(\alpha) \leqslant \alpha+1$ exists since $\alpha \in C(\alpha, \theta\alpha(\alpha+1))$.

Lemma 13. *$\mu(\alpha)$ is not a limit ordinal.*

Proof. If $\mu(\alpha)$ were a limit number, then we should have

$$\theta\alpha(\mu(\alpha)) = \sup\{\theta\alpha\eta \mid \eta < \mu(\alpha)\},$$

since by Theorem 24.9 θ_α is continuous. Then, by Lemma 6 from $\alpha \in C(\alpha, \theta\alpha(\mu(\alpha)))$ we should have $\eta < \mu(\alpha)$ such that $\alpha \in C(\alpha, \theta\alpha\eta)$. This contradicts the minimality of $\mu(\alpha)$.

Lemma 14. $\zeta \in C_n(\alpha, \beta)$, $\zeta \notin C(\gamma, \theta\gamma\delta)$, $\zeta \in C(\gamma, \theta\gamma\delta+1) \Rightarrow \delta \in C(\alpha, \beta)$.

Proof. By induction on n. By Lemma 5 c) $\zeta \notin C(\gamma, \theta\gamma\delta)$ implies $\theta\gamma\delta \leqslant \zeta$.

1. Suppose $\zeta \in C_n(\alpha, \beta)$ holds by (D1). Then $\zeta < \beta$. It follows that $\delta \leqslant \theta\gamma\delta \leqslant \zeta < \beta$ and $\delta \in C(\alpha, \beta)$.

2. Suppose $\zeta \in C_n(\alpha, \beta)$ holds by (D2). By Theorem 24.3 a) $\zeta \notin C(\gamma, \theta\gamma\delta)$ implies that there exist $\zeta_0 \in P(\zeta)$ such that $\zeta_0 \notin C(\gamma, \theta\gamma\delta)$. Hence $\zeta_0 \in C_{n-1}(\alpha, \beta)$ and $\zeta_0 \in C(\gamma, \theta\gamma\delta+1)$. By I.H. we obtain $\delta \in C(\alpha, \beta)$.

3. Suppose $\zeta \in C_n(\alpha, \beta)$ holds by (D3). Then we have $\zeta = \theta\xi\eta$ where $\xi < \alpha$, $\xi \in C(\xi, \theta\xi\eta)$, $\eta < \theta\xi\eta$ and $\xi, \eta \in C_{n-1}(\alpha, \beta)$.

3.1. Suppose $\theta\gamma\delta = \zeta$. By Theorem 24.10 $\theta\xi\eta \in \overline{\mathrm{In}}(\xi)$. Therefore from $\theta\gamma\delta = \theta\xi\eta$ we obtain $\gamma \leqslant \xi$ and by Lemma 10 a) $\delta = \eta$ or $\delta = \theta\xi\eta = \gamma$; therefore in every case $\delta \in C(\alpha, \beta)$.

3.2. Suppose $\theta\gamma\delta<\zeta$. Then $\theta\gamma\delta+1\leqslant\theta\xi\eta\in C(\gamma,\theta\gamma\delta+1)$. By Lemma 12 we obtain $\xi<\gamma$ and $\xi,\eta\in C(\gamma,\theta\gamma\delta+1)$. Then by Theorem 24.3 b) from $\theta\xi\eta\notin C(\gamma,\theta\gamma\delta)$ we obtain $\xi\notin C(\gamma,\theta\gamma\delta)$ or $\eta\notin C(\gamma,\theta\gamma\delta)$. By I.H. (for ξ or η replacing ζ) we obtain $\delta\in C(\alpha,\beta)$.

4. Suppose $\zeta\in C_n(\alpha,\beta)$ holds by (D4). Then we have $\zeta=\Omega_\xi$ where $\xi\in C_{n-1}(\alpha,\beta)$. By Theorem 24.6 a) $\xi\notin C(\gamma,\theta\gamma\delta)$ and $\xi\in C(\gamma,\theta\gamma\delta+1)$ also hold. By I.H. we obtain $\delta\in C(\alpha,\beta)$.

Lemma 15. $\gamma\in C(\alpha,\beta)\Rightarrow\mu(\gamma)\in C(\alpha,\beta)$.

Proof. If $\mu(\gamma)=0$, then $\mu(\gamma)\in C(\alpha,\beta)$ holds by Lemma 5 d). Otherwise by Lemma 13 $\mu(\gamma)=\delta+1$ where $\gamma\notin C(\gamma,\theta\gamma\delta)$ and $\gamma\in C(\gamma,\theta\gamma(\delta+1))$. By Lemma 8 we obtain $\gamma\in C(\gamma,\theta\gamma\delta+1)$. Using $\gamma\in C(\alpha,\beta)$ by Lemma 14 we obtain $\delta\in C(\alpha,\beta)$. By Lemma 6 a) from $\gamma\in C(\alpha,\beta)$ and $\gamma\notin C(\gamma,\theta\gamma\delta)$ we obtain $\gamma<\alpha$ or $\theta\gamma\delta<\beta$ and therefore $\alpha\neq0$ or $1<\beta$. If $\alpha\neq0$ then, since $0\in C(\alpha,\beta)$ by Theorem 24.3 b) $1=\theta00\in C(\alpha,\beta)$. If $1<\beta$ then $1\in C(\alpha,\beta)$ by Lemma 5 c). Therefore in every case $\delta\in C(\alpha,\beta)$ and $1\in C(\alpha,\beta)$. By Theorem 24.3 a) we obtain $\mu(\gamma)=\delta+1\in C(\alpha,\beta)$.

Definitions. 1. If $S\mu(\alpha)\leqslant S\beta$, set $r_\alpha(\beta)$ to be the unique ordinal determined by $\beta=\Omega_{S\mu(\alpha)}+r_\alpha(\beta)$.

2. If $\beta=\beta_0+n$ where $n<\omega$ and $\mu(\alpha)<\beta_0=\theta\alpha\beta_0$, then set $\iota\alpha\beta:=1$. Otherwise set $\iota\alpha\beta:=0$.

3. If $S\mu(\alpha)\leqslant S\beta$ set $\bar\theta_\alpha(\beta):=\theta\alpha(\mu(\alpha)+r_\alpha(\beta)+\iota\alpha\beta)$. We write $\bar\theta\alpha\beta$ for $\bar\theta_\alpha(\beta)$.

Theorem 24.11. *$\bar\theta_\alpha$ is an order preserving map of the set of $\xi\geqslant\Omega_{S\mu(\alpha)}$ onto the set $\overline{\text{In}}\,(\alpha)$.*

Proof. 1. Suppose $\beta\geqslant\Omega_{S\mu(\alpha)}$, then $S\mu(\alpha)\leqslant S\beta$. We prove that $\bar\theta\alpha\beta\in\overline{\text{In}}\,(\alpha)$. We have $\bar\theta\alpha\beta=\theta\alpha\bar\beta$ where

$$\bar\beta=\mu(\alpha)+r_\alpha(\beta)+\iota\alpha\beta.$$

By Lemma 6 a) from $\alpha\in C(\alpha,\theta\alpha(\mu(\alpha)))$ and $\theta\alpha(\mu(\alpha))\leqslant\theta\alpha\bar\beta$ we have

(1) $\alpha\in C(\alpha,\theta\alpha\bar\beta)$.

Assume that $\theta\alpha\bar\beta=\bar\beta$. Then $1<\bar\beta\in P$, consequently $\iota\alpha\beta=0$ and by Lemma 13 $r_\alpha(\beta)\neq0$. Hence

$$\bar\beta=\mu(\alpha)+r_\alpha(\beta)=r_\alpha(\beta).$$

Using $\Omega_{S\mu(\alpha)}\leqslant\mu(\alpha)$ we obtain $\beta=\Omega_{S\mu(\alpha)}+r_\alpha(\beta)\leqslant\bar\beta=r_\alpha(\beta)\leqslant\beta$,

and therefore $\beta=\bar\beta$, $\theta\alpha\beta=\beta$ and $\mu(\alpha)<\beta$. But in this case $\iota\alpha\beta=1$. Hence $\theta\alpha\bar\beta\neq\bar\beta$ and therefore

(2) $\bar\beta<\theta\alpha\bar\beta$.

By Theorem 24.10 from (1) and (2) we obtain $\bar\theta\alpha\beta=\theta\alpha\bar\beta\in\overline{\text{In}}\,(\alpha)$.

2. Suppose $\gamma \in \overline{\text{In}}(\alpha)$. We have to prove that there exists β such that $S\mu(\alpha) \leqslant S\beta$ and $\gamma = \bar{\theta}\alpha\beta$. By Theorem 24.10 there exists $\bar{\beta}$ such that $\gamma = \theta\alpha\bar{\beta}$, $\alpha \in C(\alpha, \theta\alpha\bar{\beta})$ and $\bar{\beta} < \theta\alpha\bar{\beta}$. By the definition of $\mu(\alpha)$ and Lemma 6 a) we obtain $\theta\alpha(\mu(\alpha)) \leqslant \theta\alpha\bar{\beta}$ and therefore $\mu(\alpha) \leqslant \bar{\beta}$. Therefore there exists β^* such that

$$\bar{\beta} = \mu(\alpha) + \beta^*.$$

If $\beta^* = \beta_0 + n$ where $\mu(\alpha) < \beta_0 = \theta\alpha\beta_0$ and $n < \omega$ then it follows from $\bar{\beta} < \theta\alpha\bar{\beta}$ that $n > 0$. In this case let $\beta := \Omega_{S\mu(\alpha)} + \beta_0 + (n-1)$. Otherwise let $\beta := \Omega_{S\mu(\alpha)} + \beta^*$. In any case we then have $S\mu(\alpha) \leqslant S\beta$ and

$$\bar{\beta} = \mu(\alpha) + r_\alpha(\beta) + \iota\alpha\beta,$$

and therefore $\gamma = \theta\alpha\bar{\beta} = \bar{\theta}\alpha\beta$.

3. Suppose $\Omega_{S\mu(\alpha)} \leqslant \beta_1 < \beta_2$. We prove that $\bar{\theta}\alpha\beta_1 < \bar{\theta}\alpha\beta_2$. We have $\bar{\theta}\alpha\beta_i = \theta\alpha\bar{\beta}_i$ where

$$\bar{\beta}_i = \mu(\alpha) + r_\alpha(\beta_i) + \iota\alpha\beta_i \quad (i = 1,2).$$

$\Omega_{S\mu(\alpha)} \leqslant \beta_1 < \beta_2$ implies $r_\alpha(\beta_1) < r_\alpha(\beta_2)$. If $\iota\alpha\beta_1 = 0$ then $\bar{\beta}_1 < \bar{\beta}_2$. If $\iota\alpha\beta_1 = 1$ and $r_\alpha(\beta_1) + 1 = r_\alpha(\beta_2)$ then we also have $\iota\alpha\beta_2 = 1$. In every case $\bar{\beta}_1 < \bar{\beta}_2$. Therefore $\bar{\theta}\alpha_1\beta_1 < \bar{\theta}\alpha_2\beta_2$.

Theorem 24.12. *If $S\mu(\xi) \leqslant S\eta$ then*

a) $\beta \leqslant \bar{\theta}\xi\eta \in C(\alpha, \beta) \Rightarrow \xi < \alpha$ *and* $\xi, \eta \in C(\alpha, \beta)$.

b) $\xi < \alpha$ *and* $\xi, \eta \in C(\alpha, \beta) \Rightarrow \bar{\theta}\xi\eta \in C(\alpha, \beta)$.

Proof. We have $\bar{\theta}\xi\eta = \theta\xi\bar{\eta}$ where $\bar{\eta} = \mu(\xi) + r_\xi(\eta) + \iota\xi\eta$. By Theorems 24.10 and 24.11 $\xi \in C(\xi, \theta\xi\bar{\eta})$ and $\bar{\eta} < \theta\xi\bar{\eta}$.

a) Suppose $\beta \leqslant \theta\xi\eta \in C(\alpha, \beta)$. By Lemma 12 we have $\xi < \alpha$ and $\xi, \bar{\eta} \in C(\alpha, \beta)$. Hence by Lemma 15 and Theorem 24.6 b) we have $\Omega_{S\mu(\xi)} \in C(\alpha, \beta)$ and by Theorem 24.3 a) $r_\xi(\eta) \in C(\alpha, \beta)$. Then by Theorem 24.3 a)

$$\eta = \Omega_{S\mu(\xi)} + r_\xi(\eta) \in C(\alpha, \beta)$$

also holds.

b) Suppose $\xi < \alpha$ and $\xi, \eta \in C(\alpha, \beta)$. Hence by Lemma 15 we have $\mu(\xi) \in C(\alpha, \beta)$ and by Theorem 24.3 a) $r_\xi(\eta) \in C(\alpha, \beta)$. Since $\alpha \neq 0$, $1 \in C(\alpha, \beta)$. By Theorem 24.3 a) we obtain

$$\bar{\eta} = \mu(\xi) + r_\xi(\eta) + \iota\alpha\beta \in C(\alpha, \beta).$$

Using $\xi < \alpha$, $\xi \in C(\xi, \theta\xi\bar{\eta})$ and $\xi \in C(\alpha, \beta)$, by Theorem 24.3 b) we obtain $\bar{\theta}\xi\eta = \theta\xi\bar{\eta} \in C(\alpha, \beta)$.

Theorem 24.13. *If $S\mu(\alpha) \leqslant S\beta$ then $\alpha \in C(\alpha, \bar{\theta}\alpha\beta)$, $\beta < \bar{\theta}\alpha\beta$ and $S\bar{\theta}\alpha\beta = S\beta$.*

Proof. We have $\bar\theta\alpha\beta=\theta\alpha\bar\beta$ where $\bar\beta=\mu(\alpha)+r_\alpha(\beta)+\iota\alpha\beta$. By Theorems 24.10 and 24.11, $\alpha\in C(\alpha,\theta\alpha\bar\beta)$ and $\bar\beta<\theta\alpha\bar\beta$. It follows that $\alpha\in C(\alpha,\bar\theta\alpha\beta)$ and

$$\beta=\Omega_{S\mu(\alpha)}+r_\alpha(\beta)\leqslant\bar\beta<\bar\theta\alpha\beta.$$

$S\mu(\alpha)\leqslant S\beta$ implies $S\bar\beta=S\beta$. Using Theorem 24.5 a) we obtain $S\bar\theta\alpha\beta=S\beta$.

Theorem 24.14. *If* $S\mu(\alpha_i)\leqslant S\beta_i$ $(i=1,2)$, *then*

a) $\bar\theta\alpha_1\beta_1=\bar\theta\alpha_2\beta_2\Rightarrow\alpha_1=\alpha_2,\ \beta_1=\beta_2$.

b) $\alpha_1<\alpha_2,\qquad \alpha_1\in C(\alpha_2,\bar\theta\alpha_2\beta_2),\qquad \beta_1<\bar\theta\alpha_2\beta_2\Rightarrow\bar\theta\alpha_1\beta_1<\bar\theta\alpha_2\beta_2$.

Proof. a) By Theorem 24.11 $\bar\theta\alpha_i\beta_i\in\overline{\text{In}}(\alpha_i)$. If $\alpha_1\neq\alpha_2$ then $\overline{\text{In}}(\alpha_1)$ and $\overline{\text{In}}(\alpha_2)$ are disjoint. Therefore from $\bar\theta\alpha_1\beta_1=\bar\theta\alpha_2\beta_2$ we obtain $\alpha_1=\alpha_2$. Then by Theorem 24.11 we also have $\beta_1=\beta_2$.

b) It follows from the hypotheses by Lemma 2 c) that $\beta_1\in C(\alpha_2,\bar\theta\alpha_2\beta_2)$ and by Theorem 24.12 b) that $\bar\theta\alpha_1\beta_1\in C(\alpha_2,\bar\theta\alpha_2\beta_2)$. By Theorem 24.13 $\beta_1<\bar\theta\alpha_2\beta_2$ implies $\bar\theta\alpha_1\beta_1<\Omega_{S\beta_2+1}$. Since $\bar\theta\alpha_2\beta_2\notin K$ by Theorem 24.8 a) we have $\bar\theta\alpha_1\beta_1<\bar\theta\alpha_2\beta_2$.

§25. A Notation System for Ordinals Based on the θ_α Functions

1. The Set $\bar\theta(\Omega)$ of Ordinals

Inductive Definition of the set $\bar\theta(\Omega)\subset\mathbb{O}^*$ and the *degree* $G\alpha\in\mathbb{N}$ of an ordinal $\alpha\in\bar\theta(\Omega)$:

1. $0\in\bar\theta(\Omega)$, $G0:=0$.
2. $0<\gamma\notin P$, $P(\gamma)\subset\bar\theta(\Omega)\Rightarrow\gamma\in\bar\theta(\Omega)$, $G\gamma:=\max\{G\gamma_i\mid\gamma_i\in P(\gamma)\}+1$.
3. $\alpha,\beta\in\bar\theta(\Omega)$, $S\mu(\alpha)\leqslant S\beta\Rightarrow\bar\theta\alpha\beta\in\bar\theta(\Omega)$, $G\bar\theta\alpha\beta:=\max(G\alpha,G\beta)+1$.
4. $0<\alpha\in\bar\theta(\Omega)\Rightarrow\Omega_\alpha\in\bar\theta(\Omega)$, $G\Omega_\alpha:=G\alpha+1$.

Obviously, for every ordinal α such that $\alpha\in\bar\theta(\Omega)$ at most one of the above four defining rules can hold. It follows by Theorem 24.14 a) that every ordinal $\alpha\in\bar\theta(\Omega)$ has a uniquely determined degree $G\alpha\in\mathbb{N}$.

2. Sets of Coefficients

We introduce finite sets of coefficients for ordinals $\gamma\in\bar\theta(\Omega)$ in such a way that $\gamma\in C(\alpha,\theta\alpha\beta)$ can be characterised in a useful recursive way.

Inductive Definition of the *coefficient sets* $K_\nu^*\gamma$ and $K_\nu\gamma$ for $\nu\in\mathbb{O}^*$ and $\gamma\in\bar\theta(\Omega)$:

1. $K_\nu^*0=K_\nu 0:=\varnothing$(empty set).
2. $0<\gamma\notin P\Rightarrow K_\nu^*\gamma:=\bigcup_{\gamma_i\in P(\gamma)}K_\nu^*\gamma_i,\ K_\nu\gamma:=\bigcup_{\gamma_i\in P(\gamma)}K_\nu\gamma_i$.

3. $\gamma \in P,\ S\gamma \leqslant \nu \Rightarrow K_\nu^*\gamma := \varnothing,\ K_\nu\gamma := \{\gamma\}$.
4. $\gamma = \bar\theta\xi\eta,\ \nu < S\gamma \Rightarrow K_\nu^*\gamma := \{\xi\} \cup K_\nu^*\xi \cup K_\nu^*\eta,\ K_\nu\gamma := K_\nu\xi \cup K_\nu\eta$.
5. $\gamma = \Omega_\xi,\ \nu < \xi = S\gamma \Rightarrow K_\nu^*\gamma := K_\nu^*\xi,\ K_\nu\gamma := K_\nu\xi$.

Corollaries. 1. *$K_\nu^*\gamma$ is a finite set of ordinals of $\bar\theta(\Omega)$ which is empty if $S\gamma \leqslant \nu$.*
2. *$K_\nu\gamma$ is a finite set of principal numbers for addition from $\bar\theta(\Omega)$ whose levels are $\leqslant \nu$.*

Notations for $M \subset \bar\theta(\Omega)$ and $\alpha \in \mathbb{O}^*$:
1. $M < \alpha$ means that $\xi < \alpha$ for all $\xi \in M$. (If M is empty then $M < \alpha$ for all $\alpha \in \mathbb{O}^*$.)
2. $\alpha \leqslant M$ means that there exists $\xi \in M$ such that $\alpha \leqslant \xi$.

Theorem 25.1. *If $\beta \notin K$ and $\gamma \in \bar\theta(\Omega)$ then $\gamma \in C(\alpha, \theta\alpha\beta)$ holds if, and only if, $K_{S\beta}^*\gamma < \alpha$ and $K_{S\beta}\gamma < \theta\alpha\beta$.*

Proof. By induction on $G\gamma$.
1. $\gamma \notin P$. Then the assertion follows from the I.H. by Theorem 24.3 a).
2. $\gamma \in P$ where $S\gamma \leqslant S\beta$. Then $K_{S\beta}^*\gamma = \varnothing < \alpha$. In this case by Theorem 24.8 a) $\gamma \in C(\alpha, \theta\alpha\beta)$ holds if, and only if, $\gamma < \theta\alpha\beta$ and therefore $K_{S\beta}\gamma = \{\gamma\} < \theta\alpha\beta$.
3. $\gamma = \bar\theta\xi\eta$ where $S\beta < S\gamma$. Then $\theta\alpha\beta < \bar\theta\xi\eta$. In this case by Theorem 24.12 a) $\gamma \in C(\alpha, \theta\alpha\beta)$ holds if, and only if, $\xi < \alpha$ and $\xi, \eta \in C(\alpha, \theta\alpha\beta)$. By I.H. this is the case if, and only if, $K_{S\beta}^*\gamma = \{\xi\} \cup K_{S\beta}^*\xi \cup K_{S\beta}^*\eta < \alpha$ and $K_{S\beta}\gamma = K_{S\beta}\xi \cup K_{S\beta}\eta < \theta\alpha\beta$.
4. $\gamma = \Omega_\xi$ where $S\beta < \xi = S\gamma$. Then the assertion follows from the I.H. by Theorem 24.6 a).

Corollary to Theorem 25.1. *If $S\mu(\alpha) \leqslant S\beta$ and $\gamma \in \bar\theta(\Omega)$ then $\gamma \in C(\alpha, \bar\theta\alpha\beta)$ if, and only if, $K_{S\beta}^*\gamma < \alpha$ and $K_{S\beta}\gamma < \bar\theta\alpha\beta$.*

Proof. There exists $\bar\beta \notin K$ such that $\bar\theta\alpha\beta = \theta\alpha\bar\beta$ and $S\beta = S\bar\beta$. Therefore the assertion holds by Theorem 25.1.

Theorem 25.2. *If $\bar\theta\alpha\beta \in \bar\theta(\Omega)$ then $K_{S\beta}^*\alpha < \alpha$ and $K_{S\beta}\alpha \cup \{\beta\} < \bar\theta\alpha\beta$.*

Proof. By Theorem 24.13 we have $\alpha \in C(\alpha, \bar\theta\alpha\beta)$ and $\beta < \bar\theta\alpha\beta$. Hence the assertion follows by the corollary to Theorem 25.1.

Theorem 25.3. *If $\alpha, \beta \in \bar\theta(\Omega)$ then $S\mu(\alpha) \leqslant S\beta$ if, and only if, $K_{S\beta}^*\alpha < \alpha$.*

Proof. 1. Suppose $S\mu(\alpha) \leqslant S\beta$. Then $\bar\theta\alpha\beta \in \bar\theta(\Omega)$. By Theorem 25.2 we obtain $K_{S\beta}^*\alpha < \alpha$.
2. Suppose $K_{S\beta}^*\alpha < \alpha$. Let γ be the maximum element of the finite set $\{\Omega_{S\beta}\} \cup K_{S\beta}\alpha$, then $K_{S\beta}\alpha < \gamma + 1 \leqslant \theta\alpha(\gamma+1)$, $S\beta = S(\gamma+1)$ and $\gamma + 1 \notin K$. By Theorem 25.1 we obtain $\alpha \in C(\alpha, \theta\alpha(\gamma+1))$. By the definition of $\mu(\alpha)$ it follows that $\mu(\alpha) \leqslant \gamma + 1$. Therefore $S\mu(\alpha) \leqslant S\beta$.

Theorem 25.4. *If $\bar{\theta}\alpha_i\beta_i \in \bar{\theta}(\Omega)$ $(i=1,2)$ where $S\beta_1 = S\beta_2 = \nu$ then $\bar{\theta}\alpha_1\beta_1 < \bar{\theta}\alpha_2\beta_2$ if, and only if, one of the following three conditions holds:*

a) $\alpha_1 < \alpha_2$, $K_\nu\alpha_1 \cup \{\beta_1\} < \bar{\theta}\alpha_2\beta_2$,

b) $\alpha_1 = \alpha_2$, $\beta_1 < \beta_2$,

c) $\alpha_2 < \alpha_1$, $\bar{\theta}\alpha_1\beta_1 \leqslant K_\nu\alpha_2 \cup \{\beta_2\}$.

Proof. Suppose condition a) holds. By Theorem 25.2 $K_\nu^*\alpha_1 < \alpha_1 < \alpha_2$. Using $K_\nu\alpha_1 < \bar{\theta}\alpha_2\beta_2$, by the corollary to Theorem 25.1, we obtain $\alpha_1 \in C(\alpha_2, \bar{\theta}\alpha_2\beta_2)$. Using $\alpha_1 < \alpha_2$ and $\beta_1 < \bar{\theta}\alpha_2\beta_2$, by Theorem 24.14 b) we obtain $\bar{\theta}\alpha_1\beta_1 < \bar{\theta}\alpha_2\beta_2$.

2. Suppose condition b) holds. Then $\bar{\theta}\alpha_1\beta_1 < \bar{\theta}\alpha_2\beta_2$ holds by Theorem 24.11.

3. Suppose condition c) holds. Then, by Theorem 25.2, $\bar{\theta}\alpha_1\beta_1 \leqslant K_\nu\alpha_2 \cup \{\beta_2\} < \bar{\theta}\alpha_2\beta_2$ and therefore $\bar{\theta}\alpha_1\beta_1 < \bar{\theta}\alpha_2\beta_2$.

4. Suppose none of the conditions holds. Then either $\bar{\theta}\alpha_1\beta_1 = \bar{\theta}\alpha_2\beta_2$ or by a)–c) (with subscripts interchanged) $\bar{\theta}\alpha_2\beta_2 < \bar{\theta}\alpha_1\beta_1$.

3. The Systems T* and OT* of Terms

Inductive Definition of the system of terms T* and of the *degree* $Gc \in \mathbb{N}$ of a term $c \in \mathrm{T}^*$.

1. $0 \in \mathrm{T}^*$, $G0 := 0$.

2. If c_i is a term $\bar{\theta}a_ib_i$ or Ω_{a_i} in T* for $i=1,\ldots,n$ $(n \geqslant 2)$ then set $c_1 + \cdots + c_n \in \mathrm{T}^*$ and $G(c_1 + \cdots + c_n) := \max(Gc_1, \ldots, Gc_n) + 1$.

3. $a, b \in \mathrm{T}^* \Rightarrow \bar{\theta}ab \in \mathrm{T}^*$, $G\bar{\theta}ab := \max(Ga, Gb) + 1$.

4. $0 \neq a \in \mathrm{T}^* \Rightarrow \Omega_a \in \mathrm{T}^*$, $G\Omega_a := Ga + 1$. (Here $0 \neq a$ means that 0 and a are terms of different forms).

We call the terms in T* of the form $\bar{\theta}ab$ or Ω_a, *principal terms*.

Inductive Definition of the *notation* $\bar{\gamma} \in \mathrm{T}^*$ for an ordinal $\gamma \in \bar{\theta}(\Omega)$:

1. $\bar{0} := 0$.

2. If $\gamma = \gamma_1 + \cdots + \gamma_n \in \bar{\theta}(\Omega)$ where $\gamma_1 \geqslant \cdots \geqslant \gamma_n$ $(n \geqslant 2)$ are principal numbers for addition, then set $\bar{\gamma} := \bar{\gamma}_1 + \cdots + \bar{\gamma}_n$.

3. If $\gamma = \bar{\theta}\alpha\beta \in \bar{\theta}(\Omega)$, then set $\bar{\gamma} := \bar{\theta}\bar{\alpha}\bar{\beta}$.

4. If $\gamma = \Omega_\alpha \in \bar{\theta}(\Omega)$ where $\alpha \neq 0$, then set $\bar{\gamma} := \Omega_{\bar{\alpha}}$.

This is an inductive definition on the degree $G\gamma$ of the ordinal $\gamma \in \bar{\theta}(\Omega)$. Obviously the term $\bar{\gamma} \in \mathrm{T}^*$ has the same degree as the ordinal γ.

The notation system OT*

Let OT* be the subsystem of T* which contains only ordinal terms $\bar{\gamma} \in \mathrm{T}^*$ which denote ordinals $\gamma \in \bar{\theta}(\Omega)$. This system OT* of terms is a system of unique notations for the set of ordinals $\bar{\theta}(\Omega)$. That is to say: Each term $\bar{\gamma} \in \mathrm{OT}^*$ denotes one and only one, ordinal $\gamma \in \bar{\theta}(\Omega)$ and each ordinal $\gamma \in \bar{\theta}(\Omega)$ is denoted by one, and only one term $\bar{\gamma} \in \bar{\theta}(\Omega)$. The remaining terms in T* which do not belong to OT* are formal strings of symbols which do not denote ordinals.

We set $\mathrm{OT}^* = \bar{\theta}(\Omega)$ whereby we identify each ordinal $\gamma \in \bar{\theta}(\Omega)$ with its notation $\bar{\gamma} \in \mathrm{OT}^*$. Also we now call the ordinals of $\bar{\theta}(\Omega)$ *ordinal terms* (because of the uniqueness of the notations).

We say that two ordinal terms $\gamma_1, \gamma_2 \in \bar{\theta}(\Omega)$ are *recursively comparable* if it can be decided whether $\gamma_1 < \gamma_2$, $\gamma_1 = \gamma_2$ or $\gamma_2 < \gamma_1$.

It is obviously decidable whether any string of symbols is a term of T^*. Below we prove that it is also decidable for every term $c \in \mathrm{T}^*$ whether c is an ordinal term of the system $\bar{\theta}(\Omega)$ and that two ordinal terms from $\bar{\theta}(\Omega)$ are recursively comparable.

If $M \subset \bar{\theta}(\Omega)$ then $GM < n$ ($GM \leqslant n$) means that $G\xi < n$ ($G\xi \leqslant n$) for all $\xi \in M$.

Lemma 1. a) $0 < \alpha \in \bar{\theta}(\Omega) \Rightarrow S\alpha \in \bar{\theta}(\Omega),\ GS\alpha < G\alpha$.

b) $v \in \mathbb{O}^*,\ \alpha \in \bar{\theta}(\Omega) \Rightarrow GK_v^*\alpha < G\alpha,\ GK_v\alpha \leqslant G\alpha$.

Proof. By induction on α.

Theorem 25.5. a) *It is decidable, for* $c \in \mathrm{T}^*$, *whether* $c \in \bar{\theta}(\Omega)$.

b) *Any two ordinal terms* $\gamma_1, \gamma_2 \in \bar{\theta}(\Omega)$ *are recursively comparable.*

Proof. of a) and b) by induction on the maximum of $2^{Gc} + 1$ and $2^{G\gamma_1} + 2^{G\gamma_2}$.

a 1) Suppose c is 0. Then $c \in \bar{\theta}(\Omega)$.

a 2) Suppose c is $c_1 + \cdots + c_n$ ($n \geqslant 2$) with principal terms $c_1, \ldots, c_n$. By I.H. it is decidable, for $i = 1, \ldots, n$, whether $c_i \in \bar{\theta}(\Omega)$. If $c_i \notin \bar{\theta}(\Omega)$, then $c \notin \bar{\theta}(\Omega)$. Now suppose $c_1, \ldots, c_n \in \bar{\theta}(\Omega)$. $2^{Gc_i} + 2^{Gc_{i+1}} < 2^{Gc} + 1$ for $i = 1, \ldots, n-1$ and therefore by I.H. c_i is recursively comparable with c_{i+1}. If $c_1 \geqslant \cdots \geqslant c_n$, then $c \in \bar{\theta}(\Omega)$. Otherwise $c \notin \bar{\theta}(\Omega)$.

a 3) Suppose c is $\bar{\theta}ab$. By I.H. it is decidable whether a and b belong to $\bar{\theta}(\Omega)$. If $a \notin \bar{\theta}(\Omega)$ or $b \notin \bar{\theta}(\Omega)$, then $c \notin \bar{\theta}(\Omega)$. Now suppose $a, b \in \bar{\theta}(\Omega)$. It then follows from I.H. that Sb and $K_{Sb}^* a$ are calculable and it is decidable whether $K_{Sb}^* a < a$. By Theorem 25.3 $c \in \bar{\theta}(\Omega)$ holds if, and only if, $K_{Sb}^* a < a$.

a 4) Suppose c is Ω_a where $a \neq 0$. By I.H. it is decidable whether $a \in \bar{\theta}(\Omega)$. $c \in \bar{\theta}(\Omega)$ holds if, and only if, $a \in \bar{\theta}(\Omega)$.

b 1) Suppose γ_1 or γ_2 is 0. If both are the ordinal term 0, then $\gamma_1 = \gamma_2$. Otherwise the ordinal term 0 is less than the other ordinal term.

b 2) Suppose γ_i is $\gamma_{i1} + \cdots + \gamma_{in_i}$ ($n_i \geqslant 1$) with principal terms $\gamma_{i1} \geqslant \cdots \geqslant \gamma_{in_i}$ ($i = 1, 2$), where $n_1 + n_2 > 2$. Let $m := \min(n_1, n_2)$. By I.H. γ_{1k} and γ_{2k} are recursively comparable for $k = 1, \ldots, m$. If there is a least subscript $j \leqslant m$ such that $\gamma_{1j} \neq \gamma_{2j}$ then $\gamma_1 < \gamma_2$ ($\gamma_2 < \gamma_1$) if $\gamma_{1j} < \gamma_{2j}$ ($\gamma_{2j} < \gamma_{1j}$). Now suppose $\gamma_{1k} = \gamma_{2k}$ for all $k = 1, \ldots, m$. Then $\gamma_1 < \gamma_2$, $\gamma_1 = \gamma_2$ or $\gamma_2 < \gamma_1$ according as $n_1 < n_2$, $n_1 = n_2$ or $n_2 < n_1$.

b 3) Suppose γ_1 and γ_2 are principal terms. Then it follows from the I.H. that $S\gamma_1$ and $S\gamma_2$ are recursively comparable. If $S\gamma_1 < S\gamma_2$ ($S\gamma_2 < S\gamma_1$), then $\gamma_1 < \gamma_2$ ($\gamma_2 < \gamma_1$). Now suppose $S\gamma_1 = S\gamma_2 = v$. If both are the ordinal term Ω_v, then $\gamma_1 = \gamma_2$. If only one of the two ordinal terms is Ω_v, then this one is less than the other ordinal term. There remains the case $\gamma_i = \bar{\theta}\alpha_i\beta_i$ where $S\gamma_i = S\beta_i = v$ ($i = 1, 2$). Then by I.H. α_1 and α_2 are recursively comparable.

b 3.1) $\alpha_1 < \alpha_2$. It follows from the I.H. that $K_\nu\alpha_1$ is calculable and it is decidable whether $K_\nu\alpha_1 \cup \{\beta_1\} < \gamma_2$ holds. By Theorem 25.4, $\gamma_1 < \gamma_2$ if $K_\nu\alpha_1 \cup \{\beta_1\} < \gamma_2$ and $\gamma_2 < \gamma_1$ if $\gamma_2 \leqslant K_\nu\alpha_1 \cup \{\beta_1\}$.

b 3.2) $\alpha_1 = \alpha_2$. By I.H. β_1 and β_2 are recursively comparable. $\gamma_1 < \gamma_2$, $\gamma_1 = \gamma_2$ or $\gamma_2 < \gamma_1$ holds according as $\beta_1 < \beta_2$, $\beta_1 = \beta_2$ or $\beta_2 < \beta_1$.

b 3.3) $\alpha_2 < \alpha_1$. Then it follows in the same way as in b 3.1) that γ_1 and γ_2 are recursively comparable.

4. The Subsystem $\bar{\theta}(\tau)$ of $\bar{\theta}(\Omega)$

Below we assume $0 < \tau \in \mathbb{O}^*$.

Inductive Definition of $\bar{\theta}(\tau) \subset \mathbb{O}^*$:

1. $0 \in \bar{\theta}(\tau)$
2. $0 < \gamma \notin P,\ P(\gamma) \subset \bar{\theta}(\tau) \Rightarrow \gamma \in \bar{\theta}(\tau)$
3. $\alpha, \beta \in \bar{\theta}(\tau),\ S\mu(\alpha) \leqslant S\beta \Rightarrow \bar{\theta}\alpha\beta \in \bar{\theta}(\tau)$
4. $0 < \alpha < \tau,\ \alpha \in \bar{\theta}(\tau) \Rightarrow \Omega_\alpha \in \bar{\theta}(\tau)$

$\bar{\theta}(\tau)$ is obviously a subsystem of $\bar{\theta}(\Omega)$. Therefore every ordinal $\alpha \in \bar{\theta}(\tau)$ has a uniquely determined degree $G\alpha \in \mathbb{N}$.

Theorem 25.6. a) *It is decidable, for* $c \in \mathrm{T}^*$, *whether* $c \in \bar{\theta}(\tau)$.

b) *Any two ordinal terms from* $\bar{\theta}(\tau)$ *are recursively comparable.*

Proof. This follows from Theorem 25.5 since $\bar{\theta}(\tau)$ is a subsystem of $\bar{\theta}(\Omega)$.

Lemma 2.

a) $\alpha \in \bar{\theta}(\tau) \Leftrightarrow P(\alpha) \subset \bar{\theta}(\tau)$

b) $\alpha \in \bar{\theta}(\tau) \Rightarrow \Omega_{S\alpha} \in \bar{\theta}(\tau)$

Proof. By induction on $G\alpha$.

Theorem 25.7. $\bar{\theta}(\tau) = C(\Omega_\tau, 0) \cap \mathbb{O}^*(\Omega_\tau)$.

Proof. 1. Suppose $\gamma \in \bar{\theta}(\tau)$. We prove that $\gamma \in C(\Omega_\tau, 0)$ and $\gamma < \Omega_\tau$ by induction on $G\gamma$.

1.1 $\gamma \notin P$. Then the assertion follows from the I.H.

1.2 $\gamma = \bar{\theta}\alpha\beta$. Then by I.H. $\alpha, \beta \in C(\Omega_\tau, 0)$, $\alpha < \Omega_\tau$ and $\beta < \Omega_\tau$ hold. It follows that $\gamma < \Omega_\tau$ and by Theorem 24.12 b) that $\gamma \in C(\Omega_\tau, 0)$.

1.3 $\gamma = \Omega_\alpha$ where $\alpha \neq 0$. Then $\alpha < \tau$ and by I.H. $\alpha \in C(\Omega_\tau, 0)$. It follows that $\gamma \in C(\Omega_\tau, 0)$ and $\gamma < \Omega_\tau$.

2. Suppose $\gamma \in C(\Omega_\tau, 0)$ and $\gamma < \Omega_\tau$. Then there exists n such that $\gamma \in C_{n+1}(\Omega_\tau, 0)$. We prove by induction on n that $\gamma \in \bar{\theta}(\tau)$.

2.1 Suppose $\gamma \in C_{n+1}(\Omega_\tau, 0)$ holds by (D2). Then the assertion follows from the I.H.

2.2 Suppose $\gamma \in C_{n+1}(\Omega_\tau, 0)$ holds by (D3). Then $\gamma = \theta\alpha\bar{\beta}$ where $\alpha < \Omega_\tau$, $\alpha \in C(\alpha, \theta\alpha\bar{\beta})$, $\bar{\beta} < \theta\alpha\bar{\beta}$ and $\alpha, \bar{\beta} \in C_n(\Omega_\tau, 0)$. By I.H. we obtain $\alpha, \bar{\beta} \in \bar{\theta}(\tau)$. By Theorems 24.10 and 24.11 there exists β such that $S\mu(\alpha) \leqslant S\beta = S\bar{\beta}$ and $\gamma = \bar{\theta}\alpha\beta$. Here

$$\bar{\beta} = \mu(\alpha) + r_\alpha(\beta) + \iota\alpha\beta.$$

It follows from $\bar{\beta} \in \bar{\theta}(\tau)$ by Lemma 2 that $\Omega_{S\beta} \in \bar{\theta}(\tau)$ and $r_\alpha(\beta) \in \bar{\theta}(\tau)$. If $S\mu(\alpha) = S\beta$, then by Lemma 2 a) we have

$$\beta = \Omega_{S\beta} + r_\alpha(\beta) \in \bar{\theta}(\tau).$$

If $S\mu(\alpha) < S\beta$ then

$$\beta = \Omega_{S\mu(\alpha)} + r_\alpha(\beta) = r_\alpha(\beta) \in \bar{\theta}(\tau).$$

In any case we have $\alpha, \beta \in \bar{\theta}(\tau)$ where $S\mu(\alpha) \leqslant S\beta$. It follows that $\gamma = \bar{\theta}\alpha\beta \in \bar{\theta}(\tau)$.

2.3 Suppose $\gamma \in C_{n+1}(\Omega_\tau, 0)$ holds by (D4). Then $\gamma = \Omega_\alpha$ where $0 < \alpha < \tau$. By I.H. we have $\alpha \in \bar{\theta}(\tau)$. Hence it follows that $\gamma \in \bar{\theta}(\tau)$.

Theorem 25.8. $\bar{\theta}(\tau) \cap \mathbb{O}^*(\Omega_1) = \mathbb{O}^*(\theta\Omega_\tau 0)$. *That is to say: The* 0 *level of* $\bar{\theta}(\tau)$ *consists of the segment of all ordinals* $< \theta\Omega_\tau 0$.

Proof. This follows from Theorems 25.7 and 24.8 b).

5. The Ordinal Λ_0

By definition $\bar{\theta}(\Omega)$ is the closure of the set $\{0\}$ under finitely many functions of ordinals. Therefore by Ax. III the set $\bar{\theta}(\Omega) \cap \mathbb{O}^*(\Omega_1)$ has cardinality $< \Omega_1$. It follows by Ax. IV that there is a least ordinal $\notin \bar{\theta}(\Omega)$ which is $< \Omega_1$. We denote this least ordinal by Λ_0. Therefore we have

$$\mathbb{O}^*(\Lambda_0) \subset \bar{\theta}(\Omega), \qquad \Lambda_0 \notin \bar{\theta}(\Omega) \quad \text{and} \quad \Lambda_0 < \Omega_1.$$

Lemma 3. If $\gamma \in \bar{\theta}(\Omega)$ there exists $\tau \neq 0$ such that $\gamma \in \bar{\theta}(\tau)$.

Proof. This follows by induction on $G\gamma$.

Theorem 25.9. $\bar{\theta}(\Omega) \cap \mathbb{O}^*(\Omega_1) = \mathbb{O}^*(\Lambda_0)$. *That is to say: The* 0 *level of* $\bar{\theta}(\Omega)$ *consists of the segment of all ordinals* $< \Lambda_0$.

Proof. Suppose $\gamma \in \bar{\theta}(\Omega) \cap \mathbb{O}^*(\Omega_1)$. By Lemma 3 there exists $\tau \neq 0$ such that $\gamma \in \bar{\theta}(\tau) \cap \mathbb{O}^*(\Omega_1)$. By Theorem 25.8 we obtain

$$\gamma \in \mathbb{O}^*(\theta\Omega_\tau 0) \subset \bar{\theta}(\tau) \subset \bar{\theta}(\Omega).$$

Since $\Lambda_0 \notin \bar{\theta}(\Omega)$, we have $\theta\Omega_\tau 0 \leqslant \Lambda_0$. Then $\gamma \in \mathbb{O}^*(\Lambda_0)$. Thus

$$\bar{\theta}(\Omega) \cap \mathbb{O}^*(\Omega_1) \subseteq \mathbb{O}^*(\Lambda_0).$$

From the properties of Λ_0 we also have

$$\mathbb{O}^*(\Lambda_0) \subseteq \bar{\theta}(\Omega) \cap \mathbb{O}^*(\Omega_1).$$

6. Relations between Cr (α) and In (α)

The set $\mathbb{O}$ of ordinals considered in §13 can be identified with $\mathbb{O}^*(\Omega_1)$. Then, for the set Cr (α) of α-critical numbers defined in §13, we have

$$\beta \in \mathrm{Cr}(\alpha) \Rightarrow \alpha, \beta \in \mathbb{O}^*(\Omega_1).$$

Lemma 4. $\alpha < \theta\Omega_1 0 \Rightarrow \alpha \in C(\alpha, \beta)$.

Proof. By Theorem 24.8 b) from $\alpha < \theta\Omega_1 0$ we have $\alpha \in C(\Omega_1, 0)$. Since Ω_1 is a limit number it follows by Lemma 3 b) of §24 that there is a least ordinal $\eta < \Omega_1$ such that $\alpha \in C(\eta, 0)$. Because of its minimality this ordinal η is not a limit number. If $\eta = 0$, then $\alpha \in C(0,0)$, hence $\alpha \in C(\alpha, \beta)$ by Lemma 3 a) of §24. Otherwise $\eta = \xi + 1$ where $\alpha \notin C(\xi, 0)$ and $\alpha \in C(\xi+1, 0)$. Then $C(\xi, 0) \neq C(\xi+1, 0)$ and consequently by Lemma 9 of §24 $\xi \in C(\xi, 0)$. Since $\xi < \Omega_1$ and $\alpha < \Omega_1$, by Theorem 24.8 b) $\xi \in C(\xi, 0)$ and $\alpha \notin C(\xi, 0)$ imply $\xi < \theta\xi 0 \leqslant \alpha$, and therefore $\xi + 1 \leqslant \alpha$. Then from $\alpha \in C(\xi+1, 0)$ it follows that $\alpha \in C(\alpha, \beta)$ by Lemma 6 a) of §24.

Lemma 5. $\xi < \alpha \leqslant \theta\Omega_1 0,\ \gamma \in \mathrm{In}(\alpha) \Rightarrow \theta\xi\gamma = \gamma$.

Proof. There exists β such that $\gamma = \theta\alpha\beta$. By Lemma 4 $\xi \in C(\xi, \theta\alpha\beta)$. By Theorem 24.7 b) we obtain $\theta\xi(\theta\alpha\beta) = \theta\alpha\beta$, and therefore $\theta\xi\gamma = \gamma$.

Theorem 25.10. *If* $\alpha \leqslant \theta\Omega_1 0$, *then:*

a) $\mathrm{Cr}(\alpha) = \mathrm{In}(\alpha) \cap \mathbb{O}^*(\Omega_1)$

b) $\beta < \Omega_1 \Rightarrow \phi\alpha\beta = \theta\alpha\beta$

Proof of a) by induction on α. By the definition of Cr (0), Cr $(0) = P \cap \mathbb{O}^*(\Omega_1)$. By Theorem 24.4 a) In $(0) = P$. Therefore the assertion holds for $\alpha = 0$. Now suppose $0 < \alpha \leqslant \theta\Omega_1 0$. If $\gamma \in \mathrm{In}(\alpha) \cap \mathbb{O}^*(\Omega_1)$, then for all $\xi < \alpha$, by Lemma 5, we have $\theta\xi\gamma = \gamma$ and consequently by I.H. $\varnothing\xi\gamma = \gamma$. Then $\gamma \in \mathrm{Cr}(\alpha)$ by the definition of Cr (α). Hence

(1) $\mathrm{In}(\alpha) \cap \mathbb{O}^*(\Omega_1) \subseteq \mathrm{Cr}(\alpha)$.

Now suppose $\gamma \in \mathrm{Cr}(\alpha)$. Then for all $\xi < \alpha$, by the definition of Cr (α) we have $\phi\xi\gamma = \gamma$ and consequently by I.H. $\theta\xi\gamma = \gamma$. It follows that $\gamma \in \bigcap_{\xi < \alpha} \mathrm{In}(\xi)$. If α is a limit number, then by Theorem 24.4 a) we have $\gamma \in \mathrm{In}(\alpha)$. Otherwise $\alpha = \alpha_0 + 1$ where $\theta\alpha_0\gamma = \gamma$. By Theorem 24.10 we have $\gamma \notin \mathrm{In}(\alpha_0)$. Since $\gamma \in \mathrm{In}(\alpha_0)$ it follows that $\gamma \in \mathrm{In}(\alpha)$. We therefore also have

(2) $\mathrm{Cr}(\alpha) \subset \mathrm{In}(\alpha)$.

The assertion follows from (1), (2) and $\mathrm{Cr}(\alpha) \subset \mathbb{O}^*(\Omega_1)$.

b) follows from a), since ϕ_α and θ_α are the ordering functions of $\mathrm{Cr}(\alpha)$ and $\mathrm{In}(\alpha)$.

Lemma 6. $\theta\Omega_1 0$ is the least ordinal η such that $\theta\eta 0 = \eta$.

Proof. 1. If $\xi < \theta\Omega_1 0$ then by Lemma 4 $\xi \in C(\xi, 0)$. It follows by Theorem 24.8 b) that $\xi < \theta\xi 0$.

2. If $\theta\Omega_1 0 \leqslant \xi < \Omega_1$, then by Theorem 24.8 a) $\xi \notin C(\Omega_1, \theta\Omega_1 0)$. By Lemma 6a) of §24 we have

$$\forall\xi(\theta\Omega_1 0 \leqslant \xi < \Omega_1 \Rightarrow \xi \notin C(\theta\Omega_1 0, \theta\Omega_1 0)).$$

By Theorem 24.7 a) we obtain $\theta(\theta\Omega_1 0)0 = \theta\Omega_1 0$.

Theorem 25.11. $\Gamma_0 = \theta\Omega_1 0$.

Proof. By definition Γ_0 is the least ordinal η such that $\phi\eta 0 = \eta$. By Theorem 25.10 b) and Lemma 6 we have $\Gamma_0 = \theta\Omega_1 0$.

Theorem 25.12. $\mathrm{OT} = \bar{\theta}(1)$.

Proof. By §13 OT denotes the segment $\mathbb{O}^*(\Gamma_0)$ of the ordinals. By Theorems 25.7 and 25.8

$$\bar{\theta}(1) = C(\Omega_1, 0) \cap \mathbb{O}^*(\Omega_1) = \mathbb{O}^*(\theta\Omega_1 0).$$

Hence, by Theorem 25.11, $\mathrm{OT} = \bar{\theta}(1)$.

§26. Level-Lowering Functions of the Ordinals

In this section we introduce some functions of ordinals which we shall use in §27 for the consistency proof of the formal system PA. Thus here we shall be chiefly concerned with functions which assign ordinal terms of lower levels to certain ordinal terms.

1. Basic Concepts

In the following we suppose $0<\sigma\leqslant\omega$. We consider ordinal terms from the notation system $\bar{\theta}(\sigma+1)$. The maximal level of these ordinal terms is σ. We also regard the ordinal terms $<\sigma$ as natural numbers.

We use the *syntactic symbols*

$\alpha, \beta, \gamma, \delta$ for principal terms of level σ in the system $\bar{\theta}(\sigma+1)$,
$\xi, \eta, \zeta, \mu, \nu$ for arbitrary ordinal terms of the system $\bar{\theta}(\sigma+1)$,
i, j, k for ordinal terms (natural numbers) $<\sigma$.

We shall also use these syntactic symbols with subscripts. For brevity we write

(ξ, η) for $\bar{\theta}\xi\eta$.

Definition of $s_{i+2}\xi\in\bar{\theta}(\sigma+1)$:

1. If $\xi<\Omega_{i+2}$, then set $s_{i+2}\xi:=0$.
2. If $\Omega_{i+2}\leqslant\xi=\xi_1+\cdots+\xi_n$ with principal terms $\xi_1\geqslant\cdots\geqslant\xi_n$ $(n\geqslant 1)$ then set $s_{i+2}\xi:=\xi_1+\cdots+\xi_m$, where m is the largest subscript $\leqslant n$ such that $\Omega_{i+2}\leqslant\xi_m$.

Corollaries. 1. *For each ξ there exists $\eta<\Omega_{i+2}$ such that $\xi=s_{i+2}\xi+\eta$.*

2. $\xi<\eta\Rightarrow s_{i+2}\xi\leqslant s_{i+2}\eta$.
3. $s_{i+2}\xi<s_{i+2}\eta\Rightarrow\xi<s_{i+2}\eta$, $s_{i+2}\xi+\Omega_{i+2}\leqslant s_{i+2}\eta$.

Inductive Definition of $h\xi$:

1. $h0=h\Omega_\sigma:=0$.
2. $0<i<\sigma\Rightarrow h\Omega_i:=i$.
3. $h(\xi_1, \xi_2):=\max\{h\xi_1, h\xi_2\}$.
4. $0<\xi\notin P\Rightarrow h\xi:=\max\{h\xi_i \mid \xi_i\in P(\xi)\}$,

that is to say: $h\xi$ is the least number such that no Ω_k occurs in ξ with $h\xi<k<\sigma$.

We say that η is a *subterm* of ξ if either $\xi\neq\eta\in P(\xi)$ or $\xi=(\eta, \xi_2)$ or $\xi=(\xi_1, \eta)$ or $\xi=\Omega_\eta$.

Lemma 1. a) $h\xi<\sigma$.

b) $S\xi<\sigma\Rightarrow S\xi\leqslant h\xi$.

c) *For every subterm η of ξ, $h\eta\leqslant h\xi$.*

d) *If η is a subterm of ξ such that $S\eta<\sigma$, then $S\eta\leqslant h\xi$.*

Proofs. These follow at once from the definition of $h\xi$.

2. Properties of the Sets of Coefficients

If $M\subset\bar{\theta}(\sigma+1)$ set $K_i^*M:=\bigcup_{\xi\in M}K_i^*\xi$ and $K_iM:=\bigcup_{\xi\in M}K_i\xi$. We denote by $k_i^*\xi$ the maximum of the finite set $K_i^*\xi\cup\{0\}$.

Lemma 2. *If* $i \leqslant k$, *then:*

a) $K_i^* K_k^* \xi \subseteq K_i^* \xi$.

b) $K_i K_k^* \xi \subseteq K_i \xi$.

c) $K_i^* \xi = K_k^* \xi \cup K_i^* K_k \xi$.

Proof. These follow by induction on $G\xi$.

Lemma 3. a) $S\eta = \sigma \Rightarrow \xi < (\xi, \eta)$.

b) $S\xi = \sigma,\ h\xi \leqslant i < \sigma \Rightarrow K_i^* \xi < \xi$.

Proof of a). If $S\eta = \sigma$ then by Theorem 25.2 $K_\sigma \xi < (\xi, \eta)$. Here $K_\sigma \xi = P(\xi)$ since $S\xi \leqslant \sigma$. Therefore $P(\xi) < (\xi, \eta)$ holds. Hence we have $\xi < (\xi, \eta)$ since (ξ, η) is a principal term.

Proof of b) by induction on $G\xi$ using Lemma 1 c), d).

1. $\xi \notin P$. For all $\xi_k \in P(\xi)$ such that $S\xi_k < \sigma$ we have $S\xi_k \leqslant h\xi \leqslant i$ and consequently $K_i^* \xi_k = \varnothing$. For all $\xi_k \in P(\xi)$ such that $S\xi_k = \sigma$ we have $h\xi_k \leqslant h\xi$ and consequently by I.H. $K_i^* \xi_k < \xi_k$. Hence we obtain $K_i^* \xi < \xi$.

2. $\xi = (\xi_1, \xi_2)$. Then $S\xi_2 = \sigma$, $h\xi_2 \leqslant h\xi$ and, by I.H., $K_i^* \xi_2 < \xi_2 < \xi$. If $S\xi_1 < \sigma$, then $\xi_1 < \xi$ and $S\xi_1 \leqslant h\xi \leqslant i$, and consequently $K_i^* \xi_1 = \varnothing$. If $S\xi_1 = \sigma$, then $h\xi_1 \leqslant h\xi$ and consequently by I.H. and a) $K_i^* \xi_1 < \xi_1 < \xi$. In every case $K_i^* \xi = \{\xi_1\} \cup K_i^* \xi_1 \cup K_i^* \xi_2 < \xi$.

3. $\xi = \Omega_\sigma$. Then $K_i^* \xi = K_i^* \sigma = \varnothing$ since $S\sigma = 0$.

Lemma 4. a) $K_i^* k_i^* \xi < k_i^* \xi$.

b) $K_i^* \xi < \xi \Rightarrow K_i^* s_{i+2} \xi < s_{i+2} \xi$.

c) $K_i^* s_{i+2} k_i^* \xi < s_{i+2} k_i^* \xi$.

Proofs. a) By Lemma 2 a) $K_i^* k_i^* \xi \subseteq K_i^* \xi$. Moreover $k_i^* \xi \notin K_i^* k_i^* \xi$. Therefore $K_i^* k_i^* \xi < k_i^* \xi$.

b) If $K_i^* \xi < s_{i+2} \xi$, then also $K_i^* s_{i+2} \xi < s_{i+2} \xi$. Now suppose $s_{i+2} \xi \leqslant K_i^* \xi$. Then since $K_i^* \xi < \xi$ it follows that there is an ordinal term η of least degree such that $s_{i+2} \xi \leqslant K_i^* \eta < \xi$. By the minimality of $G\eta$ we have $\eta \in P$ and $\eta \notin K$, and therefore $\eta = (\eta_1, \eta_2)$ where $k_i^* \eta = \eta_1$ and

$$K_i^* \eta_1 < s_{i+2} \xi \leqslant \eta_1 < \xi.$$

Then $s_{i+2} \xi = s_{i+2} \eta_1$ and consequently $K_i^* s_{i+2} \xi \subseteq K_i^* \eta_1 < s_{i+2} \xi$.

c) follows from a) and b).

Lemma 5. $K_i^* \eta < \xi + \Omega_{i+2} \Rightarrow K_i^* \eta < \xi + \eta$.

Proof by induction on $G\eta$. If $\Omega_{i+2} \leqslant \eta$, then the assertion is trivial. If $\eta \notin P$ then it follows from the I.H. If $\eta \leqslant \Omega_{i+1}$, then $K_i^* \eta = \varnothing < \xi + \eta$. There remains the case $\eta = (\eta_1, \eta_2)$ where $S\eta_2 = i+1$. Then by Lemma 2 c)

$$K_i^* \eta = \{\eta_1\} \cup K_{i+1}^* \eta_1 \cup K_i^* (K_{i+1} \eta_1 \cup \{\eta_2\}).$$

If $\xi \in K_{i+1}\eta_1 \cup \{\eta_2\}$, then $G\xi < G\eta$, $K_i^*\xi \subset K_i^*\eta < \xi + \Omega_{i+2}$ and therefore by I.H. and Theorem 25.2 $K_i^*\xi < \xi + \xi < \xi + \eta$.

By Theorem 25.2 we also have $K_{i+1}^*\eta_1 < \eta_1$. It therefore remains only to prove $\eta_1 < \xi + \eta$. If $\eta_1 \leqslant \xi$, then $\eta_1 < \xi + \eta$. Now suppose $\xi < \eta_1$. Then from

$$\eta_1 \in K_i^*\eta < \xi + \Omega_{i+2},$$

it follows that there exists $\eta_0 < \Omega_{i+2}$ such that $\eta_1 = \xi + \eta_0$. Hence we obtain $K_{i+1}\eta_0 \subset K_{i+1}\eta_1 < \eta$, where $K_{i+1}\eta_0 = P(\eta_0)$ since $\eta_0 < \Omega_{i+2}$. Therefore $P(\eta_0) < \eta$ holds. Since η is a principal term, we obtain $\eta_0 < \eta$. Then $\eta_1 = \xi + \eta_0 < \xi + \eta$ as was to be proved.

Lemma 6. a) $\xi < \Omega_{i+2}$, $K_{i+1}^*\eta < \eta$, $(\eta, \Omega_{i+1}) \leqslant \xi \Rightarrow \eta \leqslant K_i^*\xi$.
b) $K_i^*\eta < \eta$, $\xi < (\eta, \Omega_{i+1}) \Rightarrow K_i^*\xi < \eta$.

Proofs by induction on $G\xi$.
a) If $\xi \notin P$, then the assertion follows from the I.H. Otherwise $\xi = (\xi_1, \xi_2)$ where $S\xi_2 = i+1$. Then by Lemma 2 c)

$$K_i^*\xi = \{\xi_1\} \cup K_{i+1}^*\xi_1 \cup K_i^*(K_{i+1}\xi_1 \cup \{\xi_2\}).$$

If $\eta \leqslant \xi_1$, then the assertion $\eta \leqslant K_i^*\xi$ holds. If $\xi_1 < \eta$, then from $(\eta, \Omega_{i+1}) \leqslant \xi = (\xi_1, \xi_2)$, by Theorem 25.4 we obtain

$$(\eta, \Omega_{i+1}) \leqslant K_{i+1}\xi_1 \cup \{\xi_2\}.$$

Then it follows by I.H. that $\eta \leqslant K_i^*(K_{i+1}\xi_1 \cup \{\xi_2\})$ and therefore also, $\eta \leqslant K_i^*\xi$.
b) If $\xi \notin P$ the assertion follows from the I.H. If $\xi \leqslant \Omega_{i+1}$, then $K_i^*\xi = \varnothing < \eta$. There remains the case $\xi = (\xi_1, \xi_2)$ where $S\xi_2 = i+1$. Then, by Lemma 2 c)

$$K_i^*\xi = \{\xi_1\} \cup K_{i+1}^*\xi_1 \cup K_i^*(K_{i+1}\xi_1 \cup \{\xi_2\}).$$

By Theorem 25.2 $K_{i+1}^*\xi_1 < \xi_1$ and $K_{i+1}\xi_1 \cup \{\xi_2\} < \xi$. It follows by the I.H. that

$$K_i^*(K_{i+1}\xi_1 \cup \{\xi_2\}) < \eta.$$

We therefore only have to prove $\xi_1 < \eta$. Assume that $\eta \leqslant \xi_1$. Since

$$(\xi_1, \Omega_{i+1}) \leqslant (\xi_1, \xi_2) = \xi < (\eta, \Omega_{i+1}),$$

by Theorem 25.4, we obtain $(\xi_1, \Omega_{i+1}) \leqslant K_{i+1}\eta$. Then there exists $\eta_0 \in K_{i+1}\eta$ such that $\eta_0 < \Omega_{i+2}$ and $(\xi_1, \Omega_{i+1}) \leqslant \eta_0$. By a) (with η_0, ξ_1 instead of ξ, η), using $K_{i+1}^*\xi_1 < \xi_1$ we obtain $\xi_1 \leqslant K_i^*\eta_0 \subseteq K_i^*\eta < \eta$. Hence $\xi_1 < \eta$.

3. The Ordinal Term $d_i\alpha$

Now, for each number $i<\sigma$, we define a map d_i which assigns a principal term $d_i\alpha$ of level i to each principal term α of level σ. We shall use these in §27.

Inductive definition of $d_i\alpha$:

1. If $h\alpha \leqslant i < \sigma$, set $d_i\alpha := (\alpha, \Omega_i)$.
2. If $i < h\alpha$, set

$$d_i\alpha := \begin{cases} (s_{i+2}k_i^* d_{i+1}\alpha, \Omega_i), \text{ if } d_{i+1}\alpha = (s_{i+2}k_i^* d_{i+1}\alpha, \Omega_{i+1}) \\ (s_{i+2}k_i^* d_{i+1}\alpha + d_{i+1}\alpha, \Omega_i) \text{ otherwise.} \end{cases}$$

(The definition is by induction on $h\alpha - i$).

According to this definition $d_i\alpha$ is always a term of the system T*. We prove:

Theorem 26.1. *$d_i\alpha$ is a principal term of level i in the system $\bar{\theta}(\sigma+1)$.*

It suffices to prove that $d_i\alpha \in \bar{\theta}(\sigma+1)$. If $h\alpha \leqslant i < \sigma$, then by Lemma 3 b) $K_i^*\alpha < \alpha$ and consequently $d_i\alpha = (\alpha, \Omega_i) \in \bar{\theta}(\sigma+1)$. We prove the assertion by induction on $h\alpha - i$ for $i < h\alpha$. In this case $d_i\alpha = (\xi, \Omega_i)$ where

$$\xi = s_{i+2}k_i^* d_{i+1}\alpha \quad \text{or} \quad \xi = s_{i+2}k_i^* d_{i+1}\alpha + d_{i+1}\alpha.$$

Then, by I.H., $d_{i+1}\alpha \in \bar{\theta}(\sigma+1)$ and consequently $\xi \in \bar{\theta}(\sigma+1)$ too. By Lemma 4 c)

$$(1) \quad K_i^* s_{i+2}k^* d_{i+1}\alpha < s_{i+2}k_i^* d_{i+1}\alpha.$$

$$K_i^* d_{i+1}\alpha < s_{i+2}k_i^* d_{i+1}\alpha + \Omega_{i+2}$$

also holds. By Lemma 5 we obtain

$$(2) \quad K_i^* d_{i+1}\alpha < s_{i+2}k_i^* d_{i+1}\alpha + d_{i+1}\alpha.$$

From (1) and (2) we obtain $K_i^*\xi < \xi$ and therefore $d_i\alpha = (\xi, \Omega_i) \in \bar{\theta}(\sigma+1)$.

Lemma 7. $(\Omega_{i+1}, \Omega_i) \leqslant d_i\alpha$.

Proof. 1. If $h\alpha \leqslant i$, then $d_i\alpha = (\alpha, \Omega_i)$ and $\Omega_{i+1} \leqslant \alpha$. It follows that $(\Omega_{i+1}, \Omega_i) \leqslant d_i\alpha$.

2. We prove the assertion by induction on $h\alpha - i$ for $i < h\alpha$. By the definition there are the following two cases to be considered.

2.1 $d_i\alpha = (\xi, \Omega_i)$ where $d_{i+1}\alpha = (\xi, \Omega_{i+1})$. Then by I.H. $(\Omega_{i+2}, \Omega_{i+1}) \leqslant d_{i+1}\alpha = (\xi, \Omega_{i+1})$. $d_i\alpha = (\xi, \Omega_i) \in \bar{\theta}(\sigma+1)$ implies $K_i^*\xi < \xi$. By Lemma 6 from $(\xi, \Omega_{i+1}) < \Omega_{i+2}$, $K_i^*\Omega_{i+2} < \Omega_{i+2}$ and $(\Omega_{i+2}, \Omega_{i+1}) \leqslant (\xi, \Omega_{i+1})$ we therefore have $\Omega_{i+2} \leqslant K_i^*(\xi, \Omega_{i+1})$. Hence $\Omega_{i+2} \leqslant \xi$ and $(\Omega_{i+1}, \Omega_i) < (\xi, \Omega_i) = d_i\alpha$.

2.2 $d_i\alpha = (\xi, \Omega_i)$ where $\xi = s_{i+2}k_i^* d_{i+1}\alpha + d_{i+1}\alpha$. Then $\Omega_{i+1} \leqslant d_{i+1}\alpha \leqslant \xi$. Hence $(\Omega_{i+1}, \Omega_i) \leqslant d_i\alpha$.

Lemma 8. $i<j \Rightarrow K_i d_j\alpha = K_i\alpha \cup \{1\}$.

Proof. 1. If $h\alpha \leqslant j$ then

$$K_i d_j\alpha = K_i(\alpha, \Omega_j) = K_i\alpha \cup \{1\},$$

since $K_i\Omega_j = K_i j = \{1\}$.

2. We prove the assertion by induction on $h\alpha - j$ for $j < h\alpha$.

2.1. $d_j\alpha = (\xi, \Omega_j)$ where $d_{j+1}\alpha = (\xi, \Omega_{j+1})$. Then $K_i d_j\alpha = K_i d_{j+1}\alpha$ and consequently by I.H. $K_i d_j\alpha = K_i\alpha \cup \{1\}$.

2.2. $d_j\alpha = (\xi, \Omega_j)$ where $\xi = s_{j+2}k_j^* d_{j+1}\alpha + d_{j+1}\alpha$. By Lemma 2 b)

$$K_i s_{j+2} k_j^* d_{j+1}\alpha \subseteq K_i d_{j+1}\alpha.$$

It follows that $K_i\xi = K_i d_{j+1}\alpha$ and by I.H. $K_i\xi = K_i\alpha \cup \{1\}$. Then we also have $K_i d_j\alpha = K_i\alpha \cup \{1\}$.

Theorem 26.2. $K_i\alpha < d_i\alpha$.

Proof. 1. If $h\alpha \leqslant i$ then $d_i\alpha = (\alpha, \Omega_i)$ and, by Theorem 25.2, $K_i\alpha < d_i\alpha$.

2. If $i < h\alpha$ we have $d_i\alpha = (\xi, \Omega_i)$ where, by Theorem 25.2 $K_i\xi < d_i\alpha$.

2.1. $d_{i+1}\alpha = (\xi, \Omega_{i+1})$. Then, by Lemma 8,

$$K_i\alpha \cup \{1\} = K_i d_{i+1}\alpha = K_i\xi \cup \{1\}.$$

It follows from Lemma 7 that $1 < d_i\alpha$. Using $K_i\xi < d_i\alpha$ we obtain $K_i\alpha < d_i\alpha$.

2.2. $\xi = s_{i+2}k_i^* d_{i+1}\alpha + d_{i+1}\alpha$. Then, by Lemma 2 b) and 8)

$$K_i\xi = K_i d_{i+1}\alpha = K_i\alpha \cup \{1\}.$$

Therefore from $K_i\xi < d_i\alpha$ we obtain $K_i\alpha < d_i\alpha$.

Lemma 9. *If* $h\alpha < i+1 < \sigma$, *then* $d_i\alpha = (s_{i+2}k_i^* d_{i+1}\alpha, \Omega_i)$ *and* $d_{i+1}\alpha = (s_{i+2}k_i^* d_{i+1}\alpha, \Omega_{i+1})$.

Proof. If $h\alpha < i+1 < \sigma$, then $d_i\alpha = (\alpha, \Omega_i)$, $d_{i+1}\alpha = (\alpha, \Omega_{i+1})$, $K_i^* d_{i+1}\alpha = \{\alpha\} \cup K_i^*\alpha$ and by Lemma 3 b) $K_i^*\alpha < \alpha$. It follows that $k_i^* d_{i+1}\alpha = \alpha$ and, since α is a principal term of level $\sigma \geqslant i+2$, we also have $s_{i+2}k_i^* d_{i+1}\alpha = \alpha$. This yields the assertions.

Lemma 10. *If* $i+1 < \sigma$, $d_{i+1}\alpha < d_{i+1}\beta$ *and* $K_i\alpha < d_i\beta$, *then* $d_i\alpha < d_i\beta$.

Proof. By the definition of $d_i\alpha$ and Lemma 9 $d_i\alpha = (\xi, \Omega_i)$ where

$$\xi = s_{i+2}k_i^* d_{i+1}\alpha \quad \text{or} \quad \xi = s_{i+2}k_i^* d_{i+1}\alpha + d_{i+1}\alpha.$$

By Lemmata 2 b) and 8

$$K_i\xi \subseteq K_i d_{i+1}\alpha = K_i\alpha \cup \{1\}.$$

It follows from Lemma 7 that $1<d_i\beta$. Using the hypothesis $K_i\alpha<d_i\beta$ we obtain $K_i\xi<d_i\beta$. We have $d_i\beta=(\eta,\Omega_i)$. We prove $\xi<\eta$. Then using $K_i\xi<d_i\beta$ and $\Omega_i<d_i\beta$, by Theorem 25.4, the assertion $d_i\alpha<d_i\beta$ follows. By the definition of $d_i\beta$ and Lemma 9 it follows that there are two cases for η.

1. $d_{i+1}\beta=(\eta,\Omega_{i+1})$ and $\eta=s_{i+2}\eta$. Then $K_i^*\eta<\eta$. Therefore from $d_{i+1}\alpha<d_{i+1}\beta=(\eta,\Omega_{i+1})$ we obtain by Lemma 6 b) $K_i^*d_{i+1}\alpha<\eta$ and therefore $s_{i+2}k_i^*d_{i+1}\alpha<\eta$ also. Since $d_{i+1}\alpha<\Omega_{i+2}$ and $\eta=s_{i+2}\eta$ we obtain

$$\xi\leqslant s_{i+2}k_i^*d_{i+1}\alpha+d_{i+1}\alpha<\eta.$$

2. $\eta=s_{i+2}k_i^*d_{i+1}\beta+d_{i+1}\beta$.
By Lemma 4 c)

$$K_i^*s_{i+2}k_i^*d_{i+1}\alpha<s_{i+2}k_i^*d_{i+1}\alpha.$$

Were $d_{i+1}\alpha<(s_{i+2}k_i^*d_{i+1}\alpha,\Omega_{i+1})$ then by Lemma 6 b) we should obtain $K_i^*d_{i+1}\alpha<s_{i+2}k_i^*d_{i+1}\alpha$ which is false. Hence $(s_{i+2}k_i^*d_{i+1}\alpha,\Omega_{i+1})\leqslant d_{i+1}\alpha$. Using the hypothesis $d_{i+1}\alpha<d_{i+1}\beta$ we obtain $(s_{i+2}k_i^*d_{i+1}\alpha,\Omega_{i+1})<d_{i+1}\beta$. Now $d_{i+1}\beta<\Omega_{i+2}$. Therefore by Lemma 6 a) we obtain

$$s_{i+2}k_i^*d_{i+1}\alpha\leqslant K_i^*d_{i+1}\beta.$$

Then

$$s_{i+2}k_i^*d_{i+1}\alpha\leqslant s_{i+2}k_i^*d_{i+1}\beta.$$

too. Using the hypothesis $d_{i+1}\alpha<d_{i+1}\beta$ we obtain

$$\xi\leqslant s_{i+2}k_i^*d_{i+1}\alpha+d_{i+1}\alpha<s_{i+2}k_i^*d_{i+1}\beta+d_{i+1}\beta=\eta.$$

Lemma 11. *If $\alpha<\beta$ and $K_i\alpha<d_i\beta$ for all $i\geqslant j$, then $d_i\alpha<d_i\beta$ holds for all $i\geqslant j$.*

Proof. 1. $\max(j,h\alpha,h\beta)\leqslant i$. Then $d_i\alpha=(\alpha,\Omega_i)$ and $d_i\beta=(\beta,\Omega_i)$. In this case by Theorem 25.4 from the hypotheses $\alpha<\beta$ and $K_i\alpha<d_i\beta$ we obtain $d_i\alpha<d_i\beta$ by using $\Omega_i<d_i\beta$.

2. $j\leqslant i<\max(h\alpha,h\beta)$. Then we obtain $d_i\alpha<d_i\beta$ by Lemma 10 by induction on $\max(h\alpha,h\beta)-i$.

Definition. $\alpha\ll\beta$ (α is *essentially less* than β) denotes that $\alpha<\beta$ and $K_i\alpha<d_i\beta$ for all $i<\sigma$.

Theorem 26.3. *If $\alpha\ll\beta$, then $d_i\alpha<d_i\beta$ for all $i<\sigma$.*

Proof. This holds by Lemma 11 with $j=0$.

Theorem 26.4. $\alpha\ll\beta,\ \beta\ll\gamma\Rightarrow\alpha\ll\gamma$.

Proof. $\beta\ll\gamma$ implies $\beta<\gamma$ and by Theorem 26.3 $d_i\beta<d_i\gamma$ for all $i<\sigma$. Using $\alpha\ll\beta$ we obtain $\alpha\ll\gamma$.

4. The Natural Sum

Definition of the *natural sum* $\xi \# \eta$:

1. $\xi \# 0 = 0 \# \xi := \xi$.

2. If $\xi = \zeta_1 + \cdots + \zeta_m$ where $\zeta_1 \geqslant \cdots \geqslant \zeta_m$ $(m \geqslant 1)$ are principal terms and $\eta = \zeta_{m+1} + \cdots + \zeta_{m+n}$ where $\zeta_{m+1} \geqslant \cdots \geqslant \zeta_{m+n}$ $(n \geqslant 1)$ are principal terms then set

$$\xi \# \eta := \zeta_{\pi(1)} + \cdots + \zeta_{\pi(m+n)},$$

where π is a permutation of the numbers $1, \ldots, m+n$ such that $\zeta_{\pi(1)} \geqslant \cdots \geqslant \zeta_{\pi(m+n)}$.

Corollaries. 1. $\xi \# \eta = \eta \# \xi$.

2. $(\xi \# \eta) \# \zeta = \xi \# (\eta \# \zeta)$.

3. $\xi < \eta \Rightarrow \xi \# \zeta < \eta \# \zeta$.

4. $K_i^*(\xi \# \eta) = K_i^* \xi \cup K_i^* \eta$, $K_i(\xi \# \eta) = K_i \xi \cup K_i \eta$.

Theorem 26.5. a) $\alpha \ll (\nu, \alpha \# \xi)$

b) $\alpha \ll \beta \Rightarrow (\nu, \alpha \# \xi) \ll (\nu, \beta \# \xi)$.

c) $\mu < \nu < \Omega_1$, $\alpha \ll (\nu, \beta) \Rightarrow (\mu, \alpha) \ll (\nu, \beta)$.

Proofs. a) $\alpha \leqslant \alpha \# \xi < (\nu, \alpha \# \xi)$ and $K_i \alpha \subseteq K_i(\nu, \alpha \# \xi)$ hold. By Theorem 26.2 we obtain $K_i \alpha < d_i(\nu, \alpha \# \xi)$. Hence we have $\alpha \ll (\nu, \alpha \# \xi)$.

b) $\alpha < \beta$ implies $(\nu, \alpha \# \xi) < (\nu, \beta \# \xi)$. By a) $\beta \ll (\nu, \beta \# \xi)$. By Theorem 26.4, using $\alpha \ll \beta$, we obtain $\alpha \ll (\nu, \beta \# \xi)$ and therefore $K_i \alpha < d_i(\nu, \beta \# \xi)$. By Theorem 26.2 $K_i \nu \cup K_i \xi \subseteq K_i(\nu, \beta \# \xi)$ implies $K_i \nu \cup K_i \xi < d_i(\nu, \beta \# \xi)$. Hence

$$K_i(\nu, \alpha \# \xi) = K_i \nu \cup K_i \alpha \cup K_i \xi < d_i(\nu, \beta \# \xi).$$

Hence we obtain $(\nu, \alpha \# \xi) \ll (\nu, \beta \# \xi)$.

c) Using $K_\sigma \mu = P(\mu) < (\nu, \beta)$ from $\mu < \nu < \Omega_1$ and $\alpha < (\nu, \beta)$, by Theorem 25.4, we obtain $(\mu, \alpha) < (\nu, \beta)$. By Theorem 26.2 from $K_i \nu \subseteq K_i(\nu, \beta)$ we obtain $K_i \nu < d_i(\nu, \beta)$. Hence we obtain $\nu < d_i(\nu, \beta)$, since $K_i \nu = P(\nu)$ and $d_i(\nu, \beta)$ is a principal term. Using $\mu < \nu$ we obtain $K_i \mu = P(\mu) < d_i(\nu, \beta)$. Using $\alpha \ll (\nu, \beta)$ we obtain

$$K_i(\mu, \alpha) = K_i \mu \cup K_i \alpha < d_i(\nu, \beta).$$

Hence we have $(\mu, \alpha) \ll (\nu, \beta)$.

5. Deduction Functions

We define some functions which map the set of principal terms of level σ into itself. We shall use these in §27.

Inductive definition of the *deduction functions* of rank $\rho \leqslant \sigma$.

1. The identity function on the set of principal terms of level σ is a deduction function of rank ρ and degree 0.

2. If f is a deduction function of rank ρ and degree n and

$$g(\alpha)=(\nu, f(\alpha)\#\xi),$$

then g is a deduction function of rank ρ and degree $n+1$.

3. If f is a deduction function of rank ρ and degree n and

$$g(\alpha)=(0, \Omega_\sigma+d_k f(\alpha)) \quad \text{where } \rho\leqslant k<\sigma,$$

then g is a deduction function of rank ρ and degree $n+1$.

According to this definition every deduction function of rank $\rho\neq 0$ is also a deduction function of every rank $<\rho$. Therefore every deduction function is a deduction function of rank 0.

If $\rho\leqslant\sigma$ we denote the set of numbers i such that $\rho\leqslant i<\sigma$ by $[\rho, \sigma)$. (If $\rho=\sigma$ this set is empty).

Lemma 12. *If g is a deduction function of rank $\rho\leqslant\sigma$, $\alpha<\beta$ and $K_i\alpha<d_i\beta$ for all $i\in[\rho, \sigma)$, then $g(\alpha)<g(\beta)$ and $K_i g(\alpha)<d_i g(\beta)$ for all $i\in[\rho, \sigma)$.*

Proof by induction on the degree of the deduction function g.

1. $g(\alpha)=\alpha$. Then the assertion is trivial.

2. $g(\alpha)=(\nu, f(\alpha)\#\xi)$, where, by I.H., $f(\alpha)<f(\beta)$ and $K_i f(\alpha)<d_i f(\beta)$ hold for all $i\in[\rho, \sigma)$. $f(\alpha)<f(\beta)$ implies $g(\alpha)<g(\beta)$. By Theorem 26.5 a) $f(\beta)\ll g(\beta)$. Hence by Theorem 26.3 we obtain $d_i f(\beta)<d_i g(\beta)$ and by the I.H. $K_i f(\alpha)<d_i g(\beta)$ for all $i\in[\rho, \sigma)$. By Theorem 26.2 from $K_i\nu\cup K_i\xi\subseteq K_i g(\beta)$ we obtain $K_i\nu\cup K_i\xi<d_i g(\beta)$. Hence we obtain

$$K_i g(\alpha)=K_i\nu\cup K_i f(\alpha)\cup K_i\xi<d_i g(\beta)$$

for all $i\in[\rho, \sigma)$.

3. $g(\alpha)=(0, \Omega_\sigma+d_k f(\alpha))$ where $\rho\leqslant k<\sigma$ and where, by I.H., $f(\alpha)<f(\beta)$ and $K_i f(\alpha)<d_i f(\beta)$ holds for all $i\in[\rho, \sigma)$. By Lemma 11 we obtain $d_i f(\alpha)<d_i f(\beta)$ for all $i\in[\rho, \sigma)$ and therefore, in particular, $d_k f(\alpha)<d_k f(\beta)$. Hence we obtain $g(\alpha)<g(\beta)$. We now prove $K_i g(\alpha)<d_i g(\beta)$ for $i\in[\rho, \sigma)$.

3.1. $k\leqslant i<\sigma$. Then $K_i g(\beta)=K_i\sigma\cup\{d_k f(\beta)\}$. By Theorem 26.2 we obtain $K_i\sigma\cup\{d_k f(\beta)\}<d_i g(\beta)$. Using $d_k f(\alpha)<d_k f(\beta)$ we obtain

$$K_i g(\alpha)=K_i\sigma\cup\{d_k f(\alpha)\}<d_i g(\beta).$$

3.2. $\rho\leqslant i<k$. Then by Lemma 8

$$K_i g(\beta)=K_i\sigma\cup K_i d_k f(\beta)=K_i\sigma\cup K_i f(\beta)\cup\{1\}.$$

By Theorem 26.2 we obtain

$$K_i\sigma\cup K_i f(\beta)\cup\{1\}<d_i g(\beta).$$

By 3.1. $d_k f(\beta) < d_k g(\beta)$. Using $K_i f(\beta) < d_i g(\beta)$, by Lemma 10 using induction on $k-i$ we obtain $d_i f(\beta) < d_i g(\beta)$, Hence

$$K_i g(\alpha) = K_i \sigma \cup K_i f(\alpha) \cup \{1\} < d_i g(\beta)$$

Theorem 26.6. *$\alpha \ll \beta$ implies $g(\alpha) \ll g(\beta)$ for every deduction function g.*

Proof. This holds by Lemma 12 (with $\rho = 0$).

Lemma 13. *If g is a deduction function of rank $\rho \leqslant \sigma$, then*

$$K_i \alpha \subseteq K_i g(\alpha) \subseteq K_i \alpha \cup K_i g(\beta)$$

for all $i < \rho$.

Proof. By induction on the degree of the deduction function g.

1. $g(\alpha) = \alpha$. Then the assertion is trivial.
2. $g(\alpha) = (\nu, f(\alpha) \# \xi)$ where, by I.H.,

$$K_i \alpha \subseteq K_i f(\alpha) \subseteq K_i \alpha \cup K_i f(\beta)$$

holds for all $i < \rho$. In this case

$$K_i g(\alpha) = K_i \nu \cup K_i f(\alpha) \cup K_i \xi.$$

Hence the assertion follows from the I.H.

3. $g(\alpha) = (0, \Omega_\sigma + d_k f(\alpha))$ with $\rho \leqslant k < \sigma$, where, by I.H.,

$$K_i \alpha \subseteq K_i f(\alpha) \subseteq K_i \alpha \cup K_i f(\beta)$$

for all $i < \rho$. Then by Lemma 8

$$K_i g(\alpha) = K_i \sigma \cup K_i d_k f(\alpha) = K_i \sigma \cup K_i f(\alpha) \cup \{1\}$$

for $i < \rho$. Hence the assertion follows from the I.H.

We shall use the following theorem in §27.

Theorem 26.7. *If $\alpha = (\nu, (0, \Omega_\sigma + d_k \gamma) \# \delta)$, $\beta = (\nu + 1, \delta)$ and g is a deduction function of rank $k+1 \leqslant \sigma$ such that $\gamma \ll g(\beta)$, then $g(\alpha) \ll g(\beta)$.*

Proof. $d_k \gamma < \Omega_\sigma < (\nu+1, \delta)$ implies $\Omega_\sigma + d_k \gamma < (\nu+1, \delta)$. Then

$$(0, \Omega_\sigma + d_k \gamma) < (\nu+1, \delta)$$

too. Using $\delta < (\nu+1, \delta)$ we obtain

$$(0, \Omega_\sigma + d_k \gamma) \# \delta < (\nu+1, \delta) = \beta.$$

By Theorem 25.2 $K_\sigma \nu \subseteq K_\sigma(\nu+1) < (\nu+1, \delta)$. By Theorem 25.4 we obtain

$$\alpha = (\nu, (0, \Omega_\sigma + d_k\gamma) \# \delta) < (\nu+1, \delta) = \beta.$$

We now prove that $g(\alpha) < g(\beta)$ and $K_i g(\alpha) < d_i g(\beta)$ for all $i < \sigma$.

1. Suppose $k+1=\sigma$. By Lemma 12 $\alpha < \beta$ implies $g(\alpha) < g(\beta)$.

2. Suppose $k < i < \sigma$. By Theorem 26.2 from $K_i\nu \cup K_i\delta \subseteq K_i\beta$ we obtain $K_i\nu \cup K_i\delta < d_i\beta$. Using

$$K_i\sigma \cup \{d_k\gamma\} < \Omega_i < d_i\beta$$

we obtain

$$K_i\alpha = K_i\nu \cup K_i\sigma \cup \{d_k\gamma\} \cup K_i\delta < d_i\beta.$$

Since this holds for all $i \in [k+1, \sigma)$ using $\alpha < \beta$ we obtain $g(\alpha) < g(\beta)$ and $K_i g(\alpha) < d_i g(\beta)$ by Lemma 12.

3. Suppose $i \leqslant k$. Then $K_i\alpha = K_i\nu \cup K_i\sigma \cup K_i d_k\gamma \cup K_i\delta$ and consequently by Lemma 13

$$K_i g(\alpha) \subseteq K_i\nu \cup K_i\sigma \cup K_i d_k\gamma \cup K_i\delta \cup K_i g(\beta).$$

By Lemma 13 from $K_i\nu \cup K_i\delta \subseteq K_i\beta$ we obtain $K_i\nu \cup K_i\delta \subseteq K_i g(\beta)$. Thus

(1') $K_i g(\alpha) \subseteq K_i\sigma \cup K_i d_k\gamma \cup K_i g(\beta)$.

By Lemma 7 $\sigma \leqslant \omega < (\Omega_1, 0)$ implies $\sigma < d_i g(\beta)$. Thus

(2) $K_i\sigma < d_i g(\beta)$

too. From $\gamma \ll g(\beta)$ we obtain $K_i\gamma < d_i g(\beta)$ and by Theorem 26.3 we also obtain $d_k\gamma < d_k g(\beta)$. If $i=k$ we obtain

$$K_i d_k\gamma = \{d_k\gamma\} < d_i g(\beta).$$

If $i<k$, then by Lemma 8

$$K_i d_k\gamma = K_i\gamma \cup \{1\} < d_i g(\beta).$$

In every case

(3) $K_i d_k\gamma < d_i g(\beta)$.

By Theorem 26.2

(4) $K_i g(\beta) < d_i g(\beta)$.

also holds. $K_i g(\alpha) < d_i g(\beta)$ follows from (1)–(4).

§27. The Formal System GPA for a Generalized Π_1^1-Analysis

G. Takeuti [4] proved the consistency of subsystems of classical analysis in which the comprehension axiom is essentially restricted to Π_1^1-formulas. In this section we prove the consistency of a corresponding formal system GPA following the proof of Takeuti.

1. The Formal Language

We use the formal language of classical second order arithmetic described in §20.1 with the syntactic symbols listed there.

Inductive definition of the set $\mathrm{Pr}(F)$ of free predicate variables which are in the scope of a predicate quantifier:

1. If F is a prime formula or a formula $U(t)$, then $\mathrm{Pr}(F)$ is empty.
2. $\mathrm{Pr}(A \to B) := \mathrm{Pr}(A) \cup \mathrm{Pr}(B)$.
3. If F is a formula $\forall x \mathscr{F}[x]$ or $\lambda x \mathscr{F}[x](t)$, then $\mathrm{Pr}(F) := \mathrm{Pr}(\mathscr{F}[0])$.
4. If F is a formula $\forall X \mathscr{F}[X]$, then $\mathrm{Pr}(F)$ is the set of free predicate variables occurring in F.

Inductive definition of *weak formulas*.

1. Every prime formula and every formula $U(t)$ is a weak formula.
2. If A and B are weak formulas, then $(A \to B)$ is a weak formula too.
3. If $\mathscr{F}[0]$ is a weak formula in which the bound number variable x does not occur, then $\forall x \mathscr{F}[x]$ and $\lambda x \mathscr{F}[x](t)$ are weak formulas too.
4. If $\mathscr{F}[U]$ is a weak formula in which the bound predicate variable X does not occur and $U \notin \mathrm{Pr}(\mathscr{F}[U])$, then $\forall X \mathscr{F}[X]$ is a weak formula too.

G. Takeuti called the weak formulas which we have defined here *isolated.* Amongst these formulas there are, in particular, the Π_1^1-formulas that is to say the formulas $\forall X \mathscr{F}[X]$ where no bound predicate variables occur in $\mathscr{F}$.

By *weak predicators* we mean the free predicate variables and the predicators of the form $\lambda x \mathscr{F}[x]$ where $\mathscr{F}[0]$ is a weak formula.

We call formulas and predicators, which are not weak, *strong*.

Inductive definition of the *degree* $\mathrm{gr}(F)$ of a formula F:

1. If F is a prime formula or a formula $U(t)$, then set $\mathrm{gr}(F) := 0$.
2. $\mathrm{gr}(A \to B) := \max(\mathrm{gr}(A), \mathrm{gr}(B)) + 1$.
3. $\mathrm{gr}(\forall x \mathscr{F}[x]) = \mathrm{gr}(\lambda x \mathscr{F}[x](t)) := \mathrm{gr}(\mathscr{F}[0]) + 1$.
4. If $\forall X \mathscr{F}[X]$ is a weak formula, then set $\mathrm{gr}(\forall X \mathscr{F}[X]) := \mathrm{gr}(\mathscr{F}[U]) + 1$.
5. If $\forall X \mathscr{F}[X]$ is a strong formula, then set $\mathrm{gr}(\forall X \mathscr{F}[X]) := \mathrm{gr}(\mathscr{F}[U]) + \omega$.

By this definition every formula has a degree $\omega \cdot m + n$ (where $m, n \in \mathbb{N}$). We call m the *weight* of the given formula. The weight is 0 if, and only if, the formula is a weak formula.

Lemma 1. *If P is a weak predicator, then:*

a) *If $\mathscr{A}[U]$ is a weak formula, then $\mathscr{A}[P]$ is a weak formula too.*

b) *Every formula $\mathscr{F}[P]$ has the same weight as the formula $\mathscr{F}[U]$.*

c) *If $\forall X\mathscr{F}[X]$ is a strong formula, then $\mathscr{F}[P]$ has a smaller degree than $\forall X\mathscr{F}[X]$.*

Proof. a) follows by induction on the length of the nominal form $\mathscr{A}$. b) follows from a) by induction on the length of the nominal form $\mathscr{F}$. c) follows from b) since a strong formula $\forall X\mathscr{F}[X]$ has greater weight than the formula $\mathscr{F}[U]$.

Inductive definition of the *rank* rk (F) of a formula F:

1. If F is a prime formula or a formula $U(t)$, then set $\mathrm{rk}(F):=0$.
2. $\mathrm{rk}(A\to B):=\max(\mathrm{rk}(A), \mathrm{rk}(B))$.
3. $\mathrm{rk}(\forall x\mathscr{F}[x])=\mathrm{rk}(\lambda x\mathscr{F}[x](t)):=\mathrm{rk}(\mathscr{F}[0])$.
4. If $\forall X\mathscr{F}[X]$ is a weak formula, then set $\mathrm{rk}(\forall X\mathscr{F}[X]):=\mathrm{rk}(\mathscr{F}[U])+1$.
5. If $\forall X\mathscr{F}[X]$ is a strong formula, then set $\mathrm{rk}(\forall X\mathscr{F}[X]):=\omega$.

Lemma 2. a) *Every formula has a rank $\leqslant\omega$.*

b) $\mathrm{rk}(F)<\omega$ *if, and only if, F is a weak formula.*

c) *The rank of a formula is equal to the maximum of the ranks of its minimal parts.*

Proof. This follows at once from the definition of rank.

Definition of formulas $\mathscr{N}[\ \ ]$ and $\mathscr{Q}[\ \ , B]$:

1. Every N-form $\mathscr{N}$ is a nominal form $\mathscr{P}[(*_1\to C)]$ where $\mathscr{P}$ is a P-form. Now set $\mathscr{N}[\ \ ]$ to be the formula $\mathscr{P}[C]$.

2. For every NP-form $\mathscr{Q}$ $\mathscr{Q}[*_1, B]$ is an N-form $\mathscr{N}$. Now set $\mathscr{Q}[\ \ , B]$ to be the formula $\mathscr{N}[\ \ ]$.

2. Axioms, Basic Inference and Substitution Inferences

Axioms of the formal system GPA:

(Ax1) $\mathscr{P}[A]$ if A is a true constant prime formula.

(Ax2) $\mathscr{N}[A]$ if A is a false constant prime formula.

(Ax3) $\mathscr{Q}[A, B]$ if A and B are equivalent simple formulas.

(Ax4) $\mathscr{F}[a_1,\ldots,a_n]$ $(n\geqslant 1)$ if $a_1,\ldots,a_n$ do not occur in $\mathscr{F}$ and $\mathscr{F}[m_1,\ldots,m_n]$ is one of the axioms (Ax1)–(Ax3) for all n numerals $m_1,\ldots,m_n$.

The minimal parts of the axioms (Ax1)–(Ax3) indicated are said to be the *principal parts* of those axioms.

Remark. The axioms of the system GPA are formulated somewhat more generally than the axioms of the system DA in so far as the principal parts of (Ax3) in DA have degree 0 while those in GPA are only required to be simple. This has no fundamental significance but it proves to be useful for the investigations below.

Basic inferences of the formal system GPA:

(S1) $\mathscr{N}[\neg A], \mathscr{N}[B] \vdash \mathscr{N}[A \to B]$ if B is not the formula $\bot$.

(S2.0) $\mathscr{P}[\mathscr{F}[a]] \vdash \mathscr{P}[\forall x \mathscr{F}[x]]$ if a does not occur in the conclusion.

(S2.1) $\mathscr{P}[\mathscr{F}[U]] \vdash \mathscr{P}[\forall X \mathscr{F}[X]]$ if U does not occur in the conclusion.

(S3.0) $\mathscr{F}[t] \to \mathscr{N}[\forall x \mathscr{F}[x]] \vdash \mathscr{N}[\forall x \mathscr{F}[x]]$.

(S3.1) $\mathscr{F}[P] \to \mathscr{N}[\forall X \mathscr{F}[X]] \vdash \mathscr{N}[\forall X \mathscr{F}[X]]$ if P is a weak predicator or $\forall X \mathscr{F}[X]$ is a weak formula.

(S4) $\mathscr{E}[\mathscr{F}[t]] \vdash \mathscr{E}[\lambda x \mathscr{F}[x](t)]$.

(cut) $\mathscr{P}[A], \mathscr{N}[A] \vdash B$ if $\mathscr{P}[\mathscr{N}[\quad]] \models^{s} B$ holds.

(CI) $\mathscr{F}[0], \mathscr{F}[a] \to \mathscr{F}[a'] \vdash \mathscr{F}[t]$ if a does not occur in $\mathscr{F}$.

(str) $F \vdash G$ if $F \models^{s} G$ holds.

In addition to these basic inferences we use the *substitution inferences* of the following sort:

(sub) $\mathscr{F}[U] \vdash \mathscr{F}[P]$ if U does not occur in $\mathscr{F}$ and $U \notin \mathrm{Pr}(\mathscr{F}[U])$.

The way in which the formal system GPA is restricted as opposed to full classical second order arithmetic is because of the restricting condition for (S3.1)-inferences. The restriction to weak predicators P corresponds to restricting the comprehension axiom to Π_1^1-formulas. On the other hand the wider condition on (S3.1) allowing weak formulas $\forall X \mathscr{F}[X]$ and arbitrary predicators P means the formal system GPA is a strengthening of Π_1^1-analysis. (It contains Π_1^1-comprehension with bar-induction for arithmetic relations.)

We call the basic inferences (S1)–(S4) *principal inferences*. The minimal part shown in the conclusion of a principal inference is called the *principal part* of that inference.

The free variable denoted by a and U in the premises of the inferences (S2.0), (S2.1), (CI) and (sub) is called the *eigenvariable* of that inference.

The formula denoted by $\mathscr{F}[P]$ in the premise of a basic inference (S3.1) is called the *eigenformula* of that inference. The formula denoted by A in the premises of a basic inference (cut) is called the *cut formula*.

A basic inference (CI) is a *complete induction* inference and a basic inference (str) is a *structural inference*.

We set the *degree of a basic inference* (S3.1) to be the degree of its eigenformula. We set the *degree of a basic inference* (cut) to be the degree of its cut formula. We set the *degree of a basic inference* (CI) to be the degree of its first premise.

Lemma 3. *If the premise of an inference* (str) *or* (sub) *is an axiom, then the conclusion of the inference is also an axiom.*

Proof. The axioms were chosen here to be so general that Lemma 3 would hold.

Lemma 4. *If both premises of a cut are axioms which contain no free number variables and the cut-formula is simple, then the conclusion of the cut is also an axiom.*

Proof. Let $\mathscr{P}[A], \mathscr{N}[A] \vdash B$ where $\mathscr{P}[\mathscr{N}[\quad]] \models^{s} B$ be a cut satisfying the hypotheses of the Lemma. It suffices to prove that $\mathscr{P}[\mathscr{N}[\ \ ]]$ is an axiom since then by Lemma 3 B is also an axiom. If the minimal part A of $\mathscr{P}[A]$ or of $\mathscr{N}[A]$ is not a principal part of the axiom involved then obviously $\mathscr{P}[\mathscr{N}[\ \ ]]$ is also an axiom. Otherwise one of the following three cases holds.

1. A is a true constant prime formula and $\mathscr{N}$ is an N-form $\mathscr{Q}[*_1, A_2]$ where A and A_2 are equivalent. Then A_2 is also a true constant prime formula and $\mathscr{P}[\mathscr{N}[\ \ ]]$ is the (Ax1) $\mathscr{P}[\mathscr{Q}[\ \ , A_2]]$.

2. A is a false constant prime formula and $\mathscr{P}$ is a P-form $\mathscr{Q}[A_1, *_1]$ where A_1 and A are equivalent. Then A_1 is also a false constant prime formula and $\mathscr{P}[\mathscr{N}[\ \ ]]$ is the (Ax2) $\mathscr{Q}[A_1, \mathscr{N}[\ \ ]]$.

3. $\mathscr{P}$ is a P-form $\mathscr{Q}_1[A_1, *_1]$ and $\mathscr{N}$ an N-form $\mathscr{Q}_2[*_1, A_2]$ where A_1, A and A, A_2 are equivalent. Then (by Theorem 20.1) A_1 and A_2 are also equivalent and $\mathscr{P}[\mathscr{N}[\ \ ]]$ is the (Ax3) $\mathscr{Q}_1[A_1, \mathscr{Q}_2[\ \ , A_2]]$.

3. Deductions

For the system GPA *deductions* are defined in the following way as finite tree-like configurations of formulas.

Inductive definition of *deductions*:

1. If F is an axiom then the formula F forms a deduction with end formula F.

2. If H is a deduction whose end formula is the premise of an (S2.0), (S2.1), (S3.0), (S3.1), (S4), (str) or (sub) inference with conclusion F then the figure

$$\frac{H}{F}$$

forms a deduction with end formula F.

3. If H_1 and H_2 are deductions whose end formulas are the premises of an (S1), (cut) or (CI) inference with conclusion F then the figure

$$\frac{H_1 \qquad H_2}{F}$$

forms a deduction with end formula F.

We call the formulas of the deduction which do not occur as conclusions of inferences *initial formulas*. These initial formulas are axioms. However, besides the initial formulas, a deduction can contain other axioms (as conclusions of inferences).

Inductive definition of the *end piece* of a deduction:

1. The end formula of the deduction belongs to the end piece of the deduction.

2. If the conclusion of an inference (cut) belongs to the end piece of the deduction then so too do both premises of this inference.

3. If the conclusion of an inference (str) or (sub) belongs to the end piece of the deduction so too does the premise of this inference.

No other formula belongs to the end piece.

By the *inferences of the end piece* we mean those inferences of the deduction whose premises belong to the end piece. Every inference of the end piece is a (cut), (str) or (sub) inference.

By the *initial formulas of the end piece* of a deduction we mean those formulas of the end piece which do not occur as conclusions of (cut), (str) or (sub) inferences. Every initial formula of the end piece is an axiom or the conclusion of a principal inference or a basic inference (CI).

Inductive definition of the *bundle of formulas* of a minimal part M of a formula F of the end piece of a deduction:

1. The minimal part M of the formula F belongs to the bundle of formulas.
2. For every inference (cut) $\mathscr{P}[A], \mathscr{N}[A] \vdash B$ such that $\mathscr{P}[\mathscr{N}[\ \cdot]] \overset{s}{\models} B$ occurring in the end piece: If a minimal positive (negative) part of the conclusion B belongs to the bundle of formulas then so too does every identical positive (negative) part of the premise $\mathscr{P}[A]$ occurring in $\mathscr{P}$ and every identical positive (negative) part of the premise $\mathscr{N}[A]$ occurring in $\mathscr{N}$.
3. For every inference (str) $A \vdash B$ such that $A \overset{s}{\models} B$ occurring in the end piece: If a minimal positive (negative) part of the conclusion B belongs to the bundle of formulas so too does every identical positive (negative) part of the premise A.
4. For every inference (sub) $\mathscr{F}[U] \vdash \mathscr{F}[P]$, where U does not occur in $\mathscr{F}$, occurring in the end piece: If a minimal positive (negative) part $\mathscr{A}[P]$ of the conclusion $\mathscr{F}[P]$, where U does not occur in $\mathscr{A}$, belongs to the bundle of formulas then so too does each positive (negative) part $\mathscr{A}[U]$ of the premise $\mathscr{F}[U]$.

No other parts of formulas belong to the bundle of formulas.

Every bundle of formulas of the kind just defined consists either entirely of minimal positive parts or entirely of minimal negative parts of formulas of the end piece of a deduction.

Lemma 5. *If the minimal part M of a formula of the end piece of a deduction has degree 0 then the bundle of formulas of M consists solely of minimal parts of degree 0.*

Proof. This follows from the definition of bundle of formulas.

Definitions. I. A minimal part of a formula of the end piece of a deduction is said to be *explicit* if it belongs to the bundle of formulas of a minimal part of the end formula of the deduction. Otherwise it is said to be *implicit*.

II. The *rank of an inference* (sub) which occurs in the end piece of a deduction is the maximum of the ranks of the implicit minimal parts of its premise. (If the premise contains no implicit minimal parts then the inference (sub) has rank 0.)

III. A *basic deduction* is a deduction in which no inference (sub) occurs.

IV. A *normal deduction* is a deduction H for which the following holds: Every inference (sub) occurring in H is an inference of the end piece of H with a rank $<\omega$, whose eigenvariable does neither occur in the principal part of an (S2.1)-inference nor in the eigenformula of an (S3.1)-inference of H.

V. A formula F is said to be *deducible* in the formal system GPA if there is a basic deduction with the end formula F.

According to these definitions every basic deduction is a normal deduction and every initial formula of the end piece of a normal deduction is the end formula of a basic deduction.

The main task is now to prove that there is no basic deduction with the end formula $\perp$. The proof will be obtained in the way of G. Takeuti using reductions of deductions which cannot be restricted just to basic deductions but only to normal deductions.

4. Orders of Normal Deductions

We assign ordinal terms of the notation system $\bar{\theta}(\Omega)$ to the formulas of a normal deduction like the ordering functions which W. Pohlers [1] carried out using ordinal terms of the notation system Σ. For this we first assign *weights* $m<\omega$ and *ranks* $\sigma\leqslant\omega$ to normal deductions in such a way that a normal deduction of weight m and *rank* σ has the following properties:

1. Every formula of the normal deduction has a weight $\leqslant m$ and a rank $\leqslant\sigma$.

2. Every inference (sub) which occurs in the normal deduction has a rank $<\sigma$.

Obviously for each normal deduction H there is a number $m<\omega$ such that H can be regarded as a normal deduction of weight m and rank ω. If only weak formulas occur in H then H can also be regarded as a normal deduction of a rank $<\omega$.

We use principal terms of level σ of the subsystem $\bar{\theta}(\sigma+1)$ of $\bar{\theta}(\Omega)$ as *orders* of the formulas of a normal deduction of rank $\sigma\leqslant\omega$. As in §26 we write (ξ,η) for $\bar{\theta}\xi\eta$. For $k<\sigma$ we also use the function d_k introduced in §26 which assigns to every principal term of level σ a principal term of level k. In case $\sigma=0$ set $d_0\alpha:=\alpha$. If α,β are principal terms of level σ then, as in §26, $\alpha\ll\beta$ denotes that $\alpha<\beta$ and $K_i\alpha<d_i\beta$ for all $i<\sigma$. The *degrees* of formulas will be represented by ordinal terms $<\omega^2=(0,(0,0)+(0,0))$. We use ν, ν_0, ν_1, ν_2 as syntactical symbols for such degrees.

Inductive definition of the *order* of a formula in a normal deduction of weight m and rank σ:

1. Every initial formula of the normal deduction has order $(0,0)$ if $\sigma=0$ and order Ω_σ if $\sigma\neq 0$.

2. If the premises of a basic inference (S1) have orders α_1 and α_2 then the conclusion has order $(0,\alpha_1\#\alpha_2)$.

3. If the premise of a basic inference (S2.0), (S3.0), (S4) or a basic inference (S2.1) whose principal part is a weak formula has order α then the conclusion has order $(0,\alpha)$.

4. If the premise of a basic inference (S2.1) whose principal part is a strong formula has order α then the conclusion has order $(\omega\cdot(m+1),\alpha)$ (where m is the weight of the normal deduction).

5. If the premise of a basic inference (S3.1) of degree ν has order α then the conclusion has order $(\nu+1, \alpha)$.

6. If the premises of a basic inference (cut) of degree ν have orders α_1 and α_2 then the conclusion has order $(\nu, \alpha_1 \# \alpha_2)$.

7. If the premises of a basic inference (CI) of degree v have orders α_1 and α_2 then the conclusion has order $(v+1, \alpha_1 \# \alpha_2)$.

8. The conclusion of a basic inference (str) has the same order as its premise.

9. If the premise of an inference (sub) of rank $k<\sigma$ has order α then the conclusion has order $(0, \Omega_\sigma + d_k\alpha)$.

By the *order of a normal deduction* we mean the order of its end formula.

5. Transformations of Normal Deductions

Theorem 27.1. *If F and G are equivalent formulas then every normal deduction H with end formula F may be transformed into a normal deduction H' of the same weight, rank and order with the end formula G. If H is a basic deduction, so too is H'.*

Theorem 27.2. *Every normal deduction H with end formula $\mathscr{F}[a]$ where a does not occur in $\mathscr{F}$ may be transformed into a normal deduction H' of the same weight, rank and order with end formula $\mathscr{F}[t]$ for each term t. If H is a basic deduction, then so too is H'.*

Both these theorems follow by induction on the length of H.

Theorem 27.3. *If H is a basic deduction of weight m, rank ω and order α with end formula $\mathscr{F}[U]$ where U does not occur in $\mathscr{F}$ then H may be transformed into a basic deduction of the same weight, rank and with an order $\ll(\omega\cdot(m+1), \alpha)$ with end formula $\mathscr{F}[P]$ for each weak predicator P.*

Proof by induction on the length of the basic deduction H. We may assume that the eigenvariables of basic inferences (S2.0), (S2.1) and (CI) which occur in H are chosen so that they are different from U and do not occur in P. From H we then form H' in which the free predicate variable U is replaced by the weak predicator P wherever it occurs in H. It follows from Lemmas 1 and 3 that H' is also a basic deduction of weight m. H' has end formula $\mathscr{F}[P]$. We still have to prove that H', as a basic deduction of weight m and rank ω, has an order $\ll(\omega\cdot(m+1), \alpha)$.

1. Suppose $\mathscr{F}[U]$ is an initial formula of H. Then H and H' have the same order $\Omega_\omega \ll (\omega\cdot(m+1), \Omega_\omega)$.

2. Suppose $\mathscr{F}[U]$ occurs in H as the conclusion of a basic inference (str). Then the assertion follows immediately from the I.H.

3. Suppose $\mathscr{F}[U]$ occurs in H as the conclusion of a basic inference (S2.1) whose principal part is a strong formula. Then $\alpha=(\omega\cdot(m+1), \alpha_0)$ where α_0 is the order of the premise of the (S2.1)-inference. In this case $\mathscr{F}[P]$ is, in H', the conclusion of a corresponding (S2.1)-inference whose premise has, by I.H., an order $\beta_0 \ll \alpha$. Then H' has order $(\omega\cdot(m+1), \beta_0) \ll (\omega\cdot(m+1), \alpha)$.

4. Suppose $\mathscr{F}[U]$ occurs as the conclusion of a basic inference other than those under 2. and 3. Then $\alpha=(\nu_1, \alpha_1)$ or $\alpha=(\nu_1, \alpha_1 \# \alpha_2)$, where $\nu_1<\omega\cdot(m+1)$, where α_i is the order of the i-th premise of the basic inference. In this case in H', $\mathscr{F}[P]$ is the conclusion of a corresponding basic inference whose i-th premise has, by I.H., an order $\beta_i \ll (\omega\cdot(m+1), \alpha_i)$. Then there exists $\nu_2<\omega\cdot(m+1)$ such that H' has order

$$(\nu_2, \beta_1) \ll (\nu_2, (\omega\cdot(m+1), \alpha_1)) \ll (\omega\cdot(m+1), \alpha)$$

or

$$(\nu_2, \beta_1 \# \beta_2) \ll (\nu_2, (\omega\cdot(m+1), \alpha_1) \# (\omega\cdot(m+1), \alpha_2)) \ll (\omega\cdot(m+1), \alpha)$$

6. Reducible Normal Deductions

In specifying parts of a normal deduction we shall write $\vdash^{\alpha} F$ below to mean that the formula F has order α in the given place in the normal deduction.

Definition. A normal deduction of order α is said to be *reducible* if it can be transformed into a normal deduction whose order is $\ll\alpha$ while maintaining its weight, rank and end formula.

Theorem 27.4. *If a normal deduction contains a cut with cut formula* $(A \rightarrow \bot)$, *then the normal deduction is reducible.*

Proof. Suppose a normal deduction of rank σ contains the cut

$$\frac{\begin{matrix} H_1 \\ \vdash^{\alpha} \mathscr{P}[(A \rightarrow \bot)] \end{matrix} \qquad \begin{matrix} H_2 \\ \vdash^{\beta} \mathscr{N}[(A \rightarrow \bot)] \end{matrix}}{\vdash^{\gamma} B \quad \text{where } \mathscr{P}[\mathscr{N}[\;\;]] \vdash^{s} B}$$

Here H_1 and H_2 denote the subdeductions of H with end formulas $\mathscr{P}[(A \rightarrow \bot)]$ and $\mathscr{N}[(A \rightarrow \bot)]$. We replace this cut by two cuts with cut formulas $\bot$ and A as follows:

$$\frac{\begin{matrix} H_2 \\ \vdash^{\beta} \mathscr{N}[(A \rightarrow \bot)] \end{matrix} \qquad \dfrac{\begin{matrix} H_1 \\ \vdash^{\alpha} \mathscr{P}[(A \rightarrow \bot)] \end{matrix} \qquad \vdash^{\alpha_0} \mathscr{N}[\bot]}{A \rightarrow \mathscr{P}[\mathscr{N}[\;\;]]}}{\vdash^{\delta} B \quad \text{where } \mathscr{N}[(\mathscr{P}[\mathscr{N}[\;\;]] \rightarrow \bot)] \vdash^{s} B}$$

By this replacement we obtain a normal deduction H' which has the same weight, rank and end formula as H. H' has the axiom $\mathscr{N}[\bot]$ as an initial formula with order $\alpha_0=(0, 0)$ (if $\sigma=0$) or $\alpha_0=\Omega_\sigma$ (if $\sigma\neq 0$). Suppose the formula A has degree ν.

Then $(A \to \bot)$ has degree $\nu+1$ and

$$\delta = (\nu, \beta \# (0, \alpha \# \alpha_0)) \ll (\nu+1, \alpha \# \beta) = \gamma.$$

There is a deduction function g (according to the definition on p. 249), such that H has order $g(\gamma)$ and H' has order $g(\delta)$. Since $\delta \ll \gamma$ by Theorem 26.6 we have $g(\delta) \ll g(\gamma)$. Thus the normal deduction H is reducible.

Theorem 27.5. *If a normal deduction contains a basic inference* (CI) *whose conclusion contains no free number variable then the normal deduction is reducible.*

Proof. Suppose a normal deduction H contains the inference

$$\frac{\begin{array}{c} H_1 \\ \vdash^{\alpha} \mathscr{F}[0] \end{array} \qquad \begin{array}{c} H_2 \\ \vdash^{\beta} \mathscr{F}[a] \to \mathscr{F}[a'] \end{array}}{\vdash^{\gamma} \mathscr{F}[t],}$$

where a does not occur in $\mathscr{F}$ and t is a numerical term. Here H_1 and H_2 are basic deductions with orders α and β and end formulas $\mathscr{F}[0]$ and $\mathscr{F}[a] \to \mathscr{F}[a']$. By Theorem 27.2 for every numeral n, H_2 can be transformed into a basic deduction $H_{2,n}$ of order β with end formula $\mathscr{F}[n] \to \mathscr{F}[n']$ which has the same weight and rank as the normal deduction H_2. Suppose the formula $\mathscr{F}[0]$ has degree ν. Then $\gamma = (\nu+1, \alpha \# \beta)$.

We prove then

Lemma. *For every numeral n there is a basic deduction H_n' of order $\delta_n \ll \gamma$ with end formula $\mathscr{F}[n]$ which has the same weight and rank as the normal deduction H.*

The proof of this Lemma is by induction on n.

1. $n=0$. Then the assertion holds for H_1 with $\delta_0 = \alpha \ll \gamma$.

2. $n = n_0'$. By I.H. there is a basic deduction H_{n_0}' whose order $\delta_{n_0} \ll \gamma$. From this and the basic deduction $H_{2i\,n_0}$, using a cut, we form the basic deduction

$$\frac{\begin{array}{c} H_{n_0}' \\ \vdash^{\delta_{n_0}} \mathscr{F}[n_0] \end{array} \qquad \begin{array}{c} H_{2,n_0} \\ \vdash^{\beta} \mathscr{F}[n_0] \to \mathscr{F}[n] \end{array}}{\vdash^{\delta_n} \mathscr{F}[n]}$$

Here

$$\delta_n = (\nu, \delta_{n_0} \# \beta) \ll (\nu+1, \alpha \# \beta) = \gamma.$$

Thus the Lemma is proved.

The numerical term has a numeral k as its value. Then $\mathscr{F}[k]$ and $\mathscr{F}[t]$ are equivalent formulas. Therefore by Theorem 27.1 the basic deduction H_k' can be transformed into a basic deduction H' of order $\delta_k \ll \gamma$ with end formula $\mathscr{F}[t]$ where H' has the same weight and rank as the normal deduction H. Now we replace the subdeduction with end formula $\mathscr{F}[t]$ in H by the basic deduction H' and we obtain a reduction of H.

7. Singular Normal Deductions

Definitions. I. A minimal part of a formula of the end piece of a normal deduction is said to be *distinguished* if the principal part of a principal inference whose conclusion is an initial formula of the end piece belongs to its bundle of formulas.

II. A normal deduction is said to be *singular* if its end formula is not an axiom and has no distinguished minimal part.

III. A cut $\mathscr{P}[A], \mathscr{N}[A] \vdash B$ such that $\mathscr{P}[\mathscr{N}[\quad]] \overset{s}{\vdash} B$ in the end piece of a normal deduction is said to be a *suitable* cut if the following hold:

1. The negative part A of the premise $\mathscr{N}[A]$ is a distinguished minimal part.

2. If A is a simple formula then the positive part A of the premise $\mathscr{P}[A]$ is a distinguished minimal part.

Lemma 6. *Every singular normal deduction contains a cut or the conclusion of a basic inference* (CI) *in its end piece.*

Proof. Suppose H is a singular normal deduction with no cut occurring in its end piece. Then the end piece of H has precisely one initial formula A from which the end formula of H is obtained using only (str) and (sub) inferences. If A were an axiom then by Lemma 3 the end formula of H would also be an axiom. If A were the conclusion of a principal inference then the end formula of H would have a distinguished minimal part. Therefore by the singularity of H it follows that A is the conclusion of a basic inference (CI).

Theorem 27.6. *Every singular normal deduction whose end piece contains no free number variable is reducible or contains a suitable cut in its end piece.*

Proof by induction on the length of the singular normal deduction H. Since the end piece of H contains no free number variable every axiom which occurs in the end piece of H is one of the axioms (Ax1)–(Ax3). If the end piece contains the conclusion of a basic inference (CI) then the normal deduction H is reducible by Theorem 27.5. Otherwise by Lemma 6 the end piece of H contains a cut. Then suppose

$$\frac{\begin{array}{c} H_1 \\ \overset{\alpha}{\vdash} \mathscr{P}[A] \end{array} \qquad \begin{array}{c} H_2 \\ \overset{\beta}{\vdash} \mathscr{N}[A] \end{array}}{\overset{\gamma}{\vdash} B} \quad \text{where } \mathscr{P}[\mathscr{N}[\quad]] \overset{s}{\vdash} B$$

is the last cut in the end piece of H. Here H_1 and H_2 are normal deductions of orders α and β with end formulas $\mathscr{P}[A]$ and $\mathscr{N}[A]$ which occur as subdeductions in H. If H_1 or H_2 is singular then the assertion of our theorem follows from the I.H. Now assume that neither H_1 nor H_2 is singular. The end formula of H follows from the formula B by using only (str) and (sub) inferences. Were $\mathscr{P}[\mathscr{N}[\ ]]$ an axiom then by Lemma 3 the end formula of H would also be an axiom. Therefore by the singularity of H it follows that $\mathscr{P}[\mathscr{N}[\ ]]$ is not an axiom. Otherwise by the singularity of H it follows that a distinguished minimal part of $\mathscr{P}[A]$ or $\mathscr{N}[A]$ can only occur in the cut formula A.

1. Suppose A is a formula $(A_1 \to A_2)$.

1.1. Suppose A_2 is the formula $\perp$. Then by Theorem 27.4 H is reducible.

1.2. Suppose A_2 is not the formula $\perp$. Then A is a minimal negative part of $\mathscr{N}[A]$ which cannot occur as the principal part of an axiom. Since $\mathscr{P}[\mathscr{N}[\ \ ]]$ is not an axiom it follows that $\mathscr{N}[A]$ is not an axiom. Since H_2 is not singular it follows that the negative part A of $\mathscr{N}[A]$ is a distinguished minimal part. Then the cut considered is a suitable cut.

2. Suppose A is a simple formula.

2.1. Suppose neither $\mathscr{P}[A]$ not $\mathscr{N}[A]$ is an axiom. Since neither H_1 nor H_2 is singular it follows that A is a distinguished minimal part in both $\mathscr{P}[A]$ and $\mathscr{N}[A]$. Then the cut considered is a suitable cut.

2.2. Suppose $\mathscr{P}[A]$ is an axiom. Since $\mathscr{P}[\mathscr{N}[\ \ ]]$ is not an axiom it follows by Lemma 4 that $\mathscr{N}[A]$ is not an axiom. Since H_2 is not singular it follows that the negative part A in $\mathscr{N}[A]$ is a distinguished minimal part. It follows by Lemma 5 that A is not a prime formula. Since $\mathscr{P}[A]$ is an axiom and $\mathscr{P}[\mathscr{N}[\ \ ]]$ is not an axiom A is a principal part of the axiom $\mathscr{P}[A]$. Then, since A is not a prime formula, $\mathscr{P}[A]$ is an (Ax3) $\mathscr{Q}[A_1, A]$ where A_1 and A are equivalent formulas and $\mathscr{Q}[A_1, \mathscr{N}[\ \ ]] \models^{s} B$ holds. It follows that $\mathscr{N}[A_1] \models^{s} B$. By Theorem 27.1 the normal deduction H_2 can be transformed into a normal deduction with end formula $\mathscr{N}[A_1]$ while retaining the same weight, rank and order. By introducing an inference (str) we obtain a normal deduction H_2' of order $\beta \ll \gamma$ with end formula B where H_2' has the same weight and rank as H. This shows that H is reducible.

2.3. Suppose $\mathscr{N}[A]$ is an axiom. Then as under 2.2 it follows that $\mathscr{N}[A]$ is an (Ax3) $\mathscr{Q}[A, A_2]$ where A and A_2 are equivalent formulas and $\mathscr{P}[\mathscr{Q}[\ \ , A_2]] \models^{s} B$ holds. It follows that $\mathscr{P}[A_2] \models^{s} B$. Then from H_1 by Theorem 27.1 using an inference (str) we obtain a normal deduction H_1' of order $\alpha \ll \beta$ with end formula B where H_1' has the same weight and rank as H. This again shows that H is reducible.

8. Reduction of a Suitable Cut

We use the following notations for parts of a normal deduction H:

1. If F is a formula of H then set $H(F)$ to be the normal deduction with end formula F which occurs as a *subdeduction* in H.

2. A *deduction thread* $F - G$ in the end piece of H is a sequence of formulas $F_1, \ldots, F_n$ $(n \geq 1)$ of the end piece of H where F_1 is the formula F, F_i occurs in H as the premise of an inference with conclusion F_{i+1} and F_n is the formula G.

3. A *deduction part*

$$\begin{array}{c} H_1 \\ F \\ | \\ G \end{array}$$

of H consists of a deduction thread F–G of the end piece of H and the normal deduction $H_1 = H(F)$. If such a deduction part is replaced by

$$\begin{array}{c} H_1' \\ A \to F \\ | \\ B \to G \end{array}$$

then this means that H_1' is a normal deduction with end formula $A \to F$ and every formula F_i of the deduction thread F–G is replaced by the formula $A_i \to F_i$ where A_i is determined as follows: A_1 is the formula A. If F_i is a premise of an inference (cut) or (str) then A_{i+1} is the formula A_i. If $F_i \vdash F_{i+1}$ is an inference (sub) $\mathscr{F}[U] \vdash \mathscr{F}[P]$ then A_{i+1} results from A_i by replacing the free predicate variable U in A_i by the predicator P throughout. If F_n is the last formula G of the deduction thread F–G then B denotes the formula A_n.

Lemma 7. *Suppose F–G is a deduction thread in the end piece of a normal deduction H, $H_1 = H(F)$ and H_1' is a normal deduction with end formula $A \to F$. If the deduction part*

$$\begin{array}{c} H_1 \\ F \\ | \\ G \end{array} \quad \textit{in } H \textit{ is replaced by} \quad \begin{array}{c} H_1' \\ A \to F \\ | \\ B \to G, \end{array}$$

then the subdeduction $H(G)$ is transformed into a normal deduction with end formula $B \to G$ by this replacement.

Proof. This is apparent from the form of the inferences (cut), (str) and (sub) which occur as the only inferences in the end piece of H.

We make use of Lemma 7 repeatedly for the following reductions of deductions without particular mention.

Theorem 27.7. *If the end piece of a normal deduction contains a suitable cut, then the normal deduction is reducible.*

Proof. Let H be a normal deduction of weight m and rank σ whose end piece contains a suitable cut.

Case 1. Suppose $A \to B$ is the cut formula of a suitable cut. Then H has a deduction part

$$\begin{array}{c}
\qquad H_1 \qquad\qquad\qquad H_2 \\
(\mathrm{S1})\,\dfrac{\vdash^{\alpha_1} \mathcal{N}_0[\neg A_0] \qquad \vdash^{\alpha_2} \mathcal{N}_0[B_0]}{\vdash^{\beta} \mathcal{N}_0[(A_0 \to B_0)]} \\
\vdots \\
H_0 \\
(\mathrm{cut})\,\dfrac{\vdash^{\gamma_1} \mathcal{P}[(A \to B)] \qquad \vdash^{\gamma_2} \mathcal{N}[(A \to B)]}{\vdash^{\gamma} C \quad \text{where } \mathcal{P}[\mathcal{N}[\quad]] \vdash^{s} C}
\end{array}$$

Here the principal part $A_0 \to B_0$ of the (S1)-inference belongs to the bundle of formulas of the minimal negative part $A \to B$ of the second premise $\mathcal{N}[(A \to B)]$ of the suitable cut. We replace this deduction part as follows:

$$\begin{array}{c}
\begin{array}{cc}
\begin{array}{c}
H_1 \\
(\mathrm{str})\,\dfrac{\vdash^{\alpha_1} \mathcal{N}_0[\neg A_0]}{\vdash^{\alpha_1} \neg A_0 \to \mathcal{N}_0[A_0 \to B_0]} \\
\vdots \\
H_0 \\
(\mathrm{cut})\,\dfrac{\vdash^{\gamma_1} \mathcal{P}[A \to B] \qquad \vdash^{\delta_1} \neg A \to \mathcal{N}[A \to B]}{\neg A \to C}
\end{array}
&
\begin{array}{c}
H_2 \\
(\mathrm{str})\,\dfrac{\vdash^{\alpha_2} \mathcal{N}_0[B_0]}{\vdash^{\alpha_2} B_0 \to \mathcal{N}_0[A_0 \to B_0]} \\
\vdots \\
H_0 \\
H_0 \qquad (\mathrm{cut})\,\dfrac{\vdash^{\gamma_1} \mathcal{P}[A \to B] \qquad \vdash^{\delta_2} B \to \mathcal{N}[A \to B]}{B \to C} \\
(\mathrm{cut})\,\dfrac{\vdash^{\gamma_1} \mathcal{P}[A \to B] \qquad\qquad B \to C}{A \to C}
\end{array}
\end{array} \\
(\mathrm{cut})\,\dfrac{\neg A \to C \qquad\qquad\qquad A \to C}{\vdash^{\delta} C}
\end{array}$$

There is a deduction function g (see §26, p. 248) with $\gamma_2 = g(\beta)$ and $\delta_i = g(\alpha_i)$ $(i=1, 2)$. Since $\alpha_i \ll \beta$ it follows by Theorem 26.6 that $\delta_i \ll \gamma_2$ $(i=1, 2)$. Suppose the degrees of the formulas A and B are ν_1 and ν_2. Then $A \to B$ has degree $\nu := \max(\nu_1, \nu_2) + 1$. It follows that

$$\delta = (\nu_1, (\nu, \gamma_1 \# \delta_1) \# (\nu_2, \gamma_1 \# (\nu, \gamma_1 \# \delta_2))) \ll (\nu, \gamma_1 \# \gamma_2) = \gamma.$$

Thus we have that H is reduced by the given replacement.

Case 2. Suppose $\lambda x \mathcal{F}[x](t)$ is the cut formula of a suitable cut. Then H has a deduction part

$$\begin{array}{c}
\begin{array}{cc}
\begin{array}{c}
H_1 \\
(\mathrm{S4})\,\dfrac{\vdash^{\alpha_1} \mathcal{P}_0[\mathcal{F}_1[t]]}{\vdash^{\beta_1} \mathcal{P}_0[\lambda x \mathcal{F}_1[x](t)]} \\
\vdots
\end{array}
&
\begin{array}{c}
H_2 \\
(\mathrm{S4})\,\dfrac{\vdash^{\alpha_2} \mathcal{N}_0[\mathcal{F}_2[t]]}{\vdash^{\beta_2} \mathcal{N}_0[\lambda x \mathcal{F}_2[x](t)]} \\
\vdots
\end{array}
\end{array} \\
(\mathrm{cut})\,\dfrac{\vdash^{\gamma_1} \mathcal{P}[\lambda x \mathcal{F}[x](t)] \qquad\qquad \vdash^{\gamma_2} \mathcal{N}[\lambda x \mathcal{F}[x](t)]}{\vdash^{\gamma} B \quad \text{where } \mathcal{P}[\mathcal{N}[\quad]] \vdash^{s} B}
\end{array}$$

We replace this deduction part by

$$\begin{array}{c} H_1 \\ \text{(str)}\,\dfrac{\vdash^{\alpha_1} \mathscr{P}_0[\mathscr{F}_1[t]]}{\vdash^{\alpha_1} \neg\mathscr{F}_1[t] \to \mathscr{P}_0[\lambda x\mathscr{F}_1[x](t)]} \end{array} \qquad \begin{array}{c} H_1 \\ \text{(S4)}\,\dfrac{\vdash^{\alpha_1} \mathscr{P}_0[\mathscr{F}_1[t]]}{\vdash^{\beta_1} \mathscr{P}_0[\lambda x\mathscr{F}_1[x](t)]} \end{array}$$

$$\begin{array}{c} H_2 \\ \text{(S4)}\,\dfrac{\vdash^{\alpha_2} \mathscr{N}_0[\mathscr{F}_2[t]]}{\vdash^{\beta_2} \mathscr{N}_0[\lambda x\mathscr{F}_2[x](t)]} \end{array} \qquad \begin{array}{c} H_2 \\ \text{(str)}\,\dfrac{\vdash^{\alpha_2} \mathscr{N}_0[\mathscr{F}_2[t]]}{\vdash^{\alpha_2} \mathscr{F}_2[t] \to \mathscr{N}_0[\lambda x\mathscr{F}_2[x](t)]} \end{array}$$

$$\text{(cut)}\,\dfrac{\text{(cut)}\,\dfrac{\vdash^{\delta_1} \neg\mathscr{F}[t] \to \mathscr{P}[\lambda x\mathscr{F}[x](t)] \quad \vdash^{\gamma_2} \mathscr{N}[\lambda x\mathscr{F}[x](t)]}{\neg\mathscr{F}[t] \to B} \qquad \text{(cut)}\,\dfrac{\vdash^{\gamma_1} \mathscr{P}[\lambda x\mathscr{F}[x](t)] \quad \vdash^{\delta_2} \mathscr{F}[t] \to \mathscr{N}[\lambda x\mathscr{F}[x](t)]}{\mathscr{F}[t] \to B}}{\vdash^{\delta} B}$$

There are deduction functions g_i such that $g_i(\beta_i)=\gamma_i$ and $g_i(\alpha_i)=\delta_i$. Since $\alpha_i \ll \beta_i$ it follows by Theorem 26.6. that $\delta_i \ll \gamma_i$ $(i=1,2)$. Suppose the formula $\mathscr{F}[t]$ has degree ν. Then $\lambda x\mathscr{F}[x](t)$ has degree $\nu+1$. It follows that

$$\delta=(\nu,(\nu+1,\delta_1 \# \gamma_2) \# (\nu+1,\gamma_1 \# \delta_2)) \ll (\nu+1,\gamma_1 \# \gamma_2)=\gamma.$$

Consequently H is reduced by the given replacement.

Case 3. Suppose $\forall x\mathscr{F}[x]$ is the cut formula of a suitable cut. Then H has a deduction part

$$\begin{array}{c} H_1 \\ \text{(S2.0)}\,\dfrac{\vdash^{\alpha_1} \mathscr{P}_0[\mathscr{F}_1[a]]}{\vdash^{\beta_1} \mathscr{P}_0[\forall x\mathscr{F}_1[x]]} \end{array} \qquad \begin{array}{c} H_2 \\ \text{(S3.0)}\,\dfrac{\vdash^{\alpha_2} \mathscr{F}_2[t] \to \mathscr{N}_0[\forall x\mathscr{F}_2[x]]}{\vdash^{\beta_2} \mathscr{N}_0[\forall x\mathscr{F}_2[x]]} \end{array}$$

$$\text{(cut)}\,\dfrac{\vdash^{\gamma_1} \mathscr{P}[\forall x\mathscr{F}[x]] \qquad\qquad \vdash^{\gamma_2} \mathscr{N}[\forall x\mathscr{F}[x]]}{\vdash^{\gamma} B \quad \text{where } \mathscr{P}[\mathscr{N}[\quad]] \vdash^{s} B}$$

By Theorem 27.2 the basic deduction H_1 can be transformed into a basic deduction H_1' of the same weight, rank and order with end formula $\mathscr{P}_0[\mathscr{F}_1[t]]$. We replace the given deduction part as follows:

$$\begin{array}{c} H_1' \\ \text{(str)}\,\dfrac{\vdash^{\alpha_1} \mathscr{P}_0[\mathscr{F}_1[t]]}{\vdash^{\alpha_1} \neg\mathscr{F}_1[t] \to \mathscr{P}_0[\forall x\mathscr{F}_1[x]]} \end{array} \qquad \begin{array}{c} H_1 \\ \text{(S2, 0)}\,\dfrac{\vdash^{\alpha_1} \mathscr{P}_0[\mathscr{F}_1[a]]}{\vdash^{\beta_1} \mathscr{P}_0[\forall x\mathscr{F}_1[x]]} \end{array}$$

$$\begin{array}{c} H_2 \\ \text{(S3, 0)}\,\dfrac{\vdash^{\alpha_2} \mathscr{F}_2[t] \to \mathscr{N}_0[\forall x\mathscr{F}_2[x]]}{\vdash^{\beta_2} \mathscr{N}_0[\forall x\mathscr{F}_2[x]]} \end{array} \qquad \begin{array}{c} H_2 \\ \vdash^{\alpha_2} \mathscr{F}_2[t] \to \mathscr{N}_0[\forall x\mathscr{F}_2[x]] \end{array}$$

$$\text{(cut)}\,\dfrac{\text{(cut)}\,\dfrac{\vdash^{\delta_1} \neg\mathscr{F}[t] \to \mathscr{P}[\forall x\mathscr{F}[x]] \quad \vdash^{\gamma_2} \mathscr{N}[\forall x\mathscr{F}[x]]}{\neg\mathscr{F}[t] \to B} \qquad \text{(cut)}\,\dfrac{\vdash^{\gamma_1} \mathscr{P}[\forall x\mathscr{F}[x]] \quad \vdash^{\delta_2} \mathscr{F}[t] \to \mathscr{N}[\forall x\mathscr{F}[x]]}{\mathscr{F}[t] \to B}}{\vdash^{\delta} B}$$

As in case 2 we then have $\delta_i \ll \gamma_i$ $(i=1,2)$ and $\delta \ll \gamma$ so that H is reduced by the replacement.

Case 4. Suppose $\forall X \mathscr{F}[X]$ is the cut formula of a suitable cut. Then H has a deduction part

$$
\begin{array}{ccc}
H_1 & & H_2 \\
(\mathrm{S2.1})\ \dfrac{\vdash^{\alpha_1} \mathscr{P}_0[\mathscr{F}[U]]}{\vdash^{\beta_1} \mathscr{P}_0[\forall X \mathscr{F}[X]]} & & (\mathrm{S3.1})\ \dfrac{\vdash^{\alpha_2} \mathscr{F}[P] \to \mathscr{N}_0[\forall X \mathscr{F}[X]]}{\vdash^{\beta_2} \mathscr{N}_0[\forall X \mathscr{F}[X]]} \\
| & & | \\
\multicolumn{3}{c}{(\mathrm{cut})\ \dfrac{\vdash^{\gamma_1} \mathscr{P}[\forall X \mathscr{F}[X]] \qquad \vdash^{\gamma_2} \mathscr{N}[\forall X \mathscr{F}[X]]}{B \quad \text{where } \mathscr{P}[\mathscr{N}[\ \ ]] \vdash^{s} B}}
\end{array}
$$

Here the principal parts of the basic inferences (S2.1) and (S3.1) belong to the bundle of formulas of the cut formula $\forall X \mathscr{F}[X]$. The formulas in this bundle of formulas are not altered by substitution inferences since the eigenvariable of an inference (sub) does not occur governed by a predicate quantifier. Therefore this bundle of formulas consists solely of identical minimal parts $\forall X \mathscr{F}[X]$.

Case 4a. Suppose $\forall X \mathscr{F}[X]$ is a strong formula. Then $\beta_1 = (\omega \cdot (m+1), \alpha_1)$ and P is a weak predicator. In this case H has rank ω. By Theorem 27.3 the basic deduction H_1 can be transformed into a basic deduction H_1' of weight m, rank ω and order $\alpha_0 \ll (\omega \cdot (m+1), \alpha_1) = \beta_1$ with end formula $\mathscr{P}_0[\mathscr{F}[P]]$. We replace the given deduction part as follows:

$$
\begin{array}{cccc}
H_1' & & H_1 & \\
(\mathrm{str})\ \dfrac{\vdash^{\alpha_0} \mathscr{P}_0[\mathscr{F}[P]]}{\vdash^{\alpha_0} \neg \mathscr{F}[P] \to \mathscr{P}_0[\forall X \mathscr{F}[X]]} & & (\mathrm{S2.1})\ \dfrac{\vdash^{\alpha_1} \mathscr{P}_0[\mathscr{F}[U]]}{\vdash^{\beta_1} \mathscr{P}_0[\forall X \mathscr{F}[X]]} & \\
| & H_2 & | & \\
| & (\mathrm{S3.1})\ \dfrac{\vdash^{\alpha_2} \mathscr{F}[P] \to \mathscr{N}_0[\forall X \mathscr{F}[X]]}{\vdash^{\beta_2} \mathscr{N}_0[\forall X \mathscr{F}[X]]} & | & \begin{array}{c} H_2 \\ \vdash^{\alpha_2} \mathscr{F}[P] \to \mathscr{N}_0[\forall X \mathscr{F}[X]] \end{array} \\
| & | & | & | \\
\multicolumn{2}{c}{(\mathrm{cut})\ \dfrac{\vdash^{\delta_1} \neg \mathscr{F}[P] \to \mathscr{P}[\forall X \mathscr{F}[X]] \quad \vdash^{\gamma_2} \mathscr{N}[\forall X \mathscr{F}[X]]}{\neg \mathscr{F}[P] \to B}} & \multicolumn{2}{c}{(\mathrm{cut})\ \dfrac{\vdash^{\gamma_1} \mathscr{P}[\forall X \mathscr{F}[X]] \quad \vdash^{\delta_2} \mathscr{F}[P] \to \mathscr{N}[\forall X \mathscr{F}[X]]}{\mathscr{F}[P] \to B}} \\
\multicolumn{4}{c}{(\mathrm{cut})\ \dfrac{\neg \mathscr{F}[P] \to B \qquad\qquad \mathscr{F}[P] \to B}{\vdash^{\delta} B}}
\end{array}
$$

Here $\delta_i \ll \gamma_i$ $(i=1,2)$. Suppose the formula $\forall X \mathscr{F}[X]$ has degree ν. By Lemma 1c) $\mathscr{F}[P]$ has a degree $\nu_0 < \nu$. It follows that

$$\delta = (\nu_0, (\nu, \delta_1 \# \gamma_2) \# (\nu, \gamma_1 \# \delta_2)) \ll (\nu, \gamma_1 \# \gamma_2) = \gamma.$$

Hence H is reduced by the replacement.

Case 4b. Suppose $\forall X\mathscr{F}[X]$ is a weak formula. It has a rank $k+1 \leqslant \sigma$ where k is the rank of $\mathscr{F}[U]$. Suppose C is the highest formula below the premises of the suitable cut which has no implicit minimal part of rank $>k$. Such a formula C exists since the end formula H has no implicit minimal parts. Suppose C has order γ_0. We replace the given deduction part as follows:

$$
\dfrac{
\begin{array}{c}
\dfrac{\begin{array}{c} H_1 \\ \vdash^{\alpha_1} \mathscr{P}_0[\mathscr{F}[U]] \end{array}}{\vdash^{\alpha_1} \neg\mathscr{F}[U] \to \mathscr{P}_0[\forall X\mathscr{F}[X]]}\,(\text{str}) \\ \Big| \\ \neg\mathscr{F}[U] \to \mathscr{P}[\forall X\mathscr{F}[X]]
\end{array}
\qquad
\begin{array}{c}
\dfrac{\begin{array}{c} H_2 \\ \vdash^{\alpha_2} \mathscr{F}[P] \to \mathscr{N}_0[\forall X\mathscr{F}[X]] \end{array}}{\vdash^{\beta_2} \mathscr{N}_0[\forall X\mathscr{F}[X]]}\,(\text{S3.1}) \\ \Big| \\ \vdash^{\gamma_2} \mathscr{N}[\forall X\mathscr{F}[X]]
\end{array}
}{\neg\mathscr{F}[U] \to B}\,(\text{cut})
$$

$$
\begin{array}{c} \neg\mathscr{F}[U] \to B \\ \Big| \end{array}
$$

$$
\dfrac{
\begin{array}{c}
\dfrac{\begin{array}{c} H_1 \\ \vdash^{\alpha_1} \mathscr{P}_0[\mathscr{F}[U]] \end{array}}{\vdash^{\beta_1} \mathscr{P}_0[\forall X\mathscr{F}[X]]}\,(\text{S2.1}) \\ \Big| \\ \vdash^{\gamma_1} \mathscr{P}[\forall X\mathscr{F}[X]]
\end{array}
\qquad
\begin{array}{c}
\dfrac{\dfrac{\vdash^{\delta_0} \neg\mathscr{F}[U] \to C}{\neg\mathscr{F}[P] \to C}\,(\text{sub}) \qquad \begin{array}{c} H_2 \\ \vdash^{\alpha_2} \mathscr{F}[P] \to \mathscr{N}_0[\forall X\mathscr{F}[X]] \end{array}}{\vdash^{\alpha} \neg C \to \mathscr{N}_0[\forall X\mathscr{F}[X]]}\,(\text{cut}) \\ \Big| \\ \neg C \to \mathscr{N}[\forall X\mathscr{F}[X]]
\end{array}
}{\neg C \to B}\,(\text{cut})
$$

$$
\begin{array}{c} \Big| \\ \dfrac{\vdash^{\delta} \neg C \to C}{\vdash^{\delta} C}\,(\text{str}) \end{array}
$$

There is a deduction function f such that $f(\beta_1)=\gamma_0$ and $f(\alpha_1)=\delta_0$. Since $\alpha_1 \ll \beta_1$, it follows by Theorem 26.6 that

(1) $\delta_0 \ll \gamma_0$.

The newly introduced inference (sub) has rank $k<\sigma$. Suppose the formula $\mathscr{F}[P]$ has rank ν. Then

(2) $\beta_2=(\nu+1, \alpha_2)$

(3) $\alpha=(\nu, (0, \Omega_\sigma+d_k\delta_0)\#\alpha_2)$

No substitution of a rank $\leqslant k$ occurs between the formulas $\mathscr{N}_0[\forall X[X]]$ and C.

Therefore there is a deduction function g of rank $k+1$ (see §26, p. 248) with $g(\beta_2)=\gamma_0$ and $g(\alpha)=\delta$. By Theorem 26.7 from (1)–(3) we obtain

$$\delta=g(\alpha)\ll g(\beta_2)=\gamma_0.$$

Thus we have that H is reduced by the given replacement.

9. The Consistency of the System GPA

Definition. A formula F is said to be *primitive* if every minimal part of F has degree 0 and is therefore a prime formula or a formula $U(t)$.

Theorem 27.8. *A primitive formula is deducible in* GPA *only if it is an axiom.*

Proof. Suppose H is a normal deduction of order α whose end formula is a primitive formula F. We prove by induction on $d_0\alpha$ that F is an axiom.

1. Suppose F contains no free number variable. Then we may transform the normal deduction H into a normal deduction H' whose end piece contains no free number variable while retaining its weight, rank, order and end formula. By Lemma 5 the primitive formula F has no distinguished minimal part. If F were not an axiom then H' would be a singular normal deduction which, by Theorems 27.6 and 27.7, would be reducible. But then it follows from the I.H. that F is an axiom.

2. Suppose F is a formula $\mathscr{F}[a_1,\dots,a_n]$ $(n\geqslant 1)$ such that no free number variable occurs in $\mathscr{F}$. For each n numerals $m_1,\dots,m_n$ the normal deduction H can, by Theorem 27.2, be transformed into a normal deduction of the same weight, rank and order with end formula $\mathscr{F}[m_1,\dots,m_n]$. By case 1 proved above it follows that $\mathscr{F}[m_1,\dots,m_n]$ is an axiom for all n numerals $m_1,\dots,m_n$. Therefore F is an axiom (Ax4).

Corollary. *The formal system* GPA *is consistent.*

Proof. By Theorem 27.8 the formula $\perp$ is not deducible in GPA since it is a primitive formula and not an axiom.

The consistency proof we have given above depended on transfinite induction up to $((\omega^2,\Omega_\omega),0)=\theta(\theta\omega^2(\Omega_\omega+1))0$.

10. The Subsystem PA of GPA

Let PA be the subsystem of GPA where the inferences (S3.1) and (sub) are only allowed in the following restricted forms:

(S3.1) $\mathscr{F}[P]\to\mathscr{N}[\forall X\mathscr{F}[X]]\vdash\mathscr{N}[\forall X\mathscr{F}[X]]$, if P is a weak predicator,

(sub) $\mathscr{F}[U]\vdash\mathscr{F}[P]$, if U does not occur in $\mathscr{F}$, $U\notin\mathrm{Pr}(\mathscr{F}[U])$ and P is a weak predicator.

All the Lemmata proved above for GPA also hold for PA. That is to say if (S3.1) is only allowed in the restricted form then the reduction of deductions given in the proof of Theorem 27.7 only requires (sub) inferences in the restricted form.

Substitution inferences can be eliminated in PA since we have:

Theorem 27.9. *Every normal deduction in the system* PA *can be transformed into a basic deduction in the system* PA *of the same end formula.*

Proof. Suppose H is a normal deduction in the system PA with end formula F. We regard H as a normal deduction of rank ω and prove our assertion by induction on the length of H.

1. Suppose F is an initial formula of H. Then the assertion is trivial.
2. Suppose F occurs in H as the conclusion of a basic inference. Then the assertion follows from the I.H.
3. Suppose F occurs as the conclusion of an inference (sub) $\mathscr{F}[U] \vdash \mathscr{F}[P]$ where U does not occur in $\mathscr{F}$ and P is a weak predicator. Then by I.H. we have a basic deduction with end formula $\mathscr{F}[U]$. From this, by Theorem 27.3, we obtain a basic deduction with end formula F.

Theorem 27.10. *The formal system* PA *is sententially consistent. That is to say: There is no formula A such that A and $A \to \bot$ are deducible in* PA.

Proof. If A and $A \to \bot$ were deducible in PA then, by Theorem 27.9, there would be basic deductions with end formulas A and $A \to \bot$. Thence by a cut we would obtain a basic deduction with end formula $\bot$. But by Theorem 27.8 the formula $\bot$ is not deducible in PA.

§28. The Semi-Formal System PA*

We now go from the formal system PA to a corresponding semi-formal system PA* in order to eliminate cuts which are not eliminable in PA because of the basic inference (CI) which occurs there. PA is embeddable in PA*. In PA* we first eliminate those cuts which have strong cut formulas as the cuts in RA* (§22, p. 203) were eliminated. Then we show the eliminability of the other cuts using the reduction procedure of §27. In this way we obtain a consistency proof for the formal system PA using a weaker transfinite induction than we had to use in §27 for the consistency proof for GPA.

1. Axioms and Basic Inferences of the System PA*

The terms, formulas and predicators of the system PA* are the terms, formulas and predicators of the system PA which contain no free number variable. Hence every term of the system PA* is *numerical* and every prime formula is either *true* or *false*. We define *weak* and *strong* formulas and predicators and also the *degree* and *rank* of a formula in PA* just as in PA.

Axioms of the system PA*

(Ax1) and (Ax2) as in PA.
(Ax3*) $\mathscr{Q}[A, B]$, if A and B are equivalent formulas of degree 0.

Basic inferences of the system PA*

(S1), (S2.1), (S3.0), (S3.1), (S4) and (cut) as in PA.

(S2.0*) $\mathscr{P}[\mathscr{F}[n]]$ for every numeral $n \vdash \mathscr{P}[\forall x \mathscr{F}[x]]$.

All axioms and basic inferences of the system PA* are restricted to formulas in which no free number variables occur. Just as in PA the basic inferences (S3.1) are restricted in PA* to $\mathscr{F}[P] \to \mathscr{N}[\forall X \mathscr{F}[X]] \vdash \mathscr{N}[\forall X \mathscr{F}[X]]$ for weak predicators P as opposed to the corresponding basic inferences of GPA.

All the principal parts of axioms of the system PA* are formulas of degree 0 while the principal parts of (Ax3) in PA are only required to be simple. Because of this stronger restriction on the axioms of PA* we gain the permissibility of the structural inferences which we do not take as basic inferences here. Also the basic inferences (CI) occurring in PA do not occur as basic inferences in PA*. Here they arise as permissible inferences.

We count the basic inferences (S2.0*) as *principal inferences*. The *principal part* of such an inference is the positive part $\forall x \mathscr{F}[x]$ indicated in the conclusion.

As before the *degree of a cut* is defined to be the degree of its cut formula. A cut is said to be *weak* (or *strong*) if its cut formula is a weak (or strong) formula.

2. The Strength of a Formula

For eliminating strong cuts we use a notion which we call the strength of a formula. For PA* and also for PA this is defined as follows:

Inductive definition of the *strength* st (F) of a formula F:

1. If F is a weak formula, then st $(F) := 0$.
2. If F is a strong formula $(A \to B)$, then st $(F) := \max(\text{st}(A), \text{st}(B)) + 1$.
3. If F is a strong formula $\forall x \mathscr{F}[x]$ or $\lambda x \mathscr{F}[x](t)$, then st $(F) := \text{st}(\mathscr{F}[0]) + 1$.
4. If F is a strong formula $\forall X \mathscr{F}[X]$, then st $(F) := \text{st}(\mathscr{F}[U]) + 1$.

According to this definition the strength of a formula is a natural number which is 0 if, and only if, F is a weak formula.

Lemma 1. *If P is a weak predicator then any two formulas $\mathscr{F}[U]$ and $\mathscr{F}[P]$ have the same strength.*

Proof. This follows by induction on the length of the nominal form $\mathscr{F}$.

3. Basic Deductions in the System PA*

Below ξ, η, ζ (possibly with subscripts) denote ordinal terms of level 0 in the notation system $\bar{\theta}(\omega+1)$. By the *strength of a cut* we mean the strength of its cut formula.

Inductive definition of the *basic deductions* of strength $\leqslant m$.

1. If F is an axiom of the system PA* and ξ is an arbitrary ordinal term of level 0 of the system $\bar{\theta}(\omega+1)$, then the formula F forms a basic deduction of strength $\leqslant m$ and depth $\leqslant \xi$ with end formula F.

2. If H is a basic deduction of strength $\leqslant m$ and depth $\leqslant \xi$ whose end formula is the premise of a basic inference (S2.1), (S3.0), (S3.1) or (S4) of the system PA* with conclusion F and $\xi<\eta$ then

$$\frac{H}{F}$$

forms a basic deduction of strength $\leqslant m$ and depth $\leqslant \eta$ with end formula F.

3. If H_1 and H_2 are basic deductions of strengths $\leqslant m$ and depths $\leqslant \xi_i$ $(i=1, 2)$ whose end formulas are the premises of a basic inference (S1) or (cut) of strength $<m$ of the system PA* with conclusion F and $\xi_i<\eta$ $(i=1, 2)$ then

$$\frac{H_1, \qquad H_2}{F}$$

forms a basic deduction of strength $\leqslant m$ and depth $\leqslant \eta$ with end formula F.

4. If, for every natural number n, H_n is a basic deduction of strength $\leqslant m$ and depth $\leqslant \xi_n$ with end formula $\mathscr{P}[\mathscr{F}[n]]$ and $\xi_n<\eta$ for all $n \in \mathbb{N}$ then

$$\frac{\{H_n\}_{n\in\mathbb{N}}}{\mathscr{P}[\forall x \mathscr{F}[x]]}$$

forms a basic deduction of strength $\leqslant m$ and depth $\leqslant \eta$ with end formula $\mathscr{P}[\forall x \mathscr{F}[x]]$.

According to this definition, in a basic deduction of strength $\leqslant m$ no cut occurs of strength $\geqslant m$. Thus in a basic deduction of strength $\leqslant 1$ no strong cuts occur and in a basic deduction of strength $\leqslant 0$ no cuts occur at all.

By the *initial formulas* of a basic deduction H we mean, as before, those formulas in the deduction H which do not occur as conclusions of basic inferences.

PA* $\vdash^{\xi}_{m} F$ means: *There is a basic deduction of strength* $\leqslant m$ *and depth* $\leqslant \xi$ *with end formula F.*

Then by the definition of basic deduction we have:

$$\text{PA*} \vdash^{\xi_1}_{m_1} F, \qquad \xi_1 \leqslant \xi_2,\ m_1 \leqslant m_2 \Rightarrow \text{PA*} \vdash^{\xi_2}_{m_2} F.$$

Theorem 28.1. PA* $\vdash^{\omega}_{0} \mathcal{Q}[A, B]$ *holds for every formula* $\mathcal{Q}[A, B]$ *of the system* PA* *with equivalent formulas* A, B.

Proof. It follows by induction on the number of symbols $\to$, $\forall$ and λ occurring in A that PA* $\vdash^{2n}_{0} \mathcal{Q}[A, B]$ (as in the proof of Theorem 22.9). The assertion follows.

Theorem 28.2. *If F and G are equivalent formulas and* PA* $\vdash^{\xi}_{m} F$ *holds, then so too does* PA* $\vdash^{\xi}_{m} G$.

Proof. This follows by induction on ξ.

Theorem 28.3. *The following inverses of* (S1), (S2.0*), (S2.1) *and* (S4) *hold:*

(Inv1) PA* $\vdash^{\xi}_{m} \mathcal{N}[(A \to B)] \Rightarrow$ PA* $\vdash^{\xi}_{m} \mathcal{N}[\neg A]$, PA* $\vdash^{\xi}_{m} \mathcal{N}[B]$

(Inv2.0*) PA* $\vdash^{\xi}_{m} \mathcal{P}[\forall x \mathcal{F}[x]] \Rightarrow$ PA* $\vdash^{\xi}_{m} \mathcal{P}[\mathcal{F}[n]]$ *for every numeral n*

(Inv2.1) PA* $\vdash^{\xi}_{m} \mathcal{P}[\forall X \mathcal{F}[X]] \Rightarrow$ PA* $\vdash^{\xi}_{m} \mathcal{P}[\mathcal{F}[U]]$ *for every free predicate variable U.*

(Inv4) PA* $\vdash^{\xi}_{m} \mathcal{E}[\lambda x \mathcal{F}[x](t)] \Rightarrow$ PA* $\vdash^{\xi}_{m} \mathcal{E}[\mathcal{F}[t]]$.

Proof. These follow by induction on ξ since all principal parts of axioms of the system PA* have degree 0.

Theorem 28.4. *If* PA* $\vdash^{\xi}_{m} F$ *and* $F \vdash^{s} G$ *hold for formulas F, G of the system* PA* *then* PA* $\vdash^{\xi}_{m} G$ *also holds.*

Proof. This follows by induction on ξ using Theorem 28.3 (as in the proof of Theorem 4.4).

4. Embedding of PA in PA*

We define the *strength* of a cut in the system PA as for PA*. The *strength of a basic inference* (CI) of the system PA is the strength of its first premise $\mathcal{F}[0]$.

Inductive definition of PA $\vdash^{n}_{m} F$:

1. If F is an axiom of the system PA, then PA $\vdash^{1}_{m} F$.
2. If PA $\vdash^{n}_{m} F$ holds for the premise of a basic inference (S2.0), (S2.1), (S3.0),

(S3.1), (S4) or (str) of the system PA, then PA $\vdash_m^{n+1} F$ holds for the conclusion F of this basic inference.

3. If PA $\vdash_m^{n_i} F$ $(i=1, 2)$ holds for the premises F_1, F_2 of a basic inference (S1) or (cut) of strength $<m$ or (CI) of strength $<m$ in the system PA, then PA $\vdash_m^n F$ holds for the conclusion F of this inference where $n := \max(n_1, n_2)+1$.

Lemma 2. *A formula F is deducible in* PA *if, and only if, there are natural numbers $n>0$ and m such that* PA $\vdash_m^n F$.

Proof. This follows from Theorem 27.9.

Definition. A formula F^* of the system PA* is said to be a *specialization* of a formula F of the system PA if F^* arises from F by replacing all the free number variables occurring in F by numerals.

Lemma 3. *If F is an axiom of the system* PA, *then* PA* $\vdash_0^\omega F^*$ *holds for every specialization F^* of F.*

Proof. This follows from Theorem 28.1.

Theorem 28.5 (Embedding Theorem). PA $\vdash_m^n F$ *implies* PA* $\vdash_m^{\omega\cdot n} F^*$ *for every specialization F^* of F.*

Proof by induction on n.

1. Suppose $n=1$, then F is an axiom of the system PA. Then the assertion holds by Lemma 3.

2. Suppose PA $\vdash_m^n F$ by a principal inference, cut or structural inference. Then the assertion follows by the I.H. and Theorem 28.4.

3. Suppose PA $\vdash_m^n F$ is obtained by a (CI)-inference. Then F^* is a formula $\mathscr{F}^*[t^*]$ where t^* is a numerical term and we have $n_0<n$ so that by I.H. PA* $\vdash_m^{\omega\cdot n_0} \mathscr{F}^*[0]$ and PA* $\vdash_m^{\omega\cdot n_0} (\mathscr{F}^*[k] \to \mathscr{F}^*[k'])$ for every numeral k. Now $\mathscr{F}^*[k]$ has strength $<m$. Using cuts we obtain PA* $\vdash_m^{\omega\cdot n_0+k} \mathscr{F}^*[k]$ for every numeral k. The assertion PA* $\vdash_m^{\omega\cdot n} \mathscr{F}^*[t^*]$ follows by Theorem 28.2 since t^* is a numerical term and $\omega\cdot n_0+k<\omega\cdot n$.

Corollary to Theorem 28.5. *For each formula F deducible in* PA *there exist $\xi<\omega^2$ and $m\in\mathbb{N}$ such that* PA* $\vdash_m^\xi F^*$ *for every specialization F^* of F.*

Proof. This holds by Lemma 2 and Theorem 28.5.

5. Elimination of Strong Cuts in PA*

Lemma 4. *If* PA* $\vdash_m^\xi \mathscr{F}[U]$, *where U does not occur in $\mathscr{F}$ and P is a weak predicator of the system* PA*, *then* PA* $\vdash_m^{\omega+\xi} \mathscr{F}[P]$.

Proof by induction on ξ. If $\mathcal{F}[U]$ is an axiom of the system PA*, then $\mathcal{F}[P]$ is a specialization of an axiom of the system PA. Then the assertion holds by Lemma 3. In all other cases it follows by Lemma 1 from the I.H.

Lemma 5. *If A is a formula $\forall x \mathcal{F}[x]$ or $\forall X \mathcal{F}[X]$ of strength $m+1$ where* PA* $\vdash^{\xi}_{m+1} \mathcal{P}[A]$, PA* $\vdash^{\eta}_{m+1} \mathcal{N}[A]$ *and B is a formula of the system* PA* *such that $\mathcal{P}[\mathcal{N}[\ \ ]] \stackrel{s}{\vdash} B$, then* PA* $\vdash^{\omega+\xi+\eta}_{m+1} B$.

Proof by induction on η (as for Lemma 3 of §22).

1. Suppose $\mathcal{N}[A]$ is an axiom of the system PA*. Then $\mathcal{N}[\ \ ]$ is also an axiom since A is not a principal part of the axiom $\mathcal{N}[A]$. Using $\mathcal{P}[\mathcal{N}[\ \ ]] \stackrel{s}{\vdash} B$ we have that B is also an axiom of the system PA*. In this case the assertion is trivial.

2. Suppose PA* $\vdash^{\eta}_{m+1} \mathcal{N}[A]$ was obtained by a basic inference (S3.0) or (S3.1) with principal part A from PA* $\vdash^{\eta_0}_{m+1} (A_0 \rightarrow \mathcal{N}[A])$ where $\eta_0 < \eta$. Here A_0 is a formula $\mathcal{F}[t]$ or $\mathcal{F}[P]$ which has strength m by Lemma 1. By I.H. we obtain

(1) PA* $\vdash^{\omega+\xi+\eta_0}_{m+1} (A_0 \rightarrow \mathcal{P}[\mathcal{N}[\ \ ]])$.

By Theorem 28.3 and Lemma 4 from PA* $\vdash^{\xi}_{m+1} \mathcal{P}[A]$ we obtain

(2) PA* $\vdash^{\omega+\xi}_{m+1} \mathcal{P}[A_0]$.

The assertion PA* $\vdash^{\omega+\xi+\eta}_{m+1} B$ follows from (1) and (2) using a cut with cut formula A_0.

3. Suppose PA* $\vdash^{\eta}_{m+1} \mathcal{N}[A]$ by a basic inference other than those under 2. Then the assertion follows from the I.H. (as in the proof of Lemma 3 in §22).

Theorem 28.6. PA* $\vdash^{\xi}_{m+2} F$ *implies* PA* $\vdash^{\omega^{\xi}}_{m+1} F$.

Proof by induction on ξ.

1. Suppose F is an axiom of the system PA*. Then the assertion is trivial.

2. Suppose PA* $\vdash^{\xi}_{m+2} F$ was obtained by a principal inference or a cut of strength $\leq m$. Then the assertion follows from the I.H.

3. Suppose PA* $\vdash^{\xi}_{m+2} F$ was obtained by a cut with cut formula A of strength $m+1$ from PA* $\vdash^{\xi_1}_{m+2} \mathcal{P}[A]$ and PA* $\vdash^{\xi_2}_{m+2} \mathcal{N}[A]$ where $\xi_i < \xi$ $(i=1,2)$ and $\mathcal{P}[\mathcal{N}[\ \ ]] \stackrel{s}{\vdash} F$. By I.H. we then have PA* $\vdash^{\omega^{\xi_1}}_{m+1} \mathcal{P}[A]$ and PA* $\vdash^{\omega^{\xi_2}}_{m+1} \mathcal{N}[A]$.

3.1. Suppose A is a formula $(A_1 \rightarrow A_2)$. Then A_1 and A_2 have strengths $\leq m$. By Theorem 28.3 from PA* $\vdash^{\omega^{\xi_2}}_{m+1} \mathcal{N}[(A_1 \rightarrow A_2)]$ we obtain

(1) PA* $\vdash^{\omega^{\xi_2}}_{m+1} \mathcal{N}[\neg A_1]$.

(2) PA* $\vdash^{\omega^{\xi_2}}_{m+1} \mathcal{N}[A_2]$.

Suppose $\xi_0 := \max(\xi_1, \xi_2)$. From PA* $\vdash^{\omega^{\xi_1}}_{m+1} \mathcal{P}[(A_1 \rightarrow A_2)]$ and (2) we obtain

(3) PA* $\vdash^{\omega^{\xi_0}+1}_{m+1} (A_1 \rightarrow \mathcal{P}[\mathcal{N}[\ \ ]])$

by a cut with cut formula A_2. The assertion PA* $\vdash^{\omega^\xi}_{m+1} F$ follows from (1) and (3) by a cut with cut formula A_1 since $\omega^{\xi_2} < \omega^{\xi_0} + 1 < \omega^\xi$.

3.2. Suppose A is a formula $\lambda x \mathscr{F}[x](t)$. Then it follows by Theorem 28.3 that PA* $\vdash^{\omega^{\xi_1}}_{m+1} \mathscr{P}[\mathscr{F}[t]]$ and PA* $\vdash^{\omega^{\xi_2}}_{m+1} \mathscr{N}[\mathscr{F}[t]]$. Here $\mathscr{F}[t]$ has strength m. Therefore the assertion PA* $\vdash^{\omega^\xi}_{m+1} F$ follows by a cut with cut formula $\mathscr{F}[t]$.

3.3. Suppose A is a formula $\forall x \mathscr{F}[x]$ or $\forall X \mathscr{F}[X]$. If $\xi_2 = 0$, then $\mathscr{N}[A]$ is an axiom of the system PA* in which A does not occur as principal part. Then $\mathscr{N}[\ ]$ is also an axiom and since $\mathscr{P}[\mathscr{N}[\ ]] \vDash^{s} F$, F is an axiom of the system PA*. In this case the assertion is trivial. Now suppose $\xi_2 \neq 0$. Then $\omega + \omega^{\xi_1} + \omega^{\xi_2} < \omega^\xi$. In this case the assertion PA* $\vdash^{\omega^\xi}_{m+1} F$ follows from Lemma 5.

Theorem 28.7. *If F is a formula deducible in* PA *then there exists $\xi < \varepsilon_0$ such that* PA* $\vdash^{\xi}_{1} F^*$ *holds for every specialization F^* of F.*

Proof. By the Corollary to Theorem 28.5 there exist $\xi_0 < \omega^2$ and $m \in \mathbb{N}$ such that PA* $\vdash^{\xi_0}_{m} F^*$ holds for every specialization F^* of F. The assertion follows by induction on m.

6. Normal Deductions in the System PA*

By a *basic deduction in the wider sense* we mean a basic deduction in which arbitrary specializations of the axioms of the system PA may occur as initial formulas besides axioms of the system PA*.

As in PA we employ the following inferences in PA*:

(str) $F \vdash G$ if $F \vDash^{s} G$ holds.

(sub) $\mathscr{F}[U] \vdash \mathscr{F}[P]$ if U does not occur in $\mathscr{F}$, $U \notin \mathrm{Pr}(\mathscr{F}[U])$ and P is a weak predicator.

We call such an inference an inference of the system PA* if its premise and conclusion contain no free number variables.

Inductive definition of *normal deductions*:

1. Every basic deduction in the wider sense whose strength is $\leqslant 1$ is a normal deduction.

2. If H is a normal deduction whose end formula is the premise of an inference (str) or (sub) of the system PA* with conclusion F, then

$$\frac{H}{F}$$

forms a normal deduction with end formula F, if in the case of an inference (sub) the eigenvariable does not occur in the principal part of an (S2.1)-inference of H.

3. If H_1 and H_2 are normal deductions whose end formulas are premises of a weak cut of the system PA* with end formula F, then

$$\frac{H_1 \qquad H_2}{F}$$

forms a normal deduction with end formula F.

As in §27 we define the *end piece* of a normal deduction, the *initial formulas* of the end piece, the *bundle of formulas* of a minimal part of a formula of the end piece and the *explicit* and *implicit* minimal parts of the formulas of the end piece.

In PA*, as in PA, the end piece of a normal deduction consists of only finitely many formulas. But besides its end piece a normal deduction in the system PA* may contain infinitely many formulas.

As before we set the rank of an inference (sub) in a normal deduction to be the maximum of the ranks of the implicit minimal parts of its premise (or 0 if the premise has no implicit minimal part). This rank is already $<\omega$ here since no strong cut occurs in a normal deduction in the system PA* and therefore every implicit minimal part of a formula of the end piece is a weak formula.

As in PA we assign principal terms of level ω from the notation system $\bar{\theta}(\omega+1)$ to the formulas in a normal deduction. We use α, β, γ, δ (possibly with subscripts) as syntactic symbols for principal terms of level ω from the system $\bar{\theta}(\omega+1)$. As before $\alpha \ll \beta$ means that $\alpha<\beta$ and $K_i\alpha < d_i\beta$ for all $i<\omega$. Then by Theorem 26.3, $d_i\alpha < d_i\beta$ for all $i<\omega$. $\alpha \lll \beta$ means that $\alpha \ll \beta$ or $\alpha=\beta$.

Inductive definition of the *order* of a formula in a normal deduction:

1. Every initial formula of a normal deduction has order Ω_ω.

2. If the premises of a basic inference (S1) have orders α_1 and α_2, then the conclusion has order $(0, \alpha_1 \# \alpha_2)$.

3. If the premise of a basic inference (S2.1), (S3.0), (S4) or a basic inference (S3.1) whose principal part is a strong formula has order α, then the conclusion has order $(0, \alpha)$.

4. If the premise of a basic inference (S3.1) whose principal part is a weak formula has order α, then the conclusion has order $(n+1, \alpha)$ where n is the degree of the eigenformula of the inference. (This degree is finite since the principal part of the inference is a weak formula).

5. If every premise of a basic inference (S2.0*) has an order $\alpha_n \ll \alpha$ (where α is a principal term of level ω of the system $\bar{\theta}(\omega+1)$), then the conclusion has order α. (In this case the order of the conclusion is not uniquely determined by the orders of the premises but can be chosen arbitrarily as long as the condition $\alpha_n \ll \alpha$ is satisfied. Such an α always exists as the proof of Theorem 28.8 will show.)

6. If the premises of a weak cut have orders α_1 and α_2 then the conclusion has order $(n, \alpha_1 \# \alpha_2)$ where n is the degree of the cut formula.

7. If the premise of an inference (str) has order α, then the conclusion also has order α.

8. If the premise of an inference (sub) of rank k has order α, then the conclusion has order $(0, \Omega_\omega + d_k\alpha)$.

$GH \overset{\alpha}{\vdash} F$ (or $NH \overset{\alpha}{\vdash} F$) denotes that there is a basic deduction of strength $\leqslant 1$ (or a normal deduction) in which the formula F has order α. Since every basic deduction of strength $\leqslant 1$ is a normal deduction if $GH \overset{\alpha}{\vdash} F$ holds, $NH \overset{\alpha}{\vdash} F$ also holds.

Theorem 28.8. *If* PA* $\overset{\xi}{\underset{1}{\vdash}} F$ *holds, then there exists* $\alpha \ll (\omega, \Omega_\omega + \xi)$ *such that* $GH \overset{\alpha}{\vdash} F$.

Proof by induction on ξ.

1. Suppose F is an axiom of the system PA*. Then the assertion is trivial since $\Omega_\omega \ll (\omega, \Omega_\omega + \xi)$ holds.

2. Suppose PA* $\overset{\xi}{\underset{1}{\vdash}} F$ was obtained from PA* $\overset{\xi_i}{\underset{1}{\vdash}} F_i$ where $\xi_i < \xi$ $(i=1, 2)$. Then by I.H. there are $\alpha_i \ll (\omega, \Omega_\omega + \xi_i)$ such that $GH \overset{\alpha_i}{\vdash} F_i$. Using a corresponding basic inference we obtain $GH \overset{\alpha}{\vdash} F$ where $\alpha = (n, \alpha_1)$ or $\alpha = (n, \alpha_1 \# \alpha_2)$ with $n < \omega$. Since $\alpha_i \ll (\omega, \Omega_\omega + \xi_i)$ and $\xi_i < \xi$ we have $\alpha \ll (\omega, \Omega_\omega + \xi)$.

3. Suppose F is a formula $\mathscr{P}[\forall x \mathscr{F}[x]]$ and PA* $\overset{\xi_n}{\underset{1}{\vdash}} \mathscr{P}[\mathscr{F}[n]]$ where $\xi_n < \xi$ for every numeral n. Then by I.H. there exist $\alpha_n \ll (\omega, \Omega_\omega + \xi_n)$ such that $GH \overset{\alpha_n}{\vdash} \mathscr{P}[\mathscr{F}[n]]$ for every numeral n. Putting $\alpha := (\omega, \Omega_\omega + \xi)$ we have $\alpha_n \ll \alpha$ and $GH \overset{\alpha}{\vdash} F$.

7. Reducible Normal Deductions

In the same way as in §27 we define the *distinguished minimal parts* of formulas in the end piece of a normal deduction, the *suitable cuts* in the end piece of a normal deduction and *singular* and *reducible* normal deductions.

A normal deduction in the system PA* is said to be singular if its end formula is not a specialization of an axiom of the system PA and has no distinguished minimal part.

A normal deduction of order α of the system PA* is said to be *reducible* if it can be transformed into a normal deduction in the system PA* of order $\ll \alpha$ retaining the same end formula.

Theorem 28.9. *Every singular normal deduction of the system* PA* *is reducible.*

Proof. Suppose H is a singular normal deduction in the system PA*. The proofs of Theorems 27.6 and 27.7 can be immediately carried over to H. This leaves only the cases where basic inferences (CI) or strong cuts occur. Indeed in PA* the definition of order for those basic inferences (S2.1) and (S3.1) which have strong formulas as principal parts are somewhat different from those encountered in PA. But this does not affect the possibility of reductions of H since no strong cuts occur in H. H contains no free number variables. It therefore follows from the proofs of Theorems 27.6 and 27.7 that H is reducible.

8. Elimination of Cuts in PA*

Theorem 28.10. *$NH \models^{\alpha} F$ implies* PA* $\mid_{\overline{0}}^{d_0\alpha} F$.

Proof by induction on $d_0\alpha$. Suppose H is a normal deduction of order α with end formula F.

1. Suppose H is singular. Then by Theorem 28.9 there is an $\alpha_0 \ll \alpha$ where $NH \models^{\alpha_0} F$. It follows that $d_0\alpha_0 < d_0\alpha$ and by I.H. PA* $\mid_{\overline{0}}^{d_0\alpha_0} F$ and therefore PA* $\mid_{\overline{0}}^{d_0\alpha} F$.

2. Suppose F is a specialization of an axiom of the system PA. Then the assertion holds by Lemma 3 since $\omega < (\Omega_\omega, 0) = d_0\Omega_\omega \leqslant d_0\alpha$.

3. Suppose H is not singular and F is not a specialization of an axiom of the system PA. Then F has a distinguished minimal part.

3.1. Suppose F is a formula $\mathcal{N}[(A \to B)]$ with distinguished minimal negative part $(A \to B)$. Then H has a deduction part

$$\begin{array}{c} \begin{array}{cc} H_1 & H_2 \\ \models^{\beta_1} \mathcal{N}_0[\neg A_0] & \models^{\beta_2} \mathcal{N}_0[B_0] \end{array} \\ \text{(S1)} \overline{\qquad\qquad\qquad\qquad\qquad\qquad} \\ \models^{\beta} \mathcal{N}_0[(A_0 \to B_0)] \\ | \\ \models^{\alpha} \mathcal{N}[(A \to B)] \end{array}$$

Here the principal part $(A_0 \to B_0)$ of the (S1) inference belongs to the bundle of formulas of the minimal negative part $(A \to B)$ of the formula $\mathcal{N}[(A \to B)]$. From H we form two new normal deductions in which the given deduction parts are replaced by deduction parts of the following two sorts:

$$\begin{array}{c} H_1 \\ \models^{\beta_1} \mathcal{N}_0[\neg A_0] \\ \text{(str)} \overline{\qquad\qquad\qquad\qquad\qquad} \\ \models^{\beta_1} \neg A_0 \to \mathcal{N}_0[(A_0 \to B_0)] \\ | \\ \models^{\alpha_1} \neg A \to \mathcal{N}[(A \to B)] \end{array} \qquad \begin{array}{c} H_2 \\ \models^{\beta_2} \mathcal{N}_0[B_0] \\ \text{(str)} \overline{\qquad\qquad\qquad\qquad\qquad} \\ \models^{\beta_2} B_0 \to \mathcal{N}_0[(A_0 \to B_0)] \\ | \\ \models^{\alpha_2} B \to \mathcal{N}[(A \to B)] \end{array}$$

Then $\alpha_i \ll \alpha$ $(i = 1, 2)$. It follows by I.H. that PA* $\mid_{\overline{0}}^{d_0\alpha_1} (\neg A \to \mathcal{N}[(A \to B)])$ and PA* $\mid_{\overline{0}}^{d_0\alpha_2} (B \to \mathcal{N}[(A \to B)]$. Using an (S1) inference it follows that

$$\text{PA*} \mid_{\overline{0}}^{d_0\alpha} ((A \to B) \to \mathcal{N}[(A \to B)]).$$

The assertion PA* $\mid_{\overline{0}}^{d_0\alpha} \mathcal{N}[(A \to B)]$ follows by Theorem 28.4.

3.2. Suppose F is a formula $\mathcal{P}[\forall x \mathcal{F}[x]]$ with distinguished positive part $\forall x \mathcal{F}[x]$. Then H has a deduction part

$$\begin{array}{c} H_n \\ \text{(S2.0*)}\ \dfrac{\vdash^{\beta_n}\mathcal{P}_0[\mathcal{F}_0[n]] \quad \text{for every numeral } n}{\vdash^{\beta}\mathcal{P}_0[\forall x\mathcal{F}_0[x]]} \\ \Big| \\ \vdash^{\alpha}\mathcal{P}[\forall x\mathcal{F}[x]] \end{array}$$

From H, for every numeral n, we form a new normal deduction in which the given deduction part is replaced by

$$\begin{array}{c} H_n \\ \text{(str)}\ \dfrac{\vdash^{\beta_n}\mathcal{P}_0[\mathcal{F}_0[n]]}{\vdash^{\beta_n}\neg\mathcal{F}_0[n]\rightarrow\mathcal{P}_0[\forall x\mathcal{F}_0[x]]} \\ \Big| \\ \vdash^{\alpha_n}\neg\mathcal{F}[n]\rightarrow\mathcal{P}[\forall x\mathcal{F}[x]] \end{array}$$

Then $\alpha_n \ll \alpha$ for every numeral n and therefore by I.H. PA* $\vdash_0^{d_0\alpha_n}(\neg\mathcal{F}[n]\rightarrow \mathcal{P}[\forall x\mathcal{F}[x]])$. Using a basic inference (S2.0*) we obtain PA* $\vdash_0^{d_0\alpha}(\neg\forall x\mathcal{F}[x]\rightarrow \mathcal{P}[\forall x\mathcal{F}[x]])$. The assertion PA* $\vdash_0^{d_0\alpha}\mathcal{P}[\forall x\mathcal{F}[x]]$ follows by Theorem 28.4.

3.3. Suppose F is a formula $\mathcal{N}[\forall x\mathcal{F}[x]]$ or $\mathcal{N}[\forall X\mathcal{F}[X]]$ where the indicated part is distinguished. Then H has a deduction part

$$\begin{array}{c} H_0 \\ \text{(S3.0)}\ \dfrac{\vdash^{\beta_0}\mathcal{F}_0[t]\rightarrow\mathcal{N}_0[\forall x\mathcal{F}_0[x]]}{\vdash^{\beta}\mathcal{N}_0[\forall x\mathcal{F}_0[x]]} \\ \Big| \\ \vdash^{\alpha}\mathcal{N}[\forall x\mathcal{F}[x]] \end{array} \quad \text{or} \quad \begin{array}{c} H_0 \\ \text{(S3.1)}\ \dfrac{\vdash^{\beta_0}\mathcal{F}[P_0]\rightarrow\mathcal{N}_0[\forall X\mathcal{F}[X]]}{\vdash^{\beta}\mathcal{N}_0[\forall X\mathcal{F}[X]]} \\ \Big| \\ \vdash^{\alpha}\mathcal{N}[\forall X\mathcal{F}[X]] \end{array}$$

We form a new normal deduction from H in which the given deduction part is replaced by

$$\begin{array}{c} H_0 \\ \vdash^{\beta_0}\mathcal{F}_0[t]\rightarrow\mathcal{N}_0[\forall x\mathcal{F}_0[x]] \\ \Big| \\ \vdash^{\alpha_0}\mathcal{F}[t]\rightarrow\mathcal{N}[\forall x\mathcal{F}[x]] \end{array} \quad \text{or} \quad \begin{array}{c} H_0 \\ \vdash^{\beta_0}\mathcal{F}[P_0]\rightarrow\mathcal{N}_0[\forall X\mathcal{F}[X]] \\ \Big| \\ \vdash^{\alpha_0}\mathcal{F}[P]\rightarrow\mathcal{N}[\forall X\mathcal{F}[X]] \end{array}$$

Here, in the second case, P as well as P_0 is a weak predicator. In any case $\alpha_0 \ll \alpha$. By I.H. we obtain

$$\text{PA*} \vdash_0^{d_0\alpha_0}(\mathcal{F}[t]\rightarrow\mathcal{N}[\forall x\mathcal{F}[x]]) \quad \text{or} \quad \text{PA*} \vdash_0^{d_0\alpha_0}(\mathcal{F}[P]\rightarrow\mathcal{N}[\forall X\mathcal{F}[X]]).$$

The assertion PA* $\vdash_0^{d_0\alpha} F$ follows using a basic inference (S3.0) or (S3.1).

3.4. Suppose F is a formula $\mathscr{P}[\forall X\mathscr{F}[X]]$, $\mathscr{P}[\lambda x\mathscr{F}[x](t)]$ or $\mathscr{N}[\lambda x\mathscr{F}[x](t)]$ where the minimal part indicated is distinguished. Then we have in the same way as before $\alpha_0 \ll \alpha$ so by I.H. we have:

$\mathrm{PA}^* \vdash_0^{d_0\alpha_0} (\neg\mathscr{F}[U] \to \mathscr{P}[\forall X\mathscr{F}[X]]$, $\mathrm{PA}^* \vdash_0^{d_0\alpha_0} (\neg\mathscr{F}[t] \to \mathscr{P}[\lambda x\mathscr{F}[x](t)])$ or $\mathrm{PA}^* \vdash_0^{d_0\alpha_0} (\mathscr{F}[t] \to \mathscr{N}[\lambda x\mathscr{F}[x](t)]$, where, in the first case, U does not occur in $\mathscr{F}$.

Using a basic inference (S2.1) or (S4) and Theorem 28.4 we obtain the assertion $\mathrm{PA}^* \vdash_0^{d_0\alpha} F$.

Theorem 28.11. *If F is a formula deducible in* PA, *then there exists $\eta < ((\omega, \Omega_\omega + \varepsilon_0), 0)$ such that* $\mathrm{PA}^* \vdash_0^{\eta} F^*$ *holds for every specialization F^* of F.*

Proof. By Theorem 28.7 there exists $\xi < \varepsilon_0$ such that $\mathrm{PA}^* \vdash_1^{\xi} F^*$ holds for every specialization F^* of F. Then by Theorem 28.8 there exists $\alpha \ll (\omega, \Omega_\omega + \xi)$ such that $NH \models^{\alpha} F^*$. Here α may depend on the choice of the specialization F^*. It follows from Theorem 28.10 that $\mathrm{PA}^* \vdash_0^{d_0\alpha} F^*$. Then if $\eta := d_0(\omega, \Omega_\omega + \xi)$ we have

$$d_0\alpha \leqslant \eta < ((\omega, \Omega_\omega + \varepsilon_0), 0),$$

and therefore, also, $\mathrm{PA}^* \vdash_0^{\eta} F^*$ where η depends only on F.

Corollary. *The formula $\perp$ is not deducible in* PA.

Proof. This follows from Theorem 28.11 since it follows by induction up to $((\omega, \Omega_\omega + \varepsilon_0), 0)$ that there is no $\eta < ((\omega, \Omega_\omega + \varepsilon_0), 0)$ such that $\mathrm{PA}^* \vdash_0^{\eta} \perp$.

At the same time we have also proved the *sentential consistency* of the system PA since

$$A, \neg A \vdash \perp$$

is a permissible inference of the system PA.

This consistency proof uses transfinite induction up to $((\omega, \Omega_\omega + \varepsilon_0), 0)$. This is weaker than the transfinite induction we used for the consistency proof of PA in §27 since $((\omega, \Omega_\omega + \varepsilon_0), 0) < ((\omega^2, \Omega_\omega), 0)$.

Remark. The normal deductions of the system PA* we have defined here correspond to normal deductions of rank ω in the system PA. In a similar way to that for PA we may define normal deductions of rank $< \omega$ in PA* and the corresponding Theorems hold for them.

From Theorems 28.6, 28.8 and 28.10 we have that if $\mathrm{PA}^* \vdash_m^{\xi} F$ then there is an ordinal term η such that $\mathrm{PA}^* \vdash_0^{\eta} F$. If we restrict F to deducible formulas in the system PA then, by Theorem 28.5, ξ may be restricted to ordinal terms $< \omega^2$ while we can only restrict the corresponding ordinal term η so that $\eta < ((\omega, \Omega_\omega + \varepsilon_0), 0)$. But it is possible to estimate a certain ordinal term bound ζ such that to every ordinal term $\xi<\zeta$ such that $\mathrm{PA}^* \vdash^{\xi}_{m} F$ there is an ordinal term $\eta<\zeta$ such that $\mathrm{PA}^* \vdash^{\eta}_{0} F$. The smallest bound of this kind which we obtain by our theorems is $((\omega, \Omega_\omega+\Omega_1), 0)=\sup_{n\in N} \xi_n$ where $\xi_0:=0$ and $\xi_{n+1}:=((\omega, \Omega_\omega+\xi_n), 0)$. Using the next theorem we prove that the ordinal term $((\omega, \Omega_\omega+\Omega_1), 0)$ is an ordinal term bound of the type described.

Theorem 28.12. *The following three properties of a formula F are equivalent to each other.*

(1) *There exists* $m\in\mathbb{N}$ *and* $\xi<((\omega, \Omega_\omega+\Omega_1), 0)$ *such that* $\mathrm{PA}^* \vdash^{\xi}_{m} F$.

(2) *There exists* $\alpha \ll (\omega, \Omega_\omega+\Omega_1)$ *such that* $NH \models^{\alpha} F$.

(3) *There exists* $\eta<((\omega, \Omega_\omega+\Omega_1), 0)$ *such that* $\mathrm{PA}^* \vdash^{\eta}_{0} F$.

Proof of (1) $\Rightarrow$ (2). Obviously the ordinal term $((\omega, \Omega_\omega+\Omega_1), 0)$ denotes an ε-number. Therefore it follows from (1) by Theorem 28.6 using induction on m that there exists $\eta<((\omega, \Omega_\omega+\Omega_1), 0)$ such that $\mathrm{PA}^* \vdash^{\eta}_{1} F$. Then by Theorem 28.8 there exists $\alpha \ll (\omega, \Omega_\omega+\eta)$ such that $NH \models^{\alpha} F$. It follows from $\eta<((\omega, \Omega_\omega+\Omega_1), 0)$ that $(\omega, \Omega_\omega+\eta) \ll (\omega, \Omega_\omega+\Omega_1)$ and therefore $\alpha \ll (\omega, \Omega_\omega+\Omega_1)$ too.

Proof of (2) $\Rightarrow$ (3). By Theorem 28.10 it follows from (2) that $\mathrm{PA}^* \vdash^{d_0\alpha}_{0} F$. Then it follows from $\alpha \ll (\omega, \Omega_\omega+\Omega_1)$ that $d_0\alpha<((\omega, \Omega_\omega+\Omega_1), 0)$.

This completes the proof of Theorem 28.12 since (3) $\Rightarrow$ (1) holds trivially.

This theorem says that: If one defines a formula to be *deducible* in PA* if, and only if, there exist $m\in\mathbb{N}$ and $\xi<((\omega, \Omega_\omega+\Omega_1), 0)$ such that $\mathrm{PA}^* \vdash^{\xi}_{m} F$ then every formula F deducible in PA* is *deducible* with a cut free proof (that is to say, there exists $\eta<((\omega, \Omega_\omega+\Omega_1), 0)$ such that $\mathrm{PA}^* \vdash^{\eta}_{0} F$) and F is deducible in PA* if, and only if, there exists $\alpha \ll (\omega, \Omega_\omega+\Omega_1)$ such that $NH \models^{\alpha} F$. Moreover, every specialization of a formula deducible in PA is also deducible in PA*.

§29. Proofs of Well-Ordering

1. A Constructive Proof of Well-Ordering for Subsystems of $\bar{\theta}(\Omega)$

In the following we let τ be an ordinal term >0 of level 0 in the notation system $\bar{\theta}(\Omega)$.

Assumption. *The well-ordering of the ordinal terms* $<\tau$ *is constructively provable.*

Under this assumption we produce a constructive proof of well-ordering for the subsystem $\bar{\theta}(\tau)$ of $\bar{\theta}(\Omega)$. In particular this assumption is valid for $\tau \leqslant \omega+1$ so we prove constructively that the systems of ordinal terms used in §27 and §28 are well-ordered.

However the well-ordering of the full system $\bar{\theta}(\Omega)$ is not deducible using the weak methods of proof which we employ in this section. W. Buchholz [2] has produced a proof of the well-ordering of $\bar{\theta}(\Omega)$ and certain stronger systems of ordinal terms using stronger methods of proof in a certain constructive way. The proof of well-order given below uses special techniques from W. Buchholz' proof.

Notations for $M \subseteq \bar{\theta}(\Omega)$:

1. Let $M(\alpha)$ be the set of $\xi \in M$ which are $<\alpha$.
2. Let $W(M)$ be the set of $\alpha \in M$ such that $M(\alpha)$ is well-ordered.

Lemma 1. *If M, M^* are subsets of $\bar{\theta}(\Omega)$ then:*

a) $\alpha \in W(M) \Rightarrow M(\alpha) \subseteq W(M)$.

b) $\alpha \in M$, $M(\alpha) \subseteq W(M) \Rightarrow \alpha \in W(M)$.

c) $\alpha \in M(\beta) \cap W(M^*)$, $M(\beta) \subseteq M^* \Rightarrow \alpha \in W(M)$.

Proof. This follows from the definition of $W(M)$.

Inductive definition of M_σ for $\sigma < \tau$:
Let M_σ be the set of $\alpha \in \bar{\theta}(\Omega)$ such that $S\alpha \leqslant \sigma$ and $K_\nu \alpha \subseteq W(M_\nu)$ for all $\nu < \sigma$.

Lemma 2. *If $\sigma < \tau$, then:*

a) $\alpha \in M_\sigma \Rightarrow P(\alpha) \subseteq M_\sigma$

b) $\alpha, \beta \in M_\sigma \Rightarrow \alpha + \beta \in M_\sigma$

c) $\alpha + \beta \in M_\sigma \Rightarrow \beta \in M_\sigma$

d) $\alpha, \beta \in W(M_\sigma) \Rightarrow \alpha + \beta \in W(M_\sigma)$

e) $\alpha \in W(M_\sigma) \Leftrightarrow P(\alpha) \subseteq W(M_\sigma)$

Proofs. a) follows immediately from the definition of M_σ.

b) α, $\beta \in M_\sigma$ implies $S(\alpha+\beta) = \max(S\alpha, S\beta) \leqslant \sigma$ and $K_\nu(\alpha+\beta) \subseteq K_\nu\alpha \cup K_\nu\beta \subseteq W(M_\nu)$ for all $\nu < \sigma$. Hence $\alpha + \beta \in M_\sigma$.

c) By a) $\alpha + \beta \in M_\sigma$ implies $P(\beta) \subseteq P(\alpha+\beta) \subseteq M_\sigma$. Obviously $0 \in M_\sigma$. By b), using $P(\beta) \subseteq M_\sigma$ we obtain $\beta \in M_\sigma$.

d) Because of the hypothesis $\beta \in W(M_\sigma)$ we can use induction on M_σ up to β. Then we have the I.H.

$$\eta \in M_\sigma(\beta) \Rightarrow \alpha + \eta \in W(M_\sigma).$$

Now suppose $\xi \in M_\sigma(\alpha+\beta)$. If $\xi \in M_\sigma(\alpha)$, then $\xi \in W(M_\sigma)$ follows from $\alpha \in W(M_\sigma)$ by Lemma 1 a). Otherwise $\xi = \alpha + \eta \in M_\sigma$ where $\eta < \beta$. Then we obtain $\eta \in M_\sigma(\beta)$ by c) and $\xi = \alpha + \eta \in W(M_\sigma)$ by the I.H. Hence $M_\sigma(\alpha+\beta) \subseteq W(M_\sigma)$. α, $\beta \in W(M_\sigma)$ imply α, $\beta \in M_\sigma$ and by b) $\alpha + \beta \in M_\sigma$. By Lemma 1 b) we obtain $\alpha + \beta \in W(M_\sigma)$.

e) $\alpha \in W(M_\sigma)$ implies $\alpha \in M_\sigma$ and by a) $P(\alpha) \subseteq M_\sigma$. By Lemma 1 a) we obtain $P(\alpha) \subseteq W(M_\sigma)$. Obviously $0 \in W(M_\sigma)$. Therefore from $P(\alpha) \subseteq W(M_\sigma)$ we obtain $\alpha \in W(M_\sigma)$ by d).

Lemma 3. *If* $\rho \leqslant \sigma < \tau$:

a) $M_\sigma(\Omega_{\rho+1}) \subseteq M_\rho$.

b) $W(M_\rho) \subseteq W(M_\sigma)$.

Proof. a) follows immediately from the definition of M_σ.

b) is obtained as follows by induction on σ. Suppose $\alpha \in W(M_\rho)$. Then $\alpha \in M_\rho$ and $S\alpha \leqslant \rho$. If $\nu < \rho$ then $K_\nu \alpha \subseteq W(M_\nu)$. If $\rho \leqslant \nu < \sigma$ then $K_\nu \alpha = P(\alpha)$ and by I.H. $\alpha \in W(M_\nu)$ and therefore by Lemma 2 e) $K_\nu \alpha \subseteq W(M_\nu)$. Hence we have $\alpha \in M_\sigma(\Omega_{\rho+1})$. Using $\alpha \in W(M_\rho)$ it follows by a) and Lemma 1 c) that $\alpha \in W(M_\sigma)$. Hence $W(M_\rho) \subseteq W(M_\sigma)$.

Definition. $Q_\tau := \bigcup_{\sigma < \tau} W(M_\sigma)$.

Lemma 4. *If* $\sigma < \tau$ *then* $Q_\tau(\Omega_{\sigma+1}) = W(M_\sigma)$.

Proof. Suppose $\alpha \in Q_\tau(\Omega_{\sigma+1})$. Then $S\alpha \leqslant \sigma$ and there exists $v < \tau$ such that $\alpha \in W(M_\nu)$. It follows that $\alpha \in M_\nu$ and $S\alpha \leqslant \nu$. If $S\alpha < \nu$ then $P(\alpha) = K_{S\alpha}\alpha \subseteq W(M_{S\alpha})$ and by Lemma 2 e) $\alpha \in W(M_{S\alpha})$. Hence in every case $\alpha \in W(M_{S\alpha})$. Using $S\alpha \leqslant \sigma$ it follows by Lemma 3 b) that $\alpha \in W(M_\sigma)$. Therefore $Q_\tau(\Omega_{\sigma+1}) \subseteq W(M_\sigma)$ holds. By definition we also have $W(M_\sigma) \subseteq Q_\tau(\Omega_{\sigma+1})$.

Theorem 29.1. Q_τ *is well-ordered.*

Proof. If $\alpha \in Q_\tau$ then $S\alpha < \tau$ and by Lemma 4 $\alpha \in W(M_{S\alpha}) = Q_\tau(\Omega_{S\alpha+1})$. By Lemma 1 c) we have $\alpha \in W(Q_\tau)$. Hence $Q_\tau \subseteq W(Q_\tau)$ and therefore $Q_\tau$is well-ordered.

Theorem 29.2. $P(\alpha) \subseteq Q_\tau \Rightarrow \alpha \in Q_\tau$.

Proof. $P(\alpha) \subseteq Q_\tau$ implies $S\alpha < \tau$ and by Lemma 4 $P(\alpha) \subseteq W(M_{S\alpha})$. It follows by Lemma 2 e) that $\alpha \in W(M_{S\alpha}) \subseteq Q_\tau$.

Theorem 29.3. *If* $0 < \sigma < \tau$ *then* $\Omega_\sigma \in Q_\tau$.

Proof. By the assumption on τ, $\sigma \in W(M_\sigma)$. It follows from Lemmata 2 e) and 3 b) that for all $\nu < \sigma$

$$K_\nu \Omega_\sigma = K_\nu \sigma = P(\sigma) \subseteq W(M_\nu).$$

Hence $\Omega_\sigma \in M_\sigma$. $S\alpha < \sigma$ for all $\alpha \in M_\sigma(\Omega_\sigma)$ and $P(\alpha) = K_{S\alpha}\alpha \subseteq W(M_{S\alpha}$ consequently by Lemmata 2 e) and 3 b) $\alpha \in W(M_\sigma)$. Hence $M_\sigma(\Omega_\sigma) \subseteq W(M_\sigma)$. Using $\Omega_\sigma \in M_\sigma$ it follows by Lemma 1 b) that $\Omega_\sigma \in W(M_\sigma) \subseteq Q_\tau$.

Lemma 5. *If* $\bar{\theta}\alpha\beta \in \bar{\theta}(\Omega)$ *where* $S\beta < \tau$, *then:*

$$\bar{\theta}\alpha\beta \in M_{S\beta} \Rightarrow K_{S\beta}\alpha \cup \{\beta\} \subseteq M_{S\beta}.$$

Proof. If $v<S\beta$ then $K_v(K_{S\beta}\alpha)=K_v\alpha$ and consequently $K_v\bar{\theta}\alpha\beta=K_v\alpha\cup K_v\beta=K_v(K_{S\beta}\alpha\cup\{\beta\})$. Therefore it follows from $\bar{\theta}\alpha\beta\in M_{S\beta}$ that $K_v(K_{S\beta}\alpha\cup\{\beta\})\subseteq W(M_v)$ for all $v<S\beta$. Hence we obtain $K_{S\beta}\alpha\cup\{\beta\}\subseteq M_{S\beta}$.

Lemma 6. $\alpha\in Q_\tau$ *implies* $K_v\alpha\subseteq Q_\tau$ *for all* $v<\tau$.

Proof. $\alpha\in Q_\tau$ implies $S\alpha<\tau$ and by Lemma 4 we obtain $\alpha\in W(M_{S\alpha})$ and therefore $\alpha\in M_{S\alpha}$. It follows that $K_v\alpha\subseteq W(M_v)\subseteq Q_\tau$ for all $v<S\alpha$. If $S\alpha\leqslant v<\tau$ then $K_v\alpha=P(\alpha)$. Then it follows from $\alpha\in W(M_{S\alpha})$ by Lemma 2 e) that $K_v\alpha\subseteq W(M_{S\alpha})\subseteq Q_\tau$.

Lemma 7. $\alpha,\beta\in Q_\tau,\quad \bar{\theta}\alpha\beta\in\bar{\theta}(\Omega),\quad M_{S\beta}(\bar{\theta}\alpha\beta)\subseteq Q_\tau\Rightarrow\bar{\theta}\alpha\beta\in Q_\tau$.

Proof. $\alpha,\beta\in Q_\tau$ implies $S\beta<\tau$ and by Lemmas 6 and 4 we obtain

$$K_v\bar{\theta}\alpha\beta=K_v\alpha\cup K_v\beta\subseteq Q_\tau(\Omega_{v+1})=W(M_v)$$

for all $v<S\beta$. Hence we have $\bar{\theta}\alpha\beta\in M_{S\beta}$. By Lemma 4 $M_{S\beta}(\bar{\theta}\alpha\beta)\subseteq Q_\tau$ implies $M_{S\beta}(\bar{\theta}\alpha\beta)\subseteq W(M_{S\beta})$. Using $\bar{\theta}\alpha\beta\in M_{S\beta}$ we obtain $\bar{\theta}\alpha\beta\in W(M_{S\beta})\subseteq Q_\tau$ by Lemma 1 b).

Theorem 29.4. $\alpha,\beta\in Q_\tau,\ \bar{\theta}\alpha\beta\in\bar{\theta}(\Omega)\Rightarrow\bar{\theta}\alpha\beta\in Q_\tau$.

Proof. We have

(1) $\alpha,\beta\in Q_\tau$
(2) $\bar{\theta}\alpha\beta\in\bar{\theta}(\Omega)$

as hypotheses and $\bar{\theta}\alpha\beta\in Q_\tau$ as assertion. By Theorem 29.1 it follows from (1) that we can use induction over Q_τ on α. Then we have the induction hypothesis

(3) $\xi,\eta\in Q_\tau,\quad \xi<\alpha,\quad \bar{\theta}\xi\eta\in\bar{\theta}(\Omega)\Rightarrow\bar{\theta}\xi\eta\in Q_\tau$.

We may also use a subsidiary induction over Q_τ on β with induction hypothesis

$$\eta\in Q_\tau,\quad \eta<\beta,\quad \bar{\theta}\alpha\eta\in\bar{\theta}(\Omega)\Rightarrow\bar{\theta}\alpha\eta\in Q_\tau.$$

If $S\eta=S\beta$ it follows from (2) by Theorem 25.3 that $\bar{\theta}\alpha\eta\in\bar{\theta}(\Omega)$ too. Therefore we have, in particular, as induction hypothesis

(4) $\eta\in Q_\tau,\quad \Omega_{S\beta}\leqslant\eta<\beta\Rightarrow\bar{\theta}\alpha\eta\in Q_\tau$.

In order to prove the assertion $\bar{\theta}\alpha\beta\in Q_\tau$ it suffices, by (1), (2) and Lemma 7, to prove that $M_{S\beta}(\bar{\theta}\alpha\beta)\subseteq Q_\tau$. Now suppose

(5) $\gamma\in M_{S\beta}(\bar{\theta}\alpha\beta)$.

Now in order to prove our theorem it suffices to prove $\gamma\in Q_\tau$ under the hypotheses (1)–(5).

Proof of $\gamma \in Q_\tau$ by induction on the degree $G\gamma$.

1. $\gamma \notin P$. If $\delta \in P(\gamma)$ then $G\delta < G\gamma$ and by (5) and Lemma 2 a) $\delta \in M_{S\beta}(\bar{\theta}\alpha\beta)$. It follows by I.H. that $\delta \in Q_\tau$ for all $\delta \in P(\gamma)$. By Theorem 29.2 we have $\gamma \in Q_\tau$.

2. $\gamma = \Omega_\sigma$ where $0 < \sigma$. Since $\gamma < \bar{\theta}\alpha\beta$ and $\beta \in Q_\tau$, $\sigma \leqslant S\beta < \tau$ and therefore $\gamma \in Q_\tau$ by Theorem 29.3.

3. $\gamma = \bar{\theta}\gamma_1\gamma_2$ where $\gamma \leqslant K_{S\beta}\alpha \cup \{\beta\}$. It follows from (1) by Lemma 6 that $K_{S\beta}\alpha \cup \{\beta\} \subseteq Q_\tau$ and by Lemma 4 that $K_{S\beta}\alpha \cup \{\beta\} \subseteq W(M_{S\beta})$. Therefore there exists $\xi \in W(M_{S\beta})$ such that $\gamma \leqslant \xi$. Then by (5) $\gamma \in M_{S\beta}$. By Lemma 1 a) we obtain $\gamma \in W(M_{S\beta}) \subseteq Q_\tau$.

4. $\gamma = \bar{\theta}\gamma_1\gamma_2$ where $K_{S\beta}\alpha \cup \{\beta\} < \gamma$. Then $S\beta \leqslant S\gamma = S\gamma_2$. Since by (5) $\gamma < \bar{\theta}\alpha\beta$, we have $S\beta = S\gamma$. It follows by Theorem 25.2 that $K_{S\beta}\gamma_1 \cup \{\gamma_2\} < \gamma$. By I.H. we have $\delta \in Q_\tau$ for all $\delta \in K_{S\beta}\gamma_1 \cup \{\gamma_2\}$ and therefore

(6) $K_{S\beta}\gamma_1 \subseteq Q_\tau$

(7) $\gamma_2 \in Q_\tau$

4.1. $\gamma_1 = \alpha$. Now $\gamma = \bar{\theta}\gamma_1\gamma_2 < \bar{\theta}\alpha\beta$ and $S\beta = S\gamma_2$ imply $\Omega_{S\beta} \leqslant \gamma_2 < \beta$. Using (7) we therefore have, by (4), $\gamma = \bar{\theta}\alpha\gamma_2 \in Q_\tau$.

4.2. $\gamma_1 \neq \alpha$. By Theorem 25.4 $K_{S\beta}\alpha \cup \{\beta\} < \gamma = \bar{\theta}\gamma_1\gamma_2 < \bar{\theta}\alpha\beta$ implies $\gamma_1 < \alpha$. We prove $K_\sigma\gamma_1 \subseteq Q_\tau$ for all $\sigma < \tau$ by induction on σ.

a) $\sigma \leqslant S\beta$. Then $K_\sigma\gamma_1 = K_\sigma(K_{S\beta}\gamma_1)$. Using (6) we obtain $K_\sigma\gamma_1 \subseteq Q_\tau$ by Lemma 6.

b) $S\beta < \sigma < \tau$. By I.H. for all $\nu < \sigma$, we have $K_\nu(K_\sigma\gamma_1) = K_\nu\gamma_1 \subseteq Q_\tau$ and consequently by Lemma 4 $K_\nu(K_\sigma\gamma_1) \subseteq W(M_\nu)$. Hence we obtain $K_\sigma\gamma_1 \subseteq M_\sigma$. By §24 (p. 232) $\bar{\theta}\alpha\beta \in \bar{\theta}(\Omega)$ implies $S\mu(\alpha) \leqslant S\beta < \sigma$ and $\theta\alpha(\Omega_\sigma + 1) = \bar{\theta}\alpha\Omega_\sigma \in \bar{\theta}(\Omega)$. Similarly $\bar{\theta}\gamma_1\gamma_2 \in \bar{\theta}(\Omega)$ and $S\beta = S\gamma_2$ imply $\theta\gamma_1(\Omega_\sigma + 1) = \bar{\theta}\gamma_1\Omega_\sigma \in \bar{\theta}(\Omega)$. By Theorem 25.2 we obtain $K_\sigma\gamma_1 < \bar{\theta}\gamma_1\Omega_\sigma$. By Theorem 24.4 b) $\gamma_1 < \alpha$ implies $\theta\gamma_1(\Omega_\sigma + 1) \leqslant \theta\alpha(\Omega_\sigma + 1)$ and therefore $\bar{\theta}\gamma_1\Omega_\sigma \leqslant \bar{\theta}\alpha\Omega_\sigma$. Hence we have $K_\sigma\gamma_1 \subseteq M_\sigma(\bar{\theta}\alpha\Omega_\sigma)$. $\Omega_\sigma \in Q_\tau$ by Theorem 29.3. Thus the hypotheses (1)–(5) are satisfied with α, Ω_σ, $\delta \in K_\sigma\gamma_1$ replacing α, β, γ. If $\delta \in K_\sigma\gamma_1$ then $G\delta \leqslant G\gamma_1 < G\gamma$. Therefore by induction on $G\gamma$ it follows that $\delta \in Q_\tau$ for all $\delta \in K_\sigma\gamma_1$ and therefore $K_\sigma\gamma_1 \subseteq Q_\tau$.

Since $K_\sigma\gamma_1 \subseteq Q_\tau$ for all $\sigma < \tau$ and since $\gamma_1 < \alpha \in Q_\tau$ too we have in particular $P(\gamma_1) = K_{S\gamma_1}\gamma_1 \subseteq Q_\tau$. By Theorem 29.2 we have $\gamma_1 \in Q_\tau$. Using (7), $\gamma_1 < \alpha$ and $\bar{\theta}\gamma_1\gamma_2 \in \bar{\theta}(\Omega)$ by (3) we obtain $\gamma = \bar{\theta}\gamma_1\gamma_2 \in Q_\tau$.

Theorem 29.5. *$\bar{\theta}(\tau)$ is well-ordered.*

Proof. If $\alpha \in \bar{\theta}(\tau)$ we obtain $\alpha \in Q_\tau$ by induction on $G\alpha$ because of Theorems 29.2–29.4. Hence $\bar{\theta}(\tau) \subseteq Q_\tau$. By Theorem 29.1 it follows that $\bar{\theta}(\tau)$ is well-ordered.

2. The Formal System $\mathbf{ID}_n$ of n-Fold Iterated Inductive Definitions

The proof of well-ordering of $\bar{\theta}(\tau)$ above is not predicative nor can it be presented predicatively since $\bar{\theta}(1)$ already contains all ordinal terms $< \Gamma_0$. However the proof does not require full classical second order arithmetic but can be accomplished in a formal system which comes from first order arithmetic by adding generalized inductive definitions. In this way the proof of well-ordering does have a constructive character.

The proof of well-ordering of $\bar{\theta}(N)$ can be formalized in a formal system ID_N for $0<N<\omega$ where ID_n is constructed as follows for $0\leqslant n<\omega$.

The *terms* and *prime formulas* of the system ID_n are the same as those of the formal systems GPA and PA.

Inductive definition of the *formulas* and *predicators* of the system ID_n:

1. Every prime formula is a formula.
2. Every free predicate variable is a predicator.
3. If P is a predicator and t a term, then $P(t)$ is a formula.
4. If A and B are formulas so too is $(A\rightarrow B)$.
5. If $\mathscr{F}[0]$ is a formula in which the bound number variable x does not occur then $\forall x\mathscr{F}[x]$ is a formula and $\lambda x\mathscr{F}[x]$ a predicator.
6. If $n>0$ and

$$\forall x(U(x)\rightarrow V(x))\rightarrow\forall x(\mathscr{A}[U,x]\rightarrow\mathscr{A}[V,x])\ (\textit{Monotonicity condition}\text{ for }\mathscr{A})$$

where U, V are distinct free predicate variables and no free number variable and no free predicate variable occurs in $\mathscr{A}$, is a formula deducible in ID_{n-1} then $\mathscr{P}_{\mathscr{A}}$ is a predicator.

No bound predicate variables occur in ID_n. We use the same syntactic symbols for ID_n as for GPA.

Axioms of the system ID_n:

(Ax1)–(Ax4) as for GPA

(Ax5) $\mathscr{A}[\mathscr{P}_{\mathscr{A}},t]\rightarrow\mathscr{P}_{\mathscr{A}}(t)$

(Ax6) $\forall x(\mathscr{A}[P,x]\rightarrow P(x))\rightarrow(\mathscr{P}_{\mathscr{A}}(t)\rightarrow P(t))$

Basic inferences in the system ID_n:

(S1), (S2.0), (S3.0), (S4), (cut), (CI) and (str) as for GPA, (sub) $\mathscr{F}[U]\vdash\mathscr{F}[P]$ if U does not occur in $\mathscr{F}$.

All the axioms and basic inferences of the system ID_n shall consist of the formulas of ID_n defined above. P denotes an arbitrary predicator of the system ID_n and $\mathscr{P}_{\mathscr{A}}$ a predicator given by rule 6 of the definition above. By (Ax5) and (Ax6) such a predicator must represent the predicate $\mathscr{P}_{\mathscr{A}}$ such that $\mathscr{P}_{\mathscr{A}}(n)$ holds if, and only if, n is an element of the smallest set $\{x\mid P(x)\}$ such that $\forall x(\mathscr{A}[P,x]\rightarrow P(x))$ where the existence of such a smallest set is guaranteed by the monotonicity condition imposed on $\mathscr{A}$. In this way a generalized inductive definition is formalized by $\mathscr{P}_{\mathscr{A}}$.

$\mathrm{ID}_n\vdash F$ denotes that F is a deducible formula in ID_n.

Theorem 29.6. $\mathrm{ID}_n\vdash F\Rightarrow\mathrm{ID}_{n+1}\vdash F$.

Proof. This follows immediately from the definition of ID_n.

3. Formalization of the Proof of Well-Ordering of $\bar{\theta}(N)$ in ID_N

Let π be the bijective map of $\mathbb{N} \times \mathbb{N}$ onto $\mathbb{N}$ defined on p. 96. As before we set $p_0 := 2$ and p_n, for $n \neq 0$, to be the n-th odd prime.

Inductive definition of an *injective* map $\alpha \mapsto \ulcorner\alpha\urcorner$ of $\bar{\theta}(\Omega)$ into $\mathbb{N}$:

1. $\ulcorner 0 \urcorner := 1$
2. If $0 < \alpha \in \bar{\theta}(\Omega)$ set $\ulcorner \Omega_\alpha \urcorner := p_{\pi(\ulcorner\alpha\urcorner, 0)}$
3. If $\bar{\theta}\alpha\beta \in \bar{\theta}(\Omega)$ set $\ulcorner \bar{\theta}\alpha\beta \urcorner := p_{\pi(\ulcorner\alpha\urcorner, \ulcorner\beta\urcorner)}$
4. If $\alpha_1 + \cdots + \alpha_n \in \bar{\theta}(\Omega)$ where $\alpha_1 \geqslant \cdots \geqslant \alpha_n$ $(n > 1)$ are principal terms set $\ulcorner \alpha_1 + \cdots + \alpha_n \urcorner := \prod_{i=1}^{n} \ulcorner \alpha_i \urcorner$.

Corollary. *If* $\alpha \in \bar{\theta}(\Omega)$ *then* $0 < \ulcorner\alpha\urcorner \in \mathbb{N}$. *Hence it follows that* $\alpha \mapsto \ulcorner\alpha\urcorner$ *is injective.*

Definition of a *surjective* mapping $n \mapsto \hat{n}$ of $\mathbb{N}$ onto $\bar{\theta}(\Omega) \cup \{\Lambda_0\}$. ($\Lambda_0$ denotes the least ordinal which is not denoted by an ordinal term in the system $\bar{\theta}(\Omega)$.)

$$\hat{n} := \begin{cases} \alpha \in \bar{\theta}(\Omega) & \text{if } \ulcorner\alpha\urcorner = n, \\ \Lambda_0 & \text{otherwise.} \end{cases}$$

Obviously $\hat{n}$ is calculable for every natural number n.

If $m, n \in \mathbb{N}$ then $\prec (m, n)$ holds if, and only if, $\hat{m} < \hat{n}$. Therefore $\prec$ is a decidable 2-place arithmetic predicate. As before we write $s \prec t$ for $\prec(s, t)$.

If $\mathscr{F}$ is a 1-place nominal form such that $\mathscr{F}[t]$ is a formula for every term t then the progressiveness of a predicator P with respect to the relation $\prec$ on the set $\{x \mid \mathscr{F}[x]\}$ is expressed by the formula

$$\mathrm{Prog}\,[\mathscr{F}, P] := \forall x(\mathscr{F}[x] \wedge \forall y(y \prec x \wedge \mathscr{F}[y] \rightarrow P(y)) \rightarrow P(x)).$$

If

$$\mathrm{ID}_n \vdash \mathscr{F}[t] \wedge \mathrm{Pr}\,[\mathscr{F}, P] \rightarrow P(t)$$

holds for every predicator P then this means that the well-ordering of the set $\{x \mid \mathscr{F}[x]\}$ by $\prec$ up to t is provable in ID_n.

From now on we let N denote a fixed natural number >0. We define decidable arithmetic predicates as follows:

1. $T_N(m)$ holds if, and only if, $\hat{m} \in \bar{\theta}(N)$.
2. $S_i(m)$ holds if, and only if, $\hat{m} \in \bar{\theta}(\Omega)$ and $S\hat{m} \leqslant i$.
3. $K_i(m, n)$ holds if, and only if, $\hat{n} \in \bar{\theta}(\Omega)$ and $\hat{m} \in K_i\hat{n}$.

Define $(A \wedge B)$ and $(A \vee B)$ as above (p. 25). We write $A \bigwedge_{i=0}^{n} B_i$ for $A \wedge B_0 \wedge \cdots \wedge B_n$ and $\bigvee_{i=0}^{n} A_i$ for $A_0 \vee \cdots \vee A_n$.

Inductive definition of formulas $\mathscr{M}_n[t]$, $\mathscr{A}_n[U,t]$ and $\mathscr{W}_n[t]$:

1. $\mathscr{M}_0[t] := S_0(t)$
2. $\mathscr{A}_n[U,t] := \mathscr{M}_n[t] \wedge \forall y(\mathscr{M}_n[y] \wedge y < t \to U(y))$
3. $\mathscr{W}_n[t] := \mathscr{M}_n[t] \wedge \mathscr{P}_{\mathscr{A}_n}(t)$
4. $\mathscr{M}_{n+1}[t] := S_{n+1}(t) \bigwedge_{i=0}^{n} \forall z(K_i(z,t) \to \mathscr{W}_i[z])$

By induction on n one simultaneously proves:

a) $\mathscr{M}_n[t]$ is a formula of the system ID_n.
b) $\mathrm{ID}_n \vdash (\forall x(U(x) \to V(x)) \to \forall x(\mathscr{A}_n[U,x] \to \mathscr{A}_n[V,x])$.
c) $\mathscr{P}_{\mathscr{A}_n}$ is a predicator of the system ID_{n+1}.
d) $\mathscr{W}_n[t]$ is a formula of the system ID_{n+1}.

It follows that $\mathscr{M}_n[t]$, $\mathscr{P}_{\mathscr{A}_n}(t)$ and $\mathscr{W}_n[t]$ are formulas of ID_N if $n<N$. Therefore, as (Ax5) and (Ax6) hold for $n<N$

$$\mathrm{ID}_N \vdash \mathscr{M}_n[a] \wedge \forall y(\mathscr{M}_n[y] \wedge y < a \to \mathscr{P}_{\mathscr{A}_n}(y)) \to \mathscr{P}_{\mathscr{A}_n}(a)$$
$$\mathrm{ID}_N \vdash \forall x(\mathscr{M}_n[x] \wedge \forall y(\mathscr{M}_n[y] \wedge y < x \to P(y)) \to P(x)) \to (\mathscr{P}_{\mathscr{A}_n}(t) \to P(t))$$

It follows by the definition of $\mathscr{W}_n[t]$ that

$$\mathrm{ID}_N \vdash \mathrm{Prog}\,[\mathscr{M}_n, \lambda x \mathscr{W}_n[x]]$$
$$\mathrm{ID}_N \vdash \mathscr{W}_n[t] \to \mathscr{M}_n[t] \wedge (\mathrm{Prog}\,[\mathscr{M}_n, P] \to P(t))$$

From here and the definition of $\mathscr{M}_0[t]$ and $\mathscr{M}_{n+1}[t]$ we therefore see that the sets M_n and $W(M_n)$ defined on p. 282 are expressed in ID_N for $n<N$ by

$$M_n = \{x \mid \mathscr{M}_n[x]\}, \qquad W(M_n) = \{x \mid \mathscr{W}_n[x]\}.$$

Now set

$$\mathscr{Q}_N[t] := \bigvee_{i=0}^{N-1} \mathscr{W}_i[t].$$

Then $\mathscr{Q}_N[t]$ is a formula of the system ID_N such that the set Q_N defined on p. 283 is represented by

$$Q_N = \{x \mid \mathscr{Q}_N[x]\}.$$

The proof of well-ordering of $\bar{\theta}(N)$ can now be translated into ID_N in the following way:

Theorem 29.7. $\mathrm{ID}_N \vdash \mathrm{Prog}\,[T_N, P] \to \forall x(T_N(x) \to P(x))$

Proof. As for Theorem 29.1 one proves

(1) $\mathrm{ID}_N \vdash \mathscr{Q}_N[a] \wedge \mathrm{Prog}\,[\mathscr{Q}_N, P] \to P(a)$

In the same way as in the proof of Theorem 29.5 we proved $\bar{\theta}(N) \subseteq Q_N$ we now prove

(2) $\mathrm{ID}_N \vdash T_N(a) \rightarrow \mathcal{Q}_N[a]$

From (2) we have

$$\mathrm{ID}_N \vdash \mathrm{Prog}\,[T_N, P] \rightarrow \mathrm{Prog}\,[\mathcal{Q}_N, \lambda x(T_N(x) \rightarrow P(x))]$$

Since (1) also holds with $\lambda x(T_N(x) \rightarrow P(x))$ instead of P we obtain

$$\mathrm{ID}_N \vdash \mathcal{Q}_N[a] \wedge \mathrm{Prog}\,[T_N, P] \rightarrow (T_N(a) \rightarrow P(a)).$$

Using (2) we obtain

$$\mathrm{ID}_N \vdash \mathrm{Prog}\,[T_N, P] \rightarrow (T_N(a) \rightarrow P(a)).$$

The assertion follows using a basic inference (S2.0).

4. Embedding ID_n in a Subsystem of PA

If $0 < \nu \leqslant \omega$ let PA_ν be the subsystem of PA which contains only formulas of rank $< \nu$ and let PA_ν^0 be the subsystem of PA_ν in which no free predicate variable occurs inside a predicate quantifier. That is to say: The set $\mathrm{Pr}\,(F)$ defined on p. 252 is to be empty for every formula F in the system PA_ν^0. Then PA_ω is the subsystem of PA which contains only weak formulas.

If $0 < n < \nu \leqslant \omega$ then PA_n is a subsystem of PA_ν and PA_n^0 a subsystem of PA_ν^0.

Lemma 8. *If $\mathcal{F}[U]$ is a formula and P a predicator in the system* PA_ν^0 $(0 < \nu \leqslant \omega)$ *where U does not occur in $\mathcal{F}$, then*

$$\mathcal{F}[U] \vdash \mathcal{F}[P]$$

is an inference (sub) in the system PA_ν^0.

Proof. In this case $\mathrm{Pr}\,(\mathcal{F}[U])$ is empty and P is a weak predicator. Therefore the rank of $\mathcal{F}[P]$ is equal to the maximum of the ranks of $\mathcal{F}[U]$ and $P(0)$. The assertion follows.

Inductive definition of a formula $\bar{F}$ and a predicator $\bar{P}$ in the system PA_{n+1}^0 for each formula F and predicator P of the system ID_n.

1. For each prime formula F set $\bar{F} := F$.
2. For each free predicate variable U set $\overline{U} := U$.
3. For each predicator P in the system ID_n set $\overline{P(t)} := \bar{P}(t)$.
4. $\overline{(A \rightarrow B)} := (\bar{A} \rightarrow \bar{B})$.

5. If $\overline{\mathscr{F}[a]}$ is the formula $\overline{\mathscr{F}}[a]$ where a does not occur in the nominal forms $\mathscr{F}$ and $\overline{\mathscr{F}}$ then set

$$\overline{\forall x \mathscr{F}[x]} := \forall x \overline{\mathscr{F}}[x] \quad \text{and} \quad \overline{\lambda x \mathscr{F}[x]} := \lambda x \overline{\mathscr{F}}[x].$$

6. If $\mathscr{P}_{\mathscr{A}}$ is a predicator in the system ID_n and $\overline{\mathscr{A}[U, a]}$ is the formula $\overline{\mathscr{A}}[U, a]$ where U and a do not occur in the nominal forms $\mathscr{A}$ and $\overline{\mathscr{A}}$ then $n > 0$ and, because of the inductive definition, $\overline{\mathscr{A}}[U, a]$ is a formula of the system PA_n^0. Then set

$$\overline{\mathscr{P}_{\mathscr{A}}} := \lambda z \forall Y(\forall x(\overline{\mathscr{A}}[Y, x] \rightarrow Y(x)) \rightarrow Y(z)).$$

Theorem 29.8 (Embedding Theorem). *If* $\mathrm{ID}_n \vdash F$ *holds then* $\bar{F}$ *is a deducible formula in the system* PA_{n+1}^0.

Proof by induction on n with a nested induction on the length of the deduction of F in ID_n.

1. Suppose F is not an (Ax5) or (Ax6). Then the assertion follows from the subsidiary I.H. using Lemma 8.

2. Suppose F is an (Ax5) or (Ax6). Then $n > 0$ and the axiom is for a predicator $\mathscr{P}_{\mathscr{A}}$ such that

$$\mathrm{ID}_{n-1} \vdash \forall x(U(x) \rightarrow V(x)) \rightarrow \forall x(\mathscr{A}[U, x] \rightarrow \mathscr{A}[V, x]).$$

By the I.H. we obtain

(1) $\mathrm{PA}_n^0 \vdash \forall x(U(x) \rightarrow V(x)) \rightarrow \forall x(\overline{\mathscr{A}}[U, x] \rightarrow \overline{\mathscr{A}}[V, x])$.

$\overline{\mathscr{P}_{\mathscr{A}}}$ is the predicator

$$\lambda z \forall Y(\forall x(\overline{\mathscr{A}}[Y, x] \rightarrow Y(x)) \rightarrow Y(z))$$

of the system PA_{n+1}^0. For each predicator P of the system ID_n we have

$$\mathrm{PA}_{n+1}^0 \vdash \forall x(\overline{\mathscr{A}}[\bar{P}, x] \rightarrow \bar{P}(x)) \rightarrow (\forall Y(\forall x(\overline{\mathscr{A}}[Y, x] \rightarrow Y(x)) \rightarrow Y(t)) \rightarrow \bar{P}(t)).$$

Using a basic inference (S4) we obtain

(2) $\mathrm{PA}_{n+1}^0 \vdash \forall x(\overline{\mathscr{A}}[\bar{P}, x] \rightarrow \bar{P}(x)) \rightarrow (\overline{\mathscr{P}_{\mathscr{A}}}(t) \rightarrow \bar{P}(t))$.

That is to say:

(3) $\mathrm{PA}_{n+1}^0 \vdash \bar{F}$ if F is an (Ax6).

From (1) and (2) we obtain

$$\mathrm{PA}^0_{n+1} \vdash \forall x(\overline{\mathscr{P}}_{\mathscr{A}}(x) \to U(x)) \to (\overline{\mathscr{A}}[\overline{\mathscr{P}}_{\mathscr{A}}, t] \to \overline{\mathscr{A}}[U, t])$$

$$\mathrm{PA}^0_{n+1} \vdash \forall x(\overline{\mathscr{A}}[U, x] \to U(x)) \to \forall x(\overline{\mathscr{P}}_{\mathscr{A}}(x) \to U(x)).$$

From these two formulas we obtain

$$\mathrm{PA}^0_{n+1} \vdash \overline{\mathscr{A}}[\overline{\mathscr{P}}_{\mathscr{A}}, t] \to (\forall x(\overline{\mathscr{A}}[U, x] \to U(x)) \to U(t)).$$

Using basic inferences (S2.1) and (S4) we obtain

$$\mathrm{PA}^0_{n+1} \vdash \overline{\mathscr{A}}[\overline{\mathscr{P}}_{\mathscr{A}}, t] \to \overline{\mathscr{P}}_{\mathscr{A}}(t).$$

That is to say:

(4) $\mathrm{PA}^0_{n+1} \vdash \bar{F}$ if F is an (Ax5).

(3) and (4) prove the assertion.

Theorem 29.9. *If* $0 < N < \omega$, *then:*

a) *Formalized transfinite induction*

$$\mathrm{Prog}\,[T_N, P] \to \forall x(T_N(x) \to P(x))$$

over $\bar{\theta}(N)$ *is deducible in the system* PA^0_{N+1}.

b) *The consistency of the formal system* PA_{N+1} *is provable by transfinite induction up to* $((\omega, \Omega_N), 0) = \theta(\theta\omega(\Omega_N+1))0$.

Proof. a) follows from Theorems 29.7 and 29.8.

b) Every normal deduction in the system PA_{N+1} has an order $\alpha \in \bar{\theta}(N+1)$ where $\alpha \ll (\omega, \Omega_N)$. Therefore $d_0\alpha < ((\omega, \Omega_N), 0)$. The assertion therefore follows from the proofs of Theorems 27.8 and 27.9.

Remark. By W. Pohlers [2] the consistency of the formal systems ID_N and PA^0_{N+1} is provable even by transfinite induction up to $((1, \Omega_N), 0) = \theta\varepsilon_{\Omega_N+1}0$ and formalized transfinite induction is deducible in these systems up to each ordinal term $\hat{n} < ((1, \Omega_N), 0)$.

Definition. Let $T_{\omega+1}$ be the deducible arithmetic predicate which holds if, and only if, $\hat{n} \in \bar{\theta}(\omega+1)$.

Theorem 29.10. *The ordinal* $\theta\Omega_\omega 0$ *is characteristic for the formal system* PA_ω *in the following sense:*

a) *Formalized transfinite induction up to n is provable in* PA^0_ω *for every ordinal term* $\hat{n} < \theta\Omega_\omega 0$ *that is to say, the formula*

$$\text{Prog}\,[T_{\omega+1}, P] \to P(n)$$

is deducible in PA^0_ω.

b) *The consistency of the formal system* PA_ω *is provable by transfinite induction up to* $\theta\Omega_\omega 0$.

Proof. a) If $\hat{n} < \theta\Omega_\omega 0$ there is a natural number $N > 0$ such that $\hat{m} \in \bar{\theta}(N)$ for all $\hat{m} \leqslant \hat{n}$. Then it follows from Theorem 29.9 a) that the formula

$$\text{Prog}\,[T_{\omega+1}, P] \to P(n)$$

is deducible in PA^0_{N+1} and therefore in PA^0_ω too.

b) For each normal deduction H of the system PA_ω there is a natural number $N > 0$ such that H is a normal deduction of the system PA_{N+1}. Hence the assertion follows from Theorem 29.9 b) since $\theta(\theta\omega(\Omega_N + 1))0 < \theta\Omega_\omega 0$.

Remarks. 1. By W. Buchholz [1] and [6] $\theta\Omega_\omega 0$ is the order type of level 0 of the notation system Σ of H. Pfeiffer [2]. This is at the same time the order type of the union of all invariant segments of the systems $\Sigma(N)$ ($N \in \mathbb{N}$) of K. Schütte [10]. By H. Levitz [1], [2] and [3] this is simultaneously the order type of all the *ordinal diagrams of finite order* of G. Takeuti [2].

2. The proof of well-ordering given in the first part of this section for $\bar{\theta}(\tau)$ can be formalized in a system with transfinitely iterated inductive definitions in the case where $\tau = \omega + 1$. But it cannot be formalized in any system ID_n where $n < \omega$ nor in GPA. This also holds for every proof of well-ordering up to the ordinal term $((\omega^2, \Omega_\omega), 0)$ of the system $\bar{\theta}(\omega+1)$. Transfinite induction up to this ordinal was used in §27 to prove the consistency of GPA.

3. It follows from work of W. Buchholz and W. Pohlers [5] that formalized transfinite induction is deducible in GPA up to each ordinal term

$$\hat{n} < ((1, \Omega_\omega), 0) = \theta\varepsilon_{\Omega_\omega + 1} 0.$$

It can further be shown that this induction is deducible in PA up to each ordinal term $\hat{n} < ((0, \Omega_\omega + \varepsilon_0), 0) = \theta(\Omega_\omega \cdot \varepsilon_0)0$. It is possible to give consistency proofs for the systems GPA and PA with weaker transfinite inductions than were used in §§27 and 28. For by the results of W. Buchholz [4] and W. Pohlers [3] if α_{GPA} and α_{PA} are the least ordinals up to which transfinite induction is sufficient for such consistency proofs then we have

$$\alpha_{\text{GPA}} = ((1, \Omega_\omega), 0) = \theta\varepsilon_{\Omega_\omega + 1} 0,$$

$$\alpha_{\text{PA}} = ((0, \Omega_\omega + \varepsilon_0), 0) = \theta(\Omega_\omega \cdot \varepsilon_0)0.$$

Bibliography

Ackermann, W.:

[1] Zur Widerspruchsfreiheit der Zahlentheorie. Math. Ann. **117** (1940), 162–194.
[2] Widerspruchsfreier Aufbau einer typenfreien Logik. (Erweitertes System). Math. Zeitschr. **55** (1952), 364–384.
[3] Widerspruchsfreier Aufbau einer typenfreien Logik. II. Math. Zeitschr. **57** (1953), 155–166.

Aczel, P.:

[1] A new approach to the Bachmann method for describing countable ordinals. Mimeographed.

Bridge, J.:

[1] Some problems in mathematical logic. Systems of ordinal functions and ordinal notations. Thesis Oxford 1972.
[2] A simplification of the Bachmann method for generating large countable ordinals. Journal of Symbolic Logic **40** (1975), 171–185.

Buchholz, W.:

[1] Rekursive Bezeichnungssysteme für Ordinalzahlen auf der Grundlage der Feferman-Aczelschen Normalfunktionen θ_α. Dissertation München 1974.
[2] Normalfunktionen und konstruktive Systeme von Ordinalzahlen. Proof Theory Symposion Kiel 1974. Springer Lecture Notes in Math. **500** (1975), 4–25.
[3] Über Teilsysteme von $\bar{\theta}(\{g\})$. Archiv f. math. Logik u. Grundlagenf. **18** (1976), 85–98.
[4] Zur Beweisbarkeit der transfiniten Induktion in Π_1^1-CA. Erscheint im Archiv f. math. Logik u. Grundlagenf.

Buchholz, W. and Pohlers, W.:

[5] Provable wellorderings of formal theories for transfinitely iterated inductive definitions. To appear in Journal of Symbolic Logic.

Buchholz, W. und Schütte, K.:

[6] Die Beziehungen zwischen den Ordinalzahlsystemen Σ und $\bar{\theta}(\omega)$. Archiv f. math. Logik u. Grundlagenf. **17** (1976), 179–190.

Diller, J. und Schütte, K.:

[1] Simultane Rekursionen in der Theorie der Funktionale endlicher Typen. Archiv f. math. Logik u. Grundlagenf. **14** (1971), 69–74.

Feferman, S.:

[1] Systems of predicative analysis. Journal of Symbolic Logic **29** (1964), 1–30.
[2] Systems of predicative analysis, II: Representations of ordinals. Journal of Symbolic Logic **33** (1968), 193–200.
[3] Autonomous transfinite progressions and the extent of predicative mathematics. Logic, Methodology and Philosophy of Science III. Amsterdam 1968, 121–135.

[4] Hereditarely replete functionals over the ordinals. Intuitionism and Proof Theory. Amsterdam 1970, 289–301.
[5] Formal theories for transfinite iterations of generalized inductive definitions and some subsystems of analysis. Intuitionism and Proof Theory. Amsterdam 1970, 303–326.
[6] Ordinals and functionals in proof theory. Proc. of the International Congress of Math. Nice 1970. Paris 1971, Vol. 1, 229–233.

Gentzen, G.:

[1] Untersuchungen über das logische Schließen. Math. Zeitschr. **39** (1934), 176–210, 405–431.
[2] Die Widerspruchsfreiheit der reinen Zahlentheorie. Math. Ann. **112** (1936), 493–565.
[3] Neue Fassung des Widerspruchsfreiheitsbeweises für die reine Zahlentheorie. Forschungen zur Logik und zur Grundlegung der exakten Wissenschaften. Neue Folge **4** (1938), 19–44.
[4] Beweisbarkeit und Unbeweisbarkeit von Anfangsfällen der transfiniten Induktion in der reinen Zahlentheorie. Math. Ann. **119** (1943), 149–161.

Gödel, K.:

[1] Über formal unentscheidbare Sätze der Principia Mathematica und verwandter Systeme I. Monatsh. Math. Phys. **38** (1931), 173–198.
[2] Über eine bisher noch nicht benützte Erweiterung des finiten Standpunktes. Dialectica **12** (1958), 280–287.

Henkin, L.:

[1] Completeness in the theory of types. Journal of Symbolic Logic **14** (1949), 159–166.

Hilbert, D. und Bernays, P.:

[1] Grundlagen der Mathematik I, II, Die Grundlehren der math. Wissenschaften 40, 50. 2. Auflage, Springer 1968, 1970.

Howard, W.:

[1] Assignment of ordinals to terms for primitive recursive functionals of finite type. Intuitionism and Proof Theory. Amsterdam 1970, 443–458.
[2] A system of abstract constructive ordinals. Journal of Symbolic Logic **37** (1972), 355–374.

Kreisel, G.:

[1] La prédicativité. Bull. de la Soc. Math. de France **88** (1960), 371–391.
[2] A survey of proof theory. Journal of Symbolic Logic **33** (1968), 321–388.
[3] A survey of proof theory II. Proc. of the 2nd Scandinavian Logic Symposium. Amsterdam 1971, 109–170.

Levitz, H.:

[1] On a simplification of Takeuti's ordinal diagrams of finite order. Zeitschr. f. math. Logik u. Grundl. d. Math. **15** (1969), 141–154.
[2] On the relationship between Takeuti's ordinal diagrams O(n) and Schütte's system of ordinal notations Σ(n). Intuitionism and Proof Theory. Amsterdam 1970, 377–405.

Levitz, H. and Schütte, K.:

[3] A characterization of Takeuti's ordinal diagrams of finite order. Archiv f. math. Logik u. Grundlagenf. **14** (1970), 75–97.

Lorenzen, P.:

[1] Algebraische und logistische Untersuchungen über freie Verbände. Journal of Symbolic Logic **16** (1951), 81–106.

Pfeiffer, H.:

[1] Ausgezeichnete Folgen für gewisse Abschnitte der zweiten und weiterer Zahlenklassen. Dissertation Hannover 1964.

[2] Ein Bezeichnungssystem für Ordinalzahlen. Archiv f. math. Logik u. Grundlagenf. **12** (1969), 12–17.
[3] Ein Bezeichnungssystem für Ordinalzahlen. Archiv f. math. Logik u. Grundlagenf. **13** (1970), 74–90.
[4] Vergleich zweier Bezeichnungssysteme für Ordinalzahlen. Archiv f. math. Logik u. Grundlagenf. **15** (1972), 41–56.
[5] Über zwei Bezeichnungssysteme für Ordinalzahlen. Archiv f. math. Logik u. Grundlagenf. **16** (1974), 23–36.

Pohlers, W.:

[1] Eine Grenze für die Herleitbarkeit der transfiniten Induktion in einem schwachen Π_1^1-Fragment der Analysis. Dissertation München 1973.
[2] An upper bound for the provability of transfinite induction in systems with N-times iterated inductive definitions. Proof Theory Symposion Kiel 1974. Springer Lecture Notes in Math. **500** (1975), 271–289.
[3] Ordinals connected with formal theories for transfinitely iterated inductive definitions. To appear in Journal of Symbolic Logic.

Prawitz, D.:

[1] Hauptsatz for higher order logic. Journal of Symbolic Logic, **33** (1968), 452–457.
[2] Ideas and results in proof theory. Proc. of the 2nd Scandinavian Logic Symposium. Amsterdam 1971, 235–307.

Schütte, K.:

[1] Beweistheoretische Erfassung der unendlichen Induktion in der reinen Zahlentheorie. Math. Ann. **122** (1951), 369–389.
[2] Beweistheoretische Untersuchung der verzweigten Analysis. Math. Ann. **124** (1952), 123–147.
[3] Ein System des verknüpfenden Schließens. Archiv f. math. Logik u. Grundlagenf. **2** (1956), 55–67.
[4] Syntactical and semantical properties of simple type theory. Journal of Symbolic Logic **25** (1960), 305–326.
[5] Beweistheorie. Die Grundlehren der math. Wissensch. 103. Springer 1960.
[6] Logische Abgrenzungen des Transfiniten. Logik und Logikkalkül. Freiburg, Münster 1962, 105–114.
[7] Der Interpolationssatz der intuitionistischen Prädikatenlogik. Math. Ann. **148** (1962), 192–200.
[8] Eine Grenze für die Beweisbarkeit der transfiniten Induktion in der verzweigten Typenlogik. Archiv f. math. Logik u. Grundlagenf. **7** (1964), 45–60.
[9] Predicative well-orderings. Formal Systems and Recursive Functions. Proc. of the 8th Logic Colloquium Oxford 1963. Amsterdam 1965, 280–303.
[10] Ein konstruktives System von Ordinalzahlen. Archiv f. math. Logik u. Grundlagenf. **11** (1969), 126–137; **12** (1969), 3–11.
[11] Einführung der Normalfunktionen θ_α ohne Auswahlaxiom und ohne Regularitätsbedingung. Archiv f. math. Logik u. Grundlagenf. **17** (1976), 171–178.

Tait, W. W.:

[1] Infinitely long terms of transfinite type. Formal Systems and Recursive Functions. Proc. of the 8th Logic Colloquium Oxford 1963. Amsterdam 1965, 176–185.
[2] A non constructive proof of Gentzen's Hauptsatz for second order predicate logic. Bull. Amer. Math. Soc. **72** (1966), 980–983.
[3] Normal derivability in classical logic. The Syntax and Semantics of Infinitary Languages. Springer Lecture Notes in Math. **72** (1968), 204–256.

Takahashi, M.:

[1] A proof of cut-elimination theorem in simple type theory. J. Math. Soc. Japan **19** (1967), 399–410.
[2] Simple type theory of Gentzen style with the inference of extensionality. Proc. Japan Acad. **44** (1968), 43–45.
[3] A system of simple type theory of Gentzen style with inference on extensionality, and the cut-elimination in it. Comment. universitatis Sancti Pauli, Tokyo **18** (1969), 129–147.

Takeuti, G.:

[1] On a generalized logic calculus. Japan J. Math. **23** (1953), 39–96.
[2] Ordinal diagrams. J. Math. Soc. Japan **9** (1957), 386–394.

[3] Ordinal diagrams II. J. Math. Soc. Japan **12** (1960), 385–391.
[4] Consistency proofs of subsystems of classical analysis. Ann. of Math. **86** (1967), 299–348.
[5] Proof Theory. Studies in Logic **81**. Amsterdam 1975.

Troelstra, A. S.:

[1] Metamathematical investigations of intuitionistic arithmetic and analysis. Springer Lecture Notes in Math. **344** (1973).

Subject Index

λ-abstraction 121
Aczel 4, 224
addition in the system FT 120
— of ordinals 80, 223
— of ordinal terms 89, 90
additive principal numbers 80, 224
arithmetic difference 123
atomic formulas 20, 57
autonomous ordinal terms 213, 215, 217
axioms of choice 150

bar-induction 254
basic deduction 256, 271, 275
— formula of the system FT 129
— inferences 14
Bernays-Gödel set theory 74, 221
Beth 36, 55
Bridge 4
Brouwer 36
Buchholz 4, 225, 230, 282, 292
bundle of formulas 256, 276

calculable arithmetic functions 167
Cantor normal form 81
cardinality of a set of ordinals 222
cardinals 222
characteristic term of a basic formula 130
closed formulas 115
— set of ordinals 223
coefficient sets 234, 235
combinators 98
Compactness Theorem 34
complete induction inference 170, 254
Completeness Theorem 29, 35, 69
components of an ordinal 224
connectives 7
consistency 15
Consistency Theorem 17, 28, 35
consistent set of formulas 32
constant prime formulas 168
continuous ordering functions 78
Craig 51
critical ordinals and functions ϕ_α 81, 82, 83
— parts of a formula 63
Curry 98
cut 37, 170, 200, 254
Cut Elimination Theorem 203, 204
cut formula 200, 254
Cut rule 24, 62, 70

decidable arithmetic predicates 167
deducibility 14
D-chain (deduction chain) 29, 33, 63
deduction functions 248, 249
— part 263
— threat 262
degree of a basic inference 254
— of a formula 41, 252
— of an M-term 68
— of an ordinal of $\bar{\theta}(\Omega)$ 234
— of a term of T* 236
— of a type 106
D-form 40
$\mathscr{D}$-formula 27
directly derivable inferences 16
distinguished part of a formula in classical predicate calculus 29, 62, 63
— — of a formula in a normal deduction 261, 277

eigenformula of an (S3.1)-inference 254
eigenvariable of an inference 20, 41, 60, 254
elementary analysis 177
— formulas 168
— number theory 176
Embedding Theorem 175, 273, 290
end formula of a deduction 255, 275, 276
— part of a formula 40
— piece of a deduction 255, 256, 276
equality of terms of finite type 112
equivalent formulas 169
essentially less 247
explicit minimal parts of a formula in a normal deduction 256, 276
Extensionality Theorem 173, 205

Feferman 4, 84, 214, 220, 224, 225
finite ordinals 76, 221, 222
— ordinal terms 90
finitely axiomatisable theories 54
— deducible formulas 176
formal system, CS of classical sentential calculus 17
— — CP of classical predicate calculus 19
— — IP1–IP3 of intuitionistic predicate calculus 36, 38, 40
— — CT of classical simple type theory 56
— — FT of functionals of finite type 113
— — PN of pure number theory 134
— — DA of Δ_1^1-analysis 170
— — EN of elementary number theory 176
— — EA of elementary analysis 177
— — GPA of generalized Π_1^1-analysis 252
— — PA of Π_1^1-analysis 248
— — ID_n of n-fold iterated inductive definitions 286
formalized transfinite induction 179, 209, 291
functionals of finite type 99

generalized inductive definitions 286
Gentzen 3, 21, 38, 147, 220
Gödel 2, 3, 62, 98, 134, 147

height of a type 58
Henkin 66
Heyting 36
Hilbert 2, 3, 26
Howard 3, 98, 105

identity functional 121
implicit minimal parts of a formula in a normal deduction 256, 276
inaccessible ordinals and functions θ_α 224, 225
inconsistent set of formulas 32
independence of premise 150
induction on the deduction 15
Induction Theorem 75, 128, 173, 222
inductive definitions 8
initial formulas of a deduction 255, 271
Interpolation Theorem 51, 54
Interpretation Theorem 157, 207, 208
interpreting formulas 155, 156
Inversion rules 22, 61, 201, 272
isotonism 74
iterator 98

König's Lemma 30
Kreisel 55
Kripke 36

left parts of formulas and L-forms 39, 40
length of formulas and predicators 168, 198
— of a term of the set T 86
level of formulas and predicators 198
— of ordinals 224
Levitz 292
limit numbers 75, 222
limiting number of a formal system 211
Löwenheim-Skolem Theorem 29
logical consequence 176

Markov's principle 150
maximal α-critical ordinals and functions ψ_α 84, 85
maximally α-inaccessible ordinals 230, 232
metalogic, metamathematics 2
minimal parts of a formula 12, 20, 169
model of classical predicate calculus 27
M-term, M-formula, M-variant 65, 66
multiplication in the system FT 120

natural product of ordinal terms 105, 106
— sum of ordinal terms 105, 248
negative parts of formulas and N-forms 10, 11, 20
nominal forms 11, 14
normal deduction 256, 275, 276
— form of a term 104
— functions 78, 223
notation systems $\bar{\theta}(\Omega)$ and $\bar{\theta}(\tau)$ for ordinals 237, 238
NP-forms 11
number terms 134
numerals 99, 168
numerical terms 168

open terms 100
order of a deducible formula 15
— of a formula in a normal deduction 257, 276
— of a normal deduction 258
— of a term of finite type 108
ordering functions 77, 223
ordinal numbers 74, 75
ordinals of the segment $\mathfrak{O}$ 74, $\mathfrak{O}^*$ 221, 222
— of the set $\bar{\theta}(\Omega)$ 234
ordinal terms 213, 215, 217, 236
— terms of the set OT 86, 87, OT* 236, 237
$\mathfrak{O}$-segment 76, 77, $\mathfrak{O}^*$-segment 223

partial valuation 63, 64
Peano axioms 167
permissible inferences 15
$\mathcal{F}$-permissible formulas 199
Pfeiffer 292
PN-Lemma 27, 66
Pohlers 257, 291, 292

positive parts of formulas and P-forms 10, 11, 20
possible valuations 67
Prawitz 3, 62, 64, 67
predecessor functional 122
predicators 168, 198
prime formulas of second order arithmetic 168
— terms of the system CT 58
primitive formulas 63, 268
— terms of finite type 98
principal inferences 200, 254, 270
— parts of axioms 170, 200, 253
— — of basic inferences 21, 41, 60, 200, 254
Principal Semantic Lemma 29, 34, 64
— Syntactic Lemma 29, 33, 64
principal terms of T 86, T* 236
progressive 75
progressiveness $\mathscr{P}\imath[P]$ 178, $\mathscr{P}\imath[P^{\alpha}]$ 209, Prog [$\mathscr{F}$, P] 287

quantifiers 19

rank of a cut 200
— of a deduction function 248, 249
— of a formula 199, 253
— of an inference 256, 276
— of a normal deduction 257
— of a term 58, 59
recursively comparable ordinal terms 237
recursor 125
reducibility degree of a formula 18, 29
— of a term of finite type 100
reducible and irreducible formulas 12, 13, 29, 62
— normal deductions 259, 277
— parts of a formula 28, 62
reduction of terms of finite type 100, 101
regular terms of the system CT 58
regularity of a cardinal 222
Replacement rule 174, 201
right parts of formulas and R-forms 39, 40

Satisfiability Theorem 34
satisfiable set of formulas 32
Schwichtenberg 185
secondary part of an (→L)-inference 41
semantic 4
semantically equivalent sentential forms 8
semantic consequence 35
Semi-formal system, DA* of Δ_1^1-analysis 174
— — EN* of elementary number theory 177
— — EA* of elementary analysis 177
— — RA* of ramified analysis 197
— — PA* of Π_1^1-analysis 269
sentential form 8
— valuation 8

s-closed, s-complete, s-consistent, s-valid 16, 17
simple formulas of second order arithmetic 168
simultaneous primitive recursion 126
singular normal deduction 261, 277
Soundness Theorem 66
specialization of a formula 273
strength of a cut 271, 272
— of a formula 270
strong formulas and predicators 252
strongly critical ordinals 83
structural inferences $F \overset{s}{\vdash} G$ 20, 40, 169
Structural rule 23, 61, 172, 175, 201, 254
substitution inference 254
Substitution rule 22, 42, 61, 114, 136, 172
subterm of an ordinal term 242
— of a term of the system CT 58
subterm-chain 58
successor of an ordinal 75, 222
— functional 98
— number 75
suitable cut of a normal deduction 261, 277
supremum 75, 76, 222
syntactic 4
— consequence 35
— equivalence $A \overset{I}{\sim} B$ 47, $A \overset{C}{\sim} B$ 51
system of sets 65

Tait 62, 204
Takahashi 3, 62, 64
Takeuti 4, 62, 221, 252, 257, 292
terms of finite type 98
— of second order arithmetic 167, 168
— of the set T 86, T* 236
— of the system CT 57
tertium non datur 2, 3
total valuation 66
transfinite induction 75
Troelstra 151
truth functions 7
— values 7
truth-value of a constant prime formula 169
types of functions 98
— of terms of the system CT 56

valid formulas 10, 11, 27, 66
valuations 27
value of a numerical term 169

weak formulas and predicators 252
— inferences 15, 22, 23
weight of a formula 252
— of a normal deduction 257
well-ordered sets 73

Zermelo-Fraenkel set theory 74, 221

Die Grundlehren der mathematischen Wissenschaften in Einzeldarstellungen mit besonderer Berücksichtigung der Anwendungsgebiete

Eine Auswahl

---

23. Pasch: Vorlesungen über neuere Geometrie
41. Steinitz: Vorlesungen über die Theorie der Polyeder
45. Alexandroff/Hopf: Topologie. Band 1
46. Nevanlinna: Eindeutige analytische Funktionen
63. Eichler: Quadratische Formen und orthogonale Gruppen
102. Nevanlinna/Nevanlinna: Absolute Analysis
114. Mac Lane: Homology
123. Yosida: Functional Analysis
127. Hermes: Enumerability, Decidability, Computability
131. Hirzebruch: Topological Methods in Algebraic Geometry
135. Handbook for Automatic Computation. Vol. 1/Part a: Rutishauser: Description of ALGOL 60
136. Greub: Multilinear Algebra
137. Handbook for Automatic Computation. Vol. 1/Part b: Grau/Hill/Langmaack: Translation of ALGOL 60
138. Hahn: Stability of Motion
139. Mathematische Hilfsmittel des Ingenieurs. 1. Teil
140. Mathematische Hilfsmittel des Ingenieurs. 2. Teil
141. Mathematische Hilfsmittel des Ingenieurs. 3. Teil
142. Mathematische Hilfsmittel des Ingenieurs. 4. Teil
143. Schur/Grunsky: Vorlesungen über Invariantentheorie
144. Weil: Basic Number Theory
145. Butzer/Berens: Semi-Groups of Operators and Approximation
146. Treves: Locally Convex Spaces and Linear Partial Differential Equations
147. Lamotke: Semisimpliziale algebraische Topologie
148. Chandrasekharan: Introduction to Analytic Number Theory
149. Sario/Oikawa: Capacity Functions
150. Iosifescu/Theodorescu: Random Processes and Learning
151. Mandl: Analytical Treatment of One-dimensional Markov Processes
152. Hewitt/Ross: Abstract Harmonic Analysis. Vol. 2: Structure and Analysis for Compact Groups. Analysis on Locally Compact Abelian Groups
153. Federer: Geometric Measure Theory
154. Singer: Bases in Banach Spaces I
155. Müller: Foundations of the Mathematical Theory of Electromagnetic Waves
156. van der Waerden: Mathematical Statistics
157. Prohorov/Rozanov: Probability Theory. Basic Concepts. Limit Theorems. Random Processes
158. Constantinescu/Cornea: Potential Theory on Harmonic Spaces
159. Köthe: Topological Vector Spaces I
160. Agrest/Maksimov: Theory of Incomplete Cylindrical Functions and their Applications
161. Bhatia/Szegö: Stability Theory of Dynamical Systems
162. Nevanlinna: Analytic Functions
163. Stoer/Witzgall: Convexity and Optimization in Finite Dimensions I
164. Sario/Nakai: Classification Theory of Riemann Surfaces
165. Mitrinović/Vasić: Analytic Inequalities
166. Grothendieck/Dieudonné: Eléments de Géometrie Algébrique I
167. Chandrasekharan: Arithmetical Functions
168. Palamodov: Linear Differential Operators with Constant Coefficients
169. Rademacher: Topics in Analytic Number Theory
170. Lions: Optimal Control of Systems Governed by Partial Differential Equations
171. Singer: Best Approximation in Normed Linear Spaces by Elements of Linear Subspaces
172. Bühlmann: Mathematical Methods in Risk Theory
173. Maeda/Maeda: Theory of Symmetric Lattices
174. Stiefel/Scheifele: Linear and Regular Celestial Mechanics. Perturbed Two-body Motion—Numerical Methods—Canonical Theory

175. Larsen: An Introduction to the Theory of Multipliers
176. Grauert/Remmert: Analytische Stellenalgebren
177. Flügge: Practical Quantum Mechanics I
178. Flügge: Practical Quantum Mechanics II
179. Giraud: Cohomologie non abélienne
180. Landkof: Foundations of Modern Potential Theory
181. Lions/Magenes: Non-Homogeneous Boundary Value Problems and Applications I
182. Lions/Magenes: Non-Homogeneous Boundary Value Problems and Applications II
183. Lions/Magenes: Non-Homogeneous Boundary Value Problems and Applications III
184. Rosenblatt: Markov Processes. Structure and Asymptotic Behavior
185. Rubinowicz: Sommerfeldsche Polynommethode
186. Handbook for Automatic Computation. Vol. 2. Wilkinson/Reinsch: Linear Algebra
187. Siegel/Moser: Lectures on Celestial Mechanics
188. Warner: Harmonic Analysis on Semi-Simple Lie Groups I
189. Warner: Harmonic Analysis on Semi-Simple Lie Groups II
190. Faith: Algebra: Rings, Modules, and Categories I
192. Mal'cev: Algebraic Systems
193. Pólya/Szegö: Problems and Theorems in Analysis I
194. Igusa: Theta Functions
195. Berberian: Baer*-Rings
196. Athreya/Ney: Branching Processes
197. Benz: Vorlesungen über Geometrie der Algebren
198. Gaal: Linear Analysis and Representation Theory
199. Nitsche: Vorlesungen über Minimalflächen
200. Dold: Lectures on Algebraic Topology
201. Beck: Continuous Flows in the Plane
202. Schmetterer: Introduction to Mathematical Statistics
203. Schoeneberg: Elliptic Modular Functions
204. Popov: Hyperstability of Control Systems
205. Nikol'skii: Approximation of Functions of Several Variables and Imbedding Theorems
206. André: Homologie des Algèbres Commutatives
207. Donoghue: Monotone Matrix Functions and Analytic Continuation
208. Lacey: The Isometric Theory of Classical Banach Spaces
209. Ringel: Map Color Theorem
210. Gihman/Skorohod: The Theory of Stochastic Processes I
211. Comfort/Negrepontis: The Theory of Ultrafilters
212. Switzer: Algebraic Topology—Homotopy and Homology
213. Shafarevich: Basic Algebraic Geometry
214. van der Waerden: Group Theory and Quatum Mechanics
215. Schaefer: Banach Lattices and Positive Operators
216. Pólya/Szegö: Problems and Theorems in Analysis II
217. Stenström: Rings of Quotients
218. Gihman/Skorohod: The Theory of Stochastic Processes II
219. Duvaut/Lions: Inequalities in Mechanics and Physics
220. Kirillov: Elements of the Theory of Representations
221. Mumford: Algebraic Geometry I: Complex Projective Varieties
222. Lang: Introduction to Modular Forms
223. Bergh/Löfström: Interpolation Spaces. An Introduction
224. Gilbarg/Trudinger: Elliptic Partial Differential Equations of Second Order
225. Schütte: Proof Theory

MIX
Papier aus verantwortungsvollen Quellen
Paper from responsible sources
FSC® C105338

If you have any concerns about our products,
you can contact us on
ProductSafety@springernature.com

In case Publisher is established outside the EU,
the EU authorized representative is:
Springer Nature Customer Service Center GmbH
Europaplatz 3, 69115 Heidelberg, Germany

Printed by Libri Plureos GmbH
in Hamburg, Germany